Biology of European Sea Bass

Biology of European Sea Bass

Editors

F. Javier Sánchez Vázquez
Department of Physiology
University of Murcia
Murcia
Spain

José A. Muñoz-Cueto
Department of Biology
University of Cadiz
Cadiz
Spain

CRC Press
Taylor & Francis Group
Boca Raton London New York

CRC Press is an imprint of the
Taylor & Francis Group, an **informa** business

A SCIENCE PUBLISHERS BOOK

CRC Press
Taylor & Francis Group
6000 Broken Sound Parkway NW, Suite 300
Boca Raton, FL 33487-2742

First issued in paperback 2019

© 2014 Copyright reserved
CRC Press is an imprint of Taylor & Francis Group, an Informa business

No claim to original U.S. Government works

ISBN-13: 978-1-4665-9945-1 (hbk)
ISBN-13: 978-0-367-37858-5 (hbk)

**Visit the Taylor & Francis Web site at
http://www.taylorandfrancis.com**

**and the CRC Press Web site at
http://www.crcpress.com**

Preface

European sea bass, *Dicentrachus labrax*, has attracted the attention of many people: scientists, fish farmers, fishermen and other stakeholders. Since ancient times, this species has been regarded as a highly valuable table fish, and it is considered a delicacy in restaurants today. This fish has many different names. In France, it is called "loup de mer", while in the Iberian Peninsula it is known as "lubina" or "robalo". In Italy sea bass is called "branzino" or "spigola", whereas the Greeks refer to this fish as "lavraki" (from its vernacular name) and the Turks as "levrek".

European sea bass was historically cultured in coastal lagoons and tidal reservoirs, so it is not surprising that this fish species was one of the first marine fish to be farmed in Europe using modern aquaculture techniques. In fact, in Italy sea bass and seabream are the most consumed fish species. With declining fisheries, the aquaculture of European sea bass becomes increasingly important to provide a high quality product to consumers. In fact, according to FAO predictions, by 2030 over 65% of seafood will be supplied by aquaculture. Sea bass aquaculture production has been rising since the early 90s, reaching over 130,000 tonnes per year in the last few years, while fisheries catches remain stackled around 10,000 tonnes per year. Geographically, the main production of sea bass is located in two countries: Turkey (34.2%) and Greece (34.1%); although this species is also cultured in many other Mediterranean countries, such as Spain (11.4%), Italy (6.9%), Egypt (4%), Croatia (2.5%), France (2.4%), Tunisia, Cyprus, Portugal, etc.

In the last few decades, many researchers devoted their investigations to this species, publishing over 2,200 papers in scientific journals. Most of this research was conducted in three European countries: France (466), Spain (384) or Italy (361); followed by Portugal (136), Greece (133) and Turkey (99). The most popular research areas are Fisheries (including Aquaculture), Zoology and Marine & Freshwater Biology, with many significant contributions in the specific fields of (in relevance order) Physiology, Genetics, Immunology, Ecology, Endocrinology, etc. In this book, we have focused on the general Biology, Ecology, Physiology, Behavior and Pathology of the European sea bass. The book is organized in twelve chapters authored by some of the world's most renowned scientists in these fields.

The book starts with a chapter on the *"Ecology and Distribution of Dicentrarchus labrax (Linnaeus 1758)"*, written by Ángel Pérez-Ruzafa and Concepción Marcos, from the Department of Ecology and Hydrology, University of Murcia (Spain). In this chapter, the authors provide information on the distribution and genetic structure of sea bass, the main environmental factors affecting sea bass distribution and behavior and the synergetic effect of multiple factors on physiological processes, which determines the way in which this species copes with environmental variability. Other interesting aspects such as migration, reproduction, fecundity and reproductive efforts, recruitment, feeding ecology and behavior, life history strategies, response to pollution, management and conservation of sea bass are also addressed in this chapter.

Most living organisms have developed a biological clock to keep time and cope with predictable environmental changes by anticipating the forthcoming of a recurrent event in their habitat. Fish in general, and sea bass in particular, are no exception to this rule. The second chapter entitled *"The Biological Clock and Dualism"* examines this topic; it has been authored by Ana del Pozo and F. Javier Sánchez-Vázquez, from the Department of Physiology, University of Murcia (Spain) and Jack Falcón, from the Banyuls-sur-Mer Oceanological Observatory, CNRS-UPMC (France). This chapter focuses on biological rhythms and zeitgebers (food, photoperiod, temperature), the molecular clock machinery that controls these rhythms, as well as in dualism (diurnal vs. nocturnal) and the mechanisms controlling this dualism in feeding and locomotor behavior, digestive physiology, and seasonal reproductive rhythms in sea bass.

The pineal organ of fish is a photosensory and neuroendocrine epithalamic structure that plays a key role in the temporal organisation of physiological and behavioral processes. In the chapter *"The Pineal Organ of the European Sea Bass: A Neuroanatomical Approach"*, the authors present a review of their studies on the macroscopic and microscopic anatomy of the pineal organ in the European sea bass and the characterization of pineal photoreceptive and melatonin-synthesizing cells. Moreover, they summarise the bidirectional connections of the pineal with the brain through pinealofugal (efferent) and pinealopetal (afferent) projections, as well as the targets of the neuroendocrine message of the pineal organ (melatonin) by describing the tissue-specific distribution and localization melatonin receptors in the retina, the brain and somatic tissues of sea bass. This chapter has been written by Patricia Herrera-Pérez, M. Carmen Rendón-Unceta and José Antonio Muñoz-Cueto, from the Department of Biology of the University of Cadiz (Spain), Arianna Servili, from the Department of Functional Physiology of Marine Organisms, IFREMER (France) and Jack Falcón, from the Banyuls-sur-Mer Oceanological Observatory, CNRS-UPMC (France).

The following chapter is entitled *"Melatonin Rhythms"* and is authored by Luisa M. Vera and Hervé Migaud, from the Institute of Aquaculture, University of Stirling (UK). In the European sea bass, several studies have focused on the description, regulation and physiological role of melatonin rhythms. Current knowledge covers issues regarding photoreception, influence of light intensity and spectrum on melatonin production, circadian and seasonal rhythmicity, extrapineal sources of melatonin, distribution, and daily and seasonal rhythms of melatonin receptors, among others. Many of these findings are reviewed and discussed throughout this chapter.

Fish integrate environmental cues using specific sensory systems such as the pineal organ, and transduce them into a cascade of hormones along the brain-pituitary-gonadal axis, which control the timing of seasonal reproduction. The purpose of this sequence of events is to ensure that reproduction takes place at the time of year most favourable for the survival of progeny. In the next chapter, entitled *"Neuroendocrine Regulation of Reproduction in Sea Bass (Dicentrarchus Labrax)"*, the authors focus on the extraordinary complexity of the neuroendocrine control of reproductive axis of sea bass and they detail the factors modulating this reproductive axis, with both positive and negative effects, including the relevant knowledge acquired very recently in their laboratories. This chapter is authored by Arianna Servili, from the Department of Functional Physiology of Marine Organisms, IFREMER (France), Olivier Kah, from INSERM, Université de Rennes 1-CNRS (France) and Patricia Herrera-Pérez, José Antonio Paullada Salmerón and José Antonio Muñoz-Cueto, from the Department of Biology of the University of Cadiz (Spain).

During the first weeks of life, marine fish larvae undergo significant morphological and physiological modifications to acquire all the adult features by the end of the larval period. This phase is certainly the most critical in the life cycle of fish, not only in natural environments, but also under controlled aquaculture conditions. The chapter *"European Sea Bass Larval Culture"* represents the contribution of Enric Gisbert and Alicia Estévez, from IRTA Centre de Sant Carles de la Ràpita (Spain), Ignacio Fernández, from CCMAR/CIMAR, Universidade do Algarve (Portugal), Natalia Villamizar, from the Department of Physiology, University of Murcia (Spain), Maria J. Darias, from IRD-Université Montpellier II (France) and Jose L. Zambonino-Infante, from the Department of Functional Physiology of Marine Organisms, IFREMER (France). To begin, the authors focus on the embryonic and larval development of European sea bass with a comprehensive analysis of the ontogenetic changes during the early life stages of fish. Later, a description of the current larval production systems at an industrial and laboratory scale is presented with special emphasis on the optimal rearing factors affecting fry production. Finally, the authors review how some of the most important abiotic (light and temperature) and biotic (nutrition) factors affect larval performance.

Feeding by fish is dependent upon their sensory capacities to locate food, their ability to capture, handle and ingest food items, and their physiological and biochemical capacities to digest and transform ingested nutrients. In the chapter *"Foraging Behavior"*, Sandie Millot and Marie-Laure Begout from the Fishing Resources Laboratory of La Rochelle, IFREMER (France) and David Benhaïm, from LERMA, INTECHMER/CNAM, Cherbourg (France), describe the main traits of foraging behavior both in the wild and in aquaculture for the different life stages of sea bass. The chapter also presents foraging behavior in the natural environment, then details foraging in the aquaculture environment in relation with environmental factors and finally describes other related behavior and illustrates the influence of domestication and selection processes on sea bass behavioral traits.

Feeding is a key function in evolution since metabolic energy obtainment and storage make an important contribution to survival rates. The following chapter entitled *"Food Intake Regulation"*, is aimed at reviewing all the central and peripheral regulators of food intake that have been studied in sea bass. In addition, the author also reviews the relationship of food intake with some physiological processes (such as reproduction and stress response) and environmental factors (temperature, salinity, oxygen and ammonia) which are known to modulate food intake in sea bass. The author of this chapter is José Miguel Cerdá-Reverter, from the Institute of Aquaculture of Torre de la Sal, CSIC (Spain).

In fish farming, the cost of feed is the largest factor of production, and, therefore, it is important to minimize the cost of the feed while maintaining high growth rate of the fish. In this context, the chapter *"Nutrition and Dietary Selection"* represents the contribution of Rodrigo Fortes da Silva, Francisco Javier Sánchez-Vázquez and Francisco Javier Martínez-López, from the Department of Physiology of the University of Murcia (Spain). The first section of this chapter deals with basic and practical aspects, comprising protein, energy, lipid and carbohydrate nutrition in sea bass. The requirements of each macronutrient and its replacement with sustainable food sources, micronutrient requirements and the use of probiotics are also reviewed. The second section of the chapter covers dietary selection and the ability of sea bass to detect and self-select different food types. Both macro- and micro-nutrient selection are considered, as well the basic mechanisms responsible for the control of food intake and dietary selection in sea bass.

Farming practices alter the natural equilibrium between the host and the pathogen found in the wild, favoring the emergence of diseases and posing a major problem for the aquaculture industry. Farmed fish are also constantly exposed to different types of stressors such as bad or poor culture conditions, inadequate diets and other anthropogenic factors that might compromise their performance and survival. In the chapter *"Pathology"*, Ariadna Sitjà-Bobadilla, from the Institute of Aquaculture of Torre de la Sal, CSIC (Spain), Carlos Zarza, from Tassal Group Limited, Tasmania (Australia) and Belén Fouz, from the Department of Microbiology and Ecology, University of Valencia (Spain), review the main pathologies due to bacteria, virus, protozoan and metazoan parasites in the European sea bass. In addition, they also discuss other pathologies induced by aquaculture practices, disorders provoked by harmful algae blooms and jellyfish and tumors reported in sea bass populations.

The fish immune system is characterized by a multi-layered organization that provides immunity to infectious organisms, and each layer can be considered to have an increasing complexity. In this way the central challenge of fish immunology, which includes the sea bass model, is now to fully characterize the occurring immune responses. The next chapter is entitled *"Current Knowledge on the Development and Functionality of Immune Responses in the European Sea Bass (Dicentrarchus labrax)"* and is authored by Jorge Galindo-Villegas and Victoriano Mulero, from the Department of Cell Biology and Histology, University of Murcia (Spain). In this chapter, the authors briefly summarize and update the emergence, phylogenetic distribution and functionality of *D. labrax* innate and adaptive immunity, as well as the most relevant interactions between them towards host defense against pathogens.

The book ends with a chapter on *"The Response to Stressors in the Sea Bass"*. Stress is an important physiological response that all animals, including fish, experience when they are subjected to alarming situations. The stress impact in fish has been assessed in aquaculture in relation to fish transport, environmental factors, and factors related to husbandry practices, including social stressors. The issue of stress in aquaculture is highly significant, because a number of consequences often undesired but always relevant for fish performance such as growth, disease resistance, behavior and energy balance are clearly observed. Many of the findings obtained by authors in relation to the stress response in the European sea bass are reviewed and discussed in this chapter. The authors contributing to this chapter are Lluis Tort, from the Department of Cell Biology, Physiology and Immunology, Autonomous University of Barcelona, (Spain), Josep Rotllant, from the Vigo Marine Research Institute, CSIC (Spain), Michalis Pavlidis, from the Department of Biology, University of Crete (Greece), Daniel Montero, from the Aquaculture Research Group, University of Las Palmas de Gran Canaria (Spain) and Genciana Terova, from the Department of Biotechnology and Life Sciences, University of Insubria (Italy).

Contents

Section 1
General Biology and Ecology

Ecology and Distribution of *Dicentrarchus labrax* (Linnaeus 1758)

Ángel Pérez-Ruzafa[1,]* and *Concepción Marcos*[1]

Introduction

The European sea bass (*Dicentrarchus labrax* (Linnaeus 1758)) is among the most abundant and exploited fish species found off European coasts. It is also one of the more productive species in aquaculture with 125,901 T/year and a fishing yield of 10,817 T/year in 2010 (FAO 2013). This production is increasing each year; Haffray et al. (2007) reported 70,000 T/year from aquaculture and a fishing yield of around 10,000 T/year. As a consequence, it is one of the most studied species of fish. The published papers on its physiology, reproduction and feeding reach 6,754 items in ISI journals at a publication rate of 500 papers per year (Web of Knowledge 23/08/2012). However, most of these works respond to laboratory experiments and observations on cultured individuals and very scarce information can be found on wild individuals and the ecology of the species in the natural environment or its role in the marine ecosystem (see Pickett and Pawson 1994, Pawson and Pickett 1996).

The European sea bass is typically a marine species which spend most of its life in coastal lagoons and estuaries, although it has been observed occasionally in rivers. It inhabits waters ranging from hyperhaline to brackish, and shows oceanodromous behavior. Its eurihalyne and eurythermic capability permits this species to show a wide geographic

[1]Departamento de Ecología e Hidrología, Facultad de Biología, Campus de Excelencia Internacional Mare Nostrum, Universidad de Murcia, 30100 Murcia (Spain).
*Corresponding author

range of distribution and a wide depth range, occurring from shallow waters (2–10 m) to more than 100 m. It is distributed in the Eastern Atlantic, from Norway to Senegal, and in the Mediterranean and Black Sea (Fig. 1). It has been introduced for culture purposes in Israel, and more recently in Oman and the United Arab Emirates (Haffray et al. 2007).

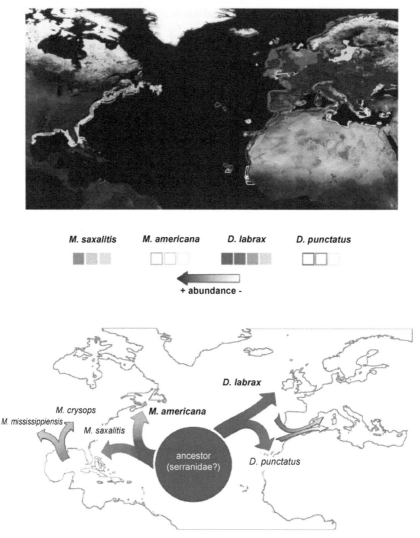

Figure 1. Distribution of Atlanto-Mediterranean species of sea bass. Arrows show possible ways of speciation and colonization or recolonization of the Mediterranean from an ancient species probably originated in the Gulf of Mexico and distributed through the North Atlantic and in the Mediterranean.

Color image of this figure appears in the color plate section at the end of the book.

This wide geographical and ecological range of distribution is related with the physiological adaptations of the species, mainly due to its tolerance to low and high temperatures and its ability to regulate osmotic stress. *D. labrax* inhabit at temperatures between 8°C–24°C (Froese and Pauly 2012), although it has been reported that they tolerate temperatures from 5°C in the northern Mediterranean coastal lagoons, such as Thau, to 27°C in the lagoon of Biban in Tunisia (Barnabé 1986), or even up to 32°C (Barnabé 1990).

Distribution and Genetic Structure

Atlanto-Mediterranean sea bass is a group of six species from the *Moronidae* family, included in *Serranidae* by earlier authors (Tortonese 1986). Two species belong to the genus *Dicentrarchus*, *D. labrax* (Linnaeus 1758) or European sea bass, and *D. punctatus* (Bloch 1792) or spotted sea bass; and four belong to the genus *Morone*: *M. saxatilis* (Walbaum 1792) or striped bass, *M. americana* (Gmelin 1789) or white perch, *M. chrysops* (Rafinesque 1820) or white bass, and *M. mississippiensis* (Jordan and Eigenmann 1887) or yellow bass. The last two are fresh water species in North America temperate and subtropical river basins, respectively.

The geographic distribution of the marine species is shown in Fig. 1. Striped bass (*Morone saxatilis*) inhabits the Atlantic coasts of North America, from the Gulf of Mexico and Florida to Labrador. The other western Atlantic marine species *M. americana* has similar distribution, but its southern limit is on North Carolina coasts.

In eastern North Atlantic, the European sea bass (*D. labrax*) is distributed from Morocco to Norway and in the Mediterranean, while spotted sea bass (*D. punctatus*), inhabits the Eastern tropical Atlantic from Senegal to Morocco and the Mediterranean (Fig. 1).

Although, according to literature, the northern limit of the distribution of *D. labrax* is Norway and it has recently been recorded in the Baltic (Bagdonas et al. 2011), the abundance of the species further north of 53° latitude (coinciding approximately with the 9°C mean annual temperature isotherm) is low, and fisheries in north-eastern England and eastern Scotland are occasional. In the Atlantic, most of the fisheries are concentrated on the Irish coasts, east of England and Wales and in the Bay of Biscay (Pickett and Pawson 1994).

The separation of these species from an ancestral predecessor, and their present genetic structure, involves the complex tectonic processes that led to the formation of the Atlantic Ocean, and probably the closure of the Mediterranean, and subsequent changes related to the saline crisis in the latter leading to intense processes of successive extinction, speciation and recolonization in this sea.

Interspecific genetic diversity shows a different pattern in the Atlantic and Mediterranean populations of *D. labrax* (Fritsch et al. 2007, Pawson et al. 2007, Coscia et al. 2012). While Atlantic populations appear genetically homogeneous over wide areas, including the Bay of Biscay and the English Channel, despite the little exchange of individuals between them, Mediterranean Sea populations appear to be genetically structured and show some differences between several sub-basins (Patarnello et al. 1993, García De León et al. 1997, Castilho and McAndrew 1998, Naciri et al. 1999, Bahri-Sfar et al. 2000, Bonhomme et al. 2002, Quéré et al. 2012). This may be related to the isolation of populations during the saline crisis, the past and present role of coastal lagoons in selecting populations adapted to local conditions (Lemaire et al. 2000, Pérez-Ruzafa et al. 2010, 2011a), and the environmental heterogeneity in the different Mediterranean sub-basins which favors the isolation of reproductive populations. However, these patterns can be altered by the introduction of individuals from other areas through aquaculture (Bahri-Sfar et al. 2005, Bagdonas et al. 2011).

In the temperate Atlanto-Mediterranean region, *D. labrax* is one of the most frequent species in coastal lagoons and estuaries, along with *Mugil cephalus* (Linnaeus 1758), *Anguilla anguilla* (Linnaeus 1758) and *Atherina boyeri* (Risso 1810). It is present in 75.0% of the 73 lagoons and 58.1% of the 30 estuaries studied by Pérez-Ruzafa et al. (2011b). This high frequency of occurrence makes *D. labrax* not associated with any trophic, hydrological, or geomorphological variable characterizing coastal lagoons (Pérez-Ruzafa et al. 2007).

Environmental Factors that Determine the Distribution and Behavior of *D. labrax*

The main environmental conditions affecting sea bass distribution, as also behavior, at geographical and small spatial scales, is temperature. Furthermore salinity, turbulence, and dissolved oxygen can also play a relevant role in some circumstances.

In the northern Atlantic coast, over a period of near 30 years, a positive correlation has been found between the abundance of sea bass and temperature (Henderson et al. 2011). This parameter can exert complex influence on sea bass populations; it directly affects the development rate of young fish, which in turn reduces mortality. At the same time however, at higher scales, temperature fluctuations such as the North Atlantic Oscillation (NAO) can affect ecosystem productivity and food availability (Henderson et al. 2011)

In sea bass under experimental conditions, with no food limitations, the major external variable influencing short term variations in growth is also water temperature (Gutiérrez and Morales-Nin 1986).

Optimum environmental conditions can show some differences in the distinct works performed to study the physiological adaptations of the species. It is basically due to the fact that different populations will be acclimated to the natural conditions in each geographical area. In this sense it must be taken into account that most of the experiments acclimate the individuals to the normal conditions in the geographic region in which the experiment is performed and the effect of salinity or temperature deviations from those considered normal is studied without acclimation to the new experimental conditions. Moreover, many works do not specify the origin of the individuals used in the experiments. Anyway, in those cases where an acclimation period is considered prior to performing the metabolic or activity measurements (e.g., Conides and Glamuzina 2006), the individuals tend to show optimums close to the natural conditions in the area.

In fact, temperature tolerance in fish differs by species, acclimation temperature as well as acclimation duration (Chatterjee et al. 2004, Das et al. 2004, Manush et al. 2004) and salinity (Jian et al. 2003, Kumlu et al. 2010). Typical seasonal acclimation allows fish to be more tolerant to higher temperature in summer and lower temperature in winter (Bevelhimer and Bennett 2000, Dülger et al. 2012).

Acclimation temperature in sea bass significantly affects the minimum and maximum Critical Thermal values of the fish (CTMin and CTMax, respectively). A 10°C increment in acclimation leads to an increase from 33.23°C to 35.94°C on CTMax and from 4.10°C to 6.77°C on CTMin (Dülger et al. 2012).

In this way, while 18°C is the water temperature suitable for growth for Atlantic sea bass populations (Russell et al. 1996), in the Mediterranean it would be 25°C (Dülger et al. 2012). Sea bass has Acclimation Response Ratio (ARR) values—that is, the relationship between the tolerance change and the total change in acclimation temperature—around 0.25–0.27, similar to some tropical species and higher than species from cold or temperate regions (Dülger et al. 2012). Higher ARR is considered an adaptation of species to greater and more frequent fluctuations of temperature that are characteristics in subtropical and tropical regions and in estuarine or coastal lagoon environments, while cooler regions show gradual long-term shifts (Re et al. 2005, Dülger et al. 2012). It is interesting to note that fluctuating temperature regimes, characteristics in coastal lagoons and estuaries, can increase not only fish's tolerance to high temperatures but also somatic growth (Bevelhimer and Bennett 2000).

The influence of temperature on metabolism is illustrated by the relationships between it and Active Metabolic Rate (AMR) and Metabolic Scope (MS) (Fig. 2). Sea bass AMR and MS increased sharply as the temperature rises from 10°C to 20°C. Maximum metabolic scope is reached between 20°C and 24°C. When temperature reaches 25°C, a certain degree

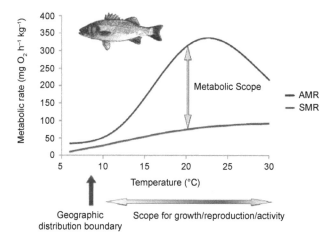

Figure 2. Metabolic scope for sea bass depending on environmental temperature according to the model from Claireaux and Lagardère 1999. AMR is the active metabolic rate and SMR the standard metabolic rate. The difference between them determines the scope for growth and reproduction effort than sea bass individuals have available. Dark arrow shows the down limit of temperature niche for the species determining the distribution range and the migratory spatial and temporal pattern.

Color image of this figure appears in the color plate section at the end of the book.

of homeostasis exists and only small changes in fish metabolic capacity take place (Claireaux and Lagardère 1999). According to the model proposed by Claireaux and Lagardère (1999), below 10°C, the metabolic scope is seriously reduced and energy budgeting conflict can arise and therefore mortality risks appear. This explains the northern distribution limit of the species and, probably, winter migrations.

Extreme temperatures can affect the behavior of the fish. Bégout and Lagardère (1998) found that below 12°C and above 22°C, the fish decrease their swimming activity which may have consequences upon feeding activity and migrations. However, these observations were made on aquaculture fish in French coasts where no acclimation of the individuals to extreme temperatures was made; therefore probably, individuals of northern and southern wild populations can show very different ranges of optimum swimming activity.

Salinity is another factor that limits the distribution of aquatic species. It constitutes the main barrier between continental and marine waters biota. Species must cope both with total salt concentration and also with ionic composition when they colonize a new aquatic environment (Herbst and Bromley 1984, Zalizniak et al. 2006). Unlike inland waters, at sea, ionic composition is constant and most marine organisms never cope with the problem of a different ionic composition, but in some coastal lagoons with strong fresh water inputs, and mainly in estuaries, the effect of different ionic composition can also be relevant. However, this aspect has received little attention.

Usually it is assumed that euryhaline species, like sea bass, spend low metabolic costs in osmotic and ionic regulation and these costs for Standard Metabolic Rate (SMR) can be slightly higher when the salinity is high, mainly in small individuals (Nordlie 1978, Von Oertzen 1984, González 2012). However, Routine Metabolic Rate (RMR) shows the reverse trend, decreasing when salinity increases, reinforcing the idea that, at least in euryhaline species, interactions with the saline environment involves more than just osmoregulation costs (Claireaux and Lagardère 1999).

The abrupt change from high to low salinity that juvenile sea bass must face when colonizing estuaries affects them relatively little. According to the experiments performed by Conides and Glamuzina (2006), the juveniles of sea bass show some reaction only during the first day after the salinity change. However, survival rate and swim bladder inflation are improved at low salinity, and, in the same manner, growth rate and conversion rate when temperature is low. Sea bass larvae, therefore, appear to show a low saline preference, which may correspond to the conditions in which juveniles live (Saillant et al. 2003) but this does not explain the case of hyperhaline coastal lagoons like the Mar Menor in the Southwestern Mediterranean and other Mediterranean coastal lagoons. In fact, Conides and Glamuzina (2006) found that the optimum salinity conditions that allow the highest values of food conversion efficiency and the highest daily growth rates of sea bass in Greece are above 28‰ and 30‰ respectively, and maintenance requirements tend to increase with decreasing salinity.

Salinity can also explain the differences in relative abundance in typical lagoonal species between different coastal lagoons. In experiments on the effects of salinity on early development and sex determination of larvae and juveniles of *D. labrax* performed by Saillant et al. (2003), survival rate ranged between 5%–15% which is common in larval rearing. During larval rearing and nursery, periods during which the greatest mortality occurs, growth was improved and survival was doubled at low salinity (15‰) when compared with a higher salinity (37‰). Such a result was also obtained by Johnson and Katavic (1986). However, as commented above, the optimum salinity conditions for sea bass growth found by Conides and Glamuzina (2006) are above 30‰ while for gilthead sea bream (*Sparus aurata*, Linnaeus 1758), they are between 18‰ and 20‰. Sea bass juveniles show higher maintenance requirements and lower growth rates than *S. aurata* at all the salinity ranges but, while growth rates decrease in *S. aurata* at salinities higher than 20‰, sea bass shows its maximum at salinities higher than 28‰ (Conides and Glamuzina 2006).

Furthermore, Saillant et al. (2003) found that although salinity is not directly involved in the sexual differentiation of sea bass and has no effect on sex-ratio when it is maintained constant, the osmotic stress generated by a salinity shock during the labile period of sexual differentiation may increase a masculinizing effect attributable to other environmental factors. In their results, the transfer from low to high salinity at 93 days post-fertilization increased the percentage of males up to 93%.

When interpreting these results we must take into account that different factors do not act independently of each other and that their effects on fish biology are the result of complex relationships affecting energy budgets (see below). In fact, Conides and Glamuzina (2006) also found that salinity around 15‰ also permits the highest growth rates in *D. labrax* juveniles, but only when temperature is also low (around 15°C).

Dissolved oxygen in the water column is another important environmental factor determining the fish ecological niche (Coutant 1986) and setting the upper limit to fish metabolic activity (Neill et al. 1994, Claireaux and Lagardère 1999). This limit depends not only on oxygen availability but also on oxygen demand which is determined by metabolic level.

Although open coastal areas are usually well oxygenated, dissolved oxygen can be a limiting factor for fish distribution in enclosed bays and estuaries with strong stratification and high inputs of organic matter. In fact, many coastal lagoons are exposed to frequent dystrophic crises (Amanieu et al. 1975, Boutiere et al. 1982, Ferrari et al. 1993, Sfriso et al. 1995, Viaroli et al. 1996, Gianmarco et al. 1997, Guyoneaud et al. 1998, Bachelet et al. 2000, Sakka Hlaili et al. 2007, Specchiulli et al. 2009, Pérez-Ruzafa et al. 2011a).

D. labrax tolerates low oxygen concentrations up to a minimum of 2 mgL^{-1}. Except in critical hypoxic concentrations, the swimming activity during the day is not affected, but high oxygen concentration (higher than 10 mgL^{-1}) increases locomotor activity during the night (Bégout Anras and Lagardère 1998).

Hypoxic conditions around 40% air saturation do not preclude sea bass growth (Thetmeyer et al. 1999) but reduce food consumption, somatic growth, and condition factor. These effects are not due to a reduction in feed conversion efficiency or variation in body size but a reduction in appetite of individuals. The latter is much less pronounced if hypoxic conditions alternate with normal conditions.

Other environmental factors, such as wind speed, atmospheric pressure, etc. can also affect swimming behavior; for example, wind speed higher than 8 ms^{-1} increases swimming activity. In general, it can be said that anticyclonic conditions and summer high temperatures reduce swimming activity while moderately low temperatures increase swimming activity. These behavioral responses to meteorological factors have been interpreted as a capability of fish to anticipate habitat changes (Bégout Anras and Lagardère 1998).

One of the effects of wind on coastal waters is the generation of waves and turbulence in the breaking zone. Moderate turbulence increases contact rates between planktonic organisms (i.e., predator and prey) (Rothschild and Osborn 1988), therefore, unless reaching a threshold level, in a prey-limited environment it is generally assumed that it enhance ingestion rates in fish larvae by increasing encounters with prey (MacKenzie et al. 1994, Saiz et al. 1992). However, turbulence, mainly in extreme events like storms, may affect not only increasing the encounter probability (Galbraith et al. 2004) but also having negative effects, increasing diffusive mixing (Rothschild and Osborn 1988), reducing the pursuit (MacKenzie and Kiørboe 2000, Mariani et al. 2005, 2007), the attack efficiency (MacKenzie and Kiørboe 1995), and the capture (MacKenzie and Kiørboe 2000) of prey. Mahjoub et al. (2012) showed that, contrary to expectations, increasing Reynolds number (Re) exponentially decreased the mean prey ingestion per individual and increased the inter-individual variability in ingestions. They conclude that turbulence exerts negative effects on ingestion in larval sea bass feeding on non-evasive prey, at a density of 200 ind.L^{-1}, commonly recorded in an estuarine environment.

Like many other fish (Montgomery et al. 2000), sea bass larvae show hydro-mechanical detection of prey (Mahjoub et al. 2011). The disadvantage of turbulence could be that it may blur hydro-mechanical signals and affect the sensory field of fish larvae (Yen et al. 2008).

Furthermore, turbulence can negatively affect the success of pursuit, reducing encounter duration (Kiørboe and MacKenzie 1995, Kiørboe and Saiz 1995, MacKenzie and Kiørboe 2000, Mariani et al. 2005, 2007). Therefore, turbulence acts as a limiting factor for feeding activity of sea bass larvae and for species distribution. As with any other environmental factor, survival in exposed areas will be the result of energy budgets. In this framework, turbulence reduces energy inputs through food and at the same time increases energy swimming costs because swimming speed and performance are affected negatively by turbulence (Pavlov et al. 2000, Enders et al. 2003, Lupandin 2005, Mahjoub et al. 2012).

Sea bass larvae can cope with high turbulence conditions in strong tidal areas or storms, displaying an intermittent feeding strategy during the short current reversal periods between flood and ebb, as reported for 0-group sea bass of the Mont Saint Michel Bay (France, English Channel) (Laffaille et al. 2001, Mahjoub et al. 2012). Furthermore, the high inter-individual variability in feeding efficiency under high turbulent conditions at velocities higher than 0.17 m s^{-1} (Mahjoub et al. 2012) would permit the segregation of populations in exposed or less turbulent areas, offering an interesting mechanism of natural selection by intra-specific competition in case there is reproductive isolation. In this context, being small is the best way to avoid high Reynolds numbers in high flow or turbulent environments. This justifies why small fish larvae are more likely than larger larvae to benefit from turbulence (Kiørboe and MacKenzie 1995) and can be a selective mechanism of larval performance in high energy coastal environments.

The Synergetic Effect of Multiple Factors

The way in which a species copes with environmental variability is the result of a complex interaction of factors and their effects on physiological processes. The resulting metabolic scope is what will allows individuals to develop their life and their function in the ecosystem, conditioning the home range, feeding activity, migrations, growth rates, fecundity, etc. Since the first law of thermodynamics must be met, if the energy that should be devoted to a function is very high, depending on the metabolic scope available in specific environmental conditions, it will be possible to perform other functions or not. For example, sea bass, and other marine species that colonize coastal lagoons and estuaries, can regulate a wide range of salinity conditions, but the energy costs involved in the osmoregulatory process prevents them from allocating enough energy to gonadal maturation and reproduction in such environments. According to Claireaux and Lagardère (1999) the metabolic scope increases by a factor of 9 between 10°C and 20°C and decreases at higher temperatures (Fig. 2). The capability to meet osmoregulatory costs is linked to metabolic activity through ventilation. This relationship was highlighted by the interaction between environmental salinity and temperature observed by Claireaux and Lagardère (1999).

The growth of sea bass is mainly affected by the combined effect of temperature/salinity. The interaction between salinity and temperature do not have linear effects on fish metabolism. Usually, contrary to what might be expected, no direct interactions have been found between metabolic rate and salinity (Hettler 1976), or between oxygenation level and salinity (Claireaux and Dutil 1992, Gonzalez and McDonald 1992). However, in sea bass, routine MO2 increases with decreasing salinity, but this raise is more pronounced in warm (T>20°C) than in cold waters (T>15°C) (Claireaux and Lagardère 1999). Conides and Glamuzina (2006) showed that when temperature is high (around 22°C), and thus metabolic rates are also high, salinity does not influence long term growth of sea bass juveniles. As a consequence, growth is higher at low salinities (around 15‰) when there are also low temperatures (around 15°C) and vice versa, when salinity is higher than 32‰ and temperature around 22°C (Alliot et al. 1983, Conides and Glamuzina 2006).

On the other hand, the relationship between temperature and metabolism can be responsible for the vulnerability of sea bass to some eutrophication events. Between 10°C and 25°C, the Standard Metabolic Rate (SMR) of sea bass increases from approximately 36 to 91 mg O_2 kg^{-1}h^{-1}. This high oxygen demand may explain why sea bass can have problems to survive or maintain normal activity when dystrophic crisis occurs in summer in coastal lagoons, affecting the levels of dissolved oxygen in the water column.

Migrations

While coastal lagoons and estuaries offer refuge and abundant food due to their intrinsic characteristics and geomorphology, at the same time their high environmental variability and the extreme and changing values of salinity and temperature involve high metabolic energy costs for the species. These costs can unbalance the energy budget when energy resources should be devoted to reproduction. In fact, *D. labrax* individuals from both sexes can accomplish gametogenesis at salinities as low as 3‰, but they do not spawn (Zanuy and Carrillo 1984), and at temperatures below 10°C the metabolic scope is seriously reduced (Claireaux and Lagardère 1999).

One of the adaptive solutions to choose the most suitable of such conditions according to the life cycle of species is maintaining ontogenic migrations between the sea and this kind of ecosystems. Like other characteristic species of coastal lagoons and estuaries, as Sparids, *Anguilla* and Mugillids, the biology and ecology of sea bass are marked by performing migrations from shallow areas in coastal lagoons and estuaries, where they grow, to deeper zones in the open coast for reproduction (Fig. 3). Pawson et al. (1987) described the distribution and movements of sea bass around England and Wales on the basis of mark-recapture studies conducted during the late 1970s and early 1980s. In the North Atlantic, annual migrations in sea bass occur from winter southern spawning areas to northern summer feeding grounds (Fritsch et al. 2007) with migratory patterns that remain largely unchanged (Pawson et al. 2007). These reproductive migrations start when they reach sexual maturity and they are approximately four years old. In autumn, between October and December, when the water is cooler, they move from the well defined inshore feeding grounds to offshore pre-spawning and spawning areas, seeking water warmer than 9°C. In spring, at the end of spawning in April/May, sea bass individuals migrate back to the feeding grounds (Pawson et al. 2007).

Temperature seems to be the main factor regulating migratory movements and the duration of spawning season. In the Mediterranean Sea, females spawn in winter (December to March) while in the Atlantic Ocean this activity is up to June (Haffray et al. 2007). In autumn the falling of sea temperature below 10°C around the British Isles is associated with the migration of sexually maturing bass from the coastal feeding zones to deeper, less cold areas (Pickett and Pawson 1994). As commented above, this critical temperature is associated with a critical reduction of metabolic scope (Claireaux and Lagardère 1999).

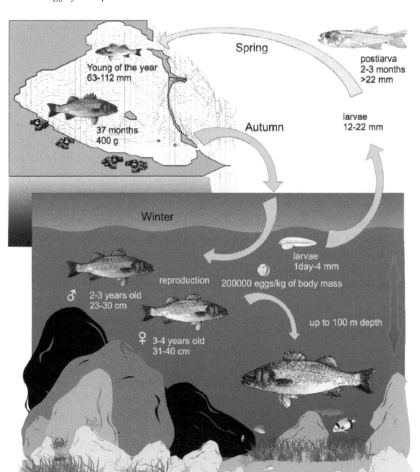

Figure 3. Migration patterns between coastal lagoons or estuaries and open coastal sea throughout the life cycle of sea bass.

Sea bass larvae recruit again in coastal zones between March and July, when they have a size between 12 mm and 22 mm, and the young of the year are invading the marshes up to November at sizes ranging between 63 mm and 112 mm. The growth rate during this period is of nearly 1.2% d^{-1} (Laffaille et al. 2001). In northwestern Mediterranean, settlement of sea bass occurred in sheltered shallow habitats from April to June, both in natural, such as coastal lagoons or estuaries, and in artificial such as marinas (Dufour et al. 2009).

The migratory juvenile behavior is innate and not altered in individuals born in captivity. In release and recapture experiments performed on larvae grown in captivity to study the possible role of artificial reefs in improving and protect sea bass populations (Grati et al. 2011), the majority of recaptured bass were concentrated in the surroundings of river mouths and harbors, suggesting that after release, sea bass migrated towards shallower and brackish waters. Only after some time, as they grew, they came back towards deeper waters and tend to aggregate around artificial structures where they fed on rocky bottoms.

Adult sea bass may move considerable distances to offshore winter spawning areas. However, there is little evidence of the spawning migrations between the North Sea and the Western Channel that were observed in the early 1980s by Pawson et al. (1987) but not in later studies (Pawson et al. 2007).

Migrations are dispersive mechanisms that maintain the genetic fluxes between distant populations. Although Pawson et al. (2008) found strong evidence that contingents of adult bass, which may share common migration routes and frequent the same spawning grounds, segregate to specific summer feeding areas, they also recognize that the extended larval drift phase (Jennings and Pawson 1992) and the wide dispersion of pre-adults (Pickett et al. 2004) precludes any genetic separation (Fritsch et al. 2007). Furthermore, around 22% of the recaptures of tagged individuals by Pickett et al. (2004) were made in fisheries outside the respective home regions, including some individuals which moved from the southeast and central south coasts of England to Wales and northwest England. However, the distribution of recruits to the coastal fisheries is not uniform, and while the Northwest of England received recruitment from nursery areas from all other regions, the southeast fishery received little recruitment from elsewhere.

Anyway, although they are usually recorded for a few individuals, migrations can also take place at larger distances from the English Channel to the Bay of Biscay and can be accomplished quite quickly (Pawson et al. 1987, Pickett et al. 2004, Fritsch et al. 2007). This could justify the substantial gene flow and the non-existence of any significant genetic population differentiation between Atlantic populations including the Bay of Biscay and the English Channel, either for juvenile or adult individuals (Fritsch et al. 2007).

The spatial range of these movements is not uniform in all the population, depending on the age of individuals. Most of the juveniles (individuals <32 cm total length) tend to be recaptured within a limited range of their respective tagging areas and show no seasonal patterns, maturing virgin or 'adolescent' bass (32–42 cm) remain inshore moving within 80 km of release areas, while adults (individuals >42 cm, the minimum size of first maturity in females) show the above mentioned extensive seasonal migrations between summer feeding areas on the west and east coasts and winter/spring spawning areas in deep water off southwest England (Pickett et al. 2004).

Reproduction, Fecundity and Reproductive Effort

The reproduction of sea bass takes place in winter, from January to March after performing the autumn Southwest migration to pre-spawning areas. Spawning starts in the British islands and in the mid-Western Channel during March (Pawson et al. 2007), when the temperature range associated with bass egg distributions is 8.5°C–11°C (Thompson and Harrop 1987), and runs through June (Froese and Pauly 2013). Eggs appear to spread east through the Channel as the surface water temperature attains 9°C (Thompson and Harrop 1987). Spawning tends to be earlier in the southern range of distribution of the species and in the Mediterranean, starting in December or January and running until March (Tortonese 1986, Froese and Pauly 2013).

Sea bass females show a high fecundity. On average, each female spawns 200,000 eggs/Kg of body mass (Haffray et al. 2007) but can reach 808,581 eggs in individuals of 57 cm of total length and around 3 kg of weight (Wassef and Emary 1989). Females start reproducing when they weigh over 2 kg and are four years old and can live upto six or seven years old (Haffray et al. 2007, Pawson et al. 2007). Eggs are pelagic and larvae are widely dispersed during the three first months of life.

As in most marine species, the high reproductive effort required to maintain the populations determines sexual dimorphism. Sea bass are not territorial and males neither take care of the territory like groupers, nor the eggs like gobies. Therefore, as most of the energy devoted to reproduction is intended for the production of gametes, the female reproductive effort is higher than that of males. This determines that females are larger. In the experiments performed by Saillant et al. (2001) males remained significantly smaller than both females and fish feminized by estradiol treatment. Size differentiation is bigger in the early stages of development, reaching 67% at 10 months of age and then decreasing to stabilize around 20%–30%. In the fourth year of life, just when they reach sexual maturity and migration starts (Pawson et al. 2007), the females grow faster in terms of both weight and length and show a relatively higher ratio of digestive tract to body weight (+26%) and start to devote energy to reproduction. When they have finished their second vitellogenic cycle, they show lower ratios of visceral fat to body weight (–49%) and muscle lipid content (–16%) (Saillant et al. 2001).

In their natural environment, the sex ratio is biased towards females, increasing from 52.0% of females in the younger fish (<30 cm total length) to 69.5% of females in the older fish (Vandeputte et al. 2012). However, in farmed populations, sex ratios are highly biased towards males (75% to 95%), which is problematic for aquaculture (Blázquez et al. 1998, Vandeputte et al. 2012). Although these authors consider that these differences could be an effect of sampling bias associated with fisheries methods and fishing grounds, taking into account that females growth faster than males and inhabit more coastal areas, this unbalanced proportion of males is in fact common in experimental conditions, and an excess of males (up to 87%) is usually found independent of the environmental conditions and salinity used in the experiment (Saillant et al. 2002, 2003). This unbalanced ratio can increase up to 93% when there is an osmotic stress produced by the transfer from low to high salinity at 93 days post-fertilization, when larvae are 44 mm long, suggesting that, the excess of males in farmed populations is actually linked to the environmental conditions during larval rearing (Vandeputte et al. 2012). As Saillant et al. (2003) recognize, the environmental component of sex determination in sea bass appears to be very complex. Temperature can also play a role in sex differentiation (Blázquez et al. 1998, Pavlidis et al. 2000, Saillant et al. 2002) but no constant temperature pattern is linked to high percentages of females except, perhaps, an increasing pattern of temperature (Saillant et al. 2003). Therefore, the level of stress during the larval period may be the major inductor of sex determination and masculinization reported in aquaculture (Saillant et al. 2003). Probably, not only the osmotic stress, but any kind of stress associated with captivity would determine preference to male differentiation in the process of sex determination.

This could be a mechanism to guarantee the maximum efficiency of the reproductive effort. Sex ratio would bias towards females mainly in environmental conditions that anticipate a good primary production in the ecosystem. This idea could be supported by the positive correlation that Vandeputte et al. (2012) found between good brood years, fishing yields, female proportion and warm springs and summers.

Recruitment

D. labrax larvae tend to occur in the upper part of the water column, and the arrival of post-larvae into sheltered bays and estuaries usually takes place during spring tides (Jennings and Pawson 1992). These authors pointed out that, in UK waters, post-larva emigrate to inshore waters when temperatures are higher than those of the surrounding sea.

Recruitment of *D. labrax* can show strong spatial and interannual variability (Cabral and Costa 2001). In the Mondego estuary, in northern Portugal, Martinho et al. (2009) found densities of new recruits going from an average of 1 ind.*1000 m^{-2} in 2007 to 19.1 ind.*1000 m^{-2} in 2003. These authors found that river runoff, and therefore precipitation, were the main factors favoring abundance of recruits, while the east-west wind component negatively affected the 0-group abundance. Lancaster et al. (1998) and Amara et al. (2000) also attribute differences in recruitment levels to variations in coastal wind direction and intensity. This negative effect of wind can be related with the effect of turbulence on food availability or catchability discussed above.

The effect of runoff on recruitment also has consequences on fisheries, so that dry years are often associated with lower catches (Meynecke et al. 2006).

Nursery areas can be concentrated in relatively small areas and from them more coastal zones can be colonized. In the British islands these areas are spread around coasts of the south of Wales and England; on the Portugal coast, one of the main recruitment sites is the Mondego estuary. Vasconcelos et al. (2008), using otolith elemental fingerprints found that about 40% of the Portuguese coastal stocks of sea bass originated in the Mondego nurseries.

Feeding Ecology and Behavior

Sea bass is a top predator, a highly adaptable and opportunistic feeder, preying on whatever prey is particularly abundant (Boulineau-Coatanea 1969, Kennedy and Fitzmaurice 1972, Arias 1980, Aranda et al. 1999, Sa et al. 2006). Figure 4 shows the main trophic relationships of the species in its natural habitats.

The feeding behavior and food preferences of *D. labrax* are related to the size of individuals. Juveniles feed mainly on Mysidacea (Moreira et al. 1992) but also on other benthic groups as copepodites, amphipods, isopods and, to a lesser extent, on bivalves (*Cerastoderma*), decapods (*Carcinus maenas* (Linnaeus, 1758)) and small fishes (*Atherina, Pomatoschistus* and *Gobius*) (Costa 1988, Mehanna Sahar et al. 2010). The composition of the food can change depending on the phase of the tide, with an increase in amphipods and polychaetes during the ebb (Cabral and Costa 2001, Laffaille et al. 2001).

The food of fish larger than 20 cm and adults consists mainly of crabs, shrimps, molluscs and fish (Sa et al. 2006, Mehanna Sahar et al. 2010, Riley et al. 2011). Crustaceans constitute up to 48% of food items and teleostei 36%. The analyses of gut content shows that they also consume polychaetes (up to 5%) such as seagrasses, seaweeds, aquatic insects and snails, but in lesser amounts (Abdel-Hakim et al. 2010). In coastal lagoons, like the Mar Menor in the south-western Mediterranean, they feed on atherinids and juvenile grey mullets.

Diet composition also varies seasonally and with body length (Abdel-Hakim et al. 2010). Crustaceans and fishes constitute the major food items all through the year. In Bardawill lagoon, crustaceans reach 88.6% in April,

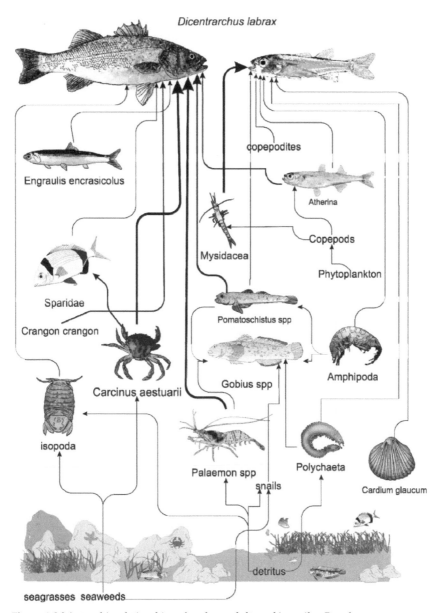

Figure 4. Main trophic relationships of sea bass adults and juveniles. Broad arrows represent the preferred food.

while fish dominate in October, contributing 53.3% by the volume of diet composition. Seagrasses can be found in small quantities in all months, but reach 4.1% in July by the volume of diet composition, while seaweeds reach 5.2% in September. Polychaeta are present only during late spring and early summer, reaching 4.8% in June. Molluscs appeared only in March, contributing 1.1% by the volume of diet composition. This means that, as mentioned before, sea bass is an opportunistic predator and consumes as food, the dominant species in each site and season.

The size of prey of all taxonomic groups increases with the growth of fish at the same time that the ratio crustaceans:fish decreases (Abdel-Hakim et al. 2010). As body size increases, the ability to capture more efficiently and to ingest a larger number of prey also increases, allowing the larvae to adjust their behavior increasing the feeding swimming activity with the density of prey (Georgalas et al. 2007).

Larvae of sea bass fall into the category of "cruise predators", showing more continuous and intense swimming than fish showing the so called "saltatory strategy", characterized by more frequent pauses and reduced swimming activity. In juvenile fish, the swimming activity increases significantly from 10 to 20 days of age, while the time spent sinking and resting decreases, coinciding with the full development of the swim bladder, and leads to a remarkable increase in prey seeking efficiency and capture (aiming posture or Sigmoid-posture and attacks) (Georgalas et al. 2007).

The adults show typical predator behavior, fishing mainly during the night and waiting for their prey in the inlets, swimming against the currents. This species is considered very voracious, to the extent that it has come to be regarded as the cause of depopulation of fish from the lagoon of Mar Menor and the dramatic decline in fish catches in the early twentieth century (Butigieg 1927).

There are no studies on the possible competition for food with other top predators in estuarine and lagoon habitats with an apparent overlap in key preys like *Anguilla* or large gobies (*Gobius cobitis*, Pallas 1814 or *Gobius paganellus*, Linnaeus 1758). Sparids can also compete for at least a wide part of their food repertory, mainly molluscs and crustaceans. Furthermore, food requirements are similar in sea bass and in *Sparus aurata* and both species show maximization of growth at feeding levels around 3% bW/day (Conides and Glamuzina 2006). Probably, the niche differentiation takes place through main food habitat (rocky in the case of sea bass and soft in the case of *Anguilla*) or prey sizes (in the case of gobies). However, more research should be done to elucidate this question. Furthermore, sea bass is the only predator that has been observed stalking leveraging inflows in the inlets at the Mar Menor coastal lagoon (pers. obs.). In open sea, despite the evident differences in habits, size and characteristics, some competence for food has been observed between sea bass and common dolphins (*Delphinus delphis,* Linnaeus 1758). In the Bay of Biscay, Spitz et al. (2013) found that sea bass primarily target small pelagic fish, as mackerel (*Scomber scombrus,* Linnaeus 1758), scads (*Trachurus* spp.), anchovy (*Engraulis encrasicolus* (Linnaeus 1758)), and sardine (*Sardina pilchardus* (Walbaum 1792)). These authors observed an important overlap with the diet of dolphins in this area, suggesting that this could be the cause of the high catches of dolphins in the bycatch of the pelagic trawl fishery devoted to sea bass fishery.

Life History Strategies

Species respond to environmental stress by adopting particular life-history strategies (e.g., *r* vs. K) (Pianka 1970, Margalef 1974). The *r*-strategy involves little care for offspring but increased reproductive effort through early age of first maturity, high fecundity, small and numerous offspring with a large dispersive capability, short lifespan and small adult body size. These traits provide a selective advantage to a species in unstable or unpredictable environments since the ability to reproduce quickly is more important in these situations than inter-specific competition.

At the other extreme, K-strategists are expected in stable and predictable environments. They are longer living species with slow maturation and reproduction at a later age, with the production of fewer offspring requiring extensive parental care. They also have higher biomass/reproductive ratios than *r*-strategists.

Sea bass spend long periods and part of their biological cycle in coastal lagoons and estuaries which are considered naturally stressed and physically controlled systems with frequent environmental disturbances and fluctuations and exposed to human impacts and land and freshwater inputs (Barnes 1980, UNESCO 1980, 1981, Kjerfve 1994, Gamito et al. 2005 Pérez-Ruzafa et al. in press). Under these conditions it is expected that inhabitants of coastal lagoons will display characteristics of *r*-selected species.

At the same time, according to traditional assumptions, immigrants to coastal lagoons, like sea bass, as they do not reproduce inside lagoons and estuaries, should mostly be K-strategists, with a competitive advantage over the *r*-strategists, at least on a temporary time scale (UNESCO 1981).

Currently, these assumptions no longer hold. As demonstrated by Pérez-Ruzafa et al. (2013), species that live and breed in coastal lagoons show mixed traits and a tendency to have reproductive strategies that are more characteristic of the K-species.

Figure 5 shows the situation of life-history strategies for sea bass regarding the main strategies of fish functional guilds (see Elliott and Dewailly 1995, Elliott et al. 2007, Franco et al. 2008 for a description of these guilds) described in Atlanto-Mediterranean coastal habitats. For most

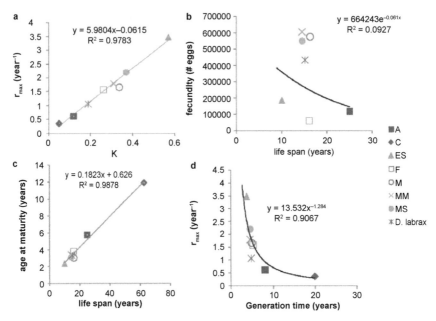

Figure 5. Life-history strategies of sea bass regarding the main strategies of fish ecological guilds in Atlanto Mediterranean fishes. M: marine species, MM: marine migrants, ES: estuarine especies, F: fresh water species, A: anadromous species (according tò data from Pérez-Ruzafa et al. in press dataset).

biological traits, sea bass is between the mean traits of anadromous and catadromous species and of fresh water, marine and marine migrant species. As Pérez-Ruzafa et al. (in press) underline, the high mortality rates of aquatic orgnisms in the larval phases condition the life strategies of species and when interpreting fish biological traits in the context of r/K-selection, they should be considered in relative terms rather than in absolute terms, especially if they are going to be compared with terrestrial organisms.

Sea bass strategy involves relatively lower intrinsic population (r_{max}) and somatic (K) growth rates, which are traits characteristic of K-strategists, than the average marine, marine migrants, fresh water and estuarine species (Fig. 5a). At the same time, fecundity per life span is lower than typically marine species, which here would have an r-strategy, but much higher than estuarine, freshwater and anadromous species which follow a K-strategy (Fig. 5b). Furthermore, sea bass also shows similar age at maturity and life span than typically marine and fresh water species, slightly higher than estuarine guild which would mark the r-strategy extreme for these traits, and lower than anadromous and catadromous fish which would determine the K-strategy extreme of the range (Fig. 5c).

Figure 5d shows the r_{max} vs. generation time relationship which is considered one of the best indices of an organism's position on the r-K continuum (Pianka 1970). The arms of the hyperbola would represent the extremes of an r/K continuum range. In the K-strategist extreme (low r_{max} and long generation time) are located the average catadromous fish and in the r-strategist extreme (high r_{max} and short generation time), the average estuarine guild. Sea bass is in an intermediate position of the curve, closer to the K extreme than typical marine fish, showing lower intrinsic population growth rates (r_{max}) than that corresponding to its relatively low generation time.

Response to Pollution

Inhabiting coastal areas and estuaries, sea bass is exposed to a large variety of potential pollutants from pesticides to crude oil spill, heavy metals or eutrophization and associated anoxic events. Although, when exposed to heavy metal contaminated sediments, this species is able to trigger a series of physiological adjustments or adaptations interfering with specific neuroendocrine control mechanisms that enable their long-term survival (De Domenico et al. 2013), pollutants can directly affect the physiology and biochemical process of the individuals or can alter their scope for growth, so that the energy that is earmarked to neutralize the effects of pollutants cannot be used for somatic growth, reproduction or the daily activity necessary for survival. Furthermore, pollutants that cannot be metabolized can accumulate in the tissues and organs of individuals.

For sea bass, food is by far the major source responsible for bioaccumulation. The high trophic level of the species makes it prone to accumulate contaminants through feeding (Loizeau et al. 2001b).

Among potential contaminants in coastal areas, and specially in coastal lagoons and estuaries with a high influence of agricultural land wastes, polychlorinated biphenyls (PCBs) are a group of organic chemicals that possess the inherent properties of compounds that bioaccumulate, i.e., high octanol/water partition coefficient (K_{ow}) and persistence (Pérez-Ruzafa et al. 2000, Loizeau et al. 2001a).

The contribution of water, via respiration, to the incorporation for most of the PCB congeners is less than 10% in sea bass and even less for the most chlorinated compound (Loizeau et al. 2001a). Therefore, feeding is the principal route for contamination, and the highest concentrations of contaminants are found in the oldest individuals (Loizeau et al. 2001b).

On the other hand, bioaccumulated substances concentrated in one individual are reduced after spawning (Loizeau et al. 2001a), but this means that contaminants are transmitted to progeny through the gametes.

Other than PCBs, in open coastal areas, many of the main spawning grounds of sea bass are close to the main routes of oil transport and where oil spills are not rare. Greater acute toxicity is generally associated with the lower molecular weight Polycyclic Aromatic Hydrocarbons (PAHs) whereas some high molecular weight PAHs form metabolites that can function as carcinogens (Gonzalez-Doncel et al. 2008).

Although sea bass individuals accumulate PAHs in lipid-rich tissues and muscles after 48 hours of petroleum exposure, this fish seems to tolerate reasonably well some levels of pollution and no prevalent mortality was observed during exposures of 28 days in the experiments performed by Kerambrun et al. (2012). Even some capacity to metabolize contaminants was observed by these researchers and a significant decrease in PAH concentrations takes place following 96 hours of exposure compared to 48 hours. After 26 days of depuration, all PAH is excreted (Kerambrun et al. 2012).

However, PAH absorption has metabolic costs and sub-lethal effects, producing a significant decrease in the sea bass specific growth rate in length, and for the RNA:DNA ratio after 28 days since exposure to light crude oil for 48 hours has ceased. After the 96-hour exposure to crude oil, specific growth rates in weight and condition index also decrease. Furthermore, PAH induced ethoxyresorufin-O-deethylase (EROD) and glutathione-S-transferase (GST) activity, even after depuration period in clean seawater (Kerambrun et al. 2012).

Reduction in growth rates can have relevant effects on population abundance. In areas where an oil spill took place, not only were 0-group bass particularly scarce in the nursery areas, probably due to increased mortality of the eggs or larvae, but also a high proportion of fish failed to reach 60 mm by the end of the summer, considered the critical length for survival through the first winter (Lancaster et al. 1998). However, these authors do not find enough evidence for the sub-lethal effect of oil contamination directly impacting the health of surviving fish once established within their nursery areas, or indirectly impacting them through effects on food supply.

Human activities and land use in coastal areas can also affect the populations of *D. labrax* mainly by altering aquatic food webs. Terrestrial herbivorous grazing activity in saltmarsh affects vegetal cover and the provision of litter to the aquatic ecosystems that sustain the populations of detrivorous amphipods. This leads to changes in species composition and relations of dominance in benthic assemblages in the aquatic ecosystem and to a reduction in food consumption by juvenile sea bass (Laffaille et al. 2000).

Management and Conservation of Sea Bass

Sea bass does not seem to be an endangered species. Although fishing effort data are not easy to obtain, total catches have been increasing through time, reaching, in 2010, 10,817 tons (FAO 2013) and do not show symptoms of overexploitation at global level. As a reference, the number of vessels fishing for sea bass in England and Wales increased from 1694 in 1991 to 2328 vessels in 2002 (Pawson 1992, ICES 2004). In the same period, total catch of sea bass in north eastern Atlantic reached 3319 tons and 4959 tons, respectively (FAO 2013).

Furthermore, at present, aquaculture production is ten times higher than fishing yields, reaching 125,901.7 tons in 2010 (FAO 2013). However, as commented above, at the local scale, populations can suffer a wide variety of impacts, including overexploitation.

The ideas and concepts on the management of exploited marine species populations have evolved rapidly in recent years. The traditional measures rested on the basis of singled-species models of population dynamics and the concept of maximum sustainable yield. They included the control of the catch and the establishment of seasonal closures and gear specifications to guarantee a minimum size of fished individuals for target species, and to ensure enough reproductive success and recruitment. However, these management measures have not produced the expected results and have given way to a more ecological approach to fishery management by developing the concept of "ecosystem approach to fisheries" (Pitcher et al. 1998, Jennings 2004) and among other management measures, to the concept of Marine Protected Areas (MPAs) (Pérez-Ruzafa et al. 2008).

In the Atlanto-Mediterranean region, MPAs have served to recover the abundance, size, structure, and genetic diversity of many target species of fisheries (Claudet et al. 2008, García-Charton et al. 2008) and these benefits are exported to the surrounding areas (Goñi et al. 2008, Harmelin-Vivien et al. 2008, Pérez-Ruzafa et al. 2008) with positive effects on the fishing yields (Higgins et al. 2008, Stelzenmüller et al. 2008, Vandeperre et al. 2011).

Although traditionally the bass fishing has been mainly concentrated in the coastal lagoons and estuaries, in the last years, as a result of protection, this species has also started to be fished around Marine Protected Areas in the Mediterranean. In this way, *D. labrax* is one of the main target species for fishermen close to many Marine Protected Areas in the Côte Bleue and it is fished using a diversity of gears like trammel net, combined net or longline (Leleu et al. 2012). In the Cabo de Palos-islas Hormigas MPA in the south western Mediterranean, although sea bass is a marginal target species, it has gone from not being fished before the creation of the MPA, to be one of the rising and conspicuous species in the catch (personal obs.). However, this could also be due to the indirect effect of fish farming plants established in the area, through the escape of fish and the potential effects of both activities, protection and aquaculture, and would deserve more detailed investigations.

Taking into account that sea bass populations share their life cycle in at least two well defined areas, spawning grounds and summer feeding areas, any protection measurement must include both areas, not only from overexploitation but also from pollution agents that could affect the larval development and food sources.

Beyond protecting populations in MPAs, other management measures can be implemented. On the basis that 55% of recaptures of tagged adult individuals (>40 cm total length) in successive years were within 16 km of their original release position, Pawson et al. (2008) suggested that mortality rates of adult bass in local populations could be reduced by around 50% if a number of carefully selected areas were designated as catch and release only for bass, thus providing a management option to improve recreational sea anglers' fishing.

References

Abdel-Hakim, N.F., S.F. Mehanna, I.A. Eisa, M.S. Hussein, D.A. Al-Azab and A.S. Ahmed. 2010. Length weight relationship, condition factor and stomach contents of the European Seabass, *D. labrax* at Bardawil Lagoon, North Sinai, Egypt. Proc. of the 3rd Global Fisheries & Aquaculture Research Conference. pp. 245–256.

Alliot, E., A. Pastoureaud and H. Thebault. 1983. Influence de la température et de la salinité sur la croissance et la composition corporelle d'alevins de *Dicentrarchus labrax*. Aquaculture 31: 181–194.

Amanieu, M., B. Baleux, O. Guelorguet and P. Michel. 1975. Etude Biologique et hydrologique d'une crise dystrophique (malaigue) dans l'étang du Prévost a Palavas (Hérault). Vie Milieu 25 B(2): 175–204.

Amara, R., F. Lagardère, Y. Desaunay and J. Marchand. 2000. Metamorphosis and estuarine colonization in the common sole, *Solea solea* (L.): implications for recruitment regulation. Oceanologica Acta 23: 469–484.

Aranda, A., F.J. Sánchez-Vázquez and J.A. Madrid. 1999. Influence of water temperature on demand-feeding rhythms in sea bass. J. Fish Biol. 55: 1029–1039.

Arias, A. 1980. Crecimiento, régimen alimentario y reproducción de la dorada (*Sparus aurata* L.) y del robalo (*Dicentrarchus labrax* L.) en los esteros de Cádiz. Investigación Pesquera, Barcelona 44: 59–83.

Bachelet, G., X. De Montaudouin, I. Auby and P.J. Labourg. 2000. Seasonal changes in macrophyte and macrozoobenthos assemblages in three coastal lagoons under varying degrees of eutrophication. Ices J. Mar. Sci. 57(5): 1495–1506.

Bagdonas, K., N. Nika, G. Bristow, R. Jankauskienė, A. Salyté and A. Kontautas. 2011. First record of *Dicentrarchus labrax* (Linnaeus 1758) from the southeastern Baltic Sea (Lithuania). J. Appl. Ichthyol. 27(6): 1390–1391.

Bahri-Sfar, L., C. Lemaire, O.K. Ben Hassine and F. Bonhomme. 2000. Fragmentation of sea bass populations in the western and eastern Mediterranean as revealed by microsatellite polymorphism. Proc. R. Soc. Lond. B 267: 929–935.

Bahri-Sfar, L., C. Lemaire, B. Chatain, P. Divanach, O.K. Ben Hassine and F. Bonhomme. 2005. Impact de l'élevage sur la structure génétique des populations méditerranéennes de *Dicentrarchus labrax*. Aquat. living Resour. 18(1): 71–76.

Barnabé, G. 1986. L'élevage du loup et de la daurade. pp. 627–666. *In*: G. Barnabé (ed.). Aquaculture. Technique et documentation-Lavoisier, Paris.

Barnabé, G. 1990. L'élevage du loup et de la daurade. pp. 627–666. *In*: G. Barnabé (ed.). Aquaculture. Technique et documentation-Lavoisier, Paris.

Barnes, R.S.K. 1980. Coastal Lagoons. Cambridge Studies in Modern Biology I. Cambridge University Press, Cambridge.

Bégout Anras, M.-L. and J.-P. Lagardère. 1998. Variabilité météorologique et hydrologique. Conséquences sur l'activité natatoire d'un poisson marin. Biologie animale CR Acad. Sci. Pris. Sciences de la vie 321: 641–648.

Bevelhimer, M. and W. Bennett. 2000. Assessing cumulative thermal stress in fish during chronic intermittent exposure to high temperatures. Environ. Sci. Technol. 3: 211–216.

Blázquez, M., S. Zanuy, M. Carrillo and F. Piferrer. 1998. Effects of rearing temperature on sex differentiation in the European Sea Bass (*Dicentrarchus labrax* L.). Journal of Experimental Zoology 281: 207–216.

Bonhomme, F., M. Naciri, L. Bahri-Sfar and C. Lemaire. 2002. Comparative analysis of genetic structure of two closely related sympatric marine fish species *Dicentrarchus labrax* and *Dicentrarchus punctatus*. Comptes Rendus Biologies 325(3): 213–220.

Boulineau-Coatanea, F. 1969. Régime alimentaire du bar (*Dicentrarchus labrax*, Serranidae)-Sur la côte atlantique bretonne. Bulletin du Museum d'Histoire Naturelle 2(41): 1106–1122.

Boutiere, H., F. Bovee, D. Delille, M. Fiala, C. Gros, G. Jacques, M. Knoepffler, J.P. Labat, M. Panouse and C. Soyer. 1982. Effect d'une crise distrophique dasn l'étang de Salses-Leucate. Oceanol. Acta. Actes Symposium International sur les lagunes cotieres 31–242.

Butigieg, J. 1927. La despoblación del Mar Menor y sus causas. 133: 251–286. *In*: Boletín de Pescas. Dirección General de Pesca del Ministerio de Marina. Instituto Español de Oceanografía.

Cabral, H. and M.J. Costa. 2001. Abundance, feeding ecology and growth of 0-group sea bass, *Dicentrarchus labrax*, within the nursery areas of the Tagus estuary. Journal of the Marine Biological Association of the United Kingdom 81(4): 679–682.

Castilho, R. and B.J. McAndrew. 1998. Population structure of sea bass in Portugal: evidence from allozymes. J. Fish Biol. 53(5): 1038–1049.

Chatterjee, N., A.K. Pal, S.M. Manush, T. Das and S.C. Mukherjee. 2004. Thermal tolerance and oxygen consumption of *Labeo rohita* and *Cyprinus carpio* early fingerlings acclimated to three different temperatures. J. Therm. Biol. 29(6): 265–270.

Claireaux, G. and J.-D. Dutil. 1992. Physiological response of the Atlantic cod, *Gadus morua*, to hypoxia at various environmental salinities. J. Exp. Biol. 163: 97–118.

Claireaux, G. and J.-P. Lagardère. 1999. Influence of temperature, oxygen and salinity on the metabolism of the European sea bass. J. Sea Res. 42: 157–168.

Claudet, J., C.W. Osenberg, L. Benedetti-Cecchi, P. Domenici, J.A. García-Charton, A. Pérez-Ruzafa, F. Badalamenti, J. Bayle-Sempere, A. Brito, F. Bulleri, J.M. Culioli, M. Dimech, J.M. Falcón, I. Guala, M. Milazzo, J. Sánchez-Meca, P.J. Somerfield, B. Stobart, F. Vandeperre, C. Valle and S. Planes. 2008. Marine reserves: Size and age do matter. Ecol. Lett. 11: 481–489.

Conides, A.J. and B. Glamuzina. 2006. Laboratory simulation of the effects of environmental salinity on acclimation, feeding and growth of wild-caught juveniles of European sea bass *Dicentrarchus labrax* and gilthead sea bream, *Sparus aurata*. Aquaculture 256: 235–245.

Coscia, I., E. Desmarais, B. Guinand and S. Mariani. 2012. Phylogeography of European sea bass in the north-east Atlantic: a correction and reanalysis of the mitochondrial DNA data from Coscia & Mariani (2011). Biol. J. Linnean Soc. 106(2): 455–458.

Costa, M.J. 1988. Écologie alimentaire des poissons de l'estuarie du Tage. Cybium 12(4): 301–320.

Coutant, C.C. 1986. Thermal niches of striped bass. Sci. Am. 254: 98–104.

Das, T., A.K. Pal, S.K. Chakraborty, S.M. Manush, N. Chatterjee and S.C. Mukherjee. 2004. Thermal tolerance and oxygen consumption of Indian Major Carps acclimated to four temperatures. J. Therm. Biol. 29(3): 157–163.

De Domenico, E., A. Mauceria, D. Giordano, M. Maisano, A. Giannetto, V. Parrino, A. Natalotto, A. D'Agata, T. Cappello and S. Fasulo. 2013. Biological responses of juvenile European sea bass (*Dicentrarchus labrax*) exposed to contaminated sediments. Ecotox. Environ. Safe. 97: 114–123.

Dülger, N., M. Kumlu, S. Türkmen, A. Olçcülü, O.T. Eroldogan, H.A. Yılmaz and N. Oçal. 2012. Thermal tolerance of European Sea Bass (*Dicentrarchus labrax*) juveniles acclimated to three temperature levels. J. Therm. Biol. 37: 79–82.

Dufour, V., M. Cantou and F. Lecomte. 2009. Identification of sea bass (*Dicentrarchus labrax*) nursery areas in the north-western Mediterranean Sea. J. Mar. Biol. Assoc. UK 89(7): 1367–1374.

Enders, E.C., D. Boisclair and A.G. Roy. 2003. The effect of turbulence on the cost of swimming for juvenile Atlantic salmon (*Salmo salar*). Can. J. Fish. Aquat. Sci. 60(9): 1149–1160.

Elliott, M. and F. Dewailly. 1995. The structure and components of European estuarine fish assemblages. Netherlands Journal of Aquatic Ecology 29(3–4): 397–417.

Elliott, M., A.K. Whitfield, I.C. Potter, S.J.M. Blaber, D.P. Cyrus, F.G. Nordlie and T.D. Harrison. 2007. The guild approach to categorizing estuarine fish assemblages: a global review. Fish Fish. 8: 241–268.

FAO. 2013. Food and Agriculture Organization of the United Nations. http://www.fao.org/fishery/species/2291/en.

Ferrari, I., V.U. Ceccherelli, M. Naldi and P. Viaroli. 1993. Planktonic and benthic communities in a shallow-water dystrophic lagoon. Verh. Int. Ver. Limnol. 25: 1043–1047.

Franco, A., M. Elliott, P. Franzoi and P. Torricelli. 2008. Life strategies of fishes in European estuaries: the functional guild approach. Mar. Ecol. Prog. Ser. 354: 219–228.

Fritsch, M., Y. Morizur, E. Lambert, F. Bonhommeb and B. Guinand. 2007. Assessment of sea bass (*Dicentrarchus labrax* L.) stock delimitation in the Bay of Biscay and the English Channel based on mark-recapture and genetic data. Fish. Res. 83: 123–132.

Froese, R. and D. Pauly. (eds.). 2012. FishBase. World Wide Web electronic publication. www.fishbase.org, version (12/2012).

Froese, R. and D. Pauly. (eds). 2013. FishBase. World Wide Web electronic publication. www.fishbase.org, version (02/2013).

Galbraith, P.S., H.I. Browman, R.G. Racca, A.B. Skiftesvik and J.F. Saint-Pierre. 2004. Effect of turbulence on the energetics of foraging in Atlantic cod *Gadus morhua* larvae. Mar. Ecol. Prog. Ser. 281: 241–257.

Gamito, S., J. Gilabert, C. Marcos and A. Pérez-Ruzafa. 2005. Effects of changing environmental conditions on lagoon ecology. pp. 193–229. *In*: I.E. Gönenç and J.P. Wolflin (eds.). Coastal Lagoons: Ecosystem Processes and Modeling for Sustainable Use and Development. CRC Press, Boca Raton, Florida.

García-Charton, J.A., A. Pérez-Ruzafa, C. Marcos, J. Claudet, F. Badalamenti, L. Benedetti-Cecchi, J.M. Falcón, M. Milazzo, P.J. Schembri, B. Stobart, F. Vandeperre, A. Brito, R. Chemello, M. Dimech, P. Domenici, I. Guala, L. Le Diréach, E. Maggi and S. Planes. 2008. Effectiveness of European Atlanto-Mediterranean MPAs: Do they accomplish the expected effects on populations, communities and ecosystems? J. Nat. Conserv. 16(4): 193–221.

Garcia De León, F.J., L. Chikhi and F. Bonhomme. 1997. Microsatellite polymorphism and population subdivision in natural populations of European sea bass *Dicentrarchus labrax* (Linnaeus, 1758). Mol. Ecol. 6: 51–62.

Georgalas, V., S. Malavasi, P. Franzoi and P. Torricelli. 2007. Swimming activity and feeding behaviour of larval European sea bass (*Dicentrarchus labrax* L.): Effects of ontogeny and increasing food density. Aquaculture 264: 418–427.

Gianmarco, G., R. Azzoni, M. Bartoli and P. Viaroli. 1997. Seasonal variations of sulphate reduction rates, sulphur pools and iron availability in the sediment of a dystrophic lagoon (Sacca di Goro, Italy). Water Air Soil Poll. 99(1–4): 363–371.

González, R.J. 2012. The physiology of hyper-salinity tolerance in teleost fish: a review. Journal of Comparative Physiology B-Biochemical Systemic and Environmental Physiology 182(3): 321–329.

González-Doncel, M., L. González, C. Fernández-Torija, J.M. Navas and V. Tarazona. 2008. Toxic effects of an oil spill on fish early life stages may not be exclusively associated to PAHs: studies with Prestige oil and medaka (*Oryzias latipes*). Aquat. Toxicol. 87: 280–288.

González, R.J. and D.G. McDonald. 1992. The relationship between oxygen consumption and ion loss in a freshwater fish. J. Exp. Biol. 163: 317–332.

Goñi, R., S. Adlerstein, D. Alvarez-Berastegui, A. Forcada, O. Reñones, G. Criquet, S. Polti, G. Cadiou, C. Valle, P. Lenfant, P. Bonhomme, A. Pérez-Ruzafa, J.L. Sánchez-Lizaso, J.A. García-Charton, G. Bernard, V. Stelzenmüller and S. Planes. 2008. Spillover from six western Mediterranean marine protected areas: evidence from artisanal fisheries. Mar. Ecol. Prog. Ser. 366: 159–174.

Grati, F., G. Scarcella, L. Bolognini and G. Fabi. 2011. Releasing of the European sea bass *Dicentrarchus labrax* (Linnaeus) in the Adriatic Sea: Large-volume versus intensively cultured juveniles. J. Exp. Mar. Biol. Ecol. 397: 144–152.

Gutiérrez, E. and B. Morales-Nin. 1986. Time series analysis of daily growth in *Dicentrarchus labrax* L. otoliths. J. Exp. Mar. Biol. Ecol. 103: 163–179.

Guyoneaud, R., R. De Wit, R. Matheron and P. Caumette. 1998. Impact of macroalgal dredging on dystrophic crises and phototrophic bacterial blooms (red waters) in a brackish coastal lagoon. Oceanologica Acta 21(4): 551–561.

Haffray, P., C.S. Tsigenopoulos, F. Bonhomme, B. Chatain, A. Magoulas, M. Rye, A. Triantafyllidis and C. Triantaphyllidis. 2007. European sea bass—*Dicentrarchus labrax*. Genimpact final scientific report. pp. 40–46.

Harmelin-Vivien, M., L. Le Diréach, J. Bayle-Sempere, E. Charbonnel, J.A. García-Charton, D. Ody, A. Pérez-Ruzafa, O. Reñones, P. Sánchez Jerez and C. Valle. 2008. Gradients of abundance and biomass across reserve boundaries in six Mediterranean marine protected areas: Evidence of fish spillover? Biol. Conserv. 141(7): 1829–1839.

Henderson, P.A., R.M.H. Seaby and J.R. Somes. 2011. Community level response to climate change: The long-term study of the fish and crustacean community of the Bristol Channel J. Exp. Mar. Biol. Ecol. 400: 78–89.

Herbst, G.N. and H.J. Bromley. 1984. Relationships between habitat stability, ionic composition, and the distribution of aquatic invertebrates in the desert regions of Israel. Limnol. Oceanogr. 29(3): 495–503.

Hettler, W.F., 1976. Influence of temperature and salinity on routine metabolic rate and growth of young Atlantic menhaden. J. Fish Biol. 8: 55–65.

Higgins, R.M., F. Vandeperre, A. Pérez-Ruzafa and R.S. Santos. 2008. Priorities for fisheries in marine protected area design and management: Implications for artisanal-type fisheries as found in southern Europe. J. Nat. Conserv. 16(4): 222–233.

ICES. 2004. Report of the Study Group on Bass, August 2003. ICES Document CM 2004/ACFM: 04. 73 pp.

Jennings, S. 2004. The ecosystem approach to fishery management: A significant step towards sustainable use of the marine environment. Mar. Ecol. Prog. Ser. 274: 279–282.

Jennings, S. and M.G. Pawson. 1992. The origin and recruitment of bass, *Dicentrarchus labrax*, larvae to nursery areas. J. Mar. Biol. Assoc. UK 72: 199–212.

Jian, C.Y., S.Y. Cheng and J.C. Chen. 2003. Temperature and salinity tolerances of yellow fin sea bream, *Acanthopagrus latus*, at different salinity and temperature levels. Aquacult. Res. 34: 175–185.

Johnson, D.V. and I. Katavic. 1986. Survival and growth of sea bass (*Dicentrarchus labrax*) larvae as influenced by temperature, salinity, and delayed initial feeding. Aquaculture 52: 11–19.

Kennedy, M. and P. Fitzmaurice. 1972. The biology of bass, *Dicentrarchus labrax*, in Irish waters. J. Mar. Biol. Assoc. UK 52: 557–597.

Kerambrun, E., S. Le Floch, W. Sanchez, H.T. Guyon, T. Meziane, F. Henry and R. Amara. 2012. Responses of juvenile sea bass, *Dicentrarchus labrax*, exposed to acute concentrations of crude oil, as assessed by molecular and physiological biomarkers. Chemosphere 87: 692–702.

Kiørboe, T. and B. MacKenzie. 1995. Turbulence-enhanced prey encounter rates in larval fish: effects of spatial scale, larval behaviour and size. J. Plankton Res. 17(12): 2319–2331.

Kiørboe, T. and E. Saiz. 1995. Planktivorous feeding in calm and turbulent environments, with emphasis on copepods. Mar. Ecol. Prog. Ser. 122(1–3): 135–145.

Kjerfve, B. (ed.). 1994. Coastal Lagoon Processes. Elsevier Oceanography Series 60. Elsevier, Amsterdam 577 pp.

Kumlu, M., M. Kumlu and S. Turkmen. 2010. Combined effects of temperature and salinity on critical thermal minima of pacific white shrimp *Litopenaeus vannamei* (Crustacea: Penaeidae). J. Therm. Biol. 35: 302–304.

Laffaille, P., J.-C. Lefeuvre and E. Feunteun. 2000. Impact of sheep grazing on juvenile sea bass, *Dicentrarchus labrax* L., in tidal salt marshes. Biol. Conserv. 96: 271–277.

Laffaille, P. J.-C. Lefeuvre, M.-T. Schricke and E. Feunteun. 2001. Feeding Ecology of 0-Group sea bass, *Dicentrarchus labrax*, in salt marshes of mont saint michel bay (France). Estuaries 24: 116–125.

Lancaster, J.E., M.G. Pawson, G.D. Pickett and S. Jennings. 1998. The Impact of the 'Sea Empress' Oil Spill on Sea bass Recruitment. Mar. Pollut. Bull. 30(9): 677–688.

Leleu, K., F. Alban, D. Pelletier, E. Charbonnel, Y. Letourneur and C.F. Boudouresque. 2012. Fishers' perceptions as indicators of the performance of Marine Protected Areas (MPAs) Marine Policy 36: 414–422.

Lemaire, C., G. Allegrucci, M. Naciri, L. Bahri-Sfar, H. Kara and F. Bonhomme. 2000. Do discrepancies between microsatellite and allozyme variation reveal differential selection between sea and lagoon in the sea bass (*Dicentrarchus labrax*)? Mol. Ecol. 9(4): 457–467.

Loizeau, V., A. Abarnou, P. Cugier, A. Jaouen-Madoulet, A.-M. Le Guellec and A. Menesguen. 2001a. A model of PCB bioaccumulation in the Sea Bass food web from the Seine Estuary (Eastern english Channel). Mar. Pollut. Bull. 43(7–12): 242–255.

Loizeau, V., A. Abarnou and A. Menesguen. 2001b. A steady-state model of PCB bioaccumulation in the sea bass (Dicentrachus labrax) food wed from the Seine estuary, France. Estuaries 24: 1074–1087.

Lupandin, A.I. 2005. Effect of flow turbulence on swimming speed of fish. Biol. Bull. 32(5): 461–466.

MacKenzie, B.R. and T. Kiørboe. 1995. Encounter rates and swimming behavior of pausetravel and cruise larval fish predators in calm and turbulent laboratory environments. Limnol. Oceanogr. 40(7): 1278–1289.

MacKenzie, B.R. and T. Kiørboe. 2000. Larval fish feeding and turbulence: a case for the downside. Limnol. Oceanogr. 45(1): 1–10.

MacKenzie, B.R., T.J. Miller, S. Cyr and W.C. Leggett. 1994. Evidence for a dome-shaped relationship between turbulence, and larval fish ingestion rates. Limnol. Oceanogr. 39(8): 1790–1799.

Mahjoub, M.-S., S. Souissi, F.G. Michalec, F.G. Schmitt and J.-S. Hwang. 2011. Swimming kinematics of *Eurytemora affinis* (Copepoda, Calanoida) reproductive stages and differential vulnerability to predation of larval *Dicentrarchus labrax* (Teleostei, Perciformes). J. Plankton Res. 33(7): 1095–1103.

Mahjoub, M.-S., R. Kumar, S. Souissi, F.G. Schmitt and J.-S. Hwang. 2012. Turbulence effects on the feeding dynamics in European sea bass (*Dicentrarchus labrax*) larvae. J. Exp. Mar. Biol. Ecol. 416–417: 61–67.

Manush, S.M., A.K. Pal, N. Chatterjee, T. Das and S.C. Mukherjee. 2004. Thermal tolerance and oxygen consumption of Macrobrachium rosenbergii acclimated to three temperatures. J. Therm. Biol. 29(1): 15–19.

Margalef, R. 1974. Ecología. Ediciones Omega, Barcelona.

Mariani, P., V. Botte and M.R. d'Alcala'. 2005. An object-oriented modelfor the prediction of turbulence effects on plankton. Deep-Sea Res. II (52): 1287–1307.

Mariani, P., B.R. MacKenzie, A.W. Visser and V. Botte. 2007. Individual-based simulations of larval fish feeding in turbulent environments. Mar. Ecol. Prog. Ser. 347: 155–169.

Martinho, F., M. Dolbeth, I. Viegas, C.M. Teixeira, H.N. Cabral and M.A. Pardal. 2009. Environmental effects on the recruitment variability of nursery species. Estuar. Coast. Shelf Sci. 83: 460–468.

Mehanna Sahar, F., A.A. El-Aiatt, M. Ameran and M. Salem. 2010. Dynamics and fisheries regulations for the European Sea bass *Dicentrarchus labrax* (Moronidae) at Bardawil lagoon, Egypt. Proc. of the 3rd Global Fisheries & Aquaculture Research Conference. pp. 199–209.

Meynecke, J.O., S.Y. Lee, N.C. Duke and J. Warken. 2006. Effect of rainfall as a component of climate change on estuarine fish production in Queensland, Australia. Estuar. Coast. Shelf Sci. 69: 491–504.

Montgomery, J., G. Carton, R. Voigt, C. Baker and C. Diebel. 2000. Sensory processing of water currents by fishes. Philos. Trans. R. Soc. Lond. B 355(1401): 1325–1327.

Moreira, F., C.A. Assis, P.R. Almeida, J.L. Costa and M.J. Costa. 1992. Trophic Relationships in the Community of the Upper Tagus Estuary (Portugal): a Preliminary Approach. Estuar. Coast. Shelf Sci. 34: 617–623.

Naciri, M., C. Lemaire, P. Borsa and F. Bonhomme. 1999. Genetic study of the Atlantic/ Mediterranean transition in sea bass (*Dicentrarchus labrax*). Journal of Heredity 90(6): 591–596.

Neill, W.H., J.M. Miller, H.W. Van der Veer and K.O. Winemiller. 1994. Ecophysiology of marine fish recruitment: a conceptual framework for understanding interannual variability. Netherlands Journal of Sea Research 32: 135–152.

Nordlie, F.G. 1978. The influence of environmental salinity on respiratory oxygen demands in the euryhaline teleost, *Ambassis interrupta* Bleeker. Comp. Biochem. Physiol. 59: 271–274.

Patarnello, T., L. Bargelloni, F. Caldara and L. Colombo. 1993. Mitochondrial DNA sequence variation in the European sea bass, *Dicentrarchus labrax* L. (Serranidae): evidence of differential haplotype distribution in natural and farmed populations. Mol. Mar. Biol. Biotechnol. 2(6): 333–7.

Pavlidis, M., G. Koumoundouris, A. Serioti, S. Somarakis, P. Divanach and M. Kentouri. 2000. Evidence of temperature-dependent sex determination in the European sea bass (*Dicentrarchus labrax* L.). J. Exp. Zool. 287: 225–232.

Pavlov, D.S., A.I. Lupandin and M.A. Skorobogatov. 2000. The effects of flow turbulence on the behavior and distribution of fish. J. Ichthyol. 40: S232–S261.

Pawson, M.G. 1992. Climatic influences on the spawning success, growth and recruitment of bass (*Dicentrarchus labrax* L.) in British waters. ICES Marine Science Symposia 195: 388–392.

Pawson, M.G. and G.D. Pickett. 1996. The annual pattern of condition and maturity in bass, *Dicentrarchus labrax*, in waters around England and Wales. J. Mar. Biol. Ass. UK 76: 107–125.

Pawson, M.G., D.F. Kelley and G.D. Pickett. 1987. The distribution and migrations of bass, *Dicentrarchus labrax* L., in waters around England and Wales as shown by tagging. J. Mar. Biol. Ass. UK 67: 183–217.

Pawson, M.G., G.D. Pickett, J. Leballeur, M. Brown and M. Fritsch. 2007. Migrations, fishery interactions, and management units of sea bass (*Dicentrarchus labrax*) in Northwest Europe. ICES Journal of Marine Science 64: 332–345.

Pawson, M.G., M. Brown, J. Leballeur and G.D. Pickett. 2008. Will philopatry in sea bass, *Dicentrarchus labrax*, facilitate the use of catch-restricted areas for management of recreational fisheries? Fisheries Research 93: 240–243.

Pérez-Ruzafa, A., S. Navarro, A. Barba, C. Marcos, M.A. Camara, F. Salas and J.M. Gutierrez. 2000. Presence of pesticides throughout trophic compartments of the food web in the Mar Menor lagoon (SE of Spain). Mar. Pollut. Bull. 40(2): 140–151.

Pérez-Ruzafa, A., M.C. Mompeán and C. Marcos. 2007. Hydrographic, geomorphologic and fish assemblage relationships in coastal lagoons. Hydrobiologia 577: 107–125.

Pérez-Ruzafa, A., E. Martín, C. Marcos, J.M. Zamarro, B. Stobart, M. Harmelin-Vivien, S. Polti, S. Planes, J.A. García-Charton and M. González-Wangüemert. 2008. Modelling spatial and temporal scales for spill-over and biomass exportation from MPAs and their potential for fisheries enhancement. J. Nat. Conserv. 16(4): 234–255.

Pérez-Ruzafa, A., C. Marcos and I.M. Pérez-Ruzafa. 2010. Mediterranean coastal lagoons in an ecosystem and aquatic resources management context. Phys. Chem. Earth 36: 160–166.

Pérez-Ruzafa, Á., C. Marcos and I.M. Pérez-Ruzafa. 2011a. Recent advances in coastal lagoons ecology: evolving old ideas and assumptions. Transit. Waters Bull. 5(1): 50–74.

Pérez-Ruzafa, A., C. Marcos, I.M. Pérez-Ruzafa and M. Pérez-Marcos. 2011b. Coastal lagoons: "transitional ecosystems" between transitional and coastal waters. J. Coast. Conserv. 15(3): 369–392.

Pérez-Ruzafa, A., C. Marcos, I.M. Pérez-Ruzafa and M. Pérez-Marcos. 2013. Are coastal lagoons physically or biologically controlled ecosystems? Revisiting r vs. K strategies in coastal lagoons and estuaries, Estuarine, Coastal and Shelf Science 132: 17–33.

Pianka, E.R. 1970. On r and K selection. Am. Nat. 104: 592–597.

Pickett, G.D. and M.G. Pawson. 1994. Sea Bass—Biology, Exploitation and Conservation. Fish and Fisheries Series 12. Chapman and Hall, London.

Pickett G.D., D.F. Kelley and M.G. Pawson. 2004. The patterns of recruitment of sea bass, *Dicentrarchus labrax* L. from nursery areas in England and Wales and implications for fisheries management. Fish. Res. 68: 329–342.

Pitcher, T.J., P.J.B. Hart and D. Pauly. (eds.). 1998. Reinventing Fisheries Management. Kluwer Academic Publishers, Dordrecht, The Netherlands.

Quéré,N., E. Desmarais, C.S. Tsigenopoulos, K. Belkhir, F. Bonhomme and B. Guinand. 2012. Gene flow at major transitional areas in sea bass (*Dicentrarchus labrax*) and the possible emergence of a hybrid swarm. Ecology and Evolution 2(12): 3061–3078.

Re, A.D., F. Diaz, E. Sierra, J. Rodriguez and E. Perez. 2005. Effect of salinity and temperature on thermal tolerance of brown shrimp *Farfantepenaeus aztecus* (Ives) (Crustacea, Penaeidae). J. Therm. Biol. 30: 618–622.

Riley, W.D., A.T. Ibbotson, W.R.C. Beaumont, M.G. Pawson, A.C. Cook and P.I. Davison. 2011. Predation of the juvenile stages of diadromous fish by sea bass (*Dicentrarchus labrax*) in the tidal reaches of an English chalk stream. Aquat. Conserv. 21(3): 307–312.

Rothschild, B.J. and T.R. Osborn. 1988. Small-scale turbulence and plankton contact rates. J. Plankton Res. 10(3): 465–474.

Russell, N.R., J.D. Fish and R.J. Wootton. 1996. Feeding and growth of juvenile sea bass: the effect of ration and temperature on growth rate and efficiency. J. Fish. Biol. 49: 206–220.

Sa, R., C. Bexiga, P. Veiga, L. Vieira and K. Erzini. 2006. Feeding ecology and trophic relationships of fish species in the lower Guadiana River Estuary and Castro Marim e Vila Realde Santo António Salt Marsh. Estuar. Coast. Shelf Sci. 70: 19–26.

Saillant, E., A. Fostier, B. Menu, P. Haffray and B. Chatain. 2001. Sexual growth dimorphism in sea bass *Dicentrarchus labrax*. Aquaculture 202: 371–387.

Saillant, E., A. Fostier, P. Haffray, B. Menu, J. Thimonier and B. Chatain. 2002. Temperature effects and genotype—temperature interactions on sex determination in the European Sea Bass (*Dicentrarchus labrax* L.). J. Exp. Zool. 292: 494–505.

Saillant, E., A. Fostier, P. Haffray, B. Menu and B. Chatain. 2003. Saline preferendum for the European sea bass, *Dicentrarchus labrax*, larvae and juveniles: effect of salinity on early development and sex determination. J. Exp. Mar. Biol. Ecol. 287: 103–117.

Saiz, E., M. Alcaraz and G.-A. Paffenhöfer. 1992. Effects of small-scale turbulence on feeding rate and gross-growth efficiency of three *Acartia* species (Copepoda: Calanoida). J. Plankton Res. 14(8): 1085–1097.

Sakka Hlaili, A., B. Grami, H. HadjMabrouk, M. Gosselin and D. Hamel. 2007. Fate of phytoplankton production in a restricted mediterranean lagoon. Rapp. Comm. int. Mer Médit. 38: 391.

Sfriso, A., B. Pavoni and A. Marcomini. 1995. Nutrient distributions in the surface sediment of the central lagoon of Venice. Sci. Total Environ. 172(1): 21–35.

Specchiulli, A., R. D'Adamo, M. Renzi, F. Vignes, A. Fabbrocini, T. Scirocco, L. Cilenti, M. Florio, P. Breber and A. Basset. 2009. Fluctuations of physicochemical characteristics in sediments and overlying water during an anoxic event: a case study from Lesina lagoon (SE Italy). Transit. Waters Bull. 3(2): 15–32.

Spitz, J., T. Chouvelon, M. Cardinaud, C. Kostecki and P. Lorance. 2013. Prey preferences of adult sea bass *Dicentrarchus labrax* in the northeastern Atlantic: implications for bycatch of common dolphin *Delphinus delphis*—ICES Journal of Marine Science, doi.10.1093/icesjms/fss200.

Stelzenmüller, V., F. Maynou, G. Bernard, G. Cadiou, M. Camilleri, R. Crech'riou, G. Criquet, M. Dimech, O. Esparza, R. Higgins, P. Lenfant and A. Pérez-Ruzafa. 2008. Spatial assessment of fishing effort around European marine reserves: Implications for a successful fisheries management. Marine Pollution Bulletin, 56(12): 2018–2026.

Thetmeyer, H., U. Waller, K.D. Black, S. Inselmann and H. Rosenthal. 1999. Growth of European sea bass (*Dicentrarchus labrax* L.) under hypoxic and oscillating oxygen conditions. Aquaculture 174: 355–367.

Thompson, B.M. and R.T. Harrop. 1987. The distribution and abundance of bass (*Dicentrarchuslabrax*) eggs and larvae in the English-channel and southern North-sea. J. Mar. Biol. Ass. UK 67(2): 263–274.

Tortonese, E. 1986. Moronidae. 2: 793–796. *In:* P.J.P. Whitehead, M.-L. Bauchot, J.-C. Hureau, J. Nielsen and E. Tortonese (eds.). Fishes of the north-eastern Atlantic and the Mediterranean. UNESCO, Paris.

UNESCO. 1980. Coastal Lagoon Survey, UNESCO Technical Papers in Marine Science 31.

UNESCO. 1981. Coastal lagoons research, present and future. UNESCO Technical Papers in Marine Science 33.

Vandeperre, F., R.M. Higgins, J. Sánchez-Meca, F. Maynou, R. Goñi, P. Martín-Sosa, A. Pérez-Ruzafa, P. Afonso, I. Bertocci, R. Crec'hriou, G. D'Anna, M. Dimech, C. Dorta, O. Esparza, J.M. Falcón, A. Forcada, I. Guala, L. Le Direach, C. Marcos, C. Ojeda-Martínez, C. Pipitone, P.J. Schembri, V. Stelzenmüller, B. Stobart and R.S. Santos. 2011. Effects of no-take area size and age of marine protected areas on fisheries yields: A meta-analytical approach. Fish Fish. 12(4): 412–426.

Vandeputte, M., E. Quillet and B. Chatain. 2012. Are sex ratios in wild European sea bass (*Dicentrarchus labrax*) populations biased? Aquat. Living Resour. 25: 77–81.

Vasconcelos, R.P., P. Reis-Santos, S. Tanner, A. Maia, C. Latkoczy, D. Günther, M.J. Costa and H. Cabral. 2008. Evidence of estuarine nursery origin of five coastalfish species along the Portuguese coast through otolith elemental fingerprints. Estuar. Coast. Shelf Sci. 79: 317–327.

Viaroli, P., M. Bartoli, C. Bondavalli, R. Christian, G. Giordani and M. Naldi. 1996. Macrophyte communities and their impact on benthic fluxes of oxygen, sulphide and nutrients in shallow eutrophic environment. Hydrobiologia 329: 93–103.

Von Oertzen, J.A. 1984. Influence of steady-state and fluctuating salinities on the oxygen consumption and activity of some brackish water shrimps and fishes. J. Exp. Mar. Biol. Ecol. 80: 29–46.

Wassef, E. and H. El Emary. 1989. Contribution to the biology of bass, *Dicentrarchus labrax* L. in the Egyptian Mediterranean waters off Alexandria. Cybium 13(4): 327–345.

Yen, J., K.D. Rasberry and D.R. Webster. 2008. Quantifying copepod kinematics in a laboratory turbulence apparatus. J. Mar. Syst. 69(3–4): 283–294.

Zalizniak, L., B. Kefford and D. Nugegoda. 2006. 'Is all salinity the same? I. The effect of ionic compositions on the salinity tolerance of five species of freshwater invertebrates'. Mar. Freshwater Res. 57: 75–82.

Zanuy, S. and M. Carrillo. 1984. La salinité: un moyen pour retarder la ponte du bar. pp. 73–80. *In*: G. Barnabé and R. Billard (eds.). L'aquaculture du bar. INRA Publication, Paris.

The Biological Clock and Dualism

Ana del Pozo,[1] Jack Falcon[2] and F. Javier Sanchez-Vazquez[1],*

Geophysical Cycles and Biological Rhythms

The earth's environment is not static, since periodic changes with different periodicities occur, driven by geophysical cycles (Fig. 1). The earth's rotation around its axis imposes daily cycles of light and darkness, while the translation around the Sun provokes annual variations (i.e., seasons). Moreover, the rotation of the moon around the earth causes tidal as well as lunar cycles. Therefore, it should not be surprising that most living organisms have developed a biological clock to keep time and cope with these predictable changes by anticipating the forthcoming of a recurrent event in their habitat. Fish in general, and sea bass in particular, are no exceptions to this rule.

One of the first evidences for an endogenous biological rhythm was observed in 1729 by De Mairan in *Mimosa* plant, which opened and closed its leaves daily (heliotropy), even indoors during constant darkness. Modern chronobiology (from Greek: *kronos* for time; *bios* for life and *logos* for study) was established in the middle of the 20th century by Colin S. Pittendrigh and Jürgen Aschoff, their work spanning from fruit fly to human being.

[1]Department of Physiology, Faculty of Biology, University of Murcia, Campus Mare Nostrum, 30100 Murcia, Spain.
[2]Centre National de la Recherche Scientifique, Unité Mixte de Recherche 7232 and Université Pierre et Marie Curie-Paris 6, Laboratoire Aragó, F-66650 Banyuls-sur-Mer, France.
*Corresponding author: javisan@um.es

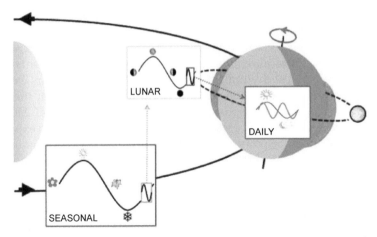

Figure 1. Seasonal, lunar and daily geophysical cycles. Geophysical phenomena, such as the earth translation around the sun, lunar translation around the earth and the earth rotation, lead different periodicities cycles in the terrestrial environment: seasonal ($\tau \approx 365$ days), mensual-lunar ($\tau \approx 28$ days), daily ($\tau \approx 24$ h) and tidal ($\tau \approx 12$ h), being contained the shortest within the longest one.

Color image of this figure appears in the color plate section at the end of the book.

Since then, a large number of biological rhythms have been described across the whole phylogenetic tree and recognized as a natural characteristic of all animals including fish, providing a temporal organization for their physiological and behavioural processes (Cymborowski 2010).

If a biological rhythm persists under constant environmental conditions, it is called an endogenous rhythm, which is generated by a self-sustainable oscillator or pacemaker. Although biological rhythms are characterized by three main parameters—period, amplitude and acrophase—they are classified in three categories depending on their periodicity (or frequency, which is the inverse relation) (Aschoff 1981):

 i) **Circadian**: when the period (T) is around one day (20h<T>28h);
 ii) **Ultradian**: when T < 20 h, for instance, tidal (T=12,4h) rhythms;
 iii) **Infradian**: when T > 28 h. **Circalunar** (T ≈ 28 days) and **circannual** (T ≈ 365 days) are classified among infradian rhythms.

Environmental time cues synchronize endogenous oscillators, which have a similar but not identical period (tau, τ) than environmental cycles (T). These (biotic or abiotic) environmental cyclic factors are called *zeitgebers* ("timegivers" in German) or **synchronizers** (Aschoff 1981). The role of a biological clock is to give an estimated internal time from the external time, and thus anticípate external changes. Therefore, biological rhythms and environmental cycles are coordinated by the following three possible ways: (i) **Synchronization**: when biological rhythms and environmental cycles display the same period and remain on a stable phase relation (phi, φ), but does not necessarily involve the endogenous rhythmic expression of the clock. (ii) **Entrainment**: synchronization where environmental cycles lead to the biological rhythm. So, when the *zeitgeber* disappears (constant conditions), the endogenous rhythm begin to **free-run** from the stable phase (phi, φ) determined by the previous *zeitgeber* (Johnson et al. 2004). (iii) **Masking**: the environmental stimulus seems to force the rhythm expression (e.g., a photophobic response), but without entraining the endogenous clock. Therefore, the biological rhythm displays the same period and appears in phase with the environmental cycle, but in constant conditions the rhythm disappears or an endogenous rhythm begins to free-run from its own phase. The anticipatory characteristic (common in entrained rhythms) is missing in masking, due to the overt rhythm responding directly to the environmental input signal without pacemaker control.

In summary, all environmental factors do not act as *zeitgebers*, even if they show a cyclic pattern. Moore-Ede et al. (1982) defined the characteristic to considerer an external signal as *zeitgeber*: when the animal is exposed to the synchronizer, (a) both periods from biological rhythms and synchronizer must coincide; (b) a stable phase relation must be established between them; when the synchronizer is removed; (c) the biological rhythm must start to free-run from the phase set by the previous synchronizer.

The Molecular Clock

Over the last decade, our understanding of the basic mechanisms driving vertebrate circadian rhythms and their synchronization to light has increased enormously. Chronobiological studies in fish have contributed most significantly to this achievement (Idda et al. 2012). Indeed, teleost fish is one of the most successful groups of vertebrates, with adaptations to live in a wide range of ecological time niches. Furthermore, the particular characteristics of the fish's molecular clock provide a unique opportunity to investigate the flexibility of circadian oscillators and their adaptation to extreme environmental conditions during evolution (Cavallari et al. 2011).

The vertebrate molecular clock comprises a set of clock genes classified into two feedback loops: a negative one, with *cryptochrome* (*Cry*) and *period* (*Per*) genes; and a positive one, with *Circadian Locomotor Output Cycles Kaput* (*Clock*) and *Brain and Muscle Aryl hydrocarbon receptor nuclear translocator (ARNT)-Like* (*Bmal*) (Fig. 2). CRY protein join PER protein to form a heterodimer complex (CRY:PER), which binds to and blocks the protein complex formed by CLOCK and BMAL (CLOCK:BMAL), inhibiting CRY and PER transcription (Iuvone et al. 2005, Okamura et al. 2002). The

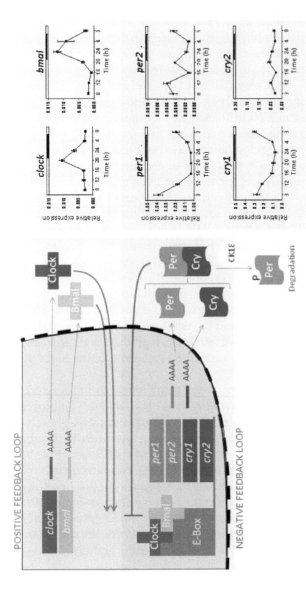

Figure 2. Sea bass molecular clock diagram based on the zebrafish one. The left panel shows the diverse sea bass molecular clock elements that are identified up to now are represented in according to the well-known zebrafish clock molecular interactions. The clock genes are organized in two feedback loop: (i) the positive one that is composed by *clock* and *bmal* genes, whose proteins form the Clock:Bmal complex to active (blue narrows) the negative loop clock genes; (ii) the negative loop with *per* and *cry* (1 and 2) genes, whose proteins inhibit their own transcription by means of blocking Clock:Bmal complex. The right panel show: the rhythmic daily expression of every clock gene (except for Cry2) in sea bass pituitary during summer under 14L/10D (7–21 h) at 20°C (MEAN±SEM, n=5; Cosinor p < 0.05). The relative expression is represented in the vertical axis and the time (in hours) in horizontal one. Black and white bars in the top mark the dark and photophase, respectively. Modified from Sánchez et al. 2010; Del Pozo et al. 2012a; Herrero et al. 2012.

Color image of this figure appears in the color plate section at the end of the book.

zebrafish's molecular clock is the best described one in fish, differing from the mammalian mechanism in the larger number of *cryptochrome* (*cry*) and *period* (*per*) genes. Several subtypes produce proteins with similar, different or still unknown function to mammalian proteins (Kobayashi et al. 2000).

In addition to the central master clock located in the brain, circadian clocks in peripheral tissues have been reported in *Drosophila*, zebrafish, mammalian cell lines and tissues, supporting the hypothesis on the existence of decentralized clocks (Tamai et al. 2005). Moreover, *Drosophila* and adult zebrafish tissues, as well as zebrafish cell lines and embryos, are directly light-responsive (Whitmore et al. 1998, Schibler and Sassone-Corsi 2002), which makes them ideal models to investigate the ontogeny of the molecular clock and its light synchronization pathways.

In sea bass, the following clock genes have been recently characterized: *per*, *cry*, *clock* and *bmal*. *Period1* (*per1*) was first cloned and its expression characterized in different tissues (brain, heart, liver, gill, muscle, digestive tract, adipose tissue, spleen and retina) (Sanchez et al. 2010), which generates a protein of 1,436 amino acids with conserved regions previously identified as a PER-ARNT-SIM (PAS) fold, a nuclear export signal (NES), carboxy terminal to PAS fold, an short-mutable domain (S/M region), and a carboxy-terminal serine/threonin-glycine (SG) repeat region. In addition, this sea bass Period 1 protein (Per1) contained several conserved Casein Kinase-I (CKI) phosphorylation regions, predictably for its degradation. The sea bass Per1 was clustered together with medaka (*Orizyas latipes*) Per1 and zebrafish (*Danio rerio*) Per1a and b. Furthermore, *per1* showed a daily rhythm of expression in brain, heart and liver, with the acrophase (time of day when the rhythm values are maximum) at the end of the night in all of them.

In a recent paper, *cry1* and *cry2* were also cloned and characterized (Del Pozo et al. 2012a), codifying two proteins of 567 and 668 aminoacids, respectively, both with two domains: FAD-binding and DNA-photolyase (SMART, $p < 0.01$). Phylogenetic analyses grouped both sea bass cryptochromes (Cry) with those of other teleost fish, separating both subtypes *Cry1* and *Cry2* in two different clusters. Moreover, *cry1* and 2 were expressed in all the analyzed sea bass tissues (brain, heart, liver, gill, muscle, intestine, spleen and retina). Daily *cry1* expression oscillated rhythmically in brain, heart and liver, peaking around Zeitgeber time (ZT) 3:15 hours, similarly to *cry2* expression in liver. However, the acrophase of *cry2* in brain was at ZT 11:08 hours, and no rhythmical daily expression was observed in heart.

Most recently, partial sequences of *clock* and *bmal* and *per2* genes have been also cloned in sea bass (Herrero et al. 2012). In this report, the daily patterns of clock gene expression of all known genes (*per1*, *per2*, *cry1*, *cry2*, *bmal* and *clock*) have been seasonally characterized in sea bass pituitary along two years cycle. The results revealed that clock genes of the positive feedback loop (*clock* and *bmal*) were higher expressed around lights-off and darkphase, while those of the negative feedback loop (*per1*, *per2* and *cry1* and *cry2*) displayed elevated levels around lights-on and photophase, independently of the season (spring, summer, autumn and winter). However, the amplitude differed between seasons according to temperature, being higher under warmest temperatures (16.0–20.0°C) and lower under coldest temperatures (12.3–13.2°C). The photoperiod seems to be less important since the acrophases were raised during the warmest (16°C) winter. Furthermore, melatonin seems to play a role in the synchronization of gene expression (Dardente et al. 2003, Falcon et al. 2011), so in sea bass pituitary melatonin modified the expression of *cry1* and *cry2* when added to the culture medium of pituitary glands kept *in vitro* or when added to the fish diet *in vivo* (Herrero et al. 2012).

Dualism (diurnal *vs.* nocturnal)

Light is the most important *zeitgeber* in nature, together with temperature (particularly among ectothermic animals, like fish). Both *zeitgebers* are often synchronized in nature, since the thermophase (warmest phase) and the chryophase (coldest phase) are closely related to the photophase and the darkphase, respectively. Most fish adjust their daily activity patterns according to these daily cycles, so they could be classified as diurnal (the greatest activity occurs during the photophase), nocturnal (the greatest activity occurs during the darkphase) or crepuscular (activity linked to dawn and dusk) (Madrid et al. 2001). However, the plasticity of the fish biological clock allows some species to change their pattern of behaviour. Thus, some individuals can be diurnal while others are nocturnal. Furthermore, the same individuals can switch from diurnal to nocturnal phase, and *vice versa*, during their life (Reebs 2002). This phenomenon is known as **dualism**, which was firstly and commonly described among fish species from high latitudes by Eriksson (1978). Since then, more and more fish species have been redefined as dual species, including fish from temperate latitudes (Lopez-Olmeda and Sanchez-Vazquez 2010).

Dualism seems to be more frequent among species considered traditionally diurnal, such as Atlantic salmon (*Salmo salar*) and sharpsnout seabream (*Diplodus puntazzo*), Nile tilapia (*Oreochromis niloticus*) for locomotor activity; rainbow trout (*Oncorhynchus mykiss*) and Artic charr (*Salvelinus alpinus*) for feeding activity; goldfish (*Carassius auratus*), gilthead seabream (*Sparus aurata*), zebrafish (*Danio rerio*) and even European sea bass (*Dicentrarchus labrax*), for both locomotor and feeding activities (Landless 1976, Fraser et al. 1995, Alanara and Brannas 1997, Sanchez-Vazquez et al. 1995a, 1996, Sanchez-Muros et al. 2003, Velazquez et al. 2004, Vera et al. 2006, 2009, Lopez-Olmeda and Sanchez-Vazquez 2009, Montoya et al. 2010, Villamizar et al. 2012). In contrast, almost no dual behaviour has been reported among traditionally considered nocturnal species, except nocturnal brown bullhead (*Ictallurus nebulosus*) which become gradually diurnal under low light intensity (Eriksson and van Veen 1980). Most nocturnal fish species never change their behaviour and remain strictly nocturnal, such as tench (*Tinca tinca*), Senegal sole (*Solea senegalensis*) or European catfish (*Silurus glanis*) (Lopez-Olmeda and Sanchez-Vazquez 2010).

Another interesting feature of the dualism is the independent phasing displayed by feeding and locomotor rhythms, which could happen during different phases (light or darkness) at the same time in the same fish (Sanchez-Vazquez et al. 1996, Del Pozo et al. 2011, Fortes-Silva et al. 2010). For instance, some goldfish and zebrafish may display diurnal locomotor activity during nocturnal feeding activity. This phenomenon appears associated with dualism as a consequence of the high flexibility of the circadian system of fish (Reebs 2002).

Behavioral patterns (feeding and locomotor)

Feeding behavior in sea bass has been more widely studied than their locomotor behaviour. The development of self-feeders for fish has supported this research, as they are a powerful tool to study feeding rhythms, meal size, as well as diet selection when given the choice to choose among different feeds. The reason is these self-feeders let fish feed freely, demanding food from a feeder whenever and as much they want. Since the pioneering self-feeder designed by Rozin and Mayer (1961), different kinds of self-feeders have been created along the last three decades. The self-feeding system is made of three main parts: (i) a feeding sensor, which should be adjusted to the fish's particular characteristic to activate it (and avoid accidental activations), being therefore, the most variable part (solenoid, rigid lever, flexible string, infrared sensor) (Adron 1972, Boujard et al. 1992, Sanchez-Vazquez et al. 1994, Rubio et al. 2004, Del Pozo et al. 2011); (ii) a food container, which delivers small amounts of food after its activation; (iii) a recording system, which monitors the feeding activity. European sea bass have very skillfully used both rigid/flexible rods and infrared sensors (Rubio et al. 2004, Del Pozo et al. 2012b), showing high learning potential and associative abilities (Rubio et al. 2003). Moreover, the individual feeding activity within a fish group could be monitored, coupling a PIT tag monitoring device with a self-feeding system (Di-Poi et al. 2008, Coves et al. 2006). Such a device has proved very useful to investigate social interactions and different diurnal/nocturnal feeding patterns in sea bass within a group (Millot and Begout 2009).

European sea bass was traditionally considered a diurnal fish species till some food demands were recorded during the dark phase (Sanchez-Vazquez et al. 1994). Later, the endogenous character of the feeding circadian rhythm was revealed in groups and in single sea bass (Fig. 3), as well as confirming the dualism under controlled and identical experimental conditions (regarding temperature, light intensity, photoperiod and salinity) (Sanchez-Vazquez et al. 1995a). That was the first time where a fish species displayed a dual behavior indoors, independent of seasonal variation in the photoperiod, changes associated with the intensity of light or temperature, or related with the season during which fish are transferred to the laboratory conditions.

The sea bass group feeding behavior is not just the sum of individual fish behavior, but the output of complex social interactions. For instance, in sea bass groups provided with a self-feeder, food demands are led by one or a few high-triggering individuals with a well defined circadian

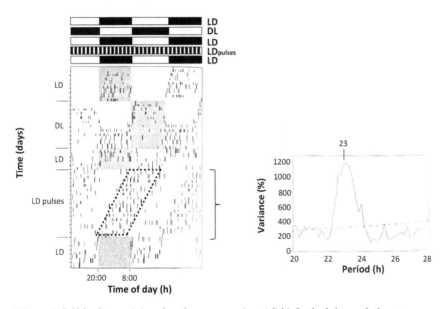

Fiigure 3. Self-feeding activity of sea bass groups (n = 4 fish). In the left panel: the actogram represents the 5 experimental phases along the time (vertical axis): (i) 12 light:12 dark (LD 12:12 h) cycles, (ii) a first photoperiod revision (doubling the light phase), (iii) a second photoperiod reversion (doubling the dark phase), (iv) LD 40:40 min pulses, evidence the endogenous character of sea bass feeding rhythm, and (v) the restoration of LD 12:12 h where phase inversion phenomenon (from nocturnal to diurnal) occurred. The horizontal axis signifies the time of day (in hours). The upper black and white bars indicate the dark and light phase for every experimental phase. In the right panel: the free-running periodigram under LD 40:40 min pulses that show a circadian rhythm with τ = 23 h. Vertical axis shows the percentage of the variety of the data that can be explained by fitting the Sokolove-Bushell periodogram method and horizontal axis time (in hours). Modified from Sanchez-Vazquez et al. (1995a).

rhythm, which feed the rest of the sea bass group (Coves et al. 2006, Millot et al. 2008, Millot and Begout 2009). The feeding activity of the leader synchronizes the food demands of the low-triggering fish, which could copy the feeding behavior of the high-triggering fish or being stimulated by the food delivered (Millot and Begout 2009). Moreover, when more than one high-triggering fish cohabited in the same tank, they activated the self-feeder following the same hourly rhythm (Coves et al. 2006). However, the feeding behavior of a sea bass group is not a single fish reflect, as the leadership may change. Transition to another leader lasted around four days (Millot et al. 2008). The reasons for a particular fish acquiring a social status within the group that changes with time remain unknown. No differences of length, mass or physiological status (muscle composition, plasma and tissues biochemistry) between high and low triggering fish, either the sex or an initial low specific growth rate (SGR) of the leader seem determine its status (Millot et al. 2008, Benhaim et al. 2012).

Regarding fish locomotor activity, it can be quantified (i) by an infrared sensor placed in a specific location in a tank, which detects the fish interruptions and sends a signal to a recording computer, or (ii) by video recording the total activity of fish in a tank, and analyzing the fish movements with specialized tracking software. Phase inversions in locomotor activity have been observed recently in sea bass. Adult sea bass showed clearly nocturnal locomotor activity (68% of daily activity occurred at night) during the spawning season (in winter and early spring), but mostly diurnal (65.5% of the daily activity occurred during the daytime) out of the studied reproductive period (April to June) (Villamizar et al. 2012). Seasonal phase inversions of both locomotor and feeding behavior in adult sea bass have been also monitored under natural conditions, finding independent phasing between feeding and locomotor rhythms (unpublished data). In that trial, three out of eight groups shifted from nocturnal to diurnal locomotor activity, while one group remained nocturnal and the other half were arrhythmic. Conversely, all sea bass groups displayed their food demands during the daytime.

Seasonal phenomenon

When first reported in fish species from a high latitude (Eriksson 1978), the dualism was considered a phenomenon typical of fish inhabiting extreme photoperiodic conditions (constant light/dark during summer/ winter, respectively), so that fish were forced to become diurnal in summer. Jorgensen and Jobling (1989, 1990) also reported seasonal changes in the daily feeding patterns of Arctic charr. The dualism was, however, also reported in fish from temperate regions, such as European sea bass (Sanchez-Vazquez et al. 1995a), therefore ruling out this "high latitude" hypothesis.

In sea bass, changes in their daily feeding pattern were reported. Several authors observed different feeding patterns depending on the season when the experiments were performed (Anthouard et al. 1993, Boujard et al. 1996). Anthouard et al. (1993) found nocturnal sea bass, which became diurnal during the course of the experiment from January to May. Discrepancies in the feeding phase also occurred when comparing April, May and June (Boujard et al. 1996). Finally, the sea bass seasonal inversions were found by Sanchez-Vazquez et al. (1998), who monitored groups of adult sea bass along a whole year under natural environmental conditions (water temperature and photoperiod). In these conditions, sea bass shifted from diurnal to nocturnal in winter and *vice versa* in spring, remaining mostly diurnal during the rest of the year. Maximum diurnal feeding activity occurred coinciding with the longest photoperiod (June), while maximum nocturnalism occurred in February (annual lowest temperature). Such changes in feeding behaviour also had practical consequences for fish performance, as observed in a follow up experiment where sea bass fed at night in winter had better growth and feed conversion rate (Azzaydi et al. 2000).

Although seasonal feeding phase inversions were reported in groups of juvenile sea bass, recent research also found such phase inversion in adult sea bass (Villamizar et al. 2012, unpublished data). It should be noted that phase inversion does not occur simultaneously in all groups, so that some fish advance/delay their switch from one type of phasing to another (Fig. 4). This fact is important because it indicates that the switch is not triggered by sudden changes in the environment (critical photoperiod or abrupt temperature changes), but points to an endogenous control.

Curiously, the pattern of macronutrient self-selection (from three diets with pair of macronutrients) also seems to have a seasonal effect (Vivas 2007). On one the hand, sea bass defended the same diet composition along the whole year (64.0% protein, 26.8% lipids and 9.3% carbohydrates). On the other hand, seasonal variability was found in the daily profile of macronutrient self-selection: sea bass demanded all macronutrients during daytime in summer, while they did so at night in winter. In autumn and spring, lipids were demanded during daytime in both seasons, while carbohydrates and proteins were demanded at the beginning of the night.

Digestive physiology

Food availability and prey/predator activity are hardly constant in the wild. Therefore, fish display feeding rhythms synchronized to a specific period of the day or night that best suit their daily way of life. As stated before, the existence of clock-controlled feeding rhythms represents an evolutionary advantage because fish can thus anticipate a meal and prepare

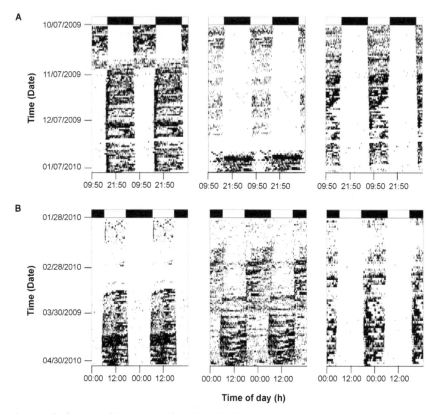

Figure 4. Both seasonal inversions of sea bass feeding behavior (A) from diurnal to nocturnal in winter and (B) *vice versa* in spring. Actograms represent the feeding activity of fish groups, which shift their feeding phase gradually (the earliest to latest inversions are painted from left to right side). Vertical axis shows the time (by date) and the horizontal one the time of day (in hours). Dark and white bars in the top indicate the dark and light phase at the trial beginning. Extracted from Del Pozo et al. 2012b.

their digestion thus allowing better nutrition utilization (Sanchez-Vazquez and Madrid 2001). Consistently, both automatic nocturnal feeding (two meals: pre-dawn and post-dusk) and (nocturnal) self-feeding in sea bass from January to April increased the specific growth rate and reduced feed conversion ratio with respect to automatic diurnal feeding (three meals: morning, noon and afternoon), the common practice in aquaculture farms (Azzaydi et al. 1998). Feeding protocols in fish farms should take into account the sea bass dualism, and the seasonal changes in feeding patterns, to improve their production rates.

Annual variations have been reported in several sea bass hormones and metabolites, such as insulin, plasma glucose (Gutierrez et al. 1987), melatonin (Garcia-Allegue et al. 2001), testosterone, estradiol (Prat et al.

1990) and plasma lipid levels (Fernandez et al. 1989). However, daily physiological differences in sea bass with diurnal and nocturnal behavioral patterns have been little explored. Blood glucose has been recently observed in diurnal and nocturnal sea bass during both seasonal inversions, when both feeding patterns coexist spontaneously (Del Pozo et al. 2012b). Seasonal differences in the daily blood glucose rhythms were found in nocturnal sea bass, presenting higher amplitude, earlier acrophase and upper mean levels in winter than in spring. Daily differences were also detected between both self-feeding patterns, starting to increase the glucose levels of both at night but peaking during the respective self-feeding phase (Fig. 5A). In another recent trial with diurnal/nocturnal restricted self-fed sea bass,

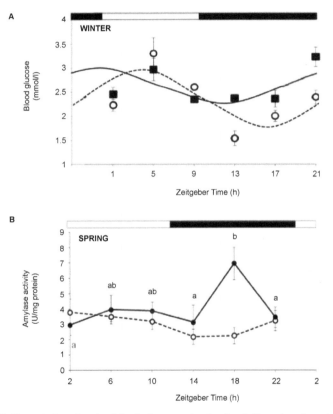

Figure 5. Daily variations in physiological parameter levels of diurnal and nocturnal sea bass groups. (A) Mean blood glucose levels (mmol/l) of five fish in winter are presented by dark squares for nocturnal fish and white circles for diurnal. The cosinor adjustment (p < 0.03) for diurnal fish is represented by dotted sinusoidal line and continuous sinusoidal line for nocturnal. (B) The mean amylase activities (U/mg protein) of seven fish in May are represented by a continuous line for the nocturnal fish and dotted line for diurnal. Different letters denote significant differences between time points in the nocturnal fish (ANOVA I, p < 0.01, followed by Tukey test). Modified from Del Pozo et al. (2012a,b).

blood glucose rose during the dark phase in both diurnal and nocturnal feeding groups, but they reached higher mean levels in diurnal ones which fitted a circadian rhythm (Del Pozo et al. 2012c). Moreover, significant differences were found in mid-intestine amylase activity of both diurnal/ nocturnal fish, which increased during the feeding phase (Fig. 5B). In short, these results revealed that self-feeding time affects physiological rhythms. Further research is necessary to understand the cause/effect relationship between behavior and physiology during the dualism.

Clock genes

As described previously, the key molecular elements involved in the sea bass biological clock have been recently reported. Most interestingly, subtle changes in the expression pattern of these rhythms have been found in diurnal/nocturnal sea bass. The impact of diurnal or nocturnal self-feedings on the daily pattern of expression of *per1* has been characterized in brain and liver (Fig. 6). Self-feeding time influences *per1* expression in the peripheral

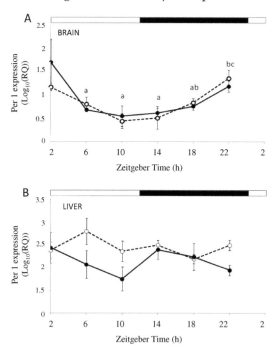

Figure 6. *Per1* relative daily expression in brain (A) and liver (B) of diurnal and nocturnal sea bass. Each point represents the mean ± SEM of 7 diurnal fish (dotted line) and nocturnal (continuous line). Letters denote significant differences between time points. Black and white bars at the top indicate the dark and light phase. The horizontal axis shows the zeitgeber time (in hours). Modified from Del Pozo et al. (2012a).

oscillator involved in digestive processes (liver), but not in the central oscillator (brain) (Del Pozo et al. 2012c). In liver, *per1* expression levels were higher in diurnal than nocturnal sea bass, without fitting a cosinor curve under any feeding condition. In contrast, rhythmic expression in the brain was identical in both diurnal and nocturnal fish, with the acrophase around the lights onset as had been previously reported in diurnally fed sea bass (Sanchez et al. 2010) and other teleost fish (Vallone et al. 2004, Lahiri et al. 2005, Velarde et al. 2009, Sanchez and Sanchez-Vazquez 2009).

Mechanisms Controlling Dualism

The exact mechanisms which control the switch from one type of phasing to another remain unknown. Nevertheless, it is known that some synchronizers affect the dualism in different ways, and ongoing research on the characterization of sea bass clock-controlled genes and reproduction, and their role during the phase inversions are shedding light on the cause/effect of dualism.

Zeitgebers

While changes in photoperiod along the year could explain by themselves the seasonal inversions in high latitude fish, other abiotic (e.g., temperature) and biotic factors (e.g., food availability), together with the photoperiod, could lead to seasonal inversions in temperate fish like sea bass (Lopez-Olmeda and Sanchez-Vazquez 2010). The following external factors which may drive the seasonal rhythms of feeding behavior in sea bass have been studied:

- The **food availability** restriction seems to be successful in leading to the feeding phase inversion in sea bass, since some individual sea bass or groups (but not all) adapted their food demands to "feeding windows" during photophase and/or darkphase, even when these windows were contrary to their previous feeding phase (Sanchez-Vazquez et al. 1995b). This evidence together with a partial coupling between food-entrained and light-entrained activity under conflicting *zeitgebers* (LD 13:13 h and restricted-feeding (RF) 4:20 h), suggest the existence of a feeding entrainable oscillator (FEO) in addition to the master light entrainable oscillator (LEO). The FEO hypothesis was supported by the higher feeding anticipatory activity (FAA) when the period of both *zeitgebers* (LD cycles and RF) was the same, instead of conflicting periods (LD 26 h and RF 24 h). However, food availability alone is not the only factor involved in the control of dualism, since some sea bass did not invert their feeding phase following the food-restriction, and moreover, the food provided in the middle of the food availability phase did not facilitate the inversion (Aranda et al. 1997).

- **Photoperiod:** There are two ways to study the photoperiod effect on the animal rhythms: (i) modifying the photoperiod length or (ii) providing the animal with information regarding dark-light transitions in skeleton photoperiods. This latter way allowed differentiated masking and total darkness effects (Pittendringh 1981). In sea bass, photoperiod length strongly synchronized the feeding activity phase, since fish confined their food demands following the contraction and expansion of daytime even under extremely short (2:22 h LD) and long (22:2 h LD) photoperiods (Aranda et al. 1999a). However, the photoperiod by itself did not control the phasing of the rhythm; thus, some individuals showed phase inversions (even "double inversions") regardless of photoperiod changes. On the other hand, either 15-minute or 1-hour light pulses separated by 12-hours, synchronized feeding rhythms in nocturnal sea bass, while only the 1-hour light pulses did so in diurnal sea bass. Later, under constant darkness, two free-running components appeared, supporting the multioscillatory circadian system.

- **Temperature:** Although seasonal inversions under natural conditions occur with temperature decreases even around 17°C (in winter), the highest percentage of nocturnalism occurred at 13.2°C (Sanchez-Vazquez et al. 1998). However, in another study, the manipulating water temperature did not drive the sea bass feeding inversions, since cool water (16°C) did not manage to shift the feeding phase from diurnal to nocturnal (Aranda et al. 1999b). Moreover, two fish of a high temperature group remained nocturnal when the temperature rose.

- **Photoperiod/Temperature:** Variations in photoperiod and water temperature are connected in nature; therefore, both factors should be manipulated together. Nevertheless, when a short photoperiod was combined with low water temperatures, and a long photoperiod with high water temperatures, sea bass failed to change their feeding phase (Aranda et al. 1999b). Fish went on demanding food during their previous active phases (photophase for diurnal and darkphase for nocturnal fish), showing a strong dependence of light conditions.

In summary, experimental manipulation of these two environmental factors (photoperiod and water temperature) failed to produce consistent changes in phasing. One explanation could be the need for gradual changes, similar to those that occur in nature along the seasons.

Reproduction rhythms and the dual oscillator hypothesis

In the wild, teleost fish from temperate waters synchronize their reproductive rhythms to the natural environmental cycles in order to ensure the best environmental conditions for their offspring (Oliveira and Sanchez-Vazquez 2010). Thus, fish have developed time-keeping systems that use cyclic oscillations of photoperiod as a reliable environmental cue to anticipate and activate gametogenesis long before spawning. Melatonin, the "time-keeping" hormone secreted by the pineal, acts as a phototransducing signal on the hypothalamic-pituitary-gonad axis, timing the production of gonadotropins, sex steroids and growth factors in the fish gonad (Amano et al. 2000, Bromage et al. 2001, Bayarri et al. 2004, Falcon et al. 2007). Actually, manipulating light and thus melatonin rhythms, has an impact on the fish biological clock and reproduction rhythms, which can be used for aquaculture purposes. Increasing/decreasing photoperiods or continuous light regimes can be used to fully inhibit reproduction and prevent precocity, although the results are species-dependent (Falcon et al. 2010). In European sea bass, light affects the daily rhythm of luteinizing hormone (LH) and also the daily melatonin rhythm, which oscillates faithfully to the seasonal changes of day length (Garcia-Allegue et. al. 2001, Bayarri et al. 2004). Furthermore, in sea bass, a given photoperiod is needed to sustain circadian oscillations in reproductive hormones and normal reproduction, because continuous light suppresses daily rhythms of key hormones such as melatonin and LH in fish, fully arresting gonad development and maturation (Bayarri et al. 2009). These issues will be further discussed in different chapters in this book.

In sea bass, maturation starts in September/October and post-vitellogenic oocytes are first observed in December. Ovulation lasts from January to mid-March, coinciding with the spawning period (Asturiano et al. 2000). Other seasonal changes have been observed regarding plasma concentration of vitellogenin and sex esteroids (E2 and T) which present a peak between December and February (Prat et al. 1990). Most interestingly, the observed seasonal phase shifts of feeding rhythms of European sea bass (nocturnal during winter, diurnal in spring-summer) (Sanchez-Vazquez et al. 1998) coincides with the spawning and resting season of the adults. This finding strongly suggests that the seasonal change of behavior is linked to reproduction and it may be based on the particular reproductive strategy of this species. Indeed in a recent study, European sea bass broodstock showed nocturnal locomotor activity during the spawning season (winter and early spring) and diurnal through the resting period (Villamizar et al. 2012). Although the role of environmental cues on the synchronization of reproduction in sea bass is clear, previous studies have also found that when this species is kept under constant photoperiod it is able to maintain the annual spawning rhythm, which suggests a strong endogenous mechanism that regulates the reproductive process (Prat et al. 1999).

Since fish synchronize reproduction to environmental factors in order to select the best season to reproduce, it is reasonable to think that the egg release and their subsequent fertilization also occur at a specific moment of the day or night. In most fish species, the discovery of daily modifications in oocyte maturation and the secretion of sexual steroids or gonadotropins have led to the study of the daily reproduction rhythms. Indeed species such as the red snapper *Lutjanus campechanus* (Jackson et al. 2006), gilthead seabream *Sparus aurata* (Meseguer et al. 2008) and the zebrafish *Danio rerio* (Blanco-Vives and Sanchez-Vazquez 2009) spawn at specific times of the day/night. In sea bass, the daily spawning rhythm has a nocturnal acrophase, so that eggs start to be released four hours after lights off, with two spawning peaks, at six and 11 hours after lights off. Further observations found that the egg viability was highest in the batches released when spawning peaked and the reproductive activity of the broodstock was positively correlated with its locomotor activity (Villamizar et al. 2012). This finding further links the dualism of sea bass with the timing of reproduction.

It should be also noted that although seasonal phase inversions occur in parallel with reproduction, juvenile sea bass also become nocturnal in winter and stay diurnal along the rest of the year (Sanchez-Vazquez et al. 1998, Del Pozo et al. 2012b). The reproductive period take place in winter, driven by annual changes in the mRNA expression and hormone levels of key elements of the reproductive axis, such as gonadotropin-releasing hormone (GnRH), three gonadotropin (GtH) subunits, namely glycoprotein α (GPα), follicle-stimulating hormone β (FSHβ) and luteinizing hormone β (LHβ), which alter their levels from the first year of sea bass life (Moles et al. 2007) in immature fish. Possibly these genetic, physiological and behavioral annual variations are controlled by a clock system, which provides sea bass with daily and seasonal information.

Finally, Aranda et al. (1999b) hypothesized that feeding rhythms are driven by an endogenous multioscillatory circadian system, by means of several oscillators interacting to form a diurnal or nocturnal "configuration". This hypothesis is supported by: (i) the low frequency of free-running rhythms, (ii) the phase instability, (iii) two free-running components appeared under continuous darkness, and (iv) the phase inversions did not occur at the same time in all fish. Expanding this hypothesis based on our current knowledge, a **dual oscillator** circadian system could lead the seasonal appearance of sea bass dualism. In this way, two oscillators: **"Daytime"** and **"Nocturnal Seasonal Reproduction (NSR)"** are coupled during almost the whole year (from spring to winter), showing a diurnal configuration held by the "Daytime", which also is suppressing the "NSR" nocturnal configuration (Fig. 7). However, during the reproductive period (from winter to spring) some external, or more probably, internal signal (i.e., a neuroendocrine compound from the reproductive axis) decouples both oscillators, allowing the "NSR" expression to switch to a nocturnal configuration. A similar hypothesis has been reported to explain the *"zugunrhue"* behavior in songbirds (Bartell and Gwinner 2005). *Zugunrhue* is the nocturnal migratory restlessness, which otherwise strictly diurnal birds show during spring and autumn migratory periods. This *zugunrhue* is endogenous, circadian, and circannual controlled and strongly coordinated with other processes such as reproduction, moult, and feeding (in accordance with nutritional status) (Coppack and Bairlein 2011, Kumar et al. 2010, Gwinner 2003). Therefore, environmental factors would act on circannual rhythms that are closely involved in the seasonal organization of animal behavior.

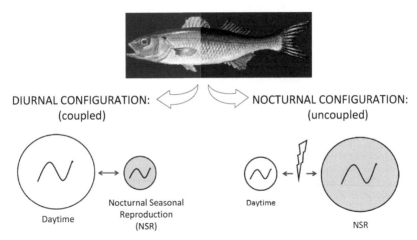

Figure 7. Diagram of dual oscillatory system, which could regulated the sea bass seasonal behavior inversions, linking to other annual phenomena, such as reproduction. Two oscillators, "daytime" and "Nocturnal Season reproduction (NSR)", would be normally (from winter to spring) coupled (as indicate the double headed narrow), so "daytime" would suppress the "NSR" nocturnal configuration and provide a diurnal configuration. Nevertheless, "NSR" would gain relevance when some external or internal signal decouples both oscillators, displaying the nocturnal configuration in winter. A reproduction signal may act as decoupler to modify the behavior configuration to nocturnal, which has place during the reproductive period, in winter.

References

Adron, J.W. 1972. Design for automatic and demand feeders for experimental. Fish. J. Cons. Perm. Int. Explor. Mer. 34: 300–305.

Alanara, A. and E. Brannas. 1997. Diurnal and nocturnal feeding activity in Artic charr (Salvelinus alpinus) and rainbow trout (*Oncorhynchus mykiss*). Can. J. Fish. Aquat. Sci. 54: 2894–2900.

Amano, M., M. Iigo, K. Ikuta, S. Kitamura, H. Yamada and K. Yamamori. 2000. Roles of melatonin in gonadal maturation of underyearling precocious male masu salmon. Gen. Comp. Endocr. 120(2): 190–197.

Anthouard, M., M. Kentouri and P. Divanach. 1993. An analysis of feeding activities of sea-bass (*Dicentrarchus labrax*, Moronidae), raised under different lighting conditions. Ichtyophysiologica Acta 16: 59–73.

Aranda, A., M. Faraco, F.J. Sanchez-Vazquez, S. Zamora and J.A. Madrid. 1997. Desfase entre actividad alimentaria y disponibilidad del alimento en lubinas. ACTAS VI CONGRESO NACIONAL ACUICULTURA. Madrid: Ministerio de Agricultura, Pesca y Alimentación. 1: 583–589.

Aranda, A., J.A. Madrid, S. Zamora and F.J. Sanchez-Vazquez. 1999a. Synchronizing effect of photoperiod on the dual phasing of demand-feeding rhythms in sea bass. Biol. Rhythm Res. 30(4): 392–406.

Aranda, A., F.J. Sánchez-Vázquez and J.A. Madrid. 1999b. Influence of water temperature on demand feeding-rhythms in sea bass. Journal of Fish Biology 55: 1029–1039.

Aschoff, J. 1981. Handbook of Behavioural Neurobiology Vol. 4: Biological Rhythms. Jurgen Aschoff. Plenum, New York.

Asturiano, J.F., L.A. Sorbera, J. Ramos, D.E. Kime, M. Carrillo and S. Zanuy. 2000. Hormonal regulation of the European sea bass reproductive cycle: an individualized female approach. J. Fish Biol. 56(5): 1155–1172.

Azzaydi, M., J.A. Madrid, S. Zamora, F.J. Sanchez-Vazquez and F.J. Martinez. 1998. Effect of three feeding strategies (automatic, *ad libitum* demand-feeding and time-restricted demand-feeding) on feeding rhythms and growth in European sea bass (*Dicentrarchus labrax* L.). Aquaculture 163(3–4): 285–296.

Azzaydi, M., F.J. Martinez, S. Zamora, F.J. Sanchez-Vazquez and J.A. Madrid. 2000. The influence of nocturnal *vs.* diurnal feeding under winter conditions on growth and feed conversion of European sea bass (*Dicentrarchus labrax* L.). Aquaculture 182(3–4): 329–338.

Bartell, P.A. and E. Gwinner. 2005. A Separate Circadian Oscillator Controls Nocturnal Migratory Restlessness in the Songbird Sylvia borin. J. Biol. Rhythm. 20(6): 538–549.

Bayarri, M.J., L. Rodriguez, S. Zanuy, J.A. Madrid, F.J. Sanchez-Vazquez, H. Kagawa, K. Okuzawa and M. Carrillo. 2004. Effect of photoperiod manipulation on the daily rhythms of melatonin and reproductive hormones in caged European sea bass (*Dicentrarchus labrax*). Gen. Comp. Endocr. 136: 72–81.

Bayarri, M.J., S. Zanuy, O. Yilmaz and M. Carrillo. 2009. Effects of continuous light on the reproductive system of european sea bass gauged by alterations of circadian variations during their first reproductive cycle. Chronobiol. Int. 26(2): 184–199.

Benhaim, D., M.L. Begout, S. Pean, B. Brisset, D. Leguay and B. Chatain. 2012. Effect of fasting on self-feeding activity in juvenile sea bass (*Dicentrarchus labrax*). Appl. Anim. Behav. Sci. 136(1): 63–73.

Blanco-Vives, B. and F.J. Sanchez-Vazquez. 2009. Synchronisation to light and feeding time of circadian rhythms of spawning and locomotor activity in zebrafish. Physiol. Behav. 98(3): 268–275.

Boujard, T., X. Dugy, D. Genner, C. Gosset and G. Grig. 1992. Description of a modular, low-cost, eater meter for the study of feeding-behavior and food preferences in fish. Physiol. Behav. 52: 1101–1106.

Boujard, T., M. Jourda, M. Kentouri and P. Divanach. 1996. Diel feeding activity and the effect of time-restricted self-feeding on growth and feed conversion in European sea bass. Aquaculture 139: 117–127.

Bromage, N., M. Porter and C. Randall. 2001. The environmental regulation of maturation in farmed finfish with special reference to the role of photoperiod and melatonin. Aquaculture 197(1–4): 63–98.

Cavallari, N., E. Frigato, D. Vallone, N. Fröhlich, A. Foà, R. Berti, J.F. Lopez-Olmeda, F.J. Sanchez-Vazquez, C. Bertolucci and N.S. Foulkes. 2011. A Blind Circadian Clock in Cavefish Reveals that Opsins Mediate Peripheral Clock Photoreception. PLOS BIOL 9 e1001142.

Coppack, T. and F. Bairlein. 2011. Circadian control of nocturnal songbird migration. J. Ornithol. 152(1): 67–73.

Coves, D., M. Beauchaud, J. Attia, G. Dutto, C. Bouchut and M.L. Begout. 2006. Long-term monitoring of individual triggering activity on a self-feeding system: An example using European sea bass (*Dicentrarchus labrax*). Aquaculture 253: 385–392.

Cymborowski, B. 2010. Introduction to circadian rhythms. pp. 155–184. *In*: E. Kulczykowska (ed.). Biological Clock in Fish. Science Publishers, Enfield, NH.

Dardente, H., J.S. Menet, V.J. Poirel, D. Streicher, F. Gauer, B. Vivien-Roels, P. Klosen, P. Pevet and M. Masson-Pévet. 2003. Melatonin induces *Cry1* expression in the *pars tuberalis* of the rat. Mol. Brain Res. 114: 101–106.

De Mairan, J.J. 1729. Observation botanique. L'Histoire de l'Academie Royal Scientifique. pp. 47–48.

Del Pozo, A., J.A. Sanchez and F.J. Sanchez-Vazquez. 2011. Circadian rhythms of self-feeding and locomotor activity in zebrafish (*Danio rerio*). Chronobiol. Int. 28(1): 39–47.

Del Pozo, A., L.M. Vera, J.A. Sanchez and F.J. Sanchez-Vazquez. 2012a. Molecular cloning, tissue distribution and daily expression of *cry1* and *cry2* clock genes in European seabass (*Dicentrarchus labrax*). Comp. Biochem. Phys. A. 163: 364–371.

Del Pozo, A., L.M. Vera, A. Montoya and F.J. Sanchez-Vazquez. 2012b. Daily rhythms of blood glucose differ in diurnal and nocturnal European sea bass (*Dicentrarchus labrax* L.) undergoing seasonal phase inversions. Fish Physiol. Biochem. DOI 10.1007/s10695-012-9732-z.

Del Pozo, A., A. Montoya, L.M. Vera and F.J. Sanchez-Vazquez. 2012c. Daily rhythms of clock gene expression, glycaemia and digestive physiology in diurnal/nocturnal European seabass. Physiol. Behav. 106: 446–450.

Di-Poi, C., M. Beauchaud, C. Bouchut, G. Dutto, D. Coves and J. Attia. 2008. Effects of high food-demand fish removal in groups of juvenile sea bass (*Dicentrarchus labrax*). Can. J. Zool. 86: 1015–1023.

Eriksson, L.O. 1978. Nocturnalism *versus* diurnalism—dualism within fish individuals. pp. 69–90. *In*: J.E. Thorpe (ed.). Rhythmic Activity of Fishes. Academic Press, London.

Eriksson, L.O. and T. van Veen. 1980. Circadian rhtyhms in the Brown bullhead, *Ictallurus nebulosus* (Teleosti). Evidence for an endogenous rhythm in feeding, locomotor, and reaction time behavior. Can. J. Zool. 58: 1899–1907.

Falcon, J., L. Besseau, S. Sauzet and G. Boeuf. 2007. Melatonin effects on the hypothalamo-pituitary axis in fish. Trends Endocrin. Met. 18(2): 81–88.

Falcon, J., H. Migaud, J.A. Munoz-Cueto and M. Carrillo. 2010. Current knowledge on the melatonin system in teleost fish. Gen. Comp. Endocr. 165(3): 469–482.

Falcon, J., L. Besseau, E. Magnanou, M.J. Herrero, M. Nagai and G. Boeuf. 2011. Melatonin, the time keeper: biosynthesis and effects in fish. Cybium. 35: 3–18.

Fernandez, J., J. Gutierrez, M. Carrillo, S. Zanuy and J. Planas. 1989. Annual cycle of plasma lipids in sea bass, *Dicentrarchus labrax* L.: Effect of environmental conditions and reproductive cycle. Comp. Biochem. Phys. A. 93: 407–412.

Fortes-Silva, R., F.J. Martínez, M. Villarroel and F.J. Sanchez-Vazquez. 2010. Daily rhythms of locomotor activity, feeding behavior and dietary selection in Nile tilapia (*Oreochromis niloticus*). Comp. Biochem. Phys. A. 156(4): 445–450.

Fraser, N.H.C., J. Heggens, N.B. Metcalfe and J.E. Thorpe. 1995. Lower summer temperatures cause juvenile Atlantic salmon to become nocturnal. Can. J. Zool. 73: 446–451.

Garcia-Allegue, R., J.A. Madrid and F.J. Sanchez-Vazquez. 2001. Melatonin rhythms in European sea bass plasma and eye: influence of seasonal photoperiod and water temperature. J. Pineal Res. 31(1): 68–75.

Gutierrez, J., J. Fernandez, M. Carillo, S. Zanuy and J. Planas. 1987. Annual cycle of plasma insulin and glucose of se bass. *Dicentrarchus labrax* L. Fish Physiol. Biochem. 4: 137–141.

Gwinner, E. 2003. Circannual rhythms in birds. Curr. Opi. Neurobiol. 13: 770–778.

Herrero, M.J., L. Besseau, M. Fuentes, F.J. Sanchez-Vazquez, A. Del Pozo, E. Isorna, L. Nisembaum and J. Falcon. 2012. Daily and seasonal expression of clock genes in the pituitary of European sea bass (*Dicentrarchus labrax*): effects of photoperiod, temperature and melatonin, XLIII Congrès de la Société Francophone de Chronobiologie. Moulin des Cordeliers, 37600 Loches.

Idda, M.L., C. Bertolucci, D. Vallone, Y. Gothilf, F.J. Sanchez-Vazquez and N.S. Foulkes. 2012. Circadian clocks: Lessons from fish. Prog. Brain Res. 199: 41–57.

Iuvone, P.M., G. Tosini, N. Pozdeyev, R. Haque, D.C. Klein and S.S. Chaurasia. 2005. Circadian clocks, networks, arylalkylamine N-acetyltransferase, and melatonin in the retina. Prog. Retin Eye Res. 24: 433–456.

Jackson, M.W., D.L. Nieland and J.H. Cowan. 2006. Diel spawning periodicity of red snapper *Lutjanus campechanus* in the northern Gulf of Mexico. J. Fish Biol. 68(3): 695–706.

Johnson, C.H., J. Elliott, R. Foster, K. Honma and R. Kronauer. 2004. Fundamental properties of circadian rhythms. pp. 67–105. *In*: J.C. Dunlap, J.J. Loros and P.J. DeCoursey (eds.). Chronobiology Biological Timekeeping. Sinauer Associates, Sunderland, MA.

Jorgensen, E.H. and M. Jobling. 1989. Patterns of food-intake in arctic charr, *Salvelinus-alpinus*, monitored by radiography. Aquaculture 81(2): 155–160.

Jorgensen, E.H. and M. Jobling. 1990. Feeding modes in arctic charr, *Salvelinus-alpinus* L.—the importance of bottom feeding for the maintenance of growth. Aquaculture 86(4): 379–385.

Kobayashi, Y., T. Ishikawa, J. Hirayama, H. Daiyasu, S. Kanai, H. Toh, I. Fukuda, T. Tsujimura, N. Terada, Y. Kamei, S. Yuba, S. Iwai and T. Todo. 2000. Molecular analysis of zebrafish photolyase/cryptochrome family two types of cryptochromes present in zebrafish. Genes Cells 5: 725–738.

Kumar, V., J.C. Wingfield, A. Dawson, M. Ramenofsky, S. Rani and P. Bartell. 2010. Biological clocks and regulation of seasonal reproduction and migration in birds. Physiol. Biochem. Zool. 83(5): 827–835.

Lahiri, K., D. Vallone, S.B. Gondi, C. Santoriello, T. Dickmeis and N.S. Foulkes. 2005. Temperature regulates transcription in the zebrafish circadian clock. PLoS Biol. 3: e351.

Landless, P.J. 1976. Demand-feeding behaviour of rainbow trout. Aquaculture 7: 11–15.

Lopez-Olmeda, J.F. and F.J. Sanchez-Vazquez. 2009. Zebrafish temperature selection and synchronization of locomotor activity circadian rhythms to ahemeral cycles of light and temperature. Chronobiol. Int. 26(2): 200–218.

Lopez-Olmeda, J.F. and F.J. Sanchez-Vazquez. 2010. Feeding rhythms in fish: from behavioral to molecular approach. pp. 155–184. *In*: E. Kulczykowska (ed.). Biological Clock in Fish. Science Publishers, Enfield, NH.

Madrid, J.A., T. Boujard and F.J. Sanchez-vazquez. 2001. Feeding rhythms. pp. 189–215. *In*: D.F. Houlihan, T. Boujard and M. Jobling (eds.). Food Intake in Fish. Blackwell Science, Oxford.

Meseguer, C., J. Ramos, M.J. Bayarri, C. Oliveira and F.J. Sanchez-Vazquez. 2008. Light synchronization of the daily spawning rhythms of gilthead sea bream (*Sparus aurata* L.) kept under different photoperiod and after shifting the LD cycle. Chronobiol. Int. 25(5): 666–679.

Millot, S. and M.L. Begout. 2009. Individual fish rhythm directs group feeding: a case study with sea bass juveniles (*Dicentrarchus labrax*) under self-demand feeding conditions. Aquat. Living Resour. 22: 363–370.

Millot, S., M.L. Begout, J. Person-Le Ruyet, G. Breuil, C. Di-Poi, J. Fievet, P. Pineau, M. Roue and A. Severe. 2008. Feed demand behavior in sea bass juveniles: Effects on individual specific growth rate variation and health (inter-individual and inter-group variation). Aquaculture 274(1): 87–95.

Moles, G., M. Carrillo, E. Mañanós, C.C. Mylonas and S. Zanuy. 2007. Temporal profile of brain and pituitary GnRHs, GnRH-R and gonadotropin mRNA expression and content during early development in European sea bass (*Dicentrarchus labrax* L.). Gen. Comp. Endocr. 150: 75–86.

Montoya, A., J.F. Lopez-Olmeda, A.B.S. Garayzar and F.J. Sanchez-Vazquez. 2010. Synchronization of daily rhythms of locomotor activity and plasma glucose, cortisol and thyroid hormones to feeding in Gilthead seabream (*Sparus aurata*) under a light-dark cycle. Physiol. Behav. 101(1): 101–107.

Moore-Ede, M.C., F.M. Sulzman and C.A. Fuller. 1982. The Clocks That Time Us: Physiology of the Circadian Timing System. Harvard University Press, Cambridge, MA/London.

Okamura, H., S. Yamaguchi and K. Yagita. 2002. Molecular machinery of the circadian clock in mammals. Cell Tissue Res. 309: 47–56.

Pittendrigh, C.S. 1981. Circadian organization and photoperiodic phenomena. pp. 1–35. B.K. Follet (ed.). Biological Clock in Reproductive Cycles. John Wright and Sons, Bristol.

Prat, F., S. Zanuy, M. Carrillo, A. Mones and A. Fostier. 1990. Seasonal changes in plasma levels of gonadal steroids of sea bass, *Dicentrarchus labrax* L. Gen. Comp. Endocr. 78: 361–373.

Prat, F., S. Zanuy, N. Bromage and M. Carrillo. 1999. Effects of constant short and long photoperiod regimes on the spawning performance and sex steroid levels of female and male sea bass. J. Fish Biol. 54(1): 125–137.

Reebs, S.G. 2002. Plasticity of diel and circadian activity rhythms in fishes. Rev. Fish Biol. Fisher. 12(4): 349–371.

Rozin, P.N. and J. Mayer. 1961. Thermal reinforcement and thermoregulatory behavior in goldfish, *Carassius auratus*. Science 134(348): 942–3.

Rubio, V.C., F.J. Sanchez-Vazquez and J.A. Madrid. 2003. Macronutrient selection through postingestive signals in sea bass fed on gelatin capsules. Physiol. Behav. 78: 795–803.

Rubio, V.C., M. Vivas, A. Sanchez-Mut, F.J. Sanchez-Vazquez, D. Coves, G. Dutto and J.A. Madrid. 2004. Self-feeding of European sea bass (*Dicentrarchus labrax* L.) under laboratory and fanning conditions using a string sensor. Aquaculture 233: 393–403.

Sanchez, J.A. and F.J. Sanchez-Vazquez. 2009. Feeding entrainment of daily rhythms of locomotor activity and clock gene expression in zebrafish brain. Chronobiol. Int. 26: 1120–1135.

Sanchez, J.A., J.A. Madrid and F.J. Sanchez-Vazquez. 2010. Molecular cloning, tissue distribution, and daily rhythms of expression of *per1* gene in European sea bass (*Dicentrarchus labrax*). Chronobiol. Int. 27(1): 19–33.

Sanchez-Muros, M.J., V. Corchete, M.D. Suarez, G. Cardenete, E. Gomez-Milan and M. de la Higuera. 2003. Effect of feeding method and protein source on Sparus aurata feeding patterns. Aquaculture 224: 83–103.

Sanchez-Vazquez, F.J. and J.A. Madrid. 2001. Feeding anticipatory activity in fish. pp. 216–232. *In*: D.F. Houlihan, T. Boujard and M. Jobling (eds.). Food Intake in Fish. Blackwell Science, Oxford.

Sanchez-Vazquez, F.J., F.J. Martinez, S. Zamora and J.A. Madrid. 1994. Design and performance of an accurate demand feeder for the study of feeding behaviour in sea bass, *Dicentrarchus labrax* L. Physiol. Behav. 56: 789–794.

Sanchez-Vazquez, F.J., J.A. Madrid and S. Zamora. 1995a. Circadian rhythms of feeding activity in sea bass, *Dicentrarchus labrax* L.: dual phasing capacity of diel demand-feeding pattern. J. Biol. Rhythm. 10: 256–266.

Sanchez-Vazquez, F.J., S. Zamora and J.A. Madrid. 1995b. Light-dark and food restriction cycles in sea bass: effect of conflicting zeitgebers on demand-feeding rhythms. Physiol. Behav. 58: 705–714.

Sanchez-Vazquez, F.J., J.A. Madrid, S. Zamora, M. Iigo and M. Tabata. 1996. Demand feeding and locomotor circadian rhythms in the goldfish, *Carassius auratus*: dual and independent phasing. Physiol. Behav. 60: 665–674.

Sanchez-Vazquez, F.J., M. Azzaydi, F.J. Martinez, S. Zamora and J.A. Madrid. 1998. Annual rhythms of demand-feeding activity in sea bass: evidence of a seasonal phase inversion of the diel feeding pattern. Chronobiol. Int. 15: 607–22.

Schibler, U. and P. Sassone-Corsi. 2002. A Web of Circadian Pacemakers. Cell 111: 919–922.

Tamai, T.K., A.J. Carr and D. Whitmore. 2005. Zebrafish circadian clocks: cell that see light. Biochem. Soc. T. 33: 962–966.

Vallone, D., S.B. Gondi, D. Whitmore and N.S. Foulkes. 2004. E-box function in a period gene repressed by light. Proc. Natl. Acad. Sci. USA 101: 4106–4111.

Velarde, E., R. Haque, P.M. Iuvone, C. Azpeleta, A.L. Alonso-Gómez and M.J. Delgado. 2009. Circadian clock genes of goldfish, *Carassius auratus*: cDNA Cloning and rhythmic expression of period and cryptochrome transcripts in retina, liver and gut. J. Biol. Rhythm. 24: 104–113.

Velazquez, M., S. Zamora and F.J. Martinez. 2004. Influence of enviromental conditions on demand-feeding behaviour of gilthead seabream (*Sparus aurata*). J. Appl. Ichthyol. 20: 536–541.

Vera, L.M., J.A. Madrid and F.J. Sanchez-Vazquez. 2006. Locomotor, feeding and melatonin daily rhythms in sharpsnout seabream (*Diplodus puntazzo*). Physiol. Behav. 88: 167–172.

Vera, L.M., L. Cairns, F.J. Sanchez-Vazquez and H. Migaud. 2009. Circadian rhythms of locomotor activity in the Nile tilapia, *Oreochromis niloticus*. Chronobiol. Int. 26(4): 1–16.

Villamizar, N., M. Herlin, M.D. Lopez and F.J. Sanchez-Vazquez. 2012. Daily spawning and locomotor activity rhythms of European sea bass broodstock (*Dicentrarchus labrax*). Aquaculture 354: 117–120.

Vivas, M. 2007. Regulación de la Auto-selección de macronutrientes en lubina y sargo picudo alimentados mediante comederos a demanda. Thesis, University of Murcia, Murcia, Spain.

Whitmore, D., N.S. Foulkes, U. Strahle and P. Sassone-Corsi. 1998. Zebrafish clock rhythmic expression reveals independent peripheral circadian oscillators. Nat. Neurosci. 1(8): 701–707.

Section 2
Physiology, Behavior and Pathology

The Pineal Organ of the European Sea Bass: A Neuroanatomical Approach

Patricia Herrera-Pérez,[1,a,]* Arianna Servili,[2,d] María del Carmen Rendón-Unceta,[1,b] Jack Falcón[3,e] and José Antonio Muñoz-Cueto[1,c,]*

Introduction

The pineal gland of vertebrates is important because it produces and releases the time-keeping hormone melatonin (Klein et al. 1981). In most species investigated, the rhythm in melatonin production results from the activity of circadian clocks synchronized by photoperiod. However, the mechanisms through which this is achieved have been profoundly modified in the course of vertebrate evolution (Collin 1971, Collin and Oksche 1981). In aquatic species, cone-like photoreceptor cells concentrate a photoreceptive unit, clock machinery and melatonin-producing unit (Falcón et al. 2007a, 2010).

[1]Departamento de Biología, Facultad de Ciencias del Mar y Ambientales, Campus de Excelencia Internacional del Mar (CEI·MAR), Universidad de Cádiz. Avenida República Saharaui s/n, Campus Río San Pedro, E11510-Puerto Real (Cádiz), Spain.
[a]Email: patricia.herrera@uca.es
[b]Email: maricarmen.rendon@uca.es
[c]Email: munoz.cueto@uca.es
[2]IFREMER, LEMAR UMR 6539 - Unit of Functional Physiology of Marine Organisms (PFOM), Plouzané, France.
[d]Email: arianna.servili@ifremer.fr
[3]CNRS, UMR7232 and Université Pierre et Marie Curie-Paris 6, Laboratoire ARAGO, Avenue Fontaulé, 66650 Banyuls-sur-Mer, France.
[e]Email: falcon@obs-banyuls.fr
*Corresponding authors

In mammals, the so-called cells of the sensory line or pinealocytes, have lost direct photosensitivity and clock properties. Instead, light caught by the eyes synchronizes the circadian clocks located in the suprachiasmatic nuclei, which in turn synchronize the pineal activity through a complex nervous pathway ending with a sympathetic innervation of the pineal (Klein et al. 1981). Intermediate situations are observed in sauropsids.

Pineal evolution has traditionally been interpreted as a long process in which the cells of the sensory line underwent a gradual transformation from typical photoreceptor cells (aquatic vertebrates), to rudimentary photoreceptors (sauropsids), and to pinealocytes *sensu stricto* (mammals and ophidians). This theory, first based on comparative ultrastructural studies of the pineal cell types organ of a wide variety of vertebrates, was further supported by the observations that during ontogeny (1) the pineal organ of amniotes develops photosensory characters before entering into regression (Araki et al. 1992, Collin 1977, Meiniel 1981, Oksche 1983, Tosini et al. 2000, Zimmerman and Tso 1975) and (2) norepinephrine suppresses photoreceptor cell characters in rat and chicken pineal (Araki and Tokunaga 1990, Araki 1992, Tosini et al. 2000). This however does not explain why early vertebrates that have diverged from the main lineage may also possess different pineal sensory cell types in one organ, from true photoreceptors to pinealocytes *sensu stricto*. This is indeed the case in lamprey (cyclostomes), which possesses typical and rudimentary photoreceptors as well as pinealocytes *sensu stricto* (Collin 1971), pike (Teleost fish) and lizards, which may possess both typical and rudimentary photoreceptor cells (Collin 1971, Falcón 1979b, Meiniel 1976) and chicken (avian) which possesses rudimentary photoreceptor cells and pinealocyte-like cells (Grève et al. 1993). Accordingly, it has been suggested that the presence of different sensory cell types within a taxa and among vertebrate species results from "changes in fate restriction within the neural lineage of the pineal field" (Ekström and Meissl 2003). The diverse types of pineal photoreceptors and pinealocytes are considered to be the result of particular combinations of regulatory mechanisms in different vertebrates (Ekström and Meissl 2003). It is clear that a higher number of species needs to be investigated in order to get a clear-cut picture.

Whatever it may be, a kind of convergent evolution may have occurred between teleost fish and mammals, at least in terms of melatonin secretion. Indeed, studies have shown that (1) pineal melatonin production by some fish pineal organ relies totally or partially on the presence of intact eyes, whereas this is not the case in others (Bayarri et al. 2003, Martinez-Chavez et al. 2008, Migaud et al. 2007); and (2) catecholamines contribute to control melatonin production in pike and carp pineal organs in culture (Falcón et al. 1991, Seth and Maitra 2011) as is the case in the avian and mammalian pineal glands (Klein et al. 1981, 1997). In this context, the pineal organ of the European sea bass, *Dicentrarchus labrax*, is interesting because both the eyes and the pineal organ appear to be necessary to sustain full night-time melatonin production by the pineal (Bayarri et al. 2003, Migaud et al. 2007); the question arise as to whether this is correlated to the existence of different photosensory cells.

Morphofunctional Characterization of the Pineal Organ

In a first attempt, we carried out an in-depth description of the anatomy and morpho-functional characteristics of the pineal organ of the European sea bass, with special attention being paid to the identification of the photoreceptor and melatonin producing cells (Herrera-Pérez et al. 2011). The determination of the anatomical and histological characteristics of sea bass pineal complex was possible using whole decalcified heads, which allowed preservation of the pineal position (Fig. 1). The sea bass pineal complex presented similarities with the organization of the pineal complex reported previously in other teleosts (Falcón 1979a, Ekström and Meissl 1997, Confente et al. 2008). Hence, it appeared composed of the pineal organ, the parapineal organ and the dorsal sac, but also showed some anatomical adaptations that differentiate it from other teleost fishes.

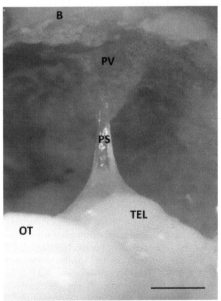

Figure 1. View of the dorsal surface of the brain of the European sea bass, *Dicentrarchus labrax*, showing the pineal gland attached to the skull bone. *Abbreviations*: B, cranial bone; OT, optic tectum; PV, pineal vesicle; PS, pineal stalk; TEL, telencephalon. Scale bar = 1 mm.

Color image of this figure appears in the color plate section at the end of the book.

Anatomical and histological analysis of the European sea bass pineal complex

The sea bass pineal organ represents a very tiny structure growing up in the midline from the junction of the telencephalic hemispheres and the optic tectum towards the top of the head (Figs. 1 and 2) (Herrera-Pérez et al. 2011). In this zone, commonly called "pineal window", the dorsal surface overlaying the cranium is translucent, being associated with light reactions of fishes (McNulty and Nafpaktitis 1977, Agha and Joy 1986). The sea bass pineal window presents a triangular shape and shows sparsely distributed melanophores (Herrera-Pérez et al. 2011). Moreover, a prominent cartilage just above the small pineal vesicle seems to act as a filter and/or a subepidermal "lens-like" structure as those described in other fish species

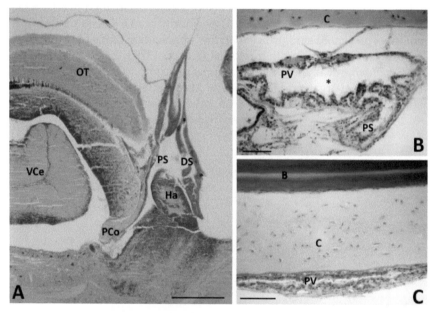

Figure 2. Histological sections of the pineal organ of the European sea bass. (**A**), Midline sagittal sections of sea bass pineal stalk emerging from the posterior commissure between the habenula and the optic tectum. (**B, C**), Transverse sections through the pineal end-vesicle of the sea bass. The pineal vesicle, which exhibits a convoluted epithelium around the pineal lumen, appears apposed to the cranium and attached to the distal pineal stalk (**B**). At the caudal end, the pineal vesicle appears firmly attached to the skull bone and exhibits a tiny lumen (**C**). *Abbreviations*: B, bone; C, cartilage; DS, dorsal sac; Ha, habenula; OT, optic tectum; PCo, posterior commissure; PS, pineal stalk; PV, pineal vesicle; VCe, valvula of the cerebellum. The asterisk in **B** indicates the pineal lumen. Scale bar = 500 μm in (A), and 100 μm in (B) and (C).

Color image of this figure appears in the color plate section at the end of the book.

(Srivastava and Srivastava 1991, Srivastava 2003). At this level, the skull bone is rather thin and flattened just above where the pineal-end vesicle is anchored. These anatomical characteristics seem to facilitate the maximal incidence of light into the underlying small-sized pineal vesicle and could reflect an adaptation to better catching the light stimuli and increase light sensitivity in the sea bass. The relative transmission of light through the skull has been quantified in different fish species including sea bass (Gern et al. 1992, Migaud et al. 2006). It has been reported that a significantly higher percentage of light penetrates the sea bass pineal window relative to salmon, with higher penetration towards the red end of the visible spectrum (Migaud et al. 2006), which could enhance pineal sensitivity in a species living in shallow waters. In fact, the pineal organ of most species investigated up to date (including sea bass) present higher sensitivity to short wavelengths (Bayarri et al. 2002, Vera et al. 2010). Furthermore, *ex vivo* and *in vivo* analyses on pineal organs demonstrated that the pineal organ of sea bass is at least 10 times more sensitive to light than the pineal of salmon (Migaud et al. 2006).

The pineal organ of sea bass has a narrow and short pineal stalk (Fig. 2A) and a pineal vesicle smaller than that of other fish species that inhabit similar geographical areas but deeper zones or sandy bottoms as the sole (Confente et al. 2008). This pineal vesicle exhibits a thin dorsal wall with a small number of pinealocytes and a thickened ventral wall that concentrates most photoreceptors (Figs. 2B, 2C) (Herrera-Pérez et al. 2011). The lower thickness of the upper parenchyma seems to facilitate the transmission of light through the organ to reach the photoreceptor cells concentrated in the ventral wall. Interestingly, the organisation of the pineal organ of a highly evolved teleost as the sea bass much more resembles the pineal organ of the sea lamprey, *Petromyzon marinus* (Cole and Youson 1982, Pombal et al. 1999, Meléndez-Ferro et al. 2002) than that of sole. Therefore, it seems that environmental photic niches have stronger influences on pineal characteristics than phylogenetic relationships.

The sea bass pineal complex also contains a minute asymmetric parapineal organ located in the left hemisphere, lying in the dorsal border of the habenular commissure, and a dorsal sac surrounding the pineal stalk, which consists of a highly folded monostratified epithelium composed of closely packed columnar cells (Herrera-Pérez et al. 2011).

Characterization of photoreceptive and melatonin-synthesizing cells in the pineal organ of sea bass

Although information concerning the photosensitive pigments present in the pineal organs of teleosts is available only in a few of the major teleost orders, the presence of both rod-like and cone-like photopigments has been reported in pineal photoreceptor cells of most species analyzed (Ekström and Meissl 1997). There is now evidence indicating that the pineal organ of adult teleosts utilizes additional photopigments. In particular, the list of photopigments has been enlarged in the last years, and includes vertebrate ancient opsin, exo-rhodopsin, UV opsin, parapinopsin, melanopsin and parietopsin (Blackshaw and Snyder 1997, Mano et al. 1999, Philp et al. 2000a,b, Kojima et al. 2000, Forsell et al. 2001, 2002, Peirson et al. 2009).

An immunohistochemical study performed in the sea bass using antibodies against cone (CERN-874) and rod (CERN-922) opsins permitted us to recognize rod opsin like and cone opsin like cells both in the pineal vesicle and stalk (Herrera-Pérez et al. 2011). Although both types of immunoreactive cells overlapped, there is some evidence suggesting that they could correspond to different types of photoreceptors cells. Thus, the immunostaining using the cone opsin antiserum was more evident in the proximity of the pineal lumen while the rod opsin antiserum labelled cells were present in more external regions of the pineal vesicle and stalk. Rod like and cone like opsins were not co-expressed in the same pinealocytes by comparing the opsin labelling of consecutive semithin sections, using similar antibodies (rod opsin CERN 858, cone opsin CERN 874) in the pineal of the Atlantic salmon (Philp et al. 2000b).

It should be noted that the sea bass pineal exhibits a differential sensitivity to lights of different wavelengths, which vary in their transmission through the skull bone (Bayarri et al. 2002, Migaud et al. 2006). These authors have suggested that the sea bass pineal organ could contain true and modified photoreceptors, as well as non-sensory pinealocytes receiving light information from the retina (Bayarri et al. 2003, Migaud et al. 2007). In an ultrastructural study developed in our laboratory, typical and modified photoreceptors were also identified in the pineal organ of the sea bass (Herrera-Pérez et al. 2004). The typical photoreceptor cells present numerous similarities with the photoreceptors of the retina, particularly with the cones, such as the cell segmentation, the regular arrangement of the discs and the polarity in the distribution of cellular organelles (Fig. 3). The modified photoreceptors exhibit low number of membrane discs, which might reflect an adaptive mechanism in pelagic species living in environments exposed to high illumination. Often, these modified photoreceptor cells have irregularly shaped outer segments that exhibit a disarranged organization and increased hyaloplasmic space between membrane lamellae (Herrera-Pérez et al. 2004). Similar modified photoreceptors have been described in other teleosts (Collin 1971, Omura and Oguri 1971, Falcón 1979b, Ekström and Meissl 1997, 2003). It has been proposed that such modified photoreceptors could represent regressed sensory cells such as those described in sauropsids (Falcón 1979b, Ekström and Meissl 2003).

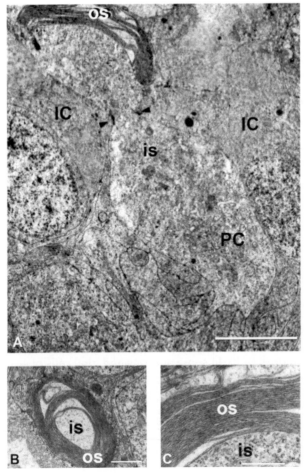

Figure 3. Ultrastructure of photoreceptor cells. (**A**), Typical pineal photoreceptor cell (PC) showing the outer segment (os) and the inner segment (is). The photoreceptor cell is surrounded by two interstitial cells (IC). (**B, C**), Photoreceptor cells with outer segments (os) of the regular type. The membrane lamellae appear compact and show a regular arrangement. Scale bars: 5000 nm in A; 100 nm in B and 500 nm in C.

Immunohistochemical studies have demonstrated that the pineal photoreceptor cells contain not only the elements of the phototransduction cascade but also from the melatonin biosynthetic pathway (Ekström and Meissl 1997, 2003). Because of the highly diffusible nature of melatonin, in a previous study from our laboratory we have identified presumed melatonin-synthesizing cells in the sea bass pineal organ by using an antibody against its precursor serotonin (5-HT) (Herrera-Pérez et al. 2011). Sea bass pineal 5-HT-immunopositive cells were present both in the pineal vesicle and stalk, and exhibited a similar pattern of distribution to that described for the opsin-immunoreactive cells, suggesting that they are also photosensitive. However, immunostaining with 5-HT antiserum was more profuse than that obtained with anti-rod opsin and anti-cone opsin sera (Herrera-Pérez et al. 2011). This result could indicate that melatonin-synthesizing photoreceptor cells coexist with non-sensory pinealocytes in the pineal organ of sea bass as it has been proposed in lampreys and some teleosts (Meiniel 1980, Ekström and Meissl 1990, 2003, Korf 2000). In fact, eye removal and optic nerve sectioning experiments in sea bass cause plasma melatonin levels to fall to approximately half of control values, suggesting the presence of different types of pinealocytes in the pineal organ (Bayarri et al. 2003, Migaud et al. 2007). These authors postulated that true and modified pineal photoreceptors would continue producing melatonin during the dark period, while pinealocytes *sensu stricto* would not receive light information from the retina, therefore causing plasma melatonin to decrease (Bayarri et al. 2003). Alternatively, 5-HT could be released from pinealocytes to the extracellular space and diffuse to other cells or, instead, could serve functions other than melatonin synthesis (Pévet 1983, Azekawa et al. 1991, Míguez et al. 1997). As in other teleosts, the pineal organ of sea bass also contains neurons, which are postsynaptic to photoreceptors and send their axons through the pineal tract, and supportive interstitial cells, which displays some cytological characteristics of glial cells (Engbretson and Linser 1991, Ekström and Meissl 1997, 2003). Therefore, the presence of 5-HT immunoreactivity in some neurons and glial cells of sea bass cannot be discarded.

In addition to the localization of presumed melatonin synthesizing cells, we have analyzed in the pineal organ of sea bass, the day-night expression pattern of two key enzymes in the melatonin-synthesizing pathway, arylalkylamine-*N*-acetyltransferase 2 (AANAT2) and hydroxyindole-*O*-methyltransferase (HIOMT) (Herrera-Pérez et al. 2011). A circadian rhythm in *Aanat2* transcript levels occurs in the pineal organ of species such as pike and zebrafish but not in species such as the trout (Bégay et al. 1998), whereas no conspicuous daily variation of *Hiomt* expression or activity appears to exist in fish (Vuilleumier et al. 2007). In our study, we found significant day-night differences in *Aanat2* mRNA expression (with higher nocturnal levels) but not in *Hiomt* transcript levels. This study was done under natural light/dark conditions, but the persistence of *in vitro* rhythmic melatonin release under constant dark conditions in sea bass pineals (Bayarri et al. 2004a) suggests that sea bass *Aanat2* could also exhibit circadian rhythms of expression as is the case of pike or zebrafish (Bégay et al. 1998).

The Targets of the Neural Message of the Pineal Organ in the European Sea Bass

The pineal organ is bidirectionally connected with the brain through pinealofugal (efferent) and pinealopetal (afferent) projections. The axons from the second-order neurons and from central projecting photoreceptor cells constitute the pineal tract that innervates specific central areas. The fish pineal organ also receives axon terminals originating from neurons located in different central cell masses (Ekström et al. 1994, Ekström and Meissl 2003, Jiménez et al. 1995, Yáñez and Anadón 1998, Pombal et al. 1999, Mandado et al. 2001, Servili et al. 2011).

Pioneer works by Holmgren (Holmgren 1918) and later, Hafeez (Hafeez 1971) reported the study of the pineal efferent tract by using methylene blue staining and silver impregnation techniques. Nowadays, the lipophilic carbocyanine dyes are preferred and currently employed as neuronal tracers. Thank to these novel tracers, the pineal projections of several fish species have already been explored (Yáñez et al. 1993, Yáñez and Anadón 1998, Pombal et al. 1999, Mandado et al. 2001, Yáñez et al. 2009). Recently, we have employed one of these neural tracers, 1,1'-dioctadecyl-3,3,3',3'-tetramethylindocarbocyanine perchlorate (DiI), to reveal the pineal neural pathway of the European sea bass (Servili et al. 2011). The DiI permitted us to effectively trace in both anterograde and retrograde directions revealing detailed pinealofugal axons as well as dendrites and cell bodies of pinealopetal neurons in sea bass (Fig. 4). This work represented the first

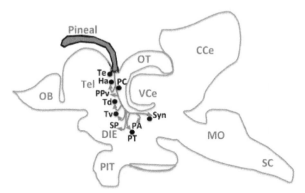

Figure 4. Schematic sagittal drawing of the brain of the European sea bass showing the efferent (red arrows) and afferent (black dots) connections of the pineal gland. *Abbreviations*: CCe, corpus of the cerebellum; DIE, diencephalon; Ha, habenula; MO, medulla oblongata; OB, olfactory bulb; OT, optic tectum; PIT, pituitary gland; PA, pretectal area; PC, posterior commissure; PPv, ventral periventricular pretectum; PT, posterior tubercle; SC, spinal cord; SP, superficial pretectum; Syn, dorsal synencephalon; Td, dorsal thalamus; Te, thalamic eminence; Tel, telencephalon; Tv, ventral thalamus; VCe, valvula of the cerebellum.

Color image of this figure appears in the color plate section at the end of the book.

report on the pineal connections in a perciform fish, which is the largest group of teleosts (Nelson 2006) and shows an extremely complex visual system (Striedter and Northcutt 1989).

Efferent projections from the pineal organ of sea bass

The DiI revealed the presence of a pineal tract in sea bass that exited from the pineal stalk and reached specific brain regions (Fig. 4), i.e., the habenula, the ventral thalamus, the periventricular pretectum, the central area, the posterior tubercle and both the medial and dorsal tegmental areas (Servili et al. 2011), which also correspond to areas receiving pinealofugal projections in other fish species (Yáñez et al. 1993, Yáñez and Anadón 1998, Mandado et al. 2001, Yáñez et al. 2009). Furthermore, many of the reported regions were identified as retino-recipient nuclei in teleost (Ekström 1984, Yáñez et al. 2009, Servili et al. 2012).

The first region receiving the pinealofugal projections is the habenula, which is located just below the junction of the pineal stalk with the brain. The habenula is described as a pinealofugal terminal region in several fish species including teleosts (Hafeez and Zerihun 1974, Ekström and van Veen 1983, Ekström 1984, Jiménez et al. 1995, Yáñez and Anadón 1996), chondrosteans (Yáñez and Anadón 1998) and elasmobranchs (Mandado et al. 2001). From this region, pineal fibers project ventrally to reach the ventral thalamus. Many nuclei of this area show pinealofugal connection in sea bass including the thalamic eminence, which is involved in higher olfactory circuits (Northcutt 1995). Interestingly, the ventral thalamic region also contains retinofugal terminals in this species (Servili and Muñoz-Cueto, unpublished). In addition, the ventral thalamus exhibits melatonin receptors (Herrera-Perez et al. 2010), GnRH receptors (González-Martínez et al. 2004), cathecolaminergic cells (Batten et al. 1993) and, in turn, it sends projections to the pituitary gland (Garcia-Robledo and Muñoz-Cueto, unpublished) in sea bass. Therefore, this nucleus could be likely involved in the mediation of the photoperiod signal to the endocrine system in this species.

A bundle of pineal efferent projections crossed the posterior commissure, giving rise to a bilateral and symmetric labelling (Servili et al. 2011). Once again, this feature seems to be conserved in fish. The stained axons proceeded in the dorso-lateral direction and entered several nuclei of the pretectal area. Some of these nuclei are sites of primary retinal and tectal terminal fields in teleosts (Presson et al. 1985, Northcutt and Butler 1991), including sea bass, and they project, in turn, to the optic tectum (Northcutt 1995). The optic tectum represents a multisensorial integrating center. Furthermore, the pretectal nuclei exhibit melatonin receptors in sea bass (Herrera-Pérez et al. 2010). Altogether, these evidences suggest a role for the pretectum in the integration of neural (via retinal and pineal connections) and also neurohormonal (via melatonin action) signals carrying the light information in this species.

Important efferent projections from the pineal organ were observed in the medial and dorsal tegmental region, notably in the medial longitudinal fascicle. This area represents in teleost a source of descending spinal axons (Oka et al. 1986) and exhibits GnRH2 cells in sea bass (González-Martínez et al. 2001, González-Martínez et al. 2002b). The GnRH2 neurons in sea bass do not innervate the pituitary gland but they project widely all over the brain (González-Martínez et al. 2002a). Therefore, these GnRH cells do not seem concerned directly in the modulation of gonadotrophin synthesis and release, although they could mediate the pineal action in the control of photo-period driven rhythmic processes (Servili et al. 2010).

Caudal pineal efferents in sea bass were also detected in the posterior tubercle, notably in the periventricular nucleus of the posterior tubercle and the paraventricular organ (Servili et al. 2011). The latter nucleus contains many serotonin- and dopamine-immunoreactive cells in this species (Batten et al. 1993). Therefore, a role for these cells in the mediation of pineal action on the central brain was suggested in sea bass (Servili et al. 2011). This connection of the pineal organ with the posterior tubercle is a shared feature between lampreys, elasmobranchs and bony fish (Hafeez and Zerihun 1974, Yáñez et al. 1993, Jiménez et al. 1995, Yáñez and Anadón 1998, Pombal et al. 1999, Mandado et al. 2001, Yáñez et al. 2009).

Afferent projections to the pineal organ of sea bass

DiI tract tracing study also revealed several pinealopetal cell masses in the sea bass brain (Fig. 4). Thus, cells projecting to the pineal organ were detected in most areas already shown to exhibit pineal efferent projections: the thalamic eminence, the habenula, the ventral and dorsal thalamus, the periventricular pretectum, the posterior tubercle and the medial tegmental area (Servili et al. 2011). For instance, the nucleus of the medial longitudinal fascicle, above described as a pineal efferent center, presented GnRH2 cells that projected to the pineal organ. This was observed by a combination of retrograde tracing and immunohistochemistry using highly specific antisera (Servili et al. 2010). Moreover, GnRH2 innervation was detected in the pineal organ, which also strongly expresses a GnRH receptor with high affinity for the GnRH2 form. In addition, both *in vivo* and *in vitro* experiments showed the stimulatory effect of GnRH2 on the nocturnal release of melatonin by the pineal organ of sea bass (Servili et al. 2010). All these evidences corroborate the specificity of the connection of this brain area and these neuroendocrine cells with the pineal organ of sea bass. In addition to GnRH projections, this photoreceptor organ also receives noradrenaline (Ekström and Korf 1986), FMRFamide (Ekström et al. 1988) and NPY (Subhedar et al. 1996, Blank et al. 1997) fibers in teleost fish. Notably, in sea bass, NPY cells were described in the thalamic eminentia, the ventral and dorsal thalamus and the periventricular pretectum (Cerdá-Reverter et al. 2000). Furthermore, catecholaminergic cells were detected in the ventromedial and dorsolateral thalamus, the periventricular pretectum, the posterior commissure and the posterior tubercle (Batten et al. 1993). The presence of kisspeptin 1 (kiss1), a neuropeptide that could be involved in the integration of the photoperiodic and environmental inputs, was revealed in cells from the sea bass habenula (Escobar et al. 2013). Interestingly, all these nuclei exhibit pinealopetal cells in sea bass (Servili et al. 2011).

Previous studies demonstrated that both ophthalmectomy and optic nerve sectioning decreased plasma melatonin levels at night. In addition, neither pinealectomy nor ophthalmectomy alone completely abolished circulating melatonin, and only when both organs were removed, daily melatonin rhythms disappeared (Bayarri et al. 2003). These results indicate that the pineal organ requires light information from the lateral eyes to normally secrete melatonin into the bloodstream in sea bass (Bayarri et al. 2003). However, no direct innervation between the retina and the pineal organ was found in sea bass (Servili et al. 2011), suggesting that putative connections between both photoreceptor organs may require intermediary neural systems conveying retinal information up to the pineal organ.

In this sense, a noteworthy observation is that the ventral thalamus of teleosts, which represents a region bidirectionally connected to the pineal organ (Servili et al. 2011), is also a retinorecipient center and a source of projections to the optic tectum (Wullimann 1998). All these evidences support the hypothesis that the ventral thalamus could play an important role in the integration and transmission of the light information. Moreover, since this region shows hypophysiotrophic neurons, it could represent a relay center in the transmission of the light information from the brain to the pituitary gland.

Our studies developed in sea bass have represented the first report of pinealofugal projections in a representative of perciform fish but also the first evidence on the existence of multiple pinealopetal connections in the brain of teleosts (Fig. 4). These studies have provided important information to better decipher how photoperiod inputs are integrated in the central nervous system of sea bass, in which the light inputs are critical to synchronize many physiological rhythmic processes.

The Targets of the Neuroendocrine Message of the Pineal Organ in Sea Bass

Melatonin represents the hormonal output of the vertebrates' circadian clocks, and exhibits a clear daily rhythm of secretion with higher nocturnal plasma levels. In ectotherms, light and temperature represent the main environmental factors modulating melatonin secretion through their actions on the activity of arylalkylamine N-acetyltransferase 2 (AANAT-2), a key enzyme in pineal melatonin rhythms (Falcón et al. 2007a, 2010). As a result, this melatonin rhythm serves as a daily and seasonal signal that can synchronize a variety of endocrine, neural and behavioral circadian and circannual rhythms (Ekström and Meissl 1997, Falcón et al. 2007a, 2007b, 2010).

The effects of melatonin are mediated through low and high affinity receptors. The low affinity melatonin receptor (MT3) identified in mammals corresponds to 'quinone reductase-2', a cytosolic enzyme that might be involved in detoxification and antioxidant processes (Mailliet et al. 2005). Three high affinity receptor subtypes have been identified to date, all belonging to the family of the seven transmembrane (TM) domains G-protein coupled receptors (GPCR) (Brydon et al. 1999, Falcón et al. 2007a). The MT1 and MT2 subtypes are found in all vertebrates investigated so far whereas the Mel1c subtype is found only in non-mammalian vertebrates.

In comparison with the huge literature concerning mammals, very few studies report on the cloning and expression of melatonin receptors in fish and very little is known on the effects that are mediated by melatonin binding to its receptors in this important group of vertebrates. In the following sections, we will summarize the studies performed by our laboratories concerning the melatonin targets in the European sea bass.

Cloning and tissue-specific distribution of melatonin receptor

In a previous study, we have reported the full length cloning of three different melatonin receptor subtypes in the sea bass *Dicentrarchus labrax*, belonging, respectively, to the MT1, MT2 and Mel1c subtypes (Sauzet et al. 2008). The three deduced amino acid sequences exhibited 350, 360 and 353 amino acids, respectively, and displayed the 7 transmembrane motifs profile as well as amino acid known to be crucial for the function of the receptors in mammals. The phylogenetic tree built after a comparative analysis of sequences further confirmed that the three clones isolated were each representative of one high affinity melatonin receptor subtype, and were therefore named dlMT1, dlMT2, and dlMel1c, respectively (Sauzet et al. 2008).

The cloning of the melatonin receptors allowed searching for the tissue specific expression of each subtype by using RT-PCR. The MT1 receptor seemed more widely distributed and more strongly expressed than the other subtypes in the sea bass brain and retina. A high MT1 expression was found in neural tissues, especially in visually-related areas as the retina and optic tectum, but an important MT1 expression was also detected in integrative centers as the cerebellum and in neuroendocrine areas as the telencephalon and hypothalamus. These results were consistent with previous studies performed in sea bass that demonstrate the presence of abundant melatonin binding sites in these central areas (Bayarri et al. 2004a,b) and also agree with previous findings on the distribution of MT1 receptor-expressing cells (Mazurais et al. 1999) or ^{125}I-melatonin binding sites (Martinoli et al. 1991, Ekström and Vanecek 1992, Davies et al. 1994, Gaildrat et al. 2002, López-Patiño et al. 2008, Sauzet et al. 2008) obtained in other teleost species. In peripheral tissues there was a conspicuous MT1 expression in the gonads (Herrera-Pérez et al. unpublished data) and the gills, and a weak expression in the muscle (Sauzet et al. 2008). The MT1 expression in gonads suggests that both ovary and testis can represent a direct target for melatonin.

In contrast to MT1, MT2 expression was strong in pituitary, weak in retina and low (optic tectum, diencephalon) or even absent (cerebellum) in central nervous tissues. No expression was detected in peripheral tissues except the liver and the blood cells (Sauzet et al. 2008). The strong MT2 expression found in extracts from sea bass pituitaries deserves special attention. The detection of melatonin receptors in the fish pituitary has been a matter of contradictory discussions in the past (Davies et al. 1994, Ekström and Vanecek 1992, Falcón et al. 2003, Gaildrat et al. 2002, Mazurais et al. 1999). Our results obtained in sea bass bring strong support to the idea that melatonin controls fish neuroendocrine functions through at least a direct action on the pituitary, mediated by MT2 receptors (Falcón et al. 2003, Gaildrat et al. 2002).

Mel1c expression in sea bass was only detected in extracts from the skin and traces were also detected in retina (Sauzet et al. 2008) whereas another study reports low levels of expression in the brain (Park et al. 2006). Recent evidence suggests that the melatonin related receptor GPR50 found in mammals is a Mel1c ortholog (Dufourny et al. 2008). In birds, *in situ* hybridization combined with immunocytochemistry for gonadotropin-inhibitory hormone (GnIH) revealed that the mRNA of Mel1c receptor subtype was expressed in GnIH-immunoreactive neurons of the paraventricular nucleus, suggesting that melatonin acts directly on GnIH neurons through this receptor to induce expression of GnIH (Ubuka et al. 2005). If this Mel1c receptor mediates similar actions in fish brain remains to be deciphered.

Melatonin receptors in cells of the retina and brain of the European sea bass

For a deeper understanding of the role played by melatonin in the different organs where receptor expression has been detected, it was necessary to more precisely identify the cell types that express these receptors. With this purpose in mind, we focused our attention in the localization of these cells by *in situ* hybridization in the sea bass retina and brain (Sauzet et al. 2008, Herrera-Pérez et al. 2010).

Localization of melatonin receptors in the sea bass retina

Sea bass MT1 and MT2 melatonin receptors were expressed in the three nuclear layers of the neural fish retina as well as in the retinal pigment epithelium. In the sea bass retina, the cells expressing the melatonin receptors were the photoreceptors of the outer nuclear layer (ONL) and most of the cell present in the ganglion cell layer (GCL), as well as yet unidentified cells located in the most inner part of the inner nuclear layer (INL), which could belong to either bipolar or amacrine or interstitial cells (Sauzet et al. 2008). The demonstration that melatonin receptors are expressed in the three different layers of the sea bass retina extends to fish previous findings obtained in frog, chicken, rodent, and human retinas (Fujieda et al. 1999, Natesan and Cassone 2002, Savaskan et al. 2002).

It is generally believed that the photoreceptor cells produce melatonin in a circadian manner and that it acts as an autocrine and paracrine modulator of retinal function (Green and Besharse 2004, Iuvone et al. 2005, Iigo et al. 2007). In fact, melatonin has been reported as an autocrine regulator of rod and cone function, including its own biosynthesis (Falcón et al. 2007a), electrical activity (ERG; (Peters and Cassone 2005, Pierce and Besharse 1985), disc shedding and photoreceptor movements (Peters and Cassone 2005, Pierce and Besharse 1985), and synchronization of circadian clocks units (Cahill and Besharse 1993, Chaurasia et al. 2006, Yu et al. 2007). However, it has also been demonstrated that retinal cells from the INL and GCL expressed the enzymes of the melatonin synthesizing pathway, the arylalkylamine N-acetyltransferase (AANAT) and/or hydroxy-indole-*O*-methyltransferase (HIOMT) in species as trout (Besseau et al. 2006) and sole (Isorna et al. 2009, 2011). In both species the melatonin-synthesizing cells occupied the same position in the retinal layers as those shown to express the melatonin receptors in sea bass. This would suggest that melatonin could also act as an autocrine-paracrine modulator in these inner layers of the sea bass retina. Interestingly, it has been demonstrated that melatonin modulates dopamine release by amacrine cells in the INL of fish and other vertebrates (Ribelayga et al. 2004), which in turn feeds back on the melatonin biosynthesis and circadian activity of the photoreceptor cells (Stella and Thoreson 2000, Yu et al. 2007). The large distribution of melatonin receptors in the INL and GCL of sea bass could reflect other functions of melatonin related to control of neurotransmitter release (Fujieda et al. 2000, Mitchell and Redburn 1991), or modulation of the electroretinogram and Purkinje shift (Peters and Cassone 2005), as reported in other species.

Localization of melatonin receptors in the sea bass brain

Despite the number of studies describing the widespread distribution of melatonin receptors in the teleost brain by means of 2-[^{125}I]-iodomelatonin binding studies (Falcón et al. 2010), there is almost no information on their precise cellular localization and only two *in situ* hybridization studies, one in rainbow trout (Mazurais et al. 1999) and the other one in sea bass (Herrera-Pérez et al. 2010) have been performed. In our sea bass study, we only focused on MT1 expression because previous RT-PCR studies indicated very low MT2 expression and no Mel1c expression in sea bass brain (Sauzet et al. 2008).

Sea bass brain MT1mRNA-expressing cells were present in the parvocellular preoptic nucleus, ventral and dorsal thalamus, lateral tuberal nucleus, periventricular layer of the optic tectum, longitudinal torus, semicircular torus, glomerular nucleus (homologue of trout posterior pretectal nucleus), and ganglionic layer (Purkinje cells) of the valvula and corpus of the cerebellum (Fig. 5), as also reported in the trout brain (Mazurais et al. 1999, Herrera-Pérez et al. 2010). However, remarkable differences were also observed between sea bass and trout because the former, but not the latter, expressed MT1 mRNAs in cells from the ventral and dorsal telencephalon, preoptic suprachiasmatic nucleus, several pretectal cell masses (including the central pretectal nucleus, superficial pretectum, periventricular pretectum, accessory pretectal nucleus, and dorsal accessory optic nucleus), mesencephalic oculomotor nucleus, hindbrain trigeminal nucleus, and medial reticular formation (Fig. 5). Conversely, the presence of weak MT1 expression in cells from hypothalamic cell masses such as the anterior tuberal nucleus, lateral nucleus of the recess, or diffuse nucleus of the inferior lobes was described in trout but not in our study in sea bass (Mazurais et al. 1999, Herrera-Pérez et al. 2010).

These discrepancies might reflect 1) different physiological status or conditions in which the animals were maintained (i.e., photoperiod, temperature, salinity); 2) different transcriptional/translational regulatory mechanisms between species; or 3) species-dependent differences, which may be related to differences in the organization of the circadian systems, as suggested recently (Falcón et al. 2007a,b, 2010, Migaud et al. 2007).

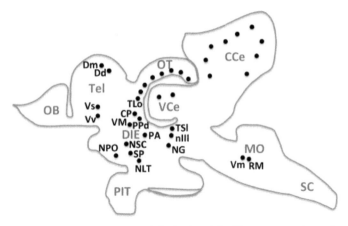

Figure 5. Schematic sagittal drawing of the brain of the European sea bass showing the distribution of type 1 melatonin receptor (MT1) mRNA-expressing cells (black dots). *Abbreviations:* CCe, corpus of the cerebellum; CP, central posterior thalamic nucleus; Dd, dorsal part of the dorsal telencephalon; DIE, diencephalon; Dm, medial part of the dorsal telencephalon; MO, medulla oblongata; nIII, oculomotor nucleus; NG, nucleus glomerulosus; NLT, lateral tuberal nucleus; NPO, parvocellular preoptic nucleus; NSC, suprachiasmatic nucleus; OB, olfactory bulb; OT, optic tectum; PA, pretectal area; PIT, pituitary gland; PPd, dorsal periventricular pretectum; RM, medial reticular formation; SC, spinal cord; SP, superficial pretectum; Tel, telencephalon; TLo, torus longitudinalis; TSl, lateral part of the torus semicircularis; VCe, valvula of the cerebellum; Vm, trigeminal nucleus; VM, ventromedial thalamic nucleus; Vs, supracommissural part of the ventral telencephalon; Vv, ventral part of the ventral telencephalon.

Melatonin targets in integrative areas related with visual and sensory information

Our *in situ* hybridization study showed that visual-system related areas such as the preoptic area, the ventral thalamus, the pretectal area and the optic tectum contained the highest amount of melatonin binding sites and MT1 expression cell masses (Herrera-Pérez et al. 2010). The overlapping of retinofugal and/or pinealofugal terminal fields (Servili et al. 2011) in MT1-expressing cell masses of sea bass brain evidence the relevant functions of these areas in the processing of light information in this species, constituting a possible pathway for exchange of information between the retina and the pineal organ. The high density of melatonin binding sites and/or MT1 expressing cells in the optic tectum, a primary target of retinal projections, is a shared characteristic among non-mammalian vertebrates (Martinoli et al. 1991, Ekström and Vanecek 1992, Davies et al. 1994, Iigo et al. 1994, 1997, Mazurais et al. 1999, López-Patiño et al. 2008, Oliveira et al. 2008) and reinforces previous results obtained in sea bass by using membrane binding assays and RT-PCR (Bayarri et al. 2004a,b, Sauzet et al. 2008). The teleostean optic tectum also receives inputs from other visual centres from the dorsal and ventral thalamus and pretectum, as well as from nonvisual sources such as the semicircular torus (Wullimann 1998). It should be noted that all these areas contained MT1-expressing cell masses in sea bass (Herrera-Pérez et al. 2010).

Our study revealed the importance of the pretectal area in the integration of pineal neuroendocrine message in sea bass because up to seven discrete cell masses from this brain region expressed MT1 gene (Herrera-Pérez et al. 2010). Thus, the clear labelling observed in the anterior and posterior subdivision of the glomerular nuclei result very interesting from the evolutionary point of view because it represents the perciform homologous of posterior pretectal nuclei of non neo-teleost like the trout, which also presents abundant melatonin receptors (Mazurais et al. 1999). Most of these visually-related areas expressing MT1 receptors also receive a pinealofugal and retinofugal inervation (Servili et al. 2011) denoting its important role in the integration of this sensorial modality. Some pretectal cell masses as the central pretectal nucleus and the accessory optic nuclei, which contain melatonin receptors in sea bass, have been implicated in detection of slowly moving objects (Friedlander 1983, Wullimann 1998).

The cells of the semicircular torus of sea bass also represent a target for melatonin in sea bass (Herrera-Pérez et al. 2010). The semicircular torus has been related with the processing of mechanosensorial/auditory stimuli, color vision, prey localization, navigation and schooling behavior (Schellart 1983, Etcheler 1984, Coombs et al. 1989). Whether melatonin plays a role in these important functions in sea bass remains to be elucidated.

Melatonin targets in behavioral, motor and integrative areas

Likewise, MT1 expression has been found in brain areas related with the control of motor and behavioral activities and integrative areas (Herrera-Pérez et al. 2010). In particular, MT1 expression has been found in the torus longitudinalis, the oculomotor nucleus and the reticular formation, which represent premotor and motor areas involved in ocular movements and eye-body motor coordination (Wulliman 1998). Two pretectal cell masses that contain melatonin receptors in sea bass, the central pretectal nucleus and the accessory optic nuclei, have also been implicated in optokinetic oculomotor reflexes (Friedlander 1983, Wullimann 1998). The torus longitudinalis, which is involved in visual activity and eye movements, represents a main relay center in premotor circuitry descending from the telencephalon to the brainstem (Northmore 1984, Wullimann 1994, Wullimann and Roth 1994). Interestingly, melatonin has been shown to modulate activity rhythms in sea bass (Herrero et al. 2007) and most of the areas connected with the torus longitudinalis express MT1 transcripts in this species, suggesting that melatonin receptors found in these brain regions could be involved in the mediation of melatonin-induced effects on locomotor activity.

The neurons of the trigeminal motor nucleus also expressed melatonin receptors in sea bass (Herrera-Pérez et al. 2010). These cells receive projections from reticular formation neurons (Luiten and van der Pers 1977) and have been functionally linked to respiratory coordination (Luiten 1976, Song and Boord 1993). Other areas that express MT1 receptors like the periventricular pretectum and the oculomotor nucleus also participate in the coordination of the rhythmic respiratory movements (Wulliman 1998). In addition, trigeminal motor neurons have been involved in feeding, aggression, sexual and brood care behaviors (Wulliman 1998).

In accordance with the reports described in other fishes (Ekström and Vanecek 1992, Davies et al. 1994, Mazurais et al. 1999, López-Patiño et al. 2008, Oliveira et al. 2008) the metencephalon of sea bass present many cells expressing melatonin receptors both in the corpus and valvula of the cerebellum (Herrera-Pérez et al. 2010). This expression is consistent with previous results obtained by radioling and assays and RT-PCR in the same species (Bayarri et al. 2004a,b, Sauzet et al. 2008). Based on physiological evidences as well as on this input-output pattern, it has been proposed that the teleostean cerebellum serves visual, motor learning and coordination functions (Kotchabhakdi 1976, Finger 1983, Pastor et al. 1997, Wullimann 1998). Complex circuits interconnect many of the brain areas implicated in the motor and behavioral control, which also represent a target for melatonin in sea bass (Herrera-Pérez et al. 2010). Thus, many of the areas involved in afferent and efferent pathways of the cerebellum also exhibit MT1 mRNAs in sea bass. Melatonin could be acting as a synchronizer of the activity of these central areas to generate coordinated responses. Moreover the cerebellum has been shown to have high proliferative capacity in sea bass (Servili et al. 2009) and other teleost species (Zupanc and Clint 2003). Neuroprotective and antiapoptotic actions of melatonin in the mouse cerebellum are mediated through its interactions with specific receptors (Jiao et al. 2004, Manda et al. 2008), and proliferative effects of melatonin have been reported in zebrafish embryos (Danilova et al. 2004). In this regard, the presence of MT1 mRNAs in the cerebellum as well as in periventricular cells of the ventral and dorsal telencephalon, which also represent active proliferative zones as reported in zebrafish (Pellegrini et al. 2007), could sustain similar neurotrophic, neuroprotective and/or proliferative effects of melatonin in these brain areas of sea bass.

Melatonin targets in neuroendocrine and hypophysiotrophic areas

The photoperiod is a key environmental factor controlling seasonal reproduction in fish. Indeed the manipulation of photoperiod is often used by the industry to advance, delay and inhibit spawning in fish species to produce a year round supply of fish or to prevent early maturation before fish reach market size (Bromage et al. 2001). The light perception system of fish has been studied in detail (Migaud et al. 2010) and the classical brain-pituitary-gonadal control axis of reproduction has been well characterized in a range of fish species (Zohar et al. 2010) but key knowledge gaps remain: what are the final links in the chain of events that connects photoperiodic information to the classical neuroendocrine and/or hypophysiotrophic systems that control the activity of the pituitary and the downstream control of reproduction?

Increasing evidence indicates that melatonin is involved in feeding, reproductive and behavioral rhythms and could mediate the effects of photoperiod on several neuroendocrine functions in teleosts (Ekström and Meissl 1997, Mayer et al. 1997, Falcón et al. 2007b, 2010). Recently, we have reported an inhibitory effect of melatonin on the nocturnal expression of GnRH-1, GnRH-3, and GnRH receptors subtypes 1c, 2a and 2b in sea bass (see chapter of Servili et al. in this book and Servili et al. 2013). In a previous study, we detected the presence of MT1 mRNA-expressing cells in some neuroendocrine and/or hypophysiotrophic nuclei of the sea bass forebrain (Herrera-Pérez et al. 2010). Thus, many cellular groups expressing MT1 gene have been evidenced in the ventral telencephalon, preoptic area (parvocellular preoptic nuclei, suprachiasmatic nucleus), ventral thalamus (ventromedial nucleus) or the hypothalamus (lateral tuberal nucleus) of sea bass. Interestingly, these brain nuclei express neuropeptides (GnRH, NPY, galanin, PACAP, GHRH, CRH, cholecystokinin, somatostatin) or amines (dopamine, serotonin) that reach the pituitary of sea bass and other fish species and have been implicated in reproduction, feeding and growth (Moons et al. 1988, 1992, Kah et al. 1991, Batten et al. 1993, 1999, Cerdá-Reverter et al. 2000, Montero et al. 2000, González-Martínez et al. 2001, 2002a). Therefore, the presence of melatonin receptors in these nuclei might be mediating photoperiod effects on these important physiological processes. Nevertheless, it should be deciphered if melatonin has direct effects on these peptidergic and aminergic cell populations or its actions are mediated by intermediary systems such as kisspeptin and gonadotropin-inhibitory hormone (GnIH) neurons. It is interesting to note that these neuropeptides seem to mediate photoperiod effects on reproductive axis in mammals and other vertebrates (Roa and Tena-Sempere 2007, Tsutsui et al. 2007, Tsutsui 2009) but these aspects have been poorly addressed in fish.

Like it occurs in birds and mammals (Cassone et al. 1986, 1987), the retino-recipient suprachiasmatic nucleus (NSC) of sea bass also express MT1 receptors (Herrera-Pérez et al. 2010). This nucleus is known for containing the biological clock that controls the circadian release of melatonin in mammals (Klein et al. 1991). In contrast, in species that lack endogenous rhythms of melatonin secretion like the trout, no MT1 mRNA-expressing cells have been found in this nucleus (Mazurais et al. 1999). These discrepancies could reflect differences in the control of melatonin secretion and in the organization of circadian system between both species, as it has been previously suggested (Migaud et al. 2007). It should be very interesting to determine if the sea bass suprachiasmatic nucleus expresses clock genes and also represents a pacemaker structure in this species.

Conclusion

In summary, our studies conducted in the European sea bass, *Dicentrarchus labrax*, have provided an in-depth description of the macroscopic and microscopic anatomy of the pineal organ and have permitted us to identify the presence of photoreceptor and melatonin-producing cells in this neural gland. In addition, we have analyzed in the pineal, the day–night expression (using quantitative real-time PCR) of two key enzymes in the melatonin-synthesizing pathway; arylalkylamine-N-acetyltransferase 2 (AANAT2) and hydroxyindole-O-methyltransferase (HIOMT), showing higher nocturnal AANAT2 transcript levels. Moreover, we have investigated the pineal outputs through the analysis of pineal projections and melatonin target sites present in sea bass brain in order to understand the regulation of time-related biological functions, specially growth, feeding and reproduction, which are of great interest for the aquaculture of this species. The efferent projections of the sea bass pineal organ reached the habenula, ventral thalamus, periventricular pretectum, central pretectal area, posterior tubercle and medial and dorsal tegmental areas. Our analysis of the pinealopetal system in sea bass demonstrated that the sea bass pineal organ receives central projections from neurons located, to a large extent, in brain areas innervated by pineal efferent projections, i.e., the thalamic eminence, habenula, ventral thalamus, dorsal thalamus, periventricular pretectum, posterior commissure, posterior tubercle and medial tegmental area. These results represented the first description of pinealofugal projections in a representative of Perciformes, which constitutes a derived order within teleosts, as well as the first evidence for the presence of pinealopetal neurons in the brain of a teleost species. Finally, we have analyzed the central and peripheral expression of melatonin receptors in sea bass and compared its brain distribution with the localization of 2-[125I]-iodomelatonin binding. Melatonin receptors were mainly expressed in visually related areas, such as the retina, pretectal area, glomerular complex, optic tectum, torus longitudinalis and thalamus. Furthermore, melatonin receptors were evident in the ganglionic cell layer of the cerebellum and the presence of iodomelatonin binding and/or MT1 mRNA-expressing cells was also observed in the hindbrain, in particular in the oculomotor and trigeminal nuclei and in the reticular formation. A conspicuous expression was also detected in neuroendocrine regions including the ventral telencephalon, preoptic area, and hypothalamus. Our results suggest an important role of MT1 in the mediation of melatonin actions in visual/light integration, mechanoreception, somatosensation, eye-body motor coordination, and integrative and neuroendocrine functions. These morpho-functional findings constitute a step forward to understand how ambient (light) information is perceived and transduced into neural (projections) and neurohormonal (melatonin) signals to regulate physiological and behavioural rhythmic functions in this species. Ongoing studies in sea bass are being directed to elucidate if the interactions between the pineal organ and the main neuroendocrine systems involved in growth, feeding and reproduction are direct or appear mediated by intermediary systems such as kisspeptins, gonadotropin-inhibitory hormone and/or other interneurons.

Acknowledgements

We thank the staff from the "Planta de Cultivos Marinos" (University of Cádiz) for their support in maintaining the experimental animals used in these studies. We thank Dr. Mairi Cowan for the critical reading of the manuscript. Grant sponsors: Regional Government of Andalusia (Junta de Andalucía, Excellence Project no. P10-AGR-05916); Spanish Ministry of Science and Innovation (Grants no. AGL2001-0593-C03-02 and HF2005-0047); European Union (Marie Curie Host Fellowships, Grant no. HPMT-CT-2000-00211); IFREMER (Grants no. 752887/00 and 753447/00); Centre National de la Recherche Scientifique (Grant no. GDR 2821).

References

Agha, A.K. and K.P. Joy. 1986. The structure of pineal window in the catfish Heteropneustes fossilis (Bloch.), with special reference to melanophore activity. Z. Mikrosk. Anat. Leipzig. 100: 104–110.

Araki, M. 1992. Cellular mechanism for norepinephrine suppression of pineal photoreceptor-like cell differentiation in rat pineal cultures. Dev. Biol. 149: 440–447.

Araki, M. and F. Tokunaga. 1990. Norepinephrine suppresses both photoreceptor and neuron-like properties expressed by cultured rat pineal glands. Cell. Differ. Dev. 31(2): 129–35.

Araki, M., Y. Fukada, Y. Shichida, T. Yoshizawa and F. Tokunaga. 1992. Differentiation of both rod and cone types of photoreceptors in the *in vivo* and *in vitro* developing pineal glands of the quail. Brain Res. Dev. Brain Res. 65(1): 85–92.

Azekawa, T., A. Sano, H. Sei and Y. Morita. 1991. Diurnal changes in pineal extracellular indoles of freely moving rats. Neurosci. Lett. 132: 93–96.

Batten, T.F., P.A. Berry, A. Maqbool, L. Moons and F. Vandesande. 1993. Immunolocalization of catecholamine enzymes, serotonin, dopamine and L-dopa in the brain of Dicentrarchus labrax (Teleostei). Brain. Res. Bull. 31: 233–252.

Batten, T.F, L. Moons and F. Vandesande. 1999. Innervation and control of the adenohypophysis by hypothalamic peptidergic neurons in teleost fishes: EM immunohistochemical evidence. Microsc. Res. Tech. 44: 19–35.

Bayarri, M.J., J.A. Madrid and F.J. Sánchez-Vázquez. 2002. Influence of light intensity, spectrum and orientation on sea bass plasma and ocular melatonin. J. Pineal. Res. 32: 34–40.

Bayarri, M.J., M.A. Rol de Lama, J.A. Madrid and F.J. Sánchez-Vázquez. 2003. Both pineal and lateral eyes are needed to sustain daily circulating melatonin rhythms in sea bass. Brain Res. 969: 175–182.

Bayarri, M.J., R. García-Allegue, J.F. López-Olmeda, J.A. Madrid and F.J. Sánchez-Vázquez. 2004a. Circadian melatonin release *in vitro* by European sea bass pineal. Fish Physiol. Biochem. 30: 87–89.

Bayarri, M.J., M. Iigo, J.A. Muñoz-Cueto, E. Isorna, M.J. Delgado, J.A. Madrid, F.J. Sánchez-Vázquez and A.L. Alonso-Gómez. 2004b. Binding characteristics and daily rhythms of melatonin receptors are distinct in the retina and the brain areas of the European sea bass (Dicentrarchus labrax). Brain Res. 1029: 241–250.

Bégay, V., J. Falcón, G.M. Cahill, D.C. Klein and S.L. Coon. 1998. Transcripts encoding two melatonin synthesis enzymes in the teleost pineal organ: circadian regulation in pike and zebrafish, but not in trout. Endocrinolology 139: 905–912.

Besseau, L., A. Benyassi, M. Moller, S.L. Coon, J.L. Weller, G. Boeuf, D.C. Klein and J. Falcón. 2006. Melatonin pathway: breaking the 'high-at-night' rule in trout retina. Exp. Eye Res. 82: 620–627.

Blackshaw, S. and S.H. Snyder. 1997. Parapinopsin, a novel catfish opsin localized to the parapineal organ, defines a new gene family. J. Neurosci. 17: 8083–8092.

Blank, H.M., B. Müller and H.W. Korf. 1997. Comparative investigations of the neuronal apparatus in the pineal organ and retina of the rainbow trout: immunocytochemical demonstration of neurofilament 200-kDa and neuropeptide Y, and tracing with DiI. Cell. Tissue Res. 288: 417–425.

Bromage, N., M. Porter and C. Randall. 2001. The environmental regulation of maturation in farmed finfish with special reference to the role of photoperiod and melatonin. Aquaculture 197: 63–98.

Brydon, L., L. Petit, P. de Coppet, P. Barrett, P.J. Morgan, A.D. Strosberg and R. Jockers. 1999. Polymorphism and signalling of melatonin receptors. Reprod. Nutr. Dev. 39(3): 315–24.

Cahill, G.M. and J.C. Besharse. 1993. Circadian clock functions localized in xenopus retinal photoreceptors. Neuron. 10: 573–577.

Cassone, V.M., M.J. Chesworth and S.M. Armstrong. 1986. Entrainment of rat circadian rhythms by daily injection of melatonin depends upon the hypothalamic supraquiasmatic nucleus. Physiol. Behav. 36: 1111–1121.

Cassone, V.M., M.H. Roberts and R.Y. Moore. 1987. Melatonin inhibits melatonin activity in the rat supraquiasmatic nuclei. Neurosci. Lett. 81: 29–34.

Cerdá-Reverter, J.M., I. Anglade, G. Martínez-Rodriguez, D. Mazurais, J.A. Muñoz-Cueto, M. Carrillo, O. Kah and S. Zanuy. 2000. Characterization of neuropeptide Y expression in the brain of a perciform fish, the sea bass (*Dicentrarchus labrax*). J. Chem. Neuroanat. 19: 197–210.

Chaurasia, S.S., N. Pozdeyev, R. Haque, A. Visser, T.N. Ivanova and P.M. Iuvone. 2006. Circadian clockwork machinery in neural retina: evidence for the presence of functional clock components in photoreceptor-enriched chick retinal cell cultures. Mol. Vis. 12: 215–223.

Cole, W.C. and J.H. Youson. 1982. Morphology of the pineal complex of the anadromous sea lamprey, *Petromyzon marinus* L. Am. J. Anat. 165(2): 131–63.

Collin, J.P. 1971. Differentiation and regression of the cells of the sensory line in the epiphysis cerebri. pp. 79–125. *In*: G.E.W. Wolstenholme and J. Knight (eds.). The Pineal Gland. Churchill Livingstone, Edinmurg-London.

Collin, J.P. 1977. La rudimentation des photorécepteurs dans l'organe pinéal des vertébrés. pp. 393–407. *In*: A. Raynaud (ed.). Mécanismes de la rudimentation des organes chez les embryons de vertébrés, Vol. 266, Coll Int CNRS, Paris.

Collin, J.P. and A. Oksche. 1981. Structural and functional relationships in the nonmammalian pineal gland. pp. 27–67. *In*: R.J. Reiter (ed.). The Pineal Gland, Vol. 1: Anatomy and Biochemistry. CRC Press, Boca Raton, FL.

Confente, F., A. El M'Rabet, A. Ouarour, P. Voisin, W.J. De Grip, M.C. Rendón and J.A. Muñoz-Cueto. 2008. The pineal complex of Senegalese sole (Solea senegalensis): anatomical, histological and immunohistochemical study. Aquaculture. 285: 207–215.

Coombs, S., P. Görner and H. Münz. 1989. The Mechanosensory Lateral Line. Neurobiology and Evolution. Springer-Verlag, New York.

Danilova, N., V.E. Krupnik, D. Sugden and I.V. Zhdanova. 2004. Melatonin stimulates cell proliferation in zebrafish embryo and accelerates its development. FASEB J. 18: 751–753.

Davies, B., L.T. Hannah, C.F. Randall, N. Bromage and L.M. Williams. 1994. Central melatonin binding sites in rainbow trout (Onchorynchus mykiss). Gen. Comp. Endocrinol. 96: 19–26.

Dufourny, L., A. Levasseur, M. Migaud, I. Callebaut, P. Pontarotti, B. Malpaux and P. Monget. 2008. GPR50 is the mammalian ortholog of Mel1c: evidence of rapid evolution in mammals. BMC. Evol. Biol. 8: 105.

Ekström, P. 1984. Central neural connections of the pineal organ and retina in the teleost. *Gasterosteus aculeatus* L. J. Comp. Neurol. 226: 321–335.

Ekström, P. and T. van Veen. 1983. Central connections of the pineal organ in the three-spined stickleback, *Gasterosteus aculeatus* L. (Teleostei). Cell. Tissue. Res. 232: 141–155.

Ekström, P. and H.W. Korf. 1986. Putative cholinergic elements in the photosensory pineal organ and retina of a teleost, *Phoxinus phoxinus* L. (Cyprinidae). Distribution of choline acetyltransferase immunoreactivity, acetylcholinesterase-positive elements and pinealofugally projecting neurons. Cell. Tissue. Res. 246: 321–329.

Ekström, P. and H. Meissl. 1990. Electron microscopic analysis of S-antigen- and serotonin-immunoreactive neural and sensory elements in the photosensory pineal organ of the salmon. J. Comp. Neurol. 292: 73–82.

Ekström, P. and J. Vanecek. 1992. Localization of 2-[I125]iodomelatonin binding sites in the brain of the Atlantic salmon. Neuroendocrinology. 55: 529–537.

Ekström, P. and H. Meissl. 1997. The pineal organ of teleost fishes. Rev. Fish Biol. Fish 7: 284.

Ekström, P. and H. Meissl. 2003. Evolution of photosensory pineal organs in new light: the fate of neuroendocrine photoreceptors. Philos. Trans. R. Soc. Lond. B: Biol. Sci. 358: 1679–1700.

Ekström, P., T. Honkanen and S.O. Ebbesson. 1988. FMRFamide-like immunoreactive neurons of the nervus terminalis of teleosts innervate both retina and pineal organ. Brain Res. 460: 68–75.

Ekström, P., T. Östholm and B.I. Holmqvist. 1994. Primary visual projections and pineal neural connections in fishes, amphibians and reptiles. pp. 1–18. *In*: M. Moller and P. Pévet (eds.). Advances in Pineal Research, Vol. 8. John Libbey, London.

Engbretson, G.A. and P.J. Linser. 1991. Glial cells of the parietal eye: structural and biochemical similarities to retinal Müller cells. J. Comp. Neurol. 314: 799–806.

Escobar, S., A. Felip, M.M. Gueguen, S. Zanuy, M. Carrillo, O. Kah and A. Servili. 2013. Expression of kisspeptins in the brain and pituitary of the European sea bass (*Dicentrarchus labrax*). J. Comp. Neurol. 521: 933–948.

Etcheler, S.M. 1984. Connections of the auditory midbrain in a teleost fish, Cyprinus carpio. J. Comp. Neurol. 230: 536–551.

Falcón, J. 1979a. L'organe pinéal du Brochet (*Esox lucius* L.). I. Etude anatomique et cytologique. Ann. Biol. Anim. Biochem. Biophys. 19: 445–465.

Falcón, J. 1979b. L'organe pinéal du brochet (*Esox lurius* L.). II. Etude en microscopie électronique de la différenciation et de la rudimentation partielle des photorécepteurs; conséquences possibles sur l'élaboration des messages photosensoriels. Ann. Biol. Anim. Bioch. Biophys. 19: 661–688.

Falcón, J., C. Thibault, C. Martin, J. Brun-Marmillon, B. Claustrat and J.P. Collin. 1991. Regulation of melatonin production by catecholamines and adenosine in a photoreceptive pineal organ. An *in vitro* study in the pike and the trout. J. Pineal. Res. 11(3–4): 123–34.

Falcón, J., Y. Gothilf, S.L. Coon, G. Boeuf and D.C. Klein. 2003. Genetic, temporal and developmental differences between melatonin rhythm generating systems in the teleost fish pineal organ and retina. 15(4): 378–82.

Falcón, J., L. Besseau and G. Boeuf. 2007a. Molecular and cellular regulation of pineal organ responses. pp. 203–406. *In*: T. Hara and B. Zielinski (eds.). Sensory Systems Neuroscience. Fish Physiology. Academic Press Elsevier, Amsterdam.

Falcón, J., L. Besseau, S. Sauzet and G. Boeuf. 2007b. Melatonin effects on the hypothalamo–pituitary axis in fish. Trends. Endocrinol. Metabolism 18: 81–88.

Falcón, J., H. Migaud, J.A. Muñoz-Cueto and M. Carrillo. 2010. Current knowledge on the melatonin system in teleost fish. Gen. Comp. Endocrinol. 165: 469–482.

Finger, T.E. 1983. Organization of the teleost cerebellum. pp. 261–284. *In*: R.E. Davis and R.G. Northcutt (eds.). Fish Neurobiology, Vol. 2: Higher Brain Areas and Functions. University of Michigan Press, Ann Arbor, MI.

Forsell, J., P. Ekström, I.N. Flamarique and B. Holmqvist. 2001. Expression of pineal ultraviolet- and green-like opsins in the pineal organ and retina of teleosts. J. Exp. Biol. 204: 2517–2525.

Forsell, J., B. Holmqvist and P. Ekström. 2002. Molecular identification and developmental expression of UV and green opsin mRNAs in the pineal organ of the Atlantic halibut. Brain Res. Dev. Brain Res. 136: 51–62.

Friedlander, M.J. 1983. The visual prosencephalon of teleosts. pp. 91–115. *In*: R.G. Northcutt and R.E. Davies (eds.). Fish Neurobiology, Vol. 2: Higher Brain Areas and Functions. University of Michigan Press, Ann Arbor, MI.

Fujieda, H., S.A. Hamadanizadeh, E. Wankiewicz, S.F. Pang and G.M. Brown. 1999. Expression of MT1 melatonin receptor in rat retina: evidence for multiple cell targets for melatonin. Neuroscience 93: 793–799.

Fujieda, H., J. Scher, S.A. Hamadanizadeh, E. Wankiewicz, S.F. Pang and G.M. Brown. 2000. Dopaminergic and GABAergic amacrine cells are direct targets of melatonin: immunocytochemical study of mt1 melatonin receptor in guinea pig retina. Vis. Neurosci. 17: 63–70.

Gaildrat, P., F. Becq and J. Falcón. 2002. First cloning and functional characterization of a melatonin receptor in fish brain: a novel one. J. Pineal. Res. 32(2): 74–84.

Gern, W.A., S.S. Greenhouse, J.M. Nervina and P.J. Gasser. 1992. The rainbow trout pineal organ: an endocrine photometer. pp. 199–218. *In*: M.A. Ali (ed.). Rhythms in Fish. Plenum, New York.

González-Martínez, D., T. Madigou, N. Zmora, I. Anglade, S. Zanuy, Y. Zohar, A. Elizur, J.A. Muñoz-Cueto and O. Kah. 2001. Differential expression of three different prepro-GnRH (gonadotrophin-releasing hormone) messengers in the brain of the European sea bass (Dicentrarchus labrax). J. Comp. Neurol. 429: 144–155.

González-Martínez, D., N. Zmora, E. Mañanos, D. Saligaut, S. Zanuy, Y. Zohar, A. Elizur, O. Kah and J.A. Muñoz-Cueto. 2002a. Immunohistochemical localization of three different prepro-GnRHs in the brain and pituitary of the European sea bass (Dicentrarchus labrax) using antibodies to the corresponding GnRH-associated peptides. J. Comp. Neurol. 446: 95–113.

González-Martínez, D., N. Zmora, S. Zanuy, C. Sarasquete, A. Elizur, O. Kah and J.A. Muñoz-Cueto. 2002b. Developmental expression of three different prepro-GnRH (gonadotrophin-releasing hormone) messengers in the brain of the European sea bass (Dicentrarchus labrax). J. Chem. Neuroanat. 23(4): 255–67.

Green, C.B. and J.C. Besharse. 2004. Retinal circadian clocks and control of retinal physiology. J. Biol. Rhythms 19: 91–102.

González-Martínez, D., T. Madigou, E. Mañanós, J.M.Cerdá-Reverter, O. Kah and J.A. Muñoz-Cueto. 2004. Cloning and expression of gonadotropin-releasing hormone receptor in the brain and pituitary of the European sea bass: an *in situ* hybridization study. Biol. Reprod. 70(5): 1380–91.

Grève, P., M. Bernard, P. Voisin, M. Cogné, J.P. Collin and J. Guerlotte. 1993. Cellular localization of hydroxyindole-O-methyltransferase mRNA in the chicken pineal gland. Neuroreport 4(6): 803–6.

Hafeez, M.A. 1971. Light microscopic studies on the pineal organ in the teleost fishes with special regard to its function. J. Morphol. 134: 281–314.

Hafeez, M.A. and L. Zerihun. 1974. Studies on central projections of the pineal nerve in rainbow trout, Salmo gairdneri Richardson, using cobalt chloride iontophoresis. Cell. Tissue Res. 154: 485–510.

Herrera-Pérez, P., A. Servili, M.C. Rendón, R. Vázquez, F.J. Sanchez-Vázquez and J.A. Muñoz-Cueto. 2004. The pineal organ of the European sea bass: a histological, immunohistochemical and ultrastructural study. 5 International Symposium on Fish Endocrinology. Castellón, Spain.

Herrera-Pérez, P., M.C. Rendón, L. Besseau, S. Sauzet, J. Falcón and J.A. Muñoz-Cueto. 2010. Melatonin receptors in the brain of the European sea bass: an *in situ* hybridization and autoradiographic study. J. Comp. Neurol. 518: 3495–3511.

Herrera-Pérez, P., A. Servili, M.C. Rendón, F.J. Sánchez-Vázquez, J. Falcón and J.A. Muñoz-Cueto. 2011. The pineal complex of the European sea bass (Dicentrarchus labrax): I. histological, immunohistochemical and qPCR study. J. Chem. Neuroanat. 41(3): 170–80.

Herrero, M.J., F.J. Martínez, J.M. Míguez and J.A. Madrid. 2007. Response of plasma and gastrointestinal melatonin, plasma cortisol and activity rhythms of European sea bass (Dicentrarchus labrax) to dietary supplementation with tryptophan and melatonin. J. Comp. Physiol. B. 177: 319–326.

Holmgren, N. 1918. Zur Frage der Epiphysen-Innervation bei Teleostiern. Folia Neurobiol. 11: 1–5.

Iigo, M., M. Kobayashi, R. Ohtani-Kanebo, M. Hara, A. Hattori, T. Suzuki and K. Aida. 1994. Characteristics, day-night changes, subcellular distribution and localization of melatonin binding sites in the goldfish brain. Brain. Res. 644: 213–220.

Iigo, M., R. Ohtani-Kanebo, M. Hara, A. Hatori, H. Takahasi, M. Tabata, T. Suzuki and K. Aida. 1997. Regulation by guanine nucleotides and cations of melatonin binding sites in the goldfish brain. Biol. Signals 6: 29–39.

Iigo, M., K. Furukawa, G. Nishi, M. Tabata and K. Aida. 2007. Ocular melatonin rhythms in teleost fish. Brain Behav. Evol. 69: 114–121.

Isorna, E., A. El M'rabet, F. Confente, J. Falcón and J.A. Muñoz-Cueto. 2009. Cloning and expression of arylalkylamine N-acetyltranferase-2 during early development and metamorphosis in the sole Solea senegalensis. Gen. Comp. Endocrinol. 161(1): 97–102.

Isorna, E., M. Aliaga-Guerrero, A. El M'Rabet, A. Servili, J. Falcón and J.A. Muñoz-Cueto. 2011. Identification of two arylalkylamine N-acetyltranferase 1 genes with different developmental expression profiles in the flatfish Solea senegalensis. J. Pineal. Res. 51(4): 434–44.

Iuvone, P.M., G. Tosini, N. Pozdeyev, R. Haque, D.C. Klein and S.S. Chaurasia. 2005. Circadian clocks, clock networks, arylalkylamine N-acetyltransferase, and melatonin in the retina. Prog. Ret. Eye Res. 24: 433–456.

Jiao, S., M.M. Wu, C.L. Hu, Z.H. Zhang and Y.A. Mei. 2004. Melatonin receptor agonist 2-iodomelatonin prevents apoptosis of cerebellar granule neurons via K(þ) current inhibition. J. Pineal. Res. 36: 109–116.

Jiménez, A.J., P. Fernández-Llébrez and J.M. Pérez-Fígares. 1995. Central projections from the goldfish pineal organ traced by HRP-immunocytochemistry. Histol. Histopathol. 10: 847–852.

Kah, O., S. Zanuy, E. Mañanos, I. Anglade and M. Carrillo. 1991. Distribution of salmon gonadotrophin releasing-hormone in the brain and pituitary of the sea bass (Dicentrarchus labrax). An immunocytochemical and immunoenzymoassay study. Cell. Tissue Res. 266: 129–136.

Klein, D.C., D.A Auerbach and J.L. Weller. 1981. Seesaw signal processing in pineal cells: homologous sensitization of adrenergic stimulation of cyclic GMP accompanies homologous desensitization of beta-adrenergic stimulation of cyclic AMP. Proc. Natl. Acad. Sci. USA 78(7): 4625–9.

Klein, D.C., R.Y. Moore and S.M. Reppert. 1991. Suprachiasmatic nucleus: the mind's clock. New York: Oxford Press.

Klein, DC., S.L. Coon, P.H. Roseboom, J.L. Weller, M. Bernard, J.A. Gastel, M. Zatz, P.M. Iuvone, I.R. Rodriguez, V. Bégay, J. Falcón, G.M. Cahill, V.M. Cassone and R. Baler. 1997. The melatonin rhythm-generating enzyme: molecular regulation of serotonin N-acetyltransferase in the pineal gland. Recent Prog. Horm. Res. 52: 307–57; discussion 357–8. Review.

Kojima, D., H. Mano and Y. Fukada. 2000. Vertebrate ancient-long opsin: a green-sensitive photoreceptive molecule present in zebrafish deep brain and retinal horizontal cells. J. Neurosci. 20: 2845–2851.

Korf, H.W. 2000. Evolution of melatonin-producing pinealocytes. pp. 17–29. In: J. Olcese (ed.). Melatonin after Four Decades. Kluwer/Plenum, New York.

Kotchabhakdi, N. 1976. Functional organization of the goldfish cerebellum. J. Comp. Physiol. 112: 75–93.

López-Patiño, M.A., A.L. Alonso-Gómez, A. Guijarro, E. Isorna and M.J. Delgado. 2008. Melatonin receptors in brain areas and ocular tissues of the teleost Tinca tinca: characterization and effect of temperature. Gen. Comp. Endocrinol. 155(3): 847–56.

Luiten, P.G. 1976. A somatotopic and functional representation of the respiratory muscles in the trigeminal and facial motor nuclei of the carp (Cyprinus carpio L.). J. Comp. Neurol. 166: 191–200.

Luiten, P.G. and J.N. van der Pers. 1977. The connections of the trigeminal and facial motor nuclei in the brain of the carp (Cyprinus carpio L.) as revealed by anterograde and retrograde transport of horseradish peroxidase. J. Comp. Neurol. 174: 575–590.

Mailliet, F., G. Ferry, F. Vella, S. Berger, F. Cogé, P. Chomarat, C. Mallet, S.P. Guénin, G. Guillaumet, M.C. Viaud-Massuard, S. Yous, P. Delagrange and J.A. Boutin. 2005. Characterization of the melatoninergic MT3 binding site on the NRH:quinone oxidoreductase 2 enzyme. Biochem. Pharmacol. 71(1–2): 74–88.

Manda, K., M. Ueno and K. Anzai. 2008. Melatonin mitigates oxidative damage and apoptosis in mouse cerebellum induced by high-LET 56Fe particle irradiation. J. Pineal. Res. 44: 189–196.

Mandado, M., P. Molist, R. Anadón and J. Yañez. 2001. A DiI-tracing study of the neural connections of the pineal organ in two elasmobranchs (Scyliorhinus canicula and Raja montagui) suggests a pineal projection to the midbrain GnRH-immunoreactive nucleus. Cell. Tissue Res. 303: 391–401.

Mano, H., D. Kojima and Y. Fukada. 1999. Exo-rhodopsin: a novel rhodopsin expressed in the zebrafish pineal gland. Mol. Brain Res. 73: 110–118.

Martinez-Chavez, C.C., S. Al-Khamees, A. Campos-Mendoza, D.J. Penman and H. Migaud. 2008. Clock-controlled endogenous melatonin rhythms in Nile tilapia (Oreochromis niloticus niloticus) and African catfish (Clarias gariepinus). Chronobiol. 25(1): 31–49.

Martinoli, M.G., L. Williams, O. Kah and G. Pelletier. 1991. Localization and characterization of melatonin-binding sites in the brain of the goldfish. Mol. Cell. Neorosci. 2: 78–85.

Mayer, I., C. Bornestaf and B. Borg. 1997. Melatonin in non-mammalian vertebrates: physiological role in reproduction? Comp. Biochem. Physiol. A. 118: 515–531.

Mazurais, D., I. Brierley, I. Anglade, J. Drew, C. Randall, N. Bromage, D. Michel, O. Kah and L.M. Williams. 1999. Central melatonin receptors in the rainbow trout: comparative distribution of ligand-binding and gene expression. J. Comp. Neurol. 409: 319–324.

McNulty, J.A. and B.G. Nafpaktitis. 1977. Morphology of the pineal complex in seven species of lantern fishes (Pisces: Myctophidae). Am. J. Anat. 150: 509–530.

Meiniel, A. 1976. L'épiphyse embryonnaire de Lacerta vivipara J. I. Différenciation des cellules de la lignée sensorielle (CLS). Le gradient morphologique. Journal of Neural Transmission 39: 139–174.

Meiniel, A. 1980. Ultrastructure of serotonin-containing cells in the pineal organ of Lampetra planeri (Petromyzontidae). A second sensory cell line from photoreceptor cell to pinealocyte. Cell Tissue Res. 207: 407–427.

Meiniel, A. 1981. New aspects of the phylogenetic evolution of sensory cell lines in the vertebrate pineal complex. pp. 27–47. *In*: A. Oksche and P. Pévet (eds.). The Pineal Organ: Photobiology, Biochronometry, Endocrinology. Elsevier/North-Holland Biomedical Press, Amsterdam.

Meléndez-Ferro, M., B. Villar-Cheda, X.M. Abalo, E. Pérez-Costas, R. Rodríguez-Muñoz, W.J. Degrip, J. Yáñez, M.C. Rodicio and R. Anadón. 2002. Early development of the retina and pineal complex in the sea lamprey: comparative immunocytochemical study. J. Comp. Neurol. 442: 250–265.

Migaud, H., J.F. Taylor, G.L. Taranger, A. Davie, J.M. Cerda-Reverter, M. Carrillo, T. Hansen and N.R. Bromage. 2006. A comparative *ex vivo* and *in vivo* study of day and night perception in teleosts species using the melatonin rhythm. J. Pineal. Res. 41: 42–52.

Migaud, H., A. Davie, C.C. Martínez-Chávez and S. Al-Khamees. 2007. Evidence for differential photic regulation of pineal melatonin synthesis in teleosts. J. Pineal Res. 43: 327–335.

Migaud, H., A. Davie and J.F. Taylor. 2010. Current knowledge on the photoneuroendocrine regulation of reproduction in temperate fish species. Journal of Fish Biology 76: 27–68.

Míguez, J.M., V. Simonneaux and P. Pévet. 1997. The role of intracellular and extracellular serotonin in the regulation of melatonin production in rat pinealocytes. J. Pineal Res. 23: 63–71.

Mitchell, C.K. and D.A. Redburn. 1991. Melatonin inhibits ACh release from rabbit retina. Vis. Neurosci. 7: 479–486.

Montero, M., L. Yon, S. Kikuyama, S. Dufour and H. Vaudry. 2000. Molecular evolution of the growth hormone-releasing hormone/pituitary adenylate cyclase-activating polypeptide gene family. Functional implication in the regulation of growth hormone secretion. J. Mol. Endocrinol. 25: 157–168.

Moons, L., M. Cambre´, S. Marivoet, T.F. Batten, J.J. Vanderhaeghen, F. Ollevier and F. Vandesande. 1988. Peptidergic innervation of the adrenocorticotropic hormone (ACTH)- and growth hormone (GH)-producing cells in the pars distalis of the sea bass (Dicentrarchus labrax). Gen. Comp. Endocrinol. 72: 171–180.

Moons, L., T.F. Batten and F. Vandesande. 1992. Comparative distribution of substance P (SP) and cholecystokinin (CCK) binding sites and immunoreactivity in the brain of the sea bass (Dicentrarchus labrax). Peptides 13: 37–46.

Natesan, A.K. and V.M. Cassone. 2002. Melatonin receptor mRNA localization and rhythmicity in the retina of the domestic chick, Gallus domesticus. Vis. Neurosci. 19: 265–274.

Nelson, J.S. 2006. Fishes of the World, ed. 4. Wiley, Hoboken, New Jersey.

Northcutt, R.G. 1995. The forebrain of gnatho stomes: in search of a morphotype. Brain Behav. Evol. 46: 275–318.

Northcutt, R.G. and A.B. Butler. 1991. Retinofugal and retinopetal projections in the green sunfish, Lepomis cyanellus. Brain Behav. Evol. 37: 333–354.

Northmore, D.P.M. 1984. Visual and saccadic activity in the goldfish torus longitudinalis. J. Comp. Physiol. A. 155: 333–340.

Oka, Y., A.D. Munro and T.J. Lam. 1986. Retinopetal projections from a subpopulation of ganglion cells of the nervus terminalis in the dwarf gourami (Colisa lalia). Brain Res. 367: 341–345.

Oliveira, C., J.F. López-Olmeda, M.J. Delgado, A.L. Alonso-Gómez and F.J. Sánchez-Vázquez. 2008. Melatonin binding sites in senegal sole: day/night changes in density and location in different regions of the brain. Chronobiol. Int. 25(4): 645–52.

Oksche, A. 1983. Aspects of evolution of the pineal organ. pp. 15–35. *In*: J. Axelrod, F. Fraschini and G.P. Velo (eds.). The Pineal Gland and its Endocrine Role. Plenum, New York.

Omura, Y. and M. Oguri. 1971. The development and degeneration of the photoreceptor outer segment of the fish pineal organ. Bull. Jap. Soc. Sci. Fish. 37: 851–860.

Park, Y.J., J.G. Park, S.J. Kim, Y.D. Lee, M. Saydur Rahman and A. Takemura. 2006. Melatonin receptor of a reef fish with lunar-related rhythmicity: cloning and daily variations. J. Pineal. Res. 41(2): 166–74.

Pastor, A.M., R.R. De la Cruz and R. Baker. 1997. Characterization of Purkinje cells in the goldfish cerebellum during eye movement and adaptive modification of the vestibulo-ocular reflex. Progr. Brain. Res. 114: 359–381.

Peirson, S.N., S. Halford and R.G. Foster. 2009. The evolution of irradiance detection: melanopsin and the non-visual opsins. Philos. Trans. R. Soc. Lond. B: Biol. Sci. 364: 2849–2865.

Pellegrini, E., K. Mouriec, I. Anglade, A. Menuet, Y. Le Page, M.M. Gueguen, M.H. Marmignon, F. Brion, F. Pakdel and O. Kah. 2007. Identification of aromatase-positive radial glial cells as progenitor cells in the ventricular layer of the forebrain in zebrafish. J. Comp. Neurol. 501: 150–167.

Peters, J.L. and V.M. Cassone. 2005. Melatonin regulates circadian electroretinogram rhythms in a dose- and time-dependent fashion J. Pineal. Res. 38: 209–215.

Pévet, P. 1983. Is 5-methoxytryptamine a pineal hormone? Psychoneuroendocrinology 8: 69–78.

Philp, A.R., J. Bellingham, J. Garcia-Fernández and R.G. Foster. 2000a. A novel rod-like opsin isolated from the extra-retinal photoreceptors of teleost fish. FEBS Lett. 468: 181–188.

Philp, A.R., J.M. García-Fernández, B.G. Soni, R.J. Lucas, J. Bellingham and R.G. Foster. 2000b. Vertebrate ancient (VA) opsin and extraretinal photoreception in the Atlantic salmon (Salmo salar). J. Exp. Biol. 203: 1925–1936.

Pierce, M.E. and J.C. Besharse. 1985. Circadian regulation of retinomotor movements. I. Interaction of melatonin and dopamine in the control of cone length. J. Gen. Physiol. 86: 671–689.

Pombal, M.A., J. Yáñez, O. Marín, A. González and R. Anadón. 1999. Cholinergic and GABAergic neuronal elements in the pineal organ of lampreys, and tract-tracing observations of differential connections of pinealofugal neurons. Cell Tissue Res. 295: 215–223.

Presson, J., R.D. Fernald and M. Max. 1985. The organization of retinal projection to the diencephalon and pretectum in the cichlid fish, Haplochromis burtoni. J. Comp. Neurol. 235: 360–374.

Ribelayga, C., Y. Wang and S.C. Mangel. 2004. A circadian clock in the fish retina regulates dopamine release via activation of melatonin receptors. J. Physiol. 554: 467–482.

Roa, J. and M. Tena-Sempere. 2007. KiSS-1 system and reproduction: comparative aspects and roles in the control of female gonadotropic axis in mammals. Gen. Comp. Endocrinol. 153(1–3): 132–40.

Sauzet, S., L. Besseau, P. Herrera-Pérez, D. Covès, B. Chatain, E. Peyric, G. Boeuf, J.A. Muñoz-Cueto and J. Falcón. 2008. Cloning and retinal expression of melatonin receptors in the European sea bass, Dicentrarchus labrax. Gen. Comp. Endocrinol. 157: 186–195.

Savaskan, E., A. Wirz-Justice, G. Olivieri, M. Pache, K. Krauchi, L. Brydon, R. Jockers, F. Muller-Spahn and P. Meyer. 2002. Distribution of melatonin MT1 receptor immunoreactivity in human retina J. Histochem. Cytochem. 50: 519–526.

Schellart, N.A. 1983. Acousticolateral and visual processing and their interaction in the torus semicircularis of the trout Salmo gairdneri. Neurosci. Lett. 42: 39–44.

Servili, A., M.R. Bufalino, R. Nishikawa, I.S. de Melo, J.A. Muñoz-Cueto and L.E. Lee. 2009. Establishment of long term cultures of neural stem cells from adult sea bass, Dicentrarchus labrax. Comp. Biochem. Physiol. A. 152: 245–254.

Servili, A., C. Lethimonier, J.J. Lareyre, J.F. López-Olmeda, F.J. Sánchez-Vázquez, O. Kah and J.A. Muñoz-Cueto. 2010. The highly conserved gonadotropin-releasing hormone-2 form acts as a melatonin-releasing factor in the pineal of a teleost fish, the european sea bass Dicentrarchus labrax. Endocrinology 151(5): 2265–75.

Servili, A., P. Herrera-Pérez, J. Yáñez and J.A. Muñoz-Cueto. 2011. Afferent and efferent connections of the pineal organ in the European sea bass Dicentrarchus labrax: a carbocyanine dye tract-tracing study. Brain Behav. Evol.78: 272–285.

Servili, A., P. Herrera-Pérez, O. Kah and J.A. Muñoz-Cueto. 2012. The retina is a target for GnRH-3 system in the European sea bass, Dicentrarchus labrax. Gen. Comp. Endocrinol. 175(3): 398–406.

Servili, A., P. Herrera-Pérez, M.C. Rendón and J.A. Muñoz-Cueto. 2013. Melatonin inhibits GnRH-1, GnRH-3 and GnRH receptor expression in the brain of the European sea bass, Dicentrarchus labrax. Int. J. Mol. Sci. 14: 7603–7616.

Seth, M. and S.K. Maitra. 2011. Neural regulation of dark-induced abundance of arylalkylamine N-acetyltransferase (AANAT) and melatonin in the carp (Catla catla) pineal: an *in vitro* study. Chronobiol. Int. 28(7): 572–85.

Song, J. and R.L. Boord. 1993. Motor components of the trigeminal nerve and organization of the mandibular arch muscles in vertebrates. Phylogenetically conservative patterns and their ontogenetic basis. Acta Anat. 148: 139–149.

Srivastava, S. 2003. Two morphological types of pineal window in catfish in relation to photophase and scotophase activity: a morphological and experimental study. J. Exp. Zool. A. 295: 17–28.

Srivastava, S. and C.B.L. Srivastava. 1991. A lens-like specialization for photic input in the pineal window of an Indian catfish, Heteropneustes fossilis. Experientia 47: 189–190.

Stella Jr., S.L. and W.B. Thoreson. 2000. Differential modulation of rod and cone calcium currents in tiger salamander retina by D2 dopamine receptors and cAMP. Eur. J. Neurosci. 12: 3537–3548.

Striedter, G.F. and R.G. Northcutt. 1989. Two distinct visual pathways through the superficial pretectum in a percomorph teleost. J. Comp. Neurol. 283: 342–354.

Subhedar, N., J. Cerdá and R.A. Fallace. 1996. Neuropeptide Y in the forebrain and retina of the killifish, Fundulus heteroclitus. Cell Tissue Res. 283: 313–323.

Tosini, G., S. Doyle, M. Geusz and M. Menaker. 2000. Induction of photosensitivity in neonatal rat pineal gland. Proc. Natl. Acad. Sci. USA 97: 11540–11544.

Tsutsui, K. 2009. A new key neurohormone controlling reproduction, gonadotropin-inhibitory hormone (GnIH): Biosynthesis, mode of action and functional significance. Prog. Neurobiol. 88(1): 76–88.

Tsutsui, K., G.E. Bentley, T. Ubuka, E. Saigoh, H. Yin, T. Osugi, K. Inoue, V.S. Chowdhury, K. Ukena, N. Ciccone, P.J. Sharp and J.C. Wingfield. 2007. The general and comparative biology of gonadotropin-inhibitory hormone (GnIH). Gen. Comp. Endocrinol. 153(1–3): 365–70.

Ubuka, T., G.E. Bentley, K. Ukena, J.C. Wingfield and K. Tsutsui. 2005. Melatonin induces the expression of gonadotropin-inhibitory hormone in the avian brain. Proc. Natl. Acad. Sci. USA 102(8): 3052–7.

Vera, L.M., A. Davie, J.F. Taylor and H. Migaud. 2010. Differential light intensity and spectral sensitivities of Atlantic salmon, European sea bass and Atlantic cod pineal glands *ex vivo*. Gen. Comp. Endocrinol. 165(1): 25–33.

Vuilleumier, R., G. Boeuf, M. Fuentes, W.J. Gehring and J. Falcón. 2007. Cloning and early expression pattern of two melatonin biosynthesis enzymes in the turbot (Scophthalmus maximus). Eur. J. Neurosci. 25: 3047–3057.

Wullimann, M.F. 1994. The teleostean torus longitudinalis: a short review on its structure, histochemistry, connectivity, possible function and phylogeny. Eur. J. Morphol. 32: 235–242.

Wullimann, M.F. 1998. The central nervous system. pp. 245–282. *In*: D.H. Evans (ed.). The Physiology of Fishes. CRC Press, Boca Raton, FL.

Wullimann, M.F. and G. Roth. 1994. Descending telencephalic information reaches longitudinal torus and cerebellum via the dorsal preglomerular nucleus in the teleost fish, Pantodon buchholzi: a case of neural preadaptation? Brain Behav. Evol. 44: 338–352.

Yáñez, J. and R. Anadón. 1996. Afferent and efferent connections of the habenula in the rainbow trout (Oncorhynchus mykiss): an indocarbocyanine dye (DiI) study. J. Comp. Neurol. 372: 529–543.

Yáñez, J. and R. Anadón. 1998. Neural connections of the pineal organ in the primitive bony fish. Acipenser baeri: a carbocyanine dye tracttracing study. J. Comp. Neurol. 398: 151–161.

Yáñez, J., R. Anadón, B.I. Holmqvist and P. Ekström. 1993. Neural projections of the pineal organ in the larval lamprey (*Petromyzon marinus* L.) revealed by indocarbocyanine dye tracing. Neurosci. Lett. 164: 213–216.

Yáñez, J., J. Busch, R. Anadón and H. Meissl. 2009. Pineal projections in the zebrafish (Danio rerio): overlap with retinal and cerebellar projections. Neuroscience 164: 1712–1720.

Yu, C.J., Y. Gao, P. Li and L. Li. 2007. Synchronizing multiphasic circadian rhythms of rhodopsin promoter expression in rod photoreceptor cells. J. Exp. Biol. 210: 676–684.

Zimmerman, B.L. and M.O. Tso. 1975. Morphologic evidence of photoreceptor differentiation of pinealocytes in the neonatal rat. Journal of Cell Biology 66: 60–75.

Zohar, Y., J.A. Muñoz-Cueto, A. Elizur and O. Kah. 2010. Neuroendocrinology of reproduction in teleost fish. Gen. Comp. Endocrinol. 165: 438–455.

Zupanc, G.K. and S.C. Clint. 2003. Potential role of radial glia in adult neurogenesis of teleost fish. Glia 43: 77–86.

Melatonin Rhythms

Luisa M. Vera[1],* and *Hervé Migaud*[2]

Introduction

Melatonin was found in the earliest life forms and is present in all organisms studied to date, ranging from bacteria to humans (Conti et al. 2002, Tan et al. 2003). In mammals and humans, melatonin acts as a sleep regulator and was reported to have a role in sleep initiation (Shochat et al. 1997, Zisapel 2007, Pandi-Perumal et al. 2008). Melatonin has been described to have other roles: it can act as a dopamine release inhibitor from the hypothalamus and retina (Zisapel 2001), it can be involved in the aging process (Reiter et al. 1998), and it can regulate blood pressure (Cavallo et al. 2004, Grossman et al. 2006) and immune response (Carrillo-Vico et al. 2006) among other roles. Melatonin is phylogenetically conserved across vertebrates and acts as a time-keeping signal but also as a direct output of the circadian clock, playing an important role in the synchronization of the circadian system (Cassone 1998). Melatonin was found to be produced mainly by the pineal gland and retina but also by many other organs and tissues including the gastrointestinal tract (Bubenik and Pang 1997), skin (Slominski et al. 2005), lymphocytes (Carrillo-Vico et al. 2004) and bone marrow (Conti et al. 2000), suggesting that melatonin is involved in a range of physiological processes. The daily cycle of illumination is the main environmental factor influencing melatonin production and hence, it is described as a "zeitgeber" or the biological time keeping hormone which entrains circadian (daily)

[1]Department of Physiology, Faculty of Biology, University of Murcia, Campus of Espinardo-30100. Murcia (Spain).
Email: lmvera@um.es
[2]Institute of Aquaculture, University of Stirling-FK9 4LA. Stirling (UK).
Email: hm7@stir.ac.uk
*Corresponding author

and circannual (seasonal) rhythms in vertebrates (Menaker et al. 1997, Falcón et al. 2006, Pandi-Perumal et al. 2006). In all the species studied so far, the synthesis of this hormone by the pineal organ takes place during the dark phase of the daily light/dark (LD) cycle, providing an internal signal of night time to the organisms, and thus timing and controlling several biological rhythms (Arendt 1995, Zachmann et al. 1992a). Daily rhythms in fish include locomotor activity, rest, food intake, vertical migration and shoaling, skin pigmentation, osmoregulation and metabolism, whereas seasonal processes include growth, reproduction and smoltification for migrating salmonids (Falcón et al. 2007, Migaud et al. 2010). However, the profile of the melatonin rhythm varies among vertebrate species and three different variants have been described (Fig. 1). The A-type is characterised by a delay of melatonin rise after the start of the dark phase, showing the

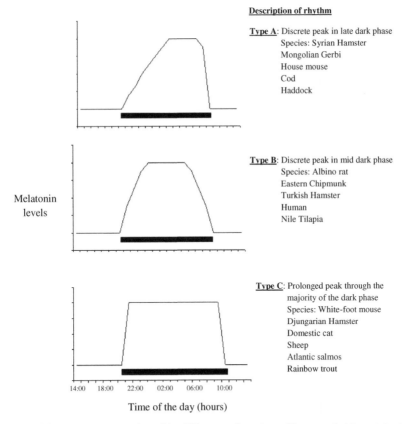

Description of rhythm

Type A: Discrete peak in late dark phase
  Species: Syrian Hamster
  Mongolian Gerbi
  House mouse
  Cod
  Haddock

Type B: Discrete peak in mid dark phase
  Species: Albino rat
  Eastern Chipmunk
  Turkish Hamster
  Human
  Nile Tilapia

Type C: Prolonged peak through the
  majority of the dark phase
  Species: White-foot mouse
  Djungarian Hamster
  Domestic cat
  Sheep
  Atlantic salmos
  Rainbow trout

Melatonin
levels

14:00 18:00 22:00 02:00 06:00 10:00

Time of the day (hours)

Figure 1. Schematic representation of the different melatonin profiles recorded in vertebrates. Examples of species which express such pattern of plasma melatonin for each profile are listed. Horizontal black bar denotes subjective dark period (Adapted from Reiter 1988, Falcón et al. 2010b).

peak towards the end of the dark phase (Reiter 1988). This pattern was found in mouse and Syrian hamster as well as in gadoid species (Atlantic cod and haddock) (Reiter 1988, Porter et al. 2001, Davie et al. 2007a,b,c). The B-type profile is characterised by a distinct peak in the middle of the dark phase as found in human and tilapia (Reiter 1988, Nikaido et al. 2009). The third profile is the C-type; it is the most common profile in vertebrates, where melatonin rises immediately following the onset of the dark phase to the maximum, which remains high and then falls rapidly to the basal level once the light phase starts. This profile is found in salmonids as well as other vertebrates (Reiter 1988, Randall et al. 1995). Also, large differences in melatonin peak levels between teleosts species has been observed (ranging from 50 to 200 pg.ml^{-1}) such as in sea bass (Bayarri et al. 2004b, Migaud et al. 2007), Atlantic cod and haddock (Bromage et al. 2001, Davie et al. 2007a,b,c), Nile tilapia (Martínez-Chávez et al. 2008) and Atlantic halibut (Davie et al. unpublished data), to 300–800 pg.ml^{-1} mainly observed in salmonid species (Migaud et al. 2010). Melatonin functions both as a clock and a calendar: the daily rhythm provides the animal information about the time of the day, since the maximum production occurs during the night, whereas the duration of the nocturnal increase would inform about the length of the photoperiod and thus about the season (Reiter 1993). In addition, in fish, the amplitude of the melatonin rhythm changes with water temperature, contributing to the seasonal variations in melatonin synthesis (Iigo and Aida 1995). Therefore, in fish, melatonin rhythms regulate the temporal coordination of many physiological processes, such as smoltification and reproduction (Ekström and Meissl 1997), growth, development, aging and cellular immunity (Yu and Reiter 1993).

The photic regulation of melatonin production is complex and involves the pineal gland, the eyes and possibly deep brain photoreceptors, with divergent circadian organisations found within teleosts. The light sensitivity of these systems would differ, possibly depending on the photoreceptors involved, their location/distribution and the photic environment inhabited by fish species (Migaud et al. 2007). However, besides light, other environmental factors such as temperature (Vera et al. 2007), feeding (Ceinos et al. 2008) or even salinity (López-Olmeda et al. 2009) appear to play a role in the regulation of the melatonin synthesis, as well as internal time-keeping system.

In the European sea bass, several studies have focused on the description, regulation and physiological role of melatonin rhythms. Current knowledge covers issues regarding photoreception, influence of light intensity and spectrum on melatonin production, circadian and seasonal rhythmicity, extrapineal sources of melatonin and melatonin receptors, among others. Many of these findings will be reviewed and discussed throughout this chapter.

Melatonin of Pineal Origin

The synthesis of melatonin occurs in different tissues of the animal's body, but the most important is the pineal gland, which is the main source releasing melatonin into the bloodstream. The pineal organ of sea bass is a photosensory and neuroendocrine structure that contains cone opsin-like and rod opsin-like photoreceptor cells. Furthermore, both *Aanat2* and *Hiomt* genes, which encode for key enzymes of the melatonin biosynthesis pathway, are expressed in sea bass pineal organ (Herrera-Pérez et al. 2011). The melatonin biosynthesis pathway takes place in the photoreceptor cells of the pineal organ and it includes four enzymatic steps (Fig. 2): 1) hydroxylation of tryptophan by means of tryptophan hydroxilase (TPHO) producing 5-hydroxytryptophan; which is then; 2) decarboxylated by means of the aromatic amino-acid decarboxylase forming serotonin; 3) serotonin is N-acetylated by arylalkylamine N-acetyltransferase (AANAT) into N-acetylserotonin which is then 4) converted into melatonin

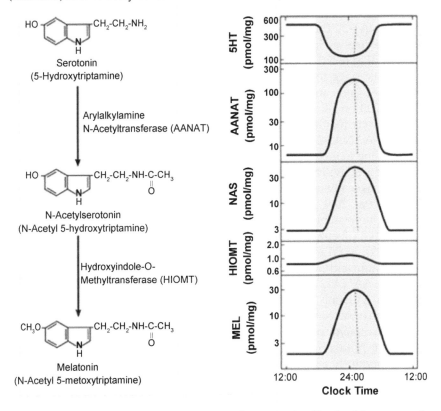

Figure 2. Melatonin pathway is shown on the left; changes in pineal levels of the compounds and enzymes are shown on the right (from Klein et al. 2002).

by the action of the hydroxyindole-O-methyltransferase enzyme (HIOMT) (Falcón 1999, Falcón et al. 2007, Falcón et al. 2010a).

The HIOMT and AANAT enzymes have been studied in a range of vertebrates as they act as the melatonin rate-limiting enzymes. AANAT is found in all vertebrates (Klein et al. 1997, 2002). In mammals, birds and anurans, only one type of AANAT has been found, while in teleosts two AANAT genes are present probably due to the genome duplication in early evolution of the teleost lineage (Coon and Klein 2006, Falcón et al. 2007). The AANAT activity increases at night with elevation in melatonin production, while the daylight results in a proteasomal degradation of the enzyme and therefore reduction in the melatonin production (Ekstrom and Meissl 1997, Falcón et al. 2010b) (Fig. 2). On the other hand HIOMT enzyme activity does not show any rhythmic changes and remains steady throughout the Light/Dark cycle (Morton and Forbes 1988, Klein et al. 2002). The HIOMT was suggested to be involved in seasonal rather than daily rhythmic oscillations in melatonin production (Ribelayga et al. 2002).

As explained above, the AANAT enzyme in fish has two forms (AANAT-1 and AANAT-2). These forms display tissue specific distribution: AANAT-1 is mainly expressed in the retina and brain, while AANAT-2 is more specific to the pineal in fish (Falcón et al. 2003, Coon and Klein 2006). Recent findings in teleosts suggested the presence of two AANAT-1 forms (1a and 1b) (Coon et al. 1999, Coon and Klein 2006, Falcón et al. 2010). At night, photoreceptor depolarization allows accumulation of cyclic AMP (cAMP) and Ca2+ entry. Both actions regulate AANAT-2 amount and activity at the cellular level through AANAT-2 protein phosphorylation (Falcón et al. 1999, 2010) and led to rising in melatonin levels. This mechanism is reversed by light, as the light triggers hyperpolarisation of photoreceptors and causing proteasomal proteolysis leading to AANAT-2 degradation and less melatonin secretion (Falcón et al. 2001, 2010), while minor part of the pineal AANAT-2 protein pool is photo-stable (Falcón et al. 2010). Contrasting to AANAT-2 protein and AANAT-2 enzyme activity, Aanat2 mRNA is not light sensitive (Coon et al. 1999), and Aanat2 gene expression is controlled by the clock machinery (Appelbaum and Gothilf 2006). A recent study studied the functional diversity of teleost AANAT2 and suggested that the AANAT2 enzyme has a conserved structure with only a few amino acid changes (Cazamea-Catalán et al. 2012). Evolution of AANAT2 has mainly been driven by phylogenetic relationships although catalytic properties (enzyme turnover and substrate affinity) are also under the influence of the respective species' normal habitat temperature.

Importantly, melatonin is a lipophilic molecule, so once it is synthesised, it easily crosses the membrane of the cells and is released into the blood and cerebrospinal fluid but is not stored within the pineal gland. Therefore, circulating melatonin levels directly reflect melatonin produced by the pineal (Falcón 1999).

Daily rhythms of melatonin: regulation by environmental factors

The existence of a daily melatonin rhythm was firstly described in sea bass by Sánchez-Vázquez and co-workers (1997). Circadian melatonin production in sea bass showed the classical pattern observed in teleost fish, under natural photoperiod with melatonin levels high during the night and low during the day (Sánchez-Vázquez et al. 1997, García-Allegue et al. 2001, Bayarri et al. 2010). Threshold of light intensity above which melatonin is suppressed depends on the species, experimental conditions (*in vitro* or *in vivo*), light spectral quality and duration (Aoki et al. 1998, Porter et al. 2000, Bayarri et al. 2002, Vera et al. 2005, Migaud et al. 2006, Oliveira et al. 2007). It depends also on the developmental stage/size of the fish, temperature profiles and previously experienced photic conditions. All these factors highlight the difficulties generally encountered when attempting to compare and review existing data in fish. Bayarri et al. (2002) investigated the light intensity threshold needed to suppress melatonin production in sea bass using white light of different intensities (ranging from 0.6 to 600 µWcm^{-2}). For this purpose sea bass were subjected to a 1-hour light pulse in the middle of the dark phase (MD). The threshold light intensity that significantly decreased plasma melatonin levels (in comparison with the MD control fish) was 6.0 µW cm^{-2}. Above this value, plasma melatonin remained low and did not differ from day levels. However, later studies showed that even a light intensity equivalent to 2 µWcm^{-2} significantly suppressed plasma melatonin levels in sea bass. A series of *in vitro* studies were also performed on cultured sea bass pineal glands and when converting the light intensities tested in pineal glands *ex vivo* into *in vivo* conditions (taking into account that the estimated pineal window penetration was 4.3% in sea bass), the theoretical threshold intensity above which melatonin is suppressed to basal day levels was found to be much lower, between 0.09 and 0.009 µW cm^{-2} (Migaud et al. 2006). However, the comparison between studies is difficult because of differences in a number of critical variables such as developmental stage/ size of the fish, temperature profiles, light spectral content and *in vivo*/ *in vitro* conditions. Importantly, acute exposure to light regimes (1-hour pulse) would not necessary reflect the true species light sensitivity. Some authors have pointed out that the minimum light intensity needed to inhibit melatonin production may be time-dependent and thresholds might decrease the duration of exposure increases (Aoki et al. 1998). The effect of light on melatonin levels is caused by changes in the mRNA expression and activity of the AANAT2 enzyme, the rate limiting enzyme in the melatonin biosynthesis pathway. This effect would be mediated by both Ca^{2+} and cAMP intracellular concentrations, which are controlled by light (Falcón 1999, Falcón et al. 2010a). In the dark, there is an accumulation of cGMP in the pineal photoreceptors which are then depolarised allowing the entry of cations (Na^+, Ca^{2+}) through the cGMP-gated channel. This increase of Ca^{2+} within the photoreceptors causes an increase in melatonin production both directly and indirectly through stimulation of cAMP synthesis (Bégay et al. 1994), since cAMP induces an increase in AANAT2 activity and protein accumulation (Falcón et al. 1992, Kroeber et al. 2000).

Although pineal melatonin production is mainly regulated by light intensity in teleosts, the spectral content of the light can also have an impact. Both light intensity and spectral content are subjected to daily variations depending on the time of the day, weather conditions and moon phase (Falcón et al. 2010a). Bayarri et al. (2002) tested the effect of three different coloured light pulses when applied at MD (half-peak bandwidth = 434–477, 498–575 and 610–687 nm for the blue, green and red lamps, respectively). The results showed that red light pulses of 2.4 μW cm^{-2} did not decrease melatonin levels whereas green and blue pulses of the same intensity did. However, a red light pulse of higher intensity was equally effective at inhibiting plasma melatonin showing that intensity threshold would depend on the spectral quality of the light. The differential effectiveness of narrow bandwidth light at reducing plasma melatonin can be explained by two facts. Firstly, the transmittance properties of the cranial bones are known to differ between species, thus modulating the amount of light reaching the pineal photoreceptors (Migaud et al. 2006, Falcón et al. 2010). The relative transmission of light at a variety of wavelengths was initially quantified in rainbow trout (Gern et al. 1992) and was refined by Migaud et al. (2007) in a range of species. Transmission was shown to differ between fish species with greater transmission of longer wavelengths through the cranial bones. Differences in light transmission through the skull in adult sea bass range from 2.5% to 6.5% depending on the spectral content of the light with longer wavelengths (650 nm–700 nm), far more effective at penetrating the skull than shorter wavelengths (400 nm–450 nm) (Migaud et al. 2006). Interestingly, no direct correlation between skin pigmentation or cranial bones thickness and the effectiveness of light transmittance was observed (Migaud et al. unpublished data). Secondly, spectral sensitivities of pineal photoreceptors differ between fish inhabiting different niches characterised by different photo-environments (Lythgoe 1980). As such, fish would have adapted and their spectral sensitivity would correspond to the available wavelengths in the fish natural environment. In the case of sea bass, a coastal marine species, the maximum reduction in melatonin was elicited by short wavelengths (blue), which is consistent with the previous theory. However, later studies carried out in sea bass pineal organs cultured *in vitro* did not reveal significant differences in the efficiency to suppress nocturnal melatonin between spectra (blue, green and red). This said, green light appeared to be more efficient (Vera et al. 2010). Nevertheless, comparisons between studies are not straightforward once again since the duration of the light exposure was different (1 hr *vs.* 12 hr), and the light sources and spectral profiles also differed. In addition, *in vitro* experiments may define the pineal spectral sensitivity when in isolation but not the animals' overall sensitivity. Further research should consider and study the pineal gland as part of an entire system/network, interacting with other photosensory structures such as the retina and extra-retinal nonpineal photoreceptors, and circadian rhythm generators. Besides the light intensity and spectrum, the light orientation also had an effect on melatonin suppression with downwelling light having a greater effect than upward-directed illumination. This may be explained by the fact that in the natural environment most sun rays reach the fish dorsally and the pineal gland is directly located beneath the skull on the top of the brain.

Other environmental factors are involved in the regulation of melatonin secretion by the pineal gland, such as water temperature. Fish are poikilothermic animals and their body temperature reflects that of their environment. Therefore, the nocturnal levels of plasma melatonin are usually lower at reduced temperatures and higher when the water temperature increases (Bromage et al. 2001). This effect appears to be mediated through the activity of the enzymes involved in the melatonin biosynthesis, particularly AANAT2 (Zachmann et al. 1992b, Falcón et al. 1994, 1996, Coon et al. 1999, Falcón 1999, Benyassi et al. 2000). Temperature would modulate the amplitude of the nocturnal rise of AANAT2 activity in the pineal organ and thus the melatonin production. Moreover, this mechanism seems to be an adaptation of fish to their natural environment since maximum activity of AANAT2 in a given species generally coincides with its preferred temperature (Falcón et al. 2009). In sea bass, a study was carried out to determine how seasonal information (taking into consideration both photoperiod and water temperature) could be transduced by the pineal organ into a daily melatonin rhythm (García-Allegue et al. 2001). To this end, plasma melatonin was determined in autumn, winter, spring and summer under natural conditions and in the summer and winter solstices under both natural and six-month-out-of-phase photoperiods (Fig. 3). The duration of the nocturnal increase reflected the length of the darkphase whereas the amplitude of the melatonin rhythm depended mostly on water temperature. However, under natural culture conditions, melatonin rhythms of sea bass were also influenced by the season, meaning that with a similar water temperature but different photoperiod, the amplitude of nocturnal melatonin varied significantly. Thus, although in spring and autumn the water temperature was similar, the melatonin rhythms presented different amplitudes as a result of the differences in photoperiod. Similar results were obtained in goldfish (Iigo and Aida 1995). Thus, the concurrent action of photoperiod, that determines the duration of the melatonin signal, and of temperature that determines its amplitude, is thought to provide accurate definitions of both the daily and annual cycles.

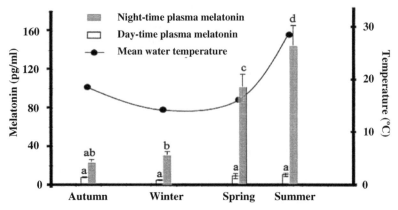

Figure 3. Average concentrations of circulating melatonin measured in night-time (full bars) and day-time (open bars) samples at different seasons of the year in sea bass. The evolution of water temperature during the year is also indicated (circles). Significant differences between mean plasma melatonin concentrations are labelled with different letters. Note that seasonal differences only appear between night melatonin levels (from García-Allegue et al. 2001).

Water salinity is another environmental factor that has been reported to play a role in the regulation of melatonin production. Sea bass is a euryhaline species able to live in a wide range of salinities, as well as in locations subjected to variations in salinity, such as estuaries, and therefore presents physiological adaptations to cope with those variations (Jensen et al. 1998). Moreover, sea bass can be classified as a diadromous species as it usually migrates in autumn and winter from tidal lagoons and estuaries to the open sea, for mating and spawning, whereas in spring fish return to freshwater environments to exploit food resources (Lemaire et al. 2000, Pawson et al. 2007). To investigate the effect of water salinity differences on nocturnal melatonin in sea bass, López-Olmeda et al. (2009) exposed fish during one week to four consecutive salinities (acclimation of 2–3 days between each salinity): 36‰ (seawater), 15‰ (isotonic water), 4‰ (brackish water) and 0‰ (freshwater). The plasma melatonin analyses revealed that all groups showed significant differences between day and night concentrations but the nocturnal levels differed depending on water salinity, the mean values being higher at lower salinities. In wild animals, the lower melatonin concentrations found in higher salinities would coincide with the winter migration, towards seawater whereas during the spring migration (back to lagoons) the melatonin levels at night would be higher. This suggests that salinity might act as a modulator and, together with photoperiod and water temperature, determine the amplitude of melatonin rhythms during the different seasons. Furthermore, osmoregulatory processes would also act on melatonin synthesis (Kulczykowska 2002).

Finally, plasma melatonin levels of sea bass can also be modified through the diet, as shown by Herrero et al. (2007). In this study the authors investigated the effect of both melatonin and tryptophan (precursor of the melatonin synthesis) diets on plasma melatonin, concluding that only the diet supplemented with melatonin resulted in an increase of sea bass physiological levels of plasma melatonin.

Circadian rhythms of melatonin

As reviewed above, melatonin rhythms are regulated by environmental factors. However, they can also be self-sustained under the control of circadian clocks. Studies from pineal glands in culture performed in both temperate and tropical teleosts have commonly demonstrated intrapineal oscillators, capable of self-sustaining melatonin rhythms *in vitro* in the absence of light stimuli (for a review see Migaud et al. 2010), in species including pike (Falcón et al. 1989), goldfish (Kezuka et al. 1989, Iigo et al. 1991), zebrafish (Cahill 1996), sailfin molly (Okimoto and Stetson 1999), ayu (Iigo et al. 2004) golden rabbitfish (Takemura et al. 2006), and sea bass (Bolliet et al. 1996, Bayarri et al. 2004a, Ron 2004, Migaud et al. 2006). However, no such endogenous rhythms have been shown to exist in salmonids (Gern and Greenhouse 1988, Migaud et al. 2006, Iigo et al. 2007) and common dentex (Pavlidis et al. 1999). Such rhythms would be sustained via endogenous clocks that regulate the expression of the enzyme AANAT through conserved elements in the gene promoter regions, and thus control the melatonin rhythm (Zilberman-Peled et al. 2006). In sea bass, the circadian rhythmicity of melatonin production has been investigated in continuous light and darkness, revealing that the circulating melatonin levels remained low in fish exposed to constant light (LL) conditions but high in constant darkness (DD), which confirms the light inhibition of the rate limiting enzymes. Interestingly, under DD, circadian rhythmicity was sustained for one subjective day cycle, but this rhythm damped during the second cycle (Iigo et al. 1997). Circadian rhythmicity has also been studied *in vitro*. For this, the pineal organs, removed from fish, were placed in a perfusion tissue culture with controlled temperature. The results showed that melatonin synthesis was inhibited by continuous light, but in continuous darkness a strong circadian rhythmicity was observed, which suggests that the sea bass pineal organ would contain an internal oscillator able to regulate melatonin production (Ron 2004). In a parallel study, Bayarri et al. (2004a) obtained similar results when they exposed sea bass pineal organs to constant darkness (DD). In this case, the circadian rhythmicity persisted for four days *in vitro*, while they damped after one subjective day *in vivo* (Iigo et al. 1997). However, contrasting results were also obtained with a clear lack of sustained melatonin production under DD in sea bass both *in vivo* and in pineal glands cultured *in vitro* (Migaud, pers. comm.). This suggests that factors such as environmental conditions, fish stocks or *in vitro* conditions may also modulate the circadian rhythmicity of melatonin production in sea bass, further highlighting the complexity of the circadian organisation in sea bass. In all teleosts investigated so far, *ex vivo* studies have shown that the pineal gland is directly photosensitive (Max and Menaker 1992, Bolliet et al. 1996, Migaud et al. 2006). Thus, the circadian rhythmicity of melatonin production is generated and controlled by a complex system composed of three elements: a photodetector (photopigment molecules), an endogenous clock and melatonin synthesizing machinery (enzymes involved in the melatonin biosynthesis) (Falcón 1999). Photopigments and the enzyme AANAT seem to be key elements of this system; however, other proteins involved in phototransduction (e.g., transducing, arrestin, phosducin), circadian clock function (e.g., clock, period and cryptochrome) and the melatoninergic pathway (e.g., tryptophan hydroxylase and hydroxyindole-O-methyltransferase) have been reported to play an important role too (Falcón et al. 2003).

Extrapineal Melatonin

It is generally accepted that the main source of melatonin is the pineal organ. However, early studies carried out in fish demonstrated that the retina also synthesizes melatonin rhythmically (Gern and Ralph 1979). Furthermore, melatonin has also been found in other tissues such as the gastrointestinal tract (Bubenik and Pang 1997), skin (Slominski et al. 2005), lymphocytes (Carrillo-Vico et al. 2004) and bone marrow (Conti et al. 2000) suggesting that melatonin is involved in a range of physiological processes. In fish, two or even three AANAT genes coexist, which indicates that genome duplication has occurred during evolution. A first round at the base of the teleost fish lineage would have resulted in the origin of the AANAT1 and AANAT2 families of enzymes whereas a second duplication event caused the appearance of AANAT1a and AANAT1b (Coon and Klein 2006). Furthermore, the expression of the different AANAT isoforms is tissue specific and their encoded proteins have different affinities for serotonin (Benyassi et al. 2000). Thus, in the pineal organ, AANAT2 is expressed whereas AANAT1a and AANAT1b are expressed in the retina and other brain regions (Coon et al. 1999).

The gastrointestinal tract (GIT) has been reported to produce large amounts of melatonin (Bubenik 2002). However, several investigations have demonstrated that circulating melatonin produced by the pineal gland or administered exogenously can be absorbed by the gut (Poeggeler et al. 2005). In sea bass, the dietary supplementation with melatonin resulted in a quick increase of this hormone concentration in the GIT. The levels observed in the intestine and bile 15 min after food intake were 40 times and 15 times higher, respectively, than those reached when a standard diet was supplied. In contrast, although tryptophan is a precursor of serotonin and melatonin, when sea bass were fed with a diet supplemented with this amino acid, melatonin levels in the GIT were only slightly increased (Herrero et al. 2007). Moreover, in this study the melatonin concentration in the bile was assessed for the first time, ranging from 803.3 and 2,287.0 pg.ml^{-1}. The physiological role of these high melatonin concentrations is unknown, although some hypotheses suggest the existence of an enterohepatic melatonin circulation (intestinal lumen, portal blood, liver, bile, intestinal lumen) that would protect the gastrointestinal mucosa against the oxidative stress of digestive secretions (Ekmekcioglu 2006).

Retinal melatonin

Fish retina has visual and non-visual photoreceptors capable of circadian entrainment and melatonin production. Falcón et al. (2010b) have suggested that the retina could directly control the pineal melatonin secretion, although further research is needed to support this hypothesis, or the retina could contribute to the circulating melatonin levels through its own production. In most vertebrate species so far investigated, melatonin production by the retina occurs at night, in parallel with that of the pineal. However, in some fish species, as observed in sea bass, this pattern is shifted. A daily rhythm in retinal melatonin was first reported in sea bass by Sánchez-Vázquez et al. (1997) who showed inverse melatonin profiles in plasma and eye, with ocular melatonin being higher during the day. Besides this daily rhythm, seasonal variations have also been found in sea bass, with ocular melatonin showing a circadian rhythm in autumn and winter (peaking during the day) but not in spring and summer (García-Allegue et al. 2001). This could suggest a role for ocular melatonin in transducing seasonal environmental information. Later investigations showed that the acute exposure of sea bass to light during the night induced an increase in ocular melatonin (Iigo et al. 1997), in contrast to the response observed in plasma, being a positive correlation between increasing light intensities and melatonin levels (Bayarri et al. 2002). Furthermore, light pulses of different wavelengths had different effects on ocular melatonin: maximum sensitivity was shifted towards shorter wavelengths (blue) as a red light pulse failed to increase melatonin concentrations, whereas green light pulses resulted in an intermediate response. Finally, the orientation of the incident light did not change the response of ocular melatonin to light, probably due to the fact that the lateral eyes can receive light from many directions, unlike the pineal (Bayarri et al. 2002). The different melatonin response from the pineal organ and lateral eyes to light could be explained by the existence of two types of AANAT, with different kinetics, which suggests different endocrine and paracrine roles for plasma and ocular melatonin (Benyassi et al. 2000). Thus, ocular melatonin might have a role in the control of retinomotor responses to light, controlling myoid contraction and pigment dispersion. However, it remains unclear whether ocular melatonin reaches the bloodstream. In sea bream, neither pinealectomy nor ophthalmectomy alone completely abolished circulating melatonin and only when both organs were removed did daily melatonin disappear. Indeed, covering the pineal (but not pinealectomy) resulted in a decrease of plasma melatonin after the exposure of fish to a light pulse during the dark phase, suggesting that light information from the eyes was able to inhibit melatonin production by the pineal and that both organs are needed to sustain melatonin rhythms in sea bass. Therefore, light information provided by the lateral eyes could affect pineal melatonin production through a neural pathway. However this is decoupled to melatonin secretion since retinal melatonin is produced at day and therefore unlikely to contribute to the circulating levels at night in this species (Bayarri et al. 2003). Moreover, other authors have suggested that a strong melatonin deacetylase activity in the retina of fish would avoid melatonin release into the bloodstream (Grace et al. 1991). The effect of ophthalmectomy on plasma melatonin levels of sea bass was later confirmed by Migaud et al. (2007) who then proposed that the regulation of pineal activity in teleosts would have evolved from an independent light sensitive pineal organ, without pacemaker activity (as reported in salmonids) to the situation observed in sea bass, where the pineal organ remains light sensitive, could possess a circadian pacemaker and is also regulated by photic information perceived by the retina (and possibly deep brain photoreceptors) (Fig. 4). Previous investigations carried out in other fish species have reported the co-existence of three different types of pinealocytes (true and modified photoreceptors and pinealocytes) (Ekström and Meissl 2003). If this is the case in sea bass, it could explain how light perceived by the retina may have an effect on melatonin production by the pineal. Further studies are needed to dissect the circadian organisation controlling melatonin production and entraining seasonality in sea bass and fish as a whole.

Melatonin Receptors

In fish, the effects of melatonin are mediated through specific high- and low-affinity membrane receptors. The melatonin receptors have been identified and classified by means of cloning techniques and pharmacokinetics studies using the radioligand 2-[^{125}I]-iodomelatonin [^{125}I]Mel (Reiter et al. 2010). Melatonin binding sites have been classified into two types, ML1 and ML2 (Dubocovich 1988). ML1 consists of a subfamily of G-protein-coupled

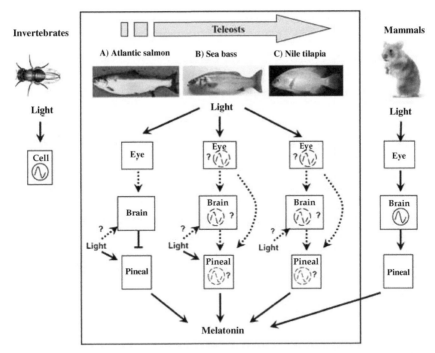

Figure 4: Suggested evolution of the regulation of pineal melatonin synthesis by the circadian axis in teleosts. In addition to the two types of circadian organisation already proposed in fish (A and B), a third type could exist where pineal light sensitivity would be dramatically reduced (C). The regulation of pineal activity would have thus evolved from an independent light sensitive pineal gland, without pacemaker activity, as seen in salmonids (A); to an intermediary state where the pineal gland remains light sensitive and could possess a circadian pacemaker, but is also regulated by photic information perceived by the retina as seen in sea bass and cod (B); to reach a more advanced system closer to higher vertebrates where light sensitivity of the pineal gland would be significantly reduced and its melatonin synthesis activity primarily regulated by a circadian pacemaker (unknown location) entrained by photic information perceived by the retina (C) (adapted from Migaud et al. 2010).

Color image of this figure appears in the color plate section at the end of the book.

receptors that includes three members: Mel1a, Mel1b and Mel1c (Reppert et al. 1996). This classification has been made according to their affinity and localisation: Mel1a (also known as MT1) has a high affinity for [^{125}I]Mel and is expressed in the brain, retina and other peripheral tissues; Mel1b (or MT2) has lower affinity for [^{125}I]Mel and has been found in brain and peripheral tissues of mammals (Masana and Dubocovich 2001); whereas Mel1c has only been found in non-mammalian vertebrates (Witt-Enderby et al. 2003). On the other hand, ML2 (currently known as MT3) receptor belongs to the family of the quinone reductase (Nosjean et al. 2000).

Distribution and daily rhythm

The distribution of melatonin binding sites in the brain of European sea bass was first described by Bayarri and co-workers (2004b), who reported an uneven distribution, with the total binding capacity (Bmax) of [^{125}I] Mel being the highest in the mesencephalic optic tectum-tegmentum and hypothalamus. Bmax was intermediate in telencephalon, cerebellum-vestibulolateral lobe and medulla oblongata-spinal cord, whereas the lowest binding levels were observed in the olfactory bulbs.

The melatonin binding sites in the brain and retina may show a daily rhythm in density and/or affinity, which could be influenced by the lighting regime, time of day and developmental or endocrine status (Vanecek 1998). However, the influence of light on the density and affinity of melatonin binding sites is not clear and contradictory results have been reported. Thus, a circadian rhythm in the binding capacity of brain receptors has been observed in avian and fish species, although their affinity did not vary (Yuan and Pang 1990, Iigo et al. 1994). However, in masu salmon, pike, goldfish, and rabbitfish brain, a circadian rhythmicity has been described in both receptor density and affinity (Gaildrat et al. 1998, Amano et al. 2003, Iigo et al. 2003, Park et al. 2007). In sea bass, the daily rhythm of melatonin receptor in the retina and brain has been described. First of all, the binding capacity of neural retina was found to be intermediate in comparison with the values obtained in the brain. Thus, the retina showed a density of melatonin receptors similar to that found in the telencephalon and cerebellum and was higher than the density observed in the medulla oblongata and olfactory bulbs (Bayarri et al. 2004c). In the same study, the authors reported that the daily rhythms in melatonin affinity and binding capacity differed in brain and retina. In all brain areas investigated, no significant variations in receptor affinity or density appeared during a 24 hour cycle. Nevertheless, these results are not in contradiction with the hypothesis about the existence of rhythmicity in the expression of receptors in sea bass brain, since in the optic tectum-thalamus, an effect of the time of day was observed on the proportions of different types of receptors: during day, both MT1 and MT2 receptors were expressed in a similar rate whereas at night the MT1 type was more abundant (75%). Unlike in the brain, a significant daily rhythm of both affinity and density has been found in the retina of sea bass. However, these rhythms were out of phase, with minimal affinity and maximal binding capacity observed at the beginning of the night. Moreover, the rhythmicity was clearer for Bmax since the maximum/minimum ratio was close to 2 whereas for the affinity rhythm this ratio was around 1.5 (Bayarri et al. 2004c). The independence of phase between the rhythm of receptor density and affinity seems to be characteristic of sea bass and has not been found in the retina (Faillace et al.

1995) or brain (Falcón et al. 1996, Amano et al. 2003) of other species so far studied. Several environmental and endocrine signals have been proposed as potential candidates controlling the daily rhythm of melatonin receptors, such as photoperiod, estrogen (Falcón et al. 1996) or melatonin (Iigo et al. 1994, Iigo et al. 2003). The low density and affinity of melatonin binding sites during the day in sea bass retina coincides with the maximum levels of ocular melatonin, so melatonin could act as a direct desensitizing factor (Bayarri et al. 2004c).

Seasonal rhythm

In most vertebrates the onset of reproduction shows a marked seasonality entrained mainly by changes in photoperiod and to some degree, temperature or food availability (Bromage et al. 2001). Cyclic variations in photoperiod are transduced by melatonin which is then considered to be the hormonal link between the circadian and reproductive pathways (Falcón et al. 2010b). In sea bass, an experiment was carried out to investigate whether melatonin binding in the optic tectum and hypothalamus differs during stages of the reproductive cycle (Ojeda et al. 2006). The results showed that both affinity and Bmax display a daily rhythm in the optic tectum (but not in hypothalamus) during spermiation, when the reproductive activity is at its peak, but not during the summer solstice, when sea bass are immature or in a resting stage, indicating that this rhythmic pattern in melatonin receptor affinity and Bmax may be involved in the regulation of reproductive hormones and fish maturation, as seen in mammals (Ojeda et al. 2006). Furthermore, under both natural photoperiod and constant light, melatonin binding changed throughout a reproductive cycle, especially in the hypothalamus, suggesting that melatonin plays a role in providing calendar information and that the brain of sea bass may contain a circannual clock, which is supported by the fact that the rhythm persisted under constant light conditions (Bayarri et al. 2010) (Fig. 5). The differences observed in melatonin binding in the optic tectum and hypothalamus throughout the reproductive cycle of sea bass suggest different roles played within the photoneuroendocrine axis. This is in tune: with the optic tectum being involved in the integration of light information while the hypothalamus is linked to reproduction and circannual rhythms (Mazurais et al. 1999, González-Martínez et al. 2002).

Effect of water salinity

As mentioned above in this chapter, sea bass is a euryhaline species that in the wild need to cope with salinity variations, as they migrate to the open sea during autumn-winter and return to coastal lagoons and estuaries

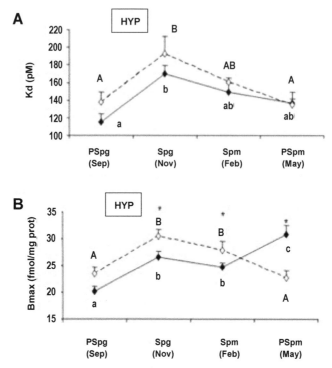

Figure 5. Kd (A) and Bmax (B) variations of melatonin binding throughout the reproductive cycle in the hypothalamus for fish maintained under normal photoperiod (black) and LL (white) conditions. Different lowercase and capital letters indicate significant variations (ANOVA, Tukey's test, $p < 0.05$) among the four study periods for fish under normal photoperiod and LL, respectively. Differences between treatments at each sampling period are represented by an asterisk. PSpg: pre-spermatogenesis; Spg: spermatogenesis; Spm: spermiation; PSpm: post-spermiation (from Bayarri et al. 2010).

during spring. The effect of water salinity on the density of melatonin binding sites has been studied by López-Olmeda and co-workers (2009) who found that this density peaked when sea bass were kept in freshwater but decreased in brackish and sea water. In fact, both plasma melatonin and binding sites density in retina and optic tectum increased with decreasing salinities. For this reason, the authors suggest that melatonin itself may regulate the expression of its receptors, although further experiments are needed to clarify this issue.

Roles of Melatonin on Sea Bass Physiology and Behaviour

In vertebrates, the daily rhythm of melatonin has been associated to circadian variations in behavioural and physiological traits. These include locomotor activity, thermal preference, rest, food intake, vertical migration, shoaling, skin pigmentation, osmoregulation, metabolism and sleep (Reiter et al. 2010). Furthermore, there is ample evidence indicating that, in fish, the pineal organ also contributes to the regulation of physiological functions displaying annual rhythmicity, such as smoltification, growth and reproduction (Falcón et al. 2007). For more details, see review by Falcón et al. (2010b).

Locomotor and feeding activity

In fish, the rhythms of locomotor and feeding activity can be considered as diurnal, nocturnal or crepuscular and this feature is species-dependent but it also can change depending on the physiological status of fish (Reebs 2002). Pinealectomy affects the display of these rhythms, which then either free-run or just disappear, depending on the species, indicating that melatonin may be involved in the regulation of behavioural rhythms. In fact, melatonin administration has been reported to induce a sleep-like state in zebrafish (Zhdanova et al. 2001) and to reduce locomotor activity and food intake in other fish species, such as goldfish and tench (López-Olmeda et al. 2006a,b). However, the effects on locomotor activity varied, depending on the time at which melatonin was administered. In sea bass showing a diurnal pattern of activity, melatonin intake produced an acute reduction in locomotor activity which was maintained until dawn, with a strong food anticipatory activity (FAA) appearing 160 min before meal time (Herrero et al. 2007). This is consistent with the effect observed on locomotor activity in other diurnal species (Murakami et al. 2001). The effect of melatonin on food intake has also been investigated in sea bass. Rubio and co-workers (2004) tested the effect of oral administration of melatonin on the selection of encapsulated macronutrients, since fish are able to choose a complete diet from separate sources of macronutrients. Sea bass were provided with gelatin capsules containing different doses of melatonin (0.1, 0.5 and 2.5 mg/kg body weight). Melatonin was absorbed and 45 min after the administration, melatonin levels were up to 26 times greater (for the highest dose) than in the control fish and a dose-dependent reduction of total food intake was observed (9%, 26% and 34%, respectively). Moreover, the pattern of macronutrient selection by fish was also altered with carbohydrate intake significantly reduced for all melatonin doses, whereas protein or fat intake was not significantly reduced.

Reproduction

It is generally accepted that photoperiod plays an important role in the control of seasonal reproduction in teleosts. In temperate regions, breeding occurs during increasing or decreasing photoperiod and although the pineal organ is involved in the transduction of the photoperiodic information, its exact role in the regulation of reproduction remains unclear. Sea bass is a short-day breeder and photoperiod manipulation has been used to delay sexual maturation of precocious males and thereby control reproduction and enhance somatic growth in the aquaculture industry (Zanuy et al. 2001). Bayarri et al. (2004d) investigated the existence of circadian rhythms of melatonin and reproductive hormones during the first year of life in sea bass kept in sea cages and under an artificially long photoperiod (18L:6D). The results showed that this long photoperiod affected several hormones, such as pituitary gonadotropin-releasing hormone (GnRH) or plasma luteinizing hormone (LH), and finally resulted in a delay of gonadal development and spawning at the time of puberty. In addition, plasma LH was higher during the night, as for melatonin. On the other hand, under LL, a photoperiod that disrupts the melatonin rhythm, an inhibition of precocity in sea bass males was observed (Bayarri et al. 2009). However, contradictory results have been obtained between fish species or even within the same species. Furthermore, the effects of melatonin have been reported to vary with gender, lighting regime and reproductive phase. Therefore, the main role of melatonin in the control of fish reproduction remains uncertain and further research is needed.

Conclusion

Melatonin is an output signal of the circadian system of vertebrates that plays a role in transducing environmental cues. Thus, the daily and circannual rhythms in melatonin production provide information about both the time of day and season. However, to date the physiological role of melatonin remains unclear in sea bass and teleosts as a whole. Although melatonin is mainly synthesised in the pineal organ and the retina, other organs can also produce melatonin. Therefore, melatonin is thought to be involved in the entrainment of tissue specific biological rhythms, which are mediated through specific receptors. Among other processes, there are strong indications that melatonin would be involved in the regulation of daily behavioural rhythms and annual functions such as reproduction, although direct evidences are lacking.

In sea bass, the photoperiod determines the length of the nocturnal melatonin production, whereas the amplitude of the daily rhythm seems to be determined by the water temperature, providing information on seasonal calendar time. Moreover, other environmental factors such as water salinity may have an impact on melatonin synthesis. In the aquaculture industry, light regimes have been frequently used to suppress the melatonin signal and hence manipulate physiological processes such as spawning, smoltification and suppress early maturation during on-growing. If artificial light regimes have been shown to be biologically efficient in a number of species (including sea bass), they are far from optimised and rely exclusively on photoperiod manipulations. The influence of light intensity, spectrum and light orientation remains to be understood and implemented within the industry. Importantly, other environmental factors (temperature, salinity, current impacting on swimming speed, etc.), fish stocks, developmental stage, and previous lighting conditions, are likely to interact.

References

Amano, M., M. Iigo, K. Ikuta, S. Kitamura and K. Yamamori. 2003. Daily variations in melatonin binding sites in the masu salmon brain. Neurosci. Lett. 350: 9–12.

Aoki, H., N. Yamada, Y. Ozeki, H. Yamane and N. Kato. 1998. Minimum light intensity required to suppress nocturnal melatonin concentration in human saliva. Neurosc. Lett. 252: 91–94.

Appelbaum, L. and Y. Gothilf. 2006. Mechanism of pineal-specific gene expression: The role of E-box and photoreceptor conserved elements. Mol. Cell. Endocrinol. 252: 27–33.

Arendt, J. 1995. Melatonin and the Mammalian Pineal Gland. Chapman & Hall, London. UK.

Bayarri, M.J., J.A. Madrid and F.J. Sánchez-Vázquez. 2002. Influence of light intensity, spectrum and orientation on sea bass plasma and ocular melatonin. J. Pineal Res. 32: 34–40.

Bayarri, M.J., M.A. Rol de Lama, J.A. Madrid and F.J. Sánchez-Vázquez. 2003. Both pineal and lateral eyes are needed to sustain daily circulating melatonin rhythms in sea bass. Brain Res. 969: 175–182.

Bayarri, M.J., R. García-Allegue, J.F. López-Olmeda, J.A. Madrid and F.J. Sánchez-Vázquez. 2004a. Circadian melatonin release *in vitro* by European sea bass pineal. Fish Physiol. Biochem. 30: 87–89.

Bayarri, M.J., R. García-Allegue, J.A. Muñoz-Cueto, J.A. Madrid, M. Tabata, F.J. Sánchez-Vázquez and M. Iigo. 2004b. Melatonin binding sites in the brain of European sea bass (*Dicentrarchus labrax*). Zool. Sci. 21: 427–434.

Bayarri, M.J., M. Iigo, J.A. Muñoz-Cueto, E. Isorna, M.J. Delgado, J.A. Madrid, F.J. Sánchez-Vázquez and A.L. Alonso-Gómez. 2004c. Binding characteristics and daily rhythms of melatonin receptors are distinct in the retina and the brain areas of the European sea bass retina (*Dicentrarchus labrax*). Brain Res. 1029: 241–250.

Bayarri, M.J., L. Rodríguez, S. Zanuy, J.A. Madrid, F.J. Sánchez-Vázquez, H. Kagawa, K. Okuzawa and M. Carrillo. 2004d. Effect of photoperiod manipulation on the daily rhythms of melatonin and reproductive hormones in caged European sea bass (*Dicentrarchus labrax*). Gen. Comp. Endocrinol. 136: 72–81.

Bayarri, M.J., S. Zanuy, O. Yilmaz and M. Carrillo. 2009. Effects of continuous light on the reproductive system of European sea bass gauged by alterations of circadian variations during their first reproductive cycle. Chronobiol. Int. 26: 184–199.

Bayarri, M.J., J. Falcón, S. Zanuy and M. Carrillo. 2010. Continuous light and melatonin: daily and seasonal variations of brain binding sites and plasma concentration during the first reproductive cycle of sea bass. Gen. Comp. Endocrinol. 169: 58–64.

Bégay, V., P. Bois, J.P. Collin, J. Lenfant and J. Falcón. 1994. Calcium and melatonin production in dissociated trout pineal photoreceptor cells in culture. Cell Calcium 16: 37–46.

Benyassi, A., C. Schwartz, S.L. Coon, D.C. Klein and J. Falcón. 2000. Melatonin synthesis: arylalkalylamine N-acetyltransferase in trout retina and pineal organ are different. Neuroreport 11: 255–258.

Bolliet, V., M.A. Ali, F.J. Lapointe and J. Falcón. 1996. Rhythmic melatonin secretion in different teleost species: an *ex vivo* study. J. Comp. Physiol. 165: 677–683.

Bromage, N.R., M.J.R. Porter and C.F. Randall. 2001. The environmental regulation of maturation in farmed finfish with special reference to the role of photoperiod and melatonin. Aquaculture 197: 63–98.

Bubenik, G.A. 2002. Gastrointestinal melatonin: localization, function, and clinical relevance. Digest. Dis. Sci. 47: 2336–2348.

Bubenik, G.A. and S.F. Pang. 1997. Melatonin levels in the gastrointestinal tissues of fish, amphibians, and a reptile. Gen. Comp. Endocrinol. 106: 415–419.

Cahill, G.M. 1996. Circadian regulation of melatonin production in cultured zebrafish pineal and retina. Brain Res. 708: 177–181.

Carrillo-Vico, A., J.R. Calvo, P. Abreu, S. García-Maurino, R.J. Reiter and J.M. Guerrero. 2004. Evidence of melatonin synthesis by human lymphocytes and its physiological significance: possible role as intracrine, autocrine, and/or paracrine substance. The FASEB Journal 18: 537–539.

Carrillo-Vico, A., R.J. Reiter, P.J. Lardone, J.L. Herrera, R. Fernandez-Montesinos, J.M. Guerrero and D. Pozo. 2006. The modulatory role of melatonin on immune responsiveness. Curr. Opin. Invest. Dr. 7: 423–431.

Cassone, V.M. 1998. Melatonin role in vertebrate circadian rhythms. Chronobiol. Int. 15: 457–473.

Cavallo, A., S.R. Daniels, L.M. Dolan, J.C. Khoury and J.A. Bean. 2004. Blood pressure response to melatonin in type 1 diabetes. Pediatr. Diabetes 5: 26–31.

Cazamea-Catalan, D., E. Magnanou, R. Helland, G. Vanegas, L. Besseau, G. Boeuf, C.H. Paulin, E.H. Jorgensen and J. Falcón. 2012. Functional diversity of Teleost arylalkylamine N-acetyltransferase-2: is the timezyme evolution driven by habitat temperature? Mol. ecol. 21: 5027–5041.

Ceinos, R.M., S. Polakof, A.R. Illamola, J.L. Soengas and J.M. Míguez. 2008. Food deprivation and refeeding effects on pineal indoles metabolism and melatonin synthesis in the rainbow trout *Oncorhynchus mykiss*. Gen. Comp. Endocrinol. 156: 410–417.

Conti, A., S. Conconi, E. Hertens, K. Skwarlo-Sonta, M. Markowska and J.M. Maestroni. 2000. Evidence for melatonin synthesis in mouse and human bone marrow cells. J. Pin. Res. 28: 193–202.

Conti, A., Ch. Tettamanti, M. Singaravel, Ch. Haldar, S.R. Pandi-Perumal and G.J.M. Maestroni. 2002. Melatonin: an ubiquitous and evolutionary hormone. pp. 105–143. *In*: Ch. Haldar, M. Singaravel and S.K. Maitra (eds.). Treatise on the Pineal Gland and Melatonin. Science Publisher, New Hampshire.

Coon, S.L. and D.C. Klein. 2006. Evolution of Arylalkylamine N-acetyltransferase: emergence and divergence. Mol. Cell. Endocrinol. 252: 2–10.

Coon, S.L., V. Bégay, D. Deurloo, J. Falcón and D.C. Klein. 1999. Two arylalkylamine N-acetyltransferase genes mediate melatonin synthesis in fish. J. Biol. Chem. 274: 9076–9082.

Davie, A., M. Porter, N. Bromage and H. Migaud. 2007a. The role of seasonally altering photoperiod in regulating physiology in Atlantic cod (*Gadus morhua*). Part I. Sexual maturation. Can. J. Fish. Aquatic Sci. 64 : 84–97.

Davie, A., M. Porter, N. Bromage and H. Migaud. 2007b. The role of seasonally altering photoperiod in regulating physiology in Atlantic cod (*Gadus morhua*). Part II. Somatic growth. Can. J. Fish. Aquatic Sci. 64: 98–112.

Davie, A., C. Mazorra de Quero, N. Bromage, J. Treasurer and H. Migaud. 2007c. Inhibition of sexual maturation in tank reared haddock (*Melanogrammus aeglefinus*) through the use of constant light photoperiods. Aquaculture 270: 379–389.

Dubocovich, M.L. 1988. Pharmacology and function of melatonin receptors. FASEB J. 2: 2765– 2773.

Ekmekcioglu, C. 2006. Melatonin receptors in humans: biological role and clinical relevance. Biomed. Pharmacother. 60: 97–108.

Ekström, P. and H. Meissl. 1997. The pineal organ of teleost fishes. Rev. Fish Biol. Fish. 7: 199–284.

Ekström, P. and H. Meissl. 2003. Evolution of photosensory pineal organs in new light: the fate of neuroendocrine photoreceptors. Phil. Trans. R. Soc. Lond. B: 1679–1700.

Faillace, M.P., M.A. De Las Heras, M.I. Keller Sarmiento and R.E. Rosenstein. 1995. Daily variations on 2-[125I]melatonin specific binding in the golden hamster retina. NeuroReport. 7: 141–144.

Falcón, J. 1999. Cellular circadian clocks in the pineal. Prog. Neurobiol. 8: 121–162.

Falcón, J., J.B. Marmillon, B. Claustrat and J.P. Collin. 1989. Regulation of melatonin secretion in a photoreceptive pineal organ: an *in vitro* study in the pike. J. Neurosci. 9: 1943–1950.

Falcón, J., V. Bégay, C. Besse, J.P. Ravault and J.P. Collin. 1992. Pineal photoreceptor cells in culture: fine structure and light control of cyclic nucleotide levels. J. Neuroendocrinol. 4: 641–651.

Falcón, J., V. Begay, J.M. Goujon, P. Voisin, J. Guerlotte and J.P. Collin. 1994. Immunicytochemical localisation of hydroxyindole-O-methyltransferase in pineal photoreceptor cells of several fish species. J. Comp. Neurol. 341: 559–566.

Falcón, J., M. Molina-Borja, J.P. Collin and S. Oaknin. 1996. Age-related changes in 2-[125I]-iodomelatonin binding sites in the brain of sea bream (*Sparus aurata* L.). Fish Physiol. Biochem. 15: 401–411.

Falcón, J., K.M. Galarneau, J.L. Weller, B. Ron, G. Chen, S.L. Coon and D.C. Klein. 2001. Regulation of arylalkylamine N-acetyltransferase-2 (AANAT2, EC 2.3.1.87) in the fish pineal organ: evidence for a role of proteasomal proteolysis. Endocrinol. 142: 1804–1813.

Falcón, J., Y. Gothilf, S.L. Coon, G. Boeuf and D.C. Klein. 2003. Genetic, temporal and developmental differences between melatonin rhythm generating systems in the teleost fish pineal organ and retina. J. Neuroendocrinol. 15: 378–382.

Falcón, J., L. Besseau and G. Bœuf. 2006. Molecular and cellular regulation of pineal organ responses. pp. 243–306. *In*: J.H.A.B. Toshiaki (ed.). Fish Physiology Sensory Systems Neuroscience. Academic Press. Salt Lake City.

Falcón, J., L. Besseau, S. Sauzet and G. Boeuf. 2007. Melatonin effects on the hypothalamo-pituitary axis in fish. Trends Endocrin. Met. 18: 81–88.

Falcón, J., L.Besseau, M. Fuentes, S. Sauzet, E. Magnanou and G. Boeuf. 2009. Structural and functional evolution of the pineal melatonin system in vertebrates. Ann. NY. Acad. Sci. 1163: 101–111.

Falcón, J., L. Besseau, E. Magnanou, S. Sauzet, M. Fuentès and G. Boeuf. 2010a. The pineal organ of fish. pp. 9–33. *In*: E. Kulczykowska, W. Popek and B.G. Kapoor (eds.). Biological Clock in Fish. CRC Press, Enfield.

Falcón, J., H. Migaud, J.A. Muñoz-Cueto and M. Carrillo. 2010b. Current knowledge on the melatonin system in teleost fish. Gen. Comp. Endocrinol. 165: 469–482.

Gaildrat, P., B. Ron and J. Falcón. 1998. Daily and circadian variations in 2-[125I]-iodomelatonin binding sites in the pike brain (*Esox lucius*). J. Neuroendocrinol. 10: 511–517.

García-Allegue, R., J.A. Madrid and F.J. Sánchez-Vázquez. 2001. Melatonin rhythms in European sea bass plasma and eye: influence of seasonal photoperiod and water temperature. J. Pineal Res. 31: 68–75.

Gern, W.A. and C.L. Ralph. 1979. Melatonin synthesis by the retina. Science 204: 183–184.

Gern, W.A. and S.S. Greenhouse. 1988. Examination of *in vitro* melatonin secretion from superfused trout (*Salmo gardnieri*) pineal organs maintained under diel illumination or continuous darkness. Gen. Comp. Endocrinol. 71: 163–174.

Gern, W.A., S.S. Greenhouse and J.M. Nervina. 1992. The rainbow trout pineal organ: an endocrine photometer. pp. 199–218. *In*: M.A. Ali (ed.). Rhythms in Fish. Plenum, New York.

González-Martínez, D., N. Zmora, E. Mañanós, D. Saligaut, S. Zanuy, Y. Zohar, A. Elizur, O. Kah and J.A. Muñoz-Cueto. 2002. Immunohistochemical localization of three different prepro-GnRHs (Gonadotrophin-releasing hormones) in the brain and pituitary of the

European sea bass (*Dicentrarchus labrax*) using antibodies against recombinant GAPs. J. Comp. Neurol. 446: 95–113.

Grace, M.S., G.M. Cahill and J.C. Besharse. 1991. Melatonin deacetylation: retinal vertebrate class distribution and *Xenopus laevis* tissue distribution. Brain Res. 559: 56–63.

Grossman, E., M. Laudon, R. Yalcin, H. Zengil, E. Peleg, Y. Sharabi, Y. Kamari, Z. Shen-Orr and N. Zisapel. 2006. Melatonin reduces night blood pressure in patients with nocturnal hypertension. Am. J. Med. 119: 898–902.

Herrera-Pérez, P., A. Servili, M.C. Rendón, F.J. Sánchez-Vázquez, J. Falcón and J.A. Muñoz-Cueto. 2011. The pineal complex of the European sea bass (*Dicentrarchus labrax*): I. Histological, immunohistochemical and qPCR study. J. Chem. Neuroanat. 41: 170–180.

Herrero, M.J., F.J. Martínez, J.M. Míguez and J.A. Madrid. 2007. Response of plasma and gastrointestinal melatonin, plasma cortisol and activity rhythms of European sea bass (Dicentrarchus labrax) to dietary supplementation with tryptophan and melatonin. J. Comp. Physiol. B. 177: 319–326.

Iigo, M. and K. Aida. 1995. Effects of season, temperature and photoperiod on plasma melatonin rhythms in the goldfish, *Carassius auratus*. J. Pineal Res. 18: 62–68.

Iigo, M., H. Kezuka, K. Aida and I. Hanyu. 1991. Circadian rhythms of melatonin secretion from superfused goldfish (*Carassius auratus*) pineal glands *in vitro*. Gen. Comp. Endocrinol. 83. 152–158.

Iigo, M., M. Kobayashi, R. Ohtani-Kaneko, M. Hara, A. Hattory, T. Suzuki and K. Aida. 1994. Characteristics, day–night changes, subcellular distribution and localization of melatonin binding sites in the goldfish brain. Brain Res. 644: 213–220.

Iigo, M., F.J. Sánchez-Vázquez, J.A. Madrid, S. Zamora and M. Tabata. 1997. Unusual responses to light and darkness of ocular melatonin in European sea bass. Neuroreport 8: 1631–1635.

Iigo, M., K. Furukawa, M. Tabata and K. Aida. 2003. Circadian variations of melatonin binding sites in the goldfish brain. Neurosci. Lett. 347: 49–52.

Iigo, M., Y. Fujimoto, M. Gunji-Suzuki, M. Yokosuka, M. Hara, R. Ohtani-Kaneko, M. Tabata, K. Aida and K. Hirata. 2004. Circadian rhythm of melatonin release from the photoreceptive pineal organ of a teleost, Ayu (*Plecoglossus altivelis*) in flow-thorough culture. J. Neuroendocrinol. 16: 45–51.

Iigo, M., T. Abe, S. Kambayashi, K. Oikawa, T. Masuda, K. Mizusawa, S. Kitamura, T. Azuma, Y. Takagi, K. Aida and T. Yanagisawa. 2007. Lack of circadian regulation of *in vitro* melatonin release from the pineal organ of salmonid teleosts. Gen. Comp. Endocrinol. 154: 91–97.

Jensen, M.K., S.S. Madsen and K. Kristiansen. 1998. Osmoregulation and salinity effects on the expression and activity of Na+,K+-ATPase in the gills of European sea bass, *Dicentrarchus labrax* (L.). J. Exp. Zool. 282: 290–300.

Kezuka, H., K. Aida and I. Hanyu. 1989. Melatonin secretion from goldfish pineal gland in organ culture. Gen. Comp. Endocrinol. 75: 217–221.

Klein, D.C., S.L. Coon, P.H. Roseboom, J.L. Weller, M. Bernard, J.A. Gastel, M. Zatz, P.M. Iuvone, M., I.R. Rodríguez, V. Bégay, J. Falcón, G.M. Cahill, V.M. Cassone and R. Baler. 1997. The melatonin rhythm-generating enzyme: molecular regulation of serotonin N-acetyltransferase in the pineal gland. Recent Prog. Horm. Res. 52: 307–357.

Klein, D.C., S. Ganguly, S.L. Coon, J.L. Weller, T. Obsil, A. Hickman and F. Dyda. 2002. Proteins and photoneuroendocrine transduction: role in controlling the daily rhythm in melatonin. Biochem. Soc. T. 30: 365–373.

Kroeber, S., H. Meissl, E. Maronde and H.W. Korf. 2000. Analyses of signal transduction cascades reveal an essential role of calcium ions for regulation of melatonin biosynthesis in the light-sensitive pineal organ of the rainbow trout (*Onchorhynchus mykiss*). J. Neurochem. 74: 2478–2489.

Kulczykowska, E. 2002. A review of the multifunctional hormone melatonin and a new hypothesis involving osmoregulation. Rev. Fish Biol. Fish. 11: 321–330.

Lemaire, C., G. Allegrucci, M. Naciri, L. Bahri-Sfar, H. Kara and F. Bonhomme. 2000. Do discrepancies between microsatellite and allozyme variation reveal differential selection between sea and lagoon in the sea bass (*Dicentrarchus labrax*)? Mol. Ecol. 9: 457–467.

López-Olmeda, J.F., M.J. Bayarri, M.A. Rol de Lama, J.A. Madrid and F.J. Sánchez-Vázquez. 2006a. Effects of melatonin administration on oxidative stress and daily locomotor activity patterns in goldfish. J. Physiol. Biochem. 62: 17–25.

López-Olmeda, J.F., J.A. Madrid and F.J. Sánchez-Vázquez. 2006b. Melatonin effects on food intake and activity rhythms in two fish species with different activity patterns: diurnal (goldfish) and nocturnal (tench). Comp. Biochem. Physiol. A. Mol. Integr. Physiol. 144: 180–187.

López-Olmeda, J.F., C. Oliveira, H. Kalamarz, E. Kulczykowska, M.J. Delgado and F.J. Sánchez-Vázquez. 2009. Effects of water salinity on melatonin levels in plasma and peripheral tissues and on melatonin binding sites in European sea bass (*Dicentrarchus labrax*). Comp. Biochem. Physiol. A. 152: 486–490.

Lythgoe, J.N. 1980. Vision in fish: ecological adaptations. pp. 1539–1550. *In*: M.A. Ali (ed.). Environmental Physiology of Fishes. Plenum Press, New York.

Martínez-Chávez, C.C., S. Al-Khamees, A. Campos-Mendoza, D. Penman and H. Migaud. 2008. Clock controlled endogenous melatonin rhythms in Nile tilapia (*Oreochromis niloticus niloticus*) and African catfish (*Clarias gariepinus*). Chronobiol. Int. 25: 31–49.

Masana, M.I. and M.L. Dubocovich. 2001. Melatonin receptor signaling: finding the path through the dark. Science's STKE 107: PE39.

Max, M. and M. Menaker. 1992. Regulation of melatonin production by light, darkness, and temperature in the trout pineal. J. Comp. Physiol. A. 170: 479–489.

Mazurais, D., I. Brierley and I. Anglade. 1999. Central melatonin receptors in the rainbow trout: comparative distribution of ligand binding and gene expression. J. Comp. Neurol. 409: 313–324.

Menaker, M., L.F. Moreira and G. Tosini. 1997. Evolution of circadian organization in vertebrates. Braz. J. Med. Biol. Res. 30: 305–313.

Migaud, H., J.F. Taylor, G.L. Taranger, A. Davie, J.M. Cerdá-Reverter, M. Carrillo, T. Hansen and N.R. Bromage. 2006. A comparative *ex vivo* and *in vivo* study of day and night perception in teleosts species using the melatonin rhythm. J. Pineal Res. 41: 42–52.

Migaud, H., A. Davie, C.C. Martínez-Chávez and S. Al-Khamees. 2007. Evidence for differential photic regulation of pineal melatonin synthesis in teleosts. J. Pineal Res. 43: 327–335.

Migaud, H., A. Davie and J.F. Taylor. 2010. Current knowledge on the photoneuroendocrine regulation of reproduction in temperate fish species. J. Fish Biol. 76: 27–68.

Morton, D.J. and H.J. Forbes. 1988. Pineal gland N-acetyltransferase and hydroxyindole-O-methyltransferase activity in the rainbow trout (*Salmo gairdneri*): seasonal variation linked to photoperiod. Neurosci. Lett. 94: 333–337.

Murakami, N., T. Kawano, K. Nakahara, T. Nasu and K. Shiota. 2001. Effect of melatonin on circadian rhythm, locomotor activity and body temperature in the intact house sparrow, Japanese quail and owl. Brain Res. 889: 220–224.

Nikaido, Y., S. Ueda and A. Takemura. 2009. Photic and circadian regulation of melatonin production in the Mozambique tilapia *Oreochromis mossambicus*. Comp. Biochem. Physiol. (A) 152: 77–82.

Nosjean, O., M. Ferros, F. Coge, P. Beauverger, J.M. Henlin, F. Lefoulon, J.L. Fauchere, P. Delagrange, E. Canet and J.A. Boutin. 2000. Identification of the melatonin binding site MT3 as the quinone reductase 2. J. Biol. Chem. 275: 31311–31317.

Ojeda, S.R., A. Lomniczi, C. Mastronardi, S. Heger, C. Roth, A.S. Parent, V. Matagne and A.E. Mungenast. 2006. The neuroendocrine regulation of puberty: is the time ripe for a systems biology approach? Endocrinol. 147: 1166–1174.

Okimoto, D.K. and M.H. Stetson. 1999. Presence of an intrapineal circadian oscillator in the teleostean family Poeciliidae. Gen. Comp. Endocrinol. 114: 304–312.

Oliveira, C., A. Ortega, J.F. López-Olmeda, L.M. Vera and F.J. Sánchez-Vázquez. 2007. Influence of constant light and darkness, light intensity, and light spectrum on plasma melatonin rhythms in senegal sole. Chronobiol. Int. 24: 615–627.

Pandi-Perumal, S.R., V. Srinivasan, G.J. Maestroni, D.P. Cardinali, B. Poeggeler and R. Hardeland. 2006. Melatonin: nature's most versatile biological signal? FEBS J. 273: 2813–2838.

Pandi-Perumal, S.R., I. Trakht, V. Sirinivasan, D.W. Spence, G.J.M. Maestroni, N. Zisapel and D.P. Cardinali. 2008. Physiological effects of melatonin: role of melatonin receptors and signal transduction pathways. Prog. Neurobiol. 85: 335–353.

Park, Y.J., J.G. Park, N. Hiyakawa, Y.D. Lee, S.J. Kim and A. Takemura. 2007. Diurnal and circadian regulation of a melatonin receptor, MT1, in the golden rabbitfish, *Siganus guttatus*. Gen. Comp. Endocrinol. 150: 253–262.

Pavlidis, M., L. Greenwood, M. Paalavuo, H. Molsa and J.T. Laitinen. 1999. The effect of photoperiod on diel rhythms in serum melatonin, cortisol, glucose, and electrolytes in the common Dentex, *Dentex dentex*. Gen. Comp. Endocrino. 113: 240–250.

Pawson, M.G., G.D. Pickett, J. Leballeur, M. Brown and M. Fritsch. 2007. Migrations, fishery interactions, and management units of sea bass (*Dicentrarchus labrax*) in Northwest Europe. J. Mar. Sci. 64: 332–345.

Poeggeler, B., G. Cornélissen, G. Huether, R. Hardeland, R. Józsa, M. Zeman, K. Stebelova, A. Oláh, G. Bubenik, W. Pan, K. Otsuka, V.O. Schwartzkop, E.E. Bakken and F. Halberg. 2005. Chronomics affirm extending scope of lead in phase of duodenal vs. pineal circadian melatonin rhythms. Biomed. Pharmacother. 59: 220–224.

Porter, M.J.R., S.O. Stefansson, G. Nyhammer, Ø. Karlsen, B. Norberg and N.R. Bromage. 2000. Environmental influences on melatonin secretion in Atlantic cod (*Gadus morhua* L.) and their relevance to commercial culture. Fish Physiol. Biochem. 23: 191–200.

Porter, M.J.R., N. Duncan, S.O. Steffansson and N.R. Bromage. 2001. Temperature, light intensity and plasma melatonin levels in juvenile Atlantic salmon. J. Biol. Rhyth. 58: 431–438.

Randall, C.F., N.R. Bromage, J.E. Thorpe, M.S. Miles and J.S. Muir. 1995. Melatonin rhythms in Atlantic salmon (*Salmo salar*) maintained under natural and out-of-phase photoperiods. Gen. Comp. Endocrinol. 98: 73–86.

Reebs, S.G. 2002. Plasticity of diel and circadian activity rhythms in fishes. Rev. Fish Biol. Fisher. 12: 349–371.

Reiter, R.J. 1988. Comparative aspects of pineal melatonin rhythms in mammals. Animal and Plant Science 1: 111–116.

Reiter, R.J. 1993. The melatonin rhythm: Both a clock and a calendar. Experientia 49: 654–664.

Reiter, R.J., J.M. Guerrero, J.J. Garcia and D. Acuna-Castroviego. 1998. Reactive oxygen intermediates, molecular damage, and aging: relation to melatonin. Ann. N.Y. Acad. Sci. 854: 410–424.

Reiter, R.J., D.X. Tan and L.C. Manchester. 2010. Melatonin in fish: circadian rhythms and functions. pp. 71–91. *In*: E. Kulczykowska, W. Popek and B.G. Kapoor (eds.). Biological Clock in Fish. Science Publishers, Enfield, NH.

Reppert, S.M., D.R. Weaver and C. Godson. 1996. Melatonin receptors step into the light: cloning and classification of subtypes. Trends Pharmacol. Sci. 17: 100–102.

Ribelayga, C., P. Pevet and I. Simonneaux. 2002. HIOMT drives the photoperiodic changes in the amplitude of the melatonin peak of the Siberian hamster. Am. J. Physiol.-Reg. I 278: 1339–1345.

Ron, B. 2004. *In vitro* melatonin rhythm reveals a clocked pineal in the European sea bass, *Dicentrarchus labrax*. Isr. J. Aquacult.-Bamid. 56: 281–285.

Rubio, V.C., F.J. Sánchez-Vázquez and J.A. Madrid. 2004. Oral administration of melatonin reduces food intake and modifies macronutrient selection in European sea bass (*Dicentrarchus labrax* L.). J. Pineal Res. 37: 42–47.

Sánchez-Vázquez, F.J., M. Iigo, J.A. Madrid, S. Zamora and M. Tabata. 1997. Daily cycles in plasma and ocular melatonin in demand-fed sea bass, *Dicentrarchus labrax* L. J. Comp. Physiol. B 167: 400–415.

Shochat, T., R. Luboshitzki and P. Lavie. 1997. Nocturnal melatonin onset is phase locked to the primary sleep gate. Am. J. Physiol.-Reg. I 273: 364–370.

Slominski, A., T.W. Fischer, M.A. Zmijewiski, J. Wortsman, I. Semak, B. Zbytek, R.M. Slominski and D.J. Tobin. 2005. On the role of melatonin in skin physiology and pathology. Endocrine 27: 137–147.

Takemura, A., S. Ueda, N. Hiyakawa and Y. Nikaido. 2006. A direct influence of moonlight intensity on changes in melatonin production by cultured pineal glands of the golden rabbitfish, *Siganus guttatus*. J. Pineal Res. 40: 236–241.

Tan, D.X., L.C. Manchester, R. Hardeland, S. Lopez-Burillo, J.C. Mayo, R.M. Sainz and R.J. Reiter. 2003. Melatonin: a hormone, a tissue factor, an autocoid, a paracoid, and an antioxidant vitamin. J. Pineal Res. 34: 75–78.

Vanecek, J. 1998. Cellular mechanisms of melatonin action. Physiol. Rev. 78: 687–721.

Vera, L.M., J.F. López-Olmeda, M.J. Bayarri, J.A. Madrid and F.J. Sánchez-Vázquez. 2005. Influence of light intensity on plasma melatonin and locomotor activity rhythms in tench. Chronobiol. Int. 22: 67–78.

Vera, L.M., C. De Oliveira, J.F. López-Olmeda, J. Ramos, E. Mañanós and F.J. Sánchez-Vázquez. 2007. Seasonal and daily plasma melatonin rhythms and reproduction in Senegal sole kept under natural photoperiod and natural or controlled water temperature. J. Pineal Res. 43: 50–55.

Vera, L.M., A. Davie, J.F. Taylor and H. Migaud. 2010. Differential light intensity and spectral sensitivities of Atlantic salmon, European sea bass and Atlantic cod pineal glands *ex vivo*. Gen. Comp. Endocrinol. 165: 25–33.

Witt-Enderby, P.A., J. Bennet, M.J. Jarzynka, S. Firestine and M.A. Melan. 2003. Melatonin receptors and their regulation: biochemical and structural mechanisms. Life Sci. 72: 2183–2198.

Yu, H.S. and R.J. Reiter. 1993. Melatonin. Biosynthesis, Physiological Effects and Clinical Application. CRC Press, Boca Raton, Florida.

Yuan, H. and S.F. Pang. 1990. [125I]Melatonin binding sites in membrane preparation of quail brain: characteristics and diurnal variations. Acta Endocrinol. 122: 633–639.

Zachmann, A., M.A. Ali and J. Falcón. 1992a. Melatonin in fish: rhythmic production by the pineal and effects; an overview. pp. 149–166. *In*: M.A. Ali (ed.). NATO Advanced Study on Rhythms in Fish. Plenum Press, New York.

Zachmann, A., J. Falcón and S.C.M. Knijff. 1992b. Effects of photoperiod and temperature on rhythmic melatonin secretion from the pineal organ of the white sucker (*Catostomus commersoni*) *in vitro*. Gen. Comp. Endocrinol. 86: 26–33.

Zanuy, S., M. Carrillo, A. Felip, L. Rodríguez, M. Blázquez, J. Ramos and F. Piferrer. 2001. Genetic, hormonal and environmental approaches for the control of reproduction in the European sea bass (*Dicentrarchus labrax* L.). Aquaculture 202: 187–203.

Zhdanova, I.V., S.Y. Wang, O.U. Leclair and N.P. Danilova. 2001. Melatonin promotes sleep-like state in zebrafish. Brain Res. 903: 263–268.

Zilberman-Peled, B., L. Appelbaum, D. Vallone, N.S. Foulkes, S. Anava, A. Anzulovich, S.L. Coon, D.C. Klein, J. Falcón, B. Ron and Y. Gothilf. 2006. Transcriptional regulation of arylalkylamine-N-acetyltransferase-2 gene in the pineal gland of the gilthead seabream. J. Neuroendocrinol. 19: 46–53.

Zisapel, N. 2001. Melatonin-dopamine interactions: from basic neurochemistry to a clinical setting. Cell. Mol. Neurobiol. 21: 605–616.

Zisapel, N. 2007. Sleep and sleep disturbances: biological basis and clinical implications. Cell. Mol. Life Sci. 64: 1174–1186.

Neuroendocrine Regulation of Reproduction in Sea Bass (*Dicentrarchus Labrax*)

Arianna Servili,[1] Patricia Herrera-Pérez,[2,a]
José Antonio Paullada Salmerón,[2,b] Olivier Kah[3,]*
and José Antonio Muñoz-Cueto[2,c,]*

Introduction

The European sea bass, *Dicentrarchus Labrax*, represents one of the most important species for marine aquaculture in Europe. The sea bass industry has strongly grown in the last decade, peaking at nearly 145,000 tonnes of total production in 2011 (FAO, 2009–2012). This remarkable production has been possible thanks to the large economical investments in the sector and the application of the scientific knowledge acquired on the biology of the reproduction, nutrition and growth of this species (for review see Scapigliati et al. 2002, Piferrer et al. 2005, Chistiakov et al. 2007, Carrillo et al. 2009).

[1]IFREMER, LEMAR UMR 6539 - Unit of Functional Physiology of Marine Organisms (PFOM), Plouzané, France.
Email: arianna.servili@ifremer.fr
[2]Departamento de Biología, Facultad de Ciencias del Mar y Ambientales, Campus de Excelencia Internacional del Mar (CEI·MAR), Universidad de Cádiz. Avenida República Saharaui s/n, Campus Río San Pedro, E11510-Puerto Real (Cádiz), Spain.
[a]Email: patricia.herrera@uca.es
[b]Email: joseantonio.paullada@uca.es
[c]Email: munoz.cueto@uca.es
[3]INSERM U1085, IRSET, Team NEED, Université de Rennes1. Case 1302, Université de Rennes1, Campus de Beaulieu, 35042 Rennes cedex, France.
Email: olivier.kah@univ-rennes1.fr
*Corresponding authors

Despite the progress achieved to date, sea bass still presents reproduction-related problems, such as early puberty in fish farming conditions (Carrillo et al. 2009). A deep understanding of the mechanisms and factors playing in the modulation and control of reproduction in sea bass would help to solve most problems related to sea bass farming.

The European sea bass belongs to the highly derived order of the Perciform, the most abundant group of Teleost. Thus, sea bass also represents a model species with both phylogenetic and comparative interest (Nelson 2006). It is a gonochoristic teleost species that exhibits a highly photoperiodic reproductive annual cycle, ending with a spawning season that takes place in winter (January–March) (Barnabé and Billard 1984). As a result, both female and male sea bass need, firstly, to achieve the adequate gonad development at the same time and, secondly, to synchronize their reproductive cycles to the environmental factors to guarantee that the birth of the offspring occurs at the most favorable season. The adequate synchronization and modulation of all these processes requires multiple and complex interactions between the external inputs and the neuroendocrine centers regulating reproduction, that take place along the pineal-brain-pituitary-gonad axis (Fig.1).

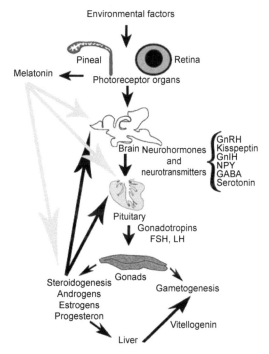

Figure 1. Schematic representation of the main interactions that assure the neuroendocrine control of the reproductive axis (pineal-brain-pituitary-gonads) in sea bass.

This is assured by the presence of specific sensory systems and receptors that are able to perceive environmental (photoperiod, temperature, etc.) and social (presence of other individuals, population density, sex ratio, etc.) stimuli. Like all fish, sea bass has two direct photoreceptor organs, the pineal gland and retina, which receive photoperiod inputs from the environment and send this information to the central nervous system, via neural (axonal projections) and hormonal (release of melatonin) signals (Ekstrom and Meissl 2003, Falcon et al. 2009). Many studies have reported the effects of the photoperiod on the reproductive physiology of this species (Carrillo et al. 1995, Bayarri et al. 2004b, Rodriguez et al. 2004, Carrillo et al. 2009), on the rhythmical melatonin secretion (Iigo et al. 1997, Sánchez-Vázquez et al. 1997, Migaud et al. 2007) and also on feeding and reproduction interaction (Sánchez-Vázquez et al. 1995).

The functions of the major endocrine organs and structures are highly preserved in vertebrates. In this sense, brain areas, neurohormonal factors as well as mechanisms underlying the neuroendocrine control of reproduction are essentially the same throughout the vertebrate phylogeny. In sea bass, as in other fish, the preoptic area and the hypothalamus are the main players of this axis because they control the pituitary activities through the secretion and releasing of different neurohormones that trigger the reproductive process. Significant differences in the way these neurohormones reach their specific targets into the pituitary gland distinguish fish from other tetrapods (Kah and Dufour 2010). In fact, the organization of the fish pituitary gland is unique among vertebrates. The pituitary gland is divided in two subunits: the adenohypophysis, which contains secreting cells releasing pituitary hormones, and the neurohypophysis, consisting of bundles of neurosecretory fibers that come from different brain areas and secrete their products in the proximity of the pituitary cells. In fish, the adenohypophysis is composed of the *pars distalis* (divided into the rostral and proximal *pars distalis*), which corresponds to the anterior lobe of terrestrial vertebrates, and the *pars intermedia*, the homologue of the intermediate lobe of terrestrial vertebrates (Olivereau and Ball 1964). The two main differences characterizing teleost pituitary involve the *pars distalis*. In fact, in teleosts, cells of the same type are grouped and concentrated in the same region, giving rise to specialized cell masses that are not found in terrestrial vertebrate pituitary. Moreover, teleosts lack median eminence and the functional vascular system that connect the hypothalamus with the pituitary, as it exists in tetrapods. Therefore, fish hypothalamic neurohormones project directly to the cells of the anterior lobe of the anodehyphophysis (Zohar et al. 2010). These neurohormones modulate the activity of the pituitary and initiate a series of endocrine cascades that lead to reproduction (Kah et al. 1993, Trudeau 1997, Fig. 1). Specifically, they stimulate or inhibit the synthesis and secretion of pituitary gonadotrophins (GTH), i.e., the follicle-stimulating hormone (FSH) and the luteinizing hormone (LH), which in turn regulate gonadal steroidogenesis and gametogenesis, together with other processes involved in reproduction.

FSH and LH belong to the glycoproteins hormone family. They are heterodimeric glycoproteins composed of a common alpha subunit (α-glycoprotein) and a hormone-specific beta subunit (βFSH or βLH). Both GTHs are steroidogenic but recent findings suggest they have different roles in the control of sea bass gonadal functions (Rocha et al. 2009, Moles et al. 2012). In fact, FSH would mostly play a role during active spermatogenesis and vitellogenesis, whereas LH would be involved in the final reproductive events, such as spermiation, oocyte maturation and ovulation (Moles et al. 2012).

In the present chapter we will focus on the extraordinary and fascinating complexity of the neuroendocrine control of reproductive axis of sea bass (Fig. 1). We will detail the factors modulating this reproductive axis, with both positive and negative effects, including the relevant knowledge achieved very recently in our laboratories.

Neuroendocrine Factors Stimulating the Reproductive Axis

Gonadotrophin releasing hormone (GnRH): forms, distribution, ontogeny and functions

It is well established that GnRH, a decapeptide, represents one of the main stimulatory factors involved in the synthesis and releasing of GTHs in fish (Zohar et al. 2010). Since the very first sets of experiments, when Breton and colleagues (Breton et al. 1971, Breton et al. 1972) demonstrated that sheep brain extracts and the mammalian form of GnRH were able to stimulate LH release in carp, many studies confirmed the presence of GnRH in fish and investigated its actions.

To date, 14 different GnRH sequences have been identified in vertebrates (Morgan and Millar 2004, Vickers et al. 2004, Guilgur et al. 2006). All forms share four of the 10 amino acid residues (at positions 1, 4, 9 and 10) and contain an N-terminal pyroglutamate and C-terminal glycinamide (Sower 1997). The overall organization of all the GnRH variants is also conserved. The typical structure of the GnRH precursor protein consists of a signal peptide at the N-terminal, the bioactive GnRH decapeptide followed by a 3 aa cleavage site (Gly-Lys-Arg) and a GnRH-associated peptide (termed GAP) at the C-terminal (Zohar et al. 2010). The function of GAP is still not completely understood and it is likely changing among species, since both amino acid and nucleotide GAP sequences are highly divergent. Nevertheless, there are evidences that show a role for GAP in providing the appropriate secondary structure for specific processing of the GnRH precursor (Sherwood et al. 1994).

Phylogenetic analysis identified that vertebrate GnRH genes cluster in three main GnRH branches including the GnRH forms termed GnRH1, GnRH2 and GnRH3 respectively, according to the novel nomenclature introduced by Fernald and White (Fernald and White 1999). In fact, it is well defined that GnRH variants in vertebrates are produced by three different paralogous genes: *gnrh1*, *gnrh2* and *gnrh3* (Fernald and White 1999).

As a modern teleost, the European sea bass expresses a variant of each GnRH group: the sea bream GnRH form, the ortholog for the GnRH1 branch; the chicken II GnRH or GnRH2; and the fish specific salmon GnRH corresponding to GnRH3 (Gonzalez-Martinez et al. 2001, Gonzalez-Martinez et al. 2002a, Zmora et al. 2002). A very precise neuroanatomical localization of these GnRHs has been performed in our laboratories at both transcript and protein levels using specific riboprobes and antibodies generated against the GAP cDNA sequences isolated in this species (Gonzalez-Martinez et al. 2001, Gonzalez-Martinez et al. 2002a, Zmora et al. 2002, Fig. 2). With this approach, we circumvented the cross-reactivity problems of probes and antisera raised directly to GnRH decapeptides. The GnRH1 cells were identified in all the classical regions described for this form in vertebrates: in the preoptic area associated with the parvocellular and anteroventral divisions of the parvocellular preotic nucleus, as well as in the ventral telencephalon (Fig. 2). Moreover, some GnRH1 cells were observed in the olfactory bulb at the terminal nerve area, and very few ones at the level of the ventrolateral hypothalamus (Fig. 2). Fibers immunoreactive to GnRH1 were localized along the ventral telencephalon, the preoptic area and the ventral hypothalamus. They were observed to profusely innervate the pituitary gland in sea bass, again a common feature of the GnRH1 branch in vertebrates. These GnRH1-immunoreactive fibers

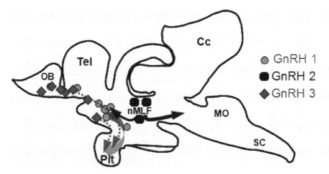

Figure 2. Schematic representation of the localization of the three GnRH systems in sea bass brain and their main projections. Circles correspond to GnRH1 cell bodies, ovals to GnRH2 neurons and diamonds to GnRH3 cells. Arrows represent the main projections of each GnRH cell population. Cc, corpus of the cerebellum; MO, medulla oblongata; nMLF, nucleus of the medial longitudinal fascicle; OB, olfactory bulb; Pit, pituitary gland; SC, spinal cord; Tel, telencephalon.

entered the neurohypophysis and reached the proximal pars distalis of the adenohypophysis and the border of the pars intermedia, ending in close proximity of GTH cells (Fig. 2).

In sea bass, as reported in most vertebrate species, GnRH2 cells were found to be restricted to the transitional area between the medial zone of the dorsal synencephalon and mesencephalic tegmentum (Fig. 2). These large cells widely project to the mid- and hindbrain of sea bass but also to the ventromedial hypothalamus, being detected near the pituitary stalk. Nevertheless, no GnRH2 immunoreactive axons were detected in the pituitary.

Lastly, the GnRH3 cells were mostly identified in the olfactory bulbs, representing the terminal nerve ganglion cells that appear at the junction of the caudal olfactory bulbs and the telencephalon. Furthermore, some GnRH3 positive cells were observed in the ventral telencephalon and in the ventral preoptic area (Fig. 2). The axons of GnRH3 cells were mainly identified in the rostral forebrain, but they were also widely distributed in more caudal regions including the hypothalamus, the ventral thalamus, the optic tectum and the rhombencephalon. The pituitary gland of sea bass also showed GnRH3 fibers entering the neurohypophysis and reaching the *proximal pars distalis* of the adenohypophysis but this innervation was markedly reduced in relation to GnRH1 (Fig. 2).

These localization studies, in which highly specific riboprobes and antibodies were used, show for the first time that GnRH1 and GnRH3 cells share an overlapping distribution from the olfactory bulbs to the preoptic region (Gonzalez-Martinez et al. 2001, Gonzalez-Martinez et al. 2002a, Fig. 2). In addition, immunoreactive fibers to both GnRH isoforms were detected in the pituitary gland of this species (Gonzalez-Martinez et al. 2002a). After this first report, several studies performed in distinct species demonstrated that the presence of two GnRH variants in the forebrain was very common in fish exhibiting three GnRH isoforms (Vickers et al. 2004, Mohamed et al. 2005, Pandolfi et al. 2005, Okubo et al. 2006).

GnRH1 and GnRH3 decapeptides present very high nucleotide and aminoacidic identity, only differing in amino acid residues located in position 7 and 8, which makes it plausible that they arise from a duplication of a single ancestral gene. This homology, together with their similar neuroanatomical localization that is maintained from developing to adult sea bass, could reflect their common embryonic origin. In fact, reports on the ontogeny of GnRH system in sea bass at protein and messenger levels revealed that both forebrain GnRH systems originate in an olfactory primordium (Gonzalez-Martinez et al. 2002b, Gonzalez-Martinez et al. 2004b). From the olfactory placode, both GnRH cell types migrate in a rostro-caudal direction during early development until they reach their final position in a continuum from the olfactory bulbs to the hypothalamus (Gonzalez-Martinez et al. 2004b).

Despite all the analogies between GnRH1 and GnRH3 systems in sea bass in terms of nucleotide/amino acid identities and embryonic origin, the patterns of projections of both systems show evident differences in brain and pituitary gland (Gonzalez-Martinez et al. 2004b), which could reflect the distinct functions for each variant. Thus, GnRH1 fibers run along the ventral forebrain and are very abundant at the level of pituitary gland, innervating the *proximal pars distalis* and the border of the *pars intermedia*, where gonadotrophic cells and also GnRH receptors were identified (Gonzalez-Martinez et al. 2004a). This localization suggests the major hypophysiotrophic role of the GnRH1, which is responsible for stimulating the synthesis and secretion of gonadotrophins in sea bass. According to these results, pituitary GnRH1 levels were 17-fold higher than GnRH3 levels in the pituitary of immature male sea bass (Rodriguez et al. 2000).

On the other hand, the GnRH3 innervation of the pituitary was much less intense, while it was much more abundant in the brain, in particular in sensory regions involved in the processing of visual and gustatory information (Kah et al. 1986, Gonzalez-Martinez et al. 2002a, Gonzalez-Martinez et al. 2002b). In fish, it has been proposed that GnRH3 might be implicated in the coordination of sensory and motivational systems (White et al. 1995) and the modulation of odorant sensitivity (Eisthen et al. 2000). Nevertheless, actions of GnRH3 in sea bass seem also to be related to the gonad differentiation, possibly via the enhancement of FSHβ gene expression (Moles et al. 2007). In sea bass the main GnRH3 neuron population was observed in the terminal nerve ganglion. In different teleost species, GnRH3 projections originating from terminal nerve cells are widely distributed throughout the brain but do not reach the pituitary gland (Oka and Ichikawa 1990). Thus, the GnRH3 cells projecting to the pituitary should be those located in the preoptic area and/or the ventral telencephalon. Nevertheless, it is nowadays assumed that this GnRH form does not serve a major role as a hypophysiotrophic hormone, but rather as a neuromodulator of diverse neural systems (Oka and Ichikawa 1990, Wayne et al. 2005). In fact, several evidences have shown that the terminal nerve ganglion cells affect the nesting comportment but not sexual behavior (Yamamoto et al. 1997) and they did not appear to be sensitive to sexual pheromones (Fujita et al. 1991). Moreover, lesions of the olfactory tract do not prevent gonadal development and ovulation, corroborating that the GnRH cells of the terminal nerve do not serve critical functions in the late reproductive events (Kobayashi et al. 1994).

In sea bass, GnRH2 cells were already detectable in the synencephalic/ mesencephalic transitional area at day 4 after hatching by both *in situ* hybridization and immunohistochemistry (Gonzalez-Martinez et al. 2002b). This result suggests that developing GnRH2 cells are not originated in the olfactory placode, as GnRH1 and GnRH3, but in the germinal zone of the third ventricle, into the synencephalon. From this region, sea bass GnRH2 neurons migrate dorsally along the midline through, the synencephalon reaching their final position at about day 30 after hatching. The number of GnRH2 neurons increases in number and size during the first weeks and decreases in adulthood. This evidence could indicate the important role of GnRH2 during early development. Since no GnRH2 projections were revealed in sea bass pituitary either in developing or in adult sea bass, the direct implication of this isoform on the secretion of gonadotrophin seems excluded, at least in this species (Gonzalez-Martinez et al. 2002a).

The GnRH2 variant is the most conserved in vertebrates and it is likely the most ancient form. The functions of the highly conserved peptide are still a matter of debate, although there are indications showing its possible role in sexual behaviour (Muske 1993) and in the regulation of feeding (Muske and Moore 1994). Furthermore, it is also proposed that GnRH2 modulates sensory-motor activity, since GnRH2 fibers profusely innervate the spinal cord, sensory areas and the cerebellum of sea bass (Gonzalez-Martinez et al. 2002b).

A recent study from our laboratories reports, for the first time in a vertebrate species, converging morphological, functional, and pharmacological evidences supporting a clear function for this highly conserved GnRH variant (Servili et al. 2010). We have shown a direct connection of GnRH2 cells with the pineal gland, a photoreceptor organ that, together with the retina, directly receives the light inputs in fish and sends them to the central nervous system. Using tract-tracing techniques coupled with immunohistochemistry, we were able to evidence that the GnRH2 cells of the synencephalon/medial tegmentum are the source of the GnRH2 innervation detected in sea bass pineal gland (Servili et al. 2010). Notably, the highly specific antibodies (Zmora et al. 2002) used in this study show that GnRH2, but neither GnRH1 nor GnRH3, reach the sea bass pineal organ. To investigate whether GnRH2 could act locally at the level of the pineal gland, the expression patterns of GnRH receptors were identified by RT-qPCR and *in situ* hybridization. Both techniques revealed that the dlGnRHR-II-2b subtype is the major form expressed in the pineal gland. This receptor is the one showing the highest biopotency for the non-hypophysiotrophic ligand GnRH2 (Gonzalez-Martinez et al. 2002a, Servili et al. 2010, Lethimonier, Lareyre and Kah, unpublished data), the only GnRH peptide found in the pineal of sea bass (Servili et al. 2010). Lower expression levels of another GnRH receptor, the dlGnRHR-II-1a form, were revealed in the sea bass pineal gland (Servili et al. 2010). This receptor subtype is strongly expressed in the pituitary (Gonzalez-Martinez et al. 2004a) and is the only receptor showing some affinity for GnRH1 and GnRH3. It has been suggested that this GnRH receptor might be evolving to improve affinity for GnRH1 and GnRH3 to better mediate their effects at the pituitary level (Kah et al. 2007, Zohar et al. 2010). Lastly, sets of *in vitro* and *in vivo* experiments performed in sea bass clearly demonstrated that GnRH2 administration consistently increased the nocturnal melatonin release and the plasma melatonin levels, respectively (Servili et al. 2010).

A similar approach was chosen for establishing a functional link between the other photoreceptive organ in sea bass, the retina, and the GnRH3 system (Servili et al. 2012). First of all, a morphological connection between the terminal nerve GnRH3 cells and the retina was observed by revealing retinopetal projections in the terminal nerve area. In fact, the GnRH3 cells located at the terminal nerve are suggested to be the source of the GnRH3 projections detected at the boundary between the inner nuclear and the inner plexiform layers as well as on the ganglion cell layer of the sea bass retina. GnRH receptors were also detected in the retina of this species by RT-qPCR and *in situ* hybridization. Interestingly, the GnRH receptor types mostly expressed in the retina corresponded to the dlGnRHR-II-2b and to the GnRHR-II-1a forms, as it occurs in the pineal gland of this species. Double-label immunostaining performed in this study showed that the GnRH3 axons present in the retina, at the border of the inner nuclear layer with the inner plexiform layer, end up in close proximity to tyrosine hydroxylase-immunoreactive (TH-ir) cells, which also express dlGnRHR-II-2b receptor (Servili et al. 2012). The TH-ir cells correspond to the dopaminergic interplexiform cells in fish retina, which are known to modulate light adaptation processes (Douglas et al. 1992). Altogether these evidences suggest a role for GnRH3 in the control of light adaptation in fish retina through its actions on dopaminergic cells (Servili et al. 2012).

Gonadotrophin releasing hormone receptors (GnRHR): forms, distribution and functions

The European sea bass exhibits five GnRHRs, which are members of the rhodopsin-like G-protein coupled receptor (GPCR) family, encoded by five different genes (Gonzalez-Martinez et al. 2004a, Lethimonier et al. 2004, Moncaut et al. 2005, Zohar et al. 2010). These sea bass GnRHRs have been named dlGnRHR-II-1a, dlGnRHR-II-1b, dlGnRHR-II-1c, dlGnRHR-II-2a, and dlGnRHR-II-2b because all of them have much more affinity for GnRH2 (Kah et al. 2007). All these GnRH receptors are detected in sea bass brain, as shown by RT-PCR studies, even if they exhibit distinct expression patterns (Moncaut et al. 2005). The exact neuroanatomical localization of one of them (dlGnRHR-II-1a) has been determined by *in situ* hybridization. These mRNAs are mainly located in the ventral telencephalon and rostral preoptic area, but also, to a lesser extent, in the dorsal telencephalon, caudal preoptic area, ventral thalamus, and periventricular hypothalamus (Gonzalez-Martinez et al. 2004a). The precise functions of the GnRHR types still need to be clarified. However, it is known that dlGnRHR-II-1a is strongly expressed in the pituitary and co-localizes with LH cells and also some FSH cells, but not with somatotrophs (Gonzalez-Martinez et al. 2004a). This receptor type is probably mediating the response of the gonadotrophs to GnRH stimulation. Pharmacologic evidences obtained by transient transfection in COS-7 cells support this hypothesis. In fact, both the IP3 and cAMP assays showed that all the sea bass receptors types are functional and all of them exhibit higher biopotency for the ligand GnRH2 (Lethimonier et al. 2004, Kah et al. 2007, Servili et al. 2010). Therefore, all five sea bass GnRH receptors belong to the "type II" GnRH receptor class, as indicated above (Kah et al. 2007). Nevertheless, the receptor type dlGnRHR-II-1a is the only one showing relative high affinity for the GnRH1 and GnRH3 ligands, the ones directly reaching the pituitary gland (Gonzalez-Martinez et al. 2004a, Lethimonier et al. 2004). There are evidences suggesting a role for GnRH1 and GnRH3 ligand in sea bass gonad differentiation that would be exerted through this dlGnRHR-II-1a receptor isoform (Moles et al. 2007). As mentioned above, this GnRH receptor is also slightly expressed in the pineal and retina of this species (Gonzalez-Martinez et al. 2004a, Servili et al. 2010), while the GnRH receptor isoform mostly associated to the photoreceptor organ activity appears to be the dlGnRHR-II-2b (Servili et al. 2010). This latter form shows the highest biopotency for the non-hypophisiotrophic ligand, GnRH2, the only one detected in sea bass pineal gland (Servili et al. 2010, Lethimonier, Lareyre, Kah, unpublished data). Thus, the dlGnRHR-II-2b receptor type could be involved in the modulation of photoperiod and environmental signals, while only the dlGnRHR-II-1a receptor is evolving in order to better mediate type 1 and type 3 GnRH ligands at the pituitary level.

It has been demonstrated that the expression levels of some GnRH receptors in sea bass brain and pituitary vary with seasonal and daily cycles (Gonzalez-Martinez et al. 2004a, Servili et al. 2013). The dlGnRHR-II-1a type exhibits the highest expression level in the anteroventral part of the parvocellular preoptic nucleus during the spawning season. On the other hand, in the sea bass pituitary, highest expression levels of this GnRH receptor were detected during late vitellogenesis (November) than during maturation (December), spawning (February) or postspawning/resting (May) periods (Gonzalez-Martinez et al. 2004a). This report considers GnRHR mRNAs and not the functional proteins, whose profiles could vary along the sea bass reproductive cycle. However, the levels of GnRH1, the main ligand of this receptor in the sea bass pituitary, are also more elevated during late vitellogenesis, at least in male sea bass (Rodriguez et al. 2000), suggesting that functional targets could also be elevated at this reproductive phase.

A recently published work from our laboratory has revealed the existence of day-night variations in sea bass brain expression of GnRH1 and GnRH3, as well as of dlGnRHR-II-1c and dlGnRHR-II-2a receptors, which exhibit higher mRNA levels at midday in relation to dusk and mid-dark periods (Servili et al. 2013). The existence of day-night variations in particular GnRH receptors suggests that they appear subjected to different regulatory mechanisms. In this study, an inhibitory effect of melatonin on the nocturnal brain expression of GnRH-1, GnRH-3, and GnRHR-II subtypes 1c, 2a and 2b was also demonstrated. To date, it remains to be determined whether daily variations in receptor transcript levels are correlated with differences in the amount of functional receptor proteins and whether this GnRH receptor mRNA fluctuation is regulated by its ligands. It appears also necessary to better define the specific expression patterns of each receptor subtype, as well as their functions. Interestingly, dlGnRHR-II-2b is the receptor subtype that shows the highest affinity for GnRH2 (Kah et al. 2007, Servili et al. 2010) and none of these genes exhibit day-night variation in transcript levels.

Kisspeptins: forms, distribution and functions

Since 2003, when two different laboratories have independently observed that inactivated mutations of GPR54 cause hypogonadotroph hypogonadism in humans (De Roux et al. 2003, Seminara et al. 2003), kisspeptins began to be recognized as a major player in the neuroendocrine control of reproduction and puberty in mammals (Colledge 2008, Kauffman 2009, Oakley et al. 2009). Indeed, several evidences unambiguously show the strong stimulatory action of kisspeptins on GnRH release and, therefore, of gonadotrophin secretion (Roa et al. 2009, Pineda et al. 2010). Kisspeptins are member of the RFamide peptide family and products of the *kiss1* gene, acting via the previously orphan receptor GPR54, which has been renamed as KISSR (Clements et al. 2001, Kotani et al. 2001, Muir et al. 2001). The precursor protein is a 145 amino acids peptide that is cleaved to yield a group of four biologically active peptides, 54, 14, 13 and 10 amino acids long (Kotani et al. 2001).

More recently, kiss system has also been investigated in other vertebrate classes. In 2009 Felip and collaborators published the description of two kisspeptin coding genes (*kiss1* and *kiss2*) within the same species, the European sea bass, and genome database searches showed that both genes were also present in non-placental vertebrate genomes (Felip et al. 2009). Nowadays, phylogenetic studies have confirmed that this phenomenon is generalized in non-placental vertebrates (Lee et al. 2009, Um et al. 2010). These two paralog genes have likely experienced further gene duplication (as in amphibians) or gene loss (as in birds) (Lee et al. 2009, Um et al. 2010). Whether the paralog genes serve similar, complementing or different functions is still a matter of debate.

To investigate the organization of these systems, several researchers looked at the expression sites of both *kiss* genes in different species. In sea bass, both *kiss1* and *kiss2* mRNAs were detected by RT-PCR into the brain and gonads (Felip et al. 2009), confirming their putative roles in the control of reproduction also in fish. We have recently reported the detailed anatomical distribution of kiss systems in this species by using *in situ* hybridization (Escobar et al. 2013, Figs. 3, 4). We have observed that *kiss1* and *kiss2* transcripts are localized in distinct cell population in the brain of sea bass (Escobar et al. 2013, Fig. 4). *Kiss1* mRNAs are expressed in the ventromedial habenula (Figs. 3A, 4A). As previously shown in medaka and zebrafish, this kiss1-expressing cell population is constantly detected independent of sex or reproductive stage (Kitahashi et al. 2009, Servili et al. 2011b, Escobar et al. 2013). However, this kiss expression site was not evident in the stripped bass (*Morone saxatilis*), another species of the family Moronidae closely related to the European sea bass (Zmora et al. 2012).

Unfortunately, to date, no specific antibodies able to recognize the sea bass kiss1 and kiss2 proteins without any cross-reactions are available. Thus, we could not trace the projections of these neurons. However, the ventromedial habenula is a highly conserved structure in vertebrates. It mainly projects to the interpeduncular and raphe nuclei via the fasciculus retroflexus, outlining the dorsal diencephalic conduction system (Guglielmotti and Cristino 2006, Hendricks and Jesuthasan 2007, Amo et al. 2010). This structure links the forebrain with the mid- and hindbrain (Bianco and Wilson 2009). In the only fish species where specific anti-preprokiss1 and anti-preprokiss2 are available, the zebrafish, the kiss1 habenular neurons were shown to project to the interpenduncular and raphe nuclei (Servili et

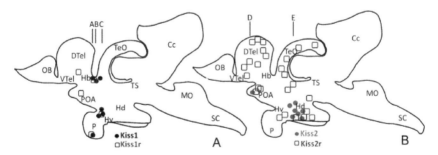

Figure 3. Schematic representation of the neuroanatomical localization of kiss and kiss receptor systems in the European sea bass. (A) Circles and squares represent the kiss1 and kiss1r expressing cells, respectively. (B) Circles and squares correspond to kiss2 and kiss2r expressing cells, respectively. Letters with lines indicate the levels of the transverse sections of the photomicrographs shown in Fig. 4. CC, corpus of the cerebellum; Hb, habenula; Hd, dorsal hypothalamus; Hv, ventral hypothalamus; MO, medulla oblongata; OB, olfactory bulb; P, pituitary gland; DTel, dorsal telencephalon; POA, preoptic area; VTel, ventral telencephalon; TeO, optic tectum; TS, torus semicircularis.

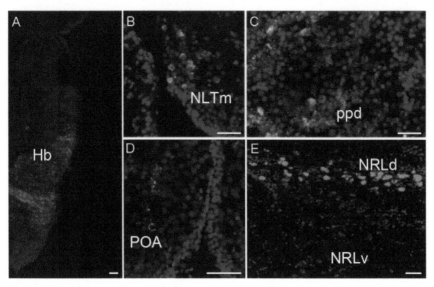

Figure 4. *In situ* hybridization on transverse sections of kiss1 (A, B, C) and kiss2 (D, E) expressing cells in the brain and pituitary gland of the European sea bass. (A) Kiss1 mRNAs were detected into the habenular nucleus (Hb) in both sexes. (B) In sea bass sacrificed during the breeding period an addition population of kiss1-expressing cells appear in the hypothalamus, at the level of the medial part of the lateral tuberal nucleus (NLTm), and (C) in the proximal pars distalis of the pituitary gland (ppd). (D) Few kiss2 expressing cells were located in the preoptic area (POA), whereas the most abundant kiss2 cell population was observed in the hypothalamus, at the level of the dorsal and ventral parts of the nucleus of the lateral recess (NRLd, NRLv). Scale bar = 25 μm.

Color image of this figure appears in the color plate section at the end of the book.

al. 2011b). On the other hand, the habenula is linked by the pineal stalk to the pineal gland, the photosensory and neuroendocrine structure involved in the reception of the environmental stimuli. Labeled pinealofugal fibers can be found in the habenula of sea bass, which also showed pinealopetal neurons (Servili et al. 2011a). In fish, this organ is the main one responsible for the release of melatonin, the hormone implicated in circadian rhythms but also in reproduction (Falcon et al. 2010, Herrera-Pérez et al. 2010). Thus, we may speculate that the habenular kiss1 population could exert the link between the photoperiod and the activation of the reproductive axis in sea bass, maybe participating in the process of sexual maturation (Bayarri et al. 2010a).

During the breeding season an additional population of *kiss1* mRNA expressing cells was revealed in sea bass mediobasal hypothalamus, just above the pituitary stalk (Figs. 3A, 4B). This kiss1 population was evident in both sea bass males and females but only in sexually mature fish (Escobar et al. 2013). This evidence suggests that gonadal steroids could modulate the hypothalamic kiss1 expression. Accordingly, the kiss1 cells located in the mediobasal hypothalamus are shown to strongly express the estrogen receptor alpha (ERα) and, to a lesser extent, also the estrogen receptor beta2 (ERβ2) (Escobar et al. 2013). This feature seems to vary among species and, in fact, in other fish like zebrafish, the hypothalamic kiss population sensible to estrogens seem to express kiss2 (Servili et al. 2011b). In mammals a similar population of KISS1 cells is observed in the arcuate nucleus that mediates the negative feedback of estrogens on LH release (Dungan et al. 2006, Tena-Sempere 2010). Mammalian arcuate nucleus also show kiss coexpression with neurokinin B, which is a major player in the neuroendocrine control of LH release (Navarro and Tena-Sempere 2011). Although recent evidences suggest a role for neurokinins in zebrafish puberty (Biran et al. 2012), they do not appear to be expressed in kiss neurons (Ogawa et al. 2012b).

In our model species, the European sea bass, a small number of kiss2-expressing cells were observed in the preoptic area in both sexes (Figs. 3B, 4D). To date coexpression with estrogen receptors has not been reported in sea bass and variation of the kiss2 expression in this area was neither observed during the reproductive cycle. Nevertheless, in goldfish the preoptic kiss2 population was described to vary with reproductive stage and to express estrogen receptors (Kanda et al. 2012).

The more conspicuous kiss2-expressing cell population was revealed at the level of the lateral recess (Figs. 3B, 4E). Kiss2 neurons appear anterior to the opening of the lateral recess extending dorsal and along the recess. This population also expands ventrally towards the lateral hypothalamic wings (Escobar et al. 2013, Figs. 3B, 4E). A similar hypothalamic kiss2 mRNA pattern is described in all fish species analyzed so far. Consistently, the kiss2 expression in this region does not seem to vary with sex or reproductive stages and positive cells do not express estrogen receptors (Kitahashi et al. 2009, Mitani et al. 2010, Servili et al. 2011b, Kanda et al. 2012, Zmora et al. 2012). The function of this kiss cell population is still unknown. Nevertheless, it is documented that the region of the lateral recess is part of the paraventricular organ and, in teleosts including sea bass, it exhibits many dopaminergic and serotoninergic neurons (Kah et al. 1978, Geffard et al. 1982, Batten et al. 1993).

Our *in situ* hybridization study revealed the expression of kiss1 mRNAs in the pituitary of mature male and female sea bass (Figs. 3A, 4C). These cells expressing kiss1 were shown to be FSHβ immunoreactive cells. No LH or other cell types appear to express kiss1 in sea bass pituitary (Escobar et al. 2013). Interestingly, recent data report that FSH increases during vitellogenesis in the European sea bass, suggesting a role for FSH in the modulation of the mid phases of oogenesis, whereas LH would be involved in the final maturation steps in females (Moles et al. 2012). Thus, our neuroanatomical observations could indicate that the kiss1 gene in the European sea bass could participate in the regulation of reproduction through the mediobasal hypothalamic kiss1 population, which is sensitive to estrogens, and via the expression in the FSH cells (Escobar et al. 2013). However, functional studies revealed that the kiss2 core decapeptide is a more potent elicitor of gonadotrophin secretion than kiss1, both in mature and immature sea bass (Felip et al. 2009). To date, ambiguous results were yielded after the exogenous administration of kiss decapeptides to several fish species. Notably, inhibitory effects of kisspeptins on LH expression were reported for the first time in the European eel (Pasquier et al. 2011). Furthermore, nowadays the general consensus is that fish produce longer mature kiss peptides like kiss1–15 and kiss2–12, which are shown to be more potent activators of kiss receptors, at least in sea bass (Felip et al. unpublished, Tena-Sempere et al. 2012). Thus, further experiments testing the effects of the administration of longer kiss forms should be addressed in fish in order to better define their role in activating gonadotrophins.

Kisspeptin receptors: forms, distribution, and functions

Kisspeptins bind and activate the GPR54 receptor, a member of G Protein-Coupled Receptor (GPCR) family. As in the case of kiss ligands, multiple kiss receptor genes have been identified in non-mammalian species. All these receptor forms exhibit a conserved genomic organization, a possible evidence of the fact that each of them originated from a common ancestral gene (Zohar et al. 2010). To date, two paralog kiss receptor genes have been isolated and cloned in the European sea bass (Felip et al. unpublished, Tena-Sempere et al. 2012), as well as in several fish species (Biran et al. 2008, Kanda et al. 2008, Kitahashi et al. 2009, Li et al. 2009, Zmora et al. 2012).

The neuroanatomical localization of kiss1r- and kiss2r-expressing cells has been recently addressed in the European sea bass (Escobar et al. 2012, Fig. 3). Interestingly, kiss1r and kiss2r were distributed mainly in the same regions as kiss1 and kiss2 ligands, respectively (Fig. 3). For instance, kiss1r-expressing cells were detected in the habenular region in the same area where kiss1 mRNAs were observed (Escobar et al. 2012, Fig. 3A). The same correspondence between kiss1 and kiss1r-expressing cells were identified in the habenula of zebrafish. In this latter species, it has been shown that habenular kiss1 cells actually express kiss1r (Servili et al. 2011b, Ogawa et al. 2012a). This fact could be an evidence of an autocrine regulation of kiss1-expressing cells via kiss1r in both species. Surprisingly, recent findings in zebrafish reported that kiss1r gene produces several splice forms that generate truncated proteins, including one able to enter the nucleus and to exert transactivation activity (Onuma and Duan 2012). Therefore, we cannot exclude that kiss1r variant could modulate gene expression in habenular kiss1 neurons. Further expression sites for kiss1r gene in sea bass include the mediobasal hypothalamus as well as the pituitary gland (Fig. 3A), where kiss1-expressing cells were also observed in breeding animals (Figs. 3A, 4B). This observation could further support a possible autocrine regulation of kiss1 through the kiss1r (Escobar et al. 2012).

In our *in situ* hybridization analysis, there were remarkably high expression levels of the kiss2r mRNA-expressing cells, which were widely distributed in forebrain and hindbrain (Escobar et al. 2012, Fig. 3B). The main kiss2r mRNA-containing cell populations were evidenced in the telencephalon, in the preoptic area, all along the lateral recess and also in the posterior recess region, close but likely not overlapping with kiss2 neurons (Escobar et al. 2012, Fig. 3B). In the preoptic area we have found very few kiss2r cells (Fig. 3B) that co-express GnRH1. As we mentioned above, these preoptic GnRH1 neurons are responsible for the synthesis and release of gonadotrophins (Gonzalez-Martinez et al. 2001, Gonzalez-Martinez et al. 2002a). Due to the scarce number of GnRH cells identified as a target for kisspeptin actions via kiss receptor in sea bass and the fact that so far no study has reported a very clear direct effect of kisspeptins on GnRH neurons in fish, we wonder whether non identified interneurons could be connecting kiss and GnRH systems in fish. One possible candidate for this role could be the nitric oxide (NO) neurons. Recent data obtained in mammals show that neuronal nitric oxide synthase (nNOS)-expressing cells from the preoptic area are in proximity to GnRH neurons, appear surrounded by kispeptins fibers and express KISS1R (Hanchate et al. 2012).

Our very preliminary data also suggest that kisspeptins could be involved in functions other than reproduction in sea bass. Indeed, different populations of kiss2r-expressing cells were observed to express tyrosine hydroxylase, neuropeptide Y, somatostatin, or leptin receptor in sea bass (Escobar et al. 2012).

The results obtained in sea bass are in line with the evidence that kiss1 is more efficiently activated by kiss1–10 than by kiss2–10, whereas kiss2r is activated at comparable levels by kiss1–10 and kiss2–10 (Felip et al. unpublished, Tena-Sempere et al. 2012). The same ligand-receptor affinities were also observed for the two kisspeptin forms longer that 10 amino acids (kiss1–15 and kiss2–12), which were more effective than their corresponding kiss-10 peptides in activating kiss receptors in sea bass (Felip et al. unpublished, Tena-Sempere et al. 2012).

Inhibition of the Reproductive Axis

Dopamine

While GnRH and kisspeptin systems are considered the major hypothalamic factors stimulating pituitary gonadotrophin release in most vertebrates, studies in teleost fish have revealed the existence of an inhibitory control of the reproductive axis exerted by dopamine (Dufour et al. 2010). Pioneer studies performed in goldfish identified a brain gonadotrophin-release inhibitory factor (GRIF) involved in the neuroendocrine control of the last steps of gametogenesis (Peter and Paulencu 1980). Afterward, pharmacological experiments showed that the inhibition of LH release and ovulation could be also achieved using dopamine (DA) agonist and prevented by the administration of DA antagonist (Chang and Peter 1983, Peter et al. 1986, Peter et al. 1991).

Dopamine is a neurotransmitter synthesized from tyrosine thank to the actions of two different limiting enzymes, tyrosine hydroxylase and DOPA-decarboxylase (Dufour et al. 2010, Zohar et al. 2010). Dopamine exerts its action by binding dopamine receptors, which belong to the GPCR family. All DA receptor variants are grouped in two main classes, D1 and D2 receptors, according to their ability to inhibit or activate the enzyme adenylate cyclase (Kebabian and Calne 1979, Cardinaud et al. 1998).

To date it is known that DA exerts a variety of effects on the brain and the pituitary of vertebrates, including its action on the release of prolactin in mammals. In fish, this neurotransmitter acts directly on pituitary gonadotrophs, mostly inhibiting the hormonal release in LH cells, which express D2 receptors (Chang and Peter 1983, Dufour et al. 2005). DA can also inhibit GnRH secretion from hypothalamic GnRH neurons by using D2 receptors present on GnRH pituitary terminals and D1 receptors existing on GnRH cell bodies (Yu et al. 1991).

After the discovery of the dopamine inhibition of reproductive axis in goldfish, several fish species were investigated to find out whether a similar DA action was present. Nowadays, the general consensus is that many teleosts, but not all, exhibit a DA inhibition of gonadotrophin secretion. Hence, within Perciformes, an order of modern teleosts, only a few species like the tilapia (*Oreochromis mossambicus*) showed DA inhibitory action (Yaron et al. 2003). In contrast, other species of the same order and relative to the European sea bass, as the gilthead seabream (*Sparus aurata*, Zohar et al. 2005), the red seabream (*Pagrus major*, Kumakura et al. 2003) and the striped bass (*Morone saxatilis*, Holland et al. 1998), do not appear to exhibit such a DA inhibition. Notably, this effect could not be observed in many marine species (Dufour et al. 2005, Dufour et al. 2010). In sea bass, very few studies are available in literature regarding the DA actions on gonadotrophin release. Prat and co-authors have shown that co-treatment of GnRH analog (GnRHa) with DA antagonists such as pimozide (PIM), did not improve GnRHa-induced oocyte maturation or spawning (Prat et al. 2001). Therefore, it is very likely that DA does not exert a relevant inhibitory effect on the reproductive axis of sea bass. These variations among teleost species in terms of DA inhibition could reflect the large diversity and plasticity of teleost reproductive cycles.

In the European sea bass, brain tyrosine hydroxylase (TH)-immunoreactive (ir) cells and projections have been mapped (Batten et al. 1993). The TH is the rate-limiting enzyme in the biosynthesis of different catecholamines (DA, adrenaline and noradrenaline) but its localization, together with the immunoreactivity against DA antisera, can be considered as markers for DA cells since many of the catecholaminergic cells present in teleost forebrain and midbrain are dopaminergic (Kaslin and Panula 2001, Sebert et al. 2008). The TH-ir and DA-ir cells are mainly located in the olfactory bulb, ventral/central telencephalon, periventricular preoptic area, suprachiasmatic area, dorsolateral and ventromedial thalamus, posterior tuberal nucleus and paraventricular organ. Broadly distributed DA projections were identified in the sea bass brain, also reaching the pituitary (Batten et al. 1993). The DA cells from the preoptic area, an important hypophysiotrophic region containing both GnRH and kiss neurons, seem to express kiss2r in sea bass (Escobar et al. 2012). Moreover, the most evident DA cell population was observed within the paraventricular organ (Batten et al. 1993), which also contains the main kiss 2 population and exhibit kiss2 receptors in this species (Escobar et al. 2012, 2013). Taken together, all these evidences suggest that dopaminergic neurons could represent targets for kisspeptin actions. However, the putative role of DA in sea bass reproduction remains undetermined.

DA acts also at the level of the retina in fish (Grens et al. 2005, Maruska and Tricas 2007). In sea bass TH-ir cells were observed at the border of the inner nuclear layer with the inner plexiform layer. These cells were in morphological vicinity with GnRH3 fibers coming from the terminal nerve GnRH3 neurons. Moreover, the dopaminergic interplexiform retinal cells express GnRH receptors (Servili et al. 2012). The existence of contacts between terminal nerve axons and the perikarya of dopaminergic interplexiform cells has been previously demonstrated in the teleost retina at electron microscopy (Zucker and Dowling 1987). In fish, DA is involved in the modulation of the retinal adaptation to light. This monoamine is associated with the retinomotor cone contraction, the dispersion of melanosomes in the pigment epithelial cells, and the formation of the spinules on horizontal cell dendrites (Wagner 1980, Kirsch et al. 1990, Douglas et al. 1992, Wagner and Behrens 1993). Previous studies in fish showed that GnRH of the terminal nerve projections stimulate dopamine release by acting directly on the interplexiform dopaminergic cells or indirectly via amacrine cells (Umino and Dowling 1991). Our data also suggest that the activity of DA cells in sea bass retina might be modulated by the GnRH3 system. However, the mechanisms underlying these interactions need further investigations to be clarified.

Gonadotrophin-inhibitory hormone (GnIH)

In 2000, a novel hypothalamic RFamide neuropeptide, able to inhibit the gonadotrophin release from the pituitary was discovered in birds (Tsutsui et al. 2000). As a result, it was named as gonadotrophin-inhibitory hormone or GnIH. GnIH is mostly expressed in the hypothalamus and the septal region in the avian brain and its inhibitory effect on the synthesis and release of gonadotrophins is directly exerted at pituitary levels via a novel G protein-coupled GnIH receptor (GnIHR, Tsutsui and Ukena 2006, Tsutsui et al. 2007). Moreover, GnIH seems to act also on GnRH cells since preoptic GnRH neurons and GnRH fibers of the median eminence receive an evident GnIH innervation and GnRH cells exhibit GnIH binding sites (Bentley et al. 2006, Bentley et al. 2008). Furthermore, a possible effect of GnIH on gonads has been described, notably inhibiting steroidogenesis and development (Tsutsui et al. 2007).

After the discovery in birds, ortologous genes to GnIH were identified in several vertebrate groups such as mammals (RFRPs, Fukusumi et al. 2001, Ukena et al. 2002), amphibians (fGRP and fGRP-RPs; R-RFa, Chartrel et al. 2002, Koda et al. 2002, Sawada et al. 2002a, Ukena et al. 2003) and teleosts (gfLPXRFa, Sawada et al. 2002b, Zhang et al. 2010, Shahjahan et al. 2011). All the GnIH variants exhibit a C-terminal Leu-Pro-Xaa-Arg-Phe-NH$_2$ (with Xaa being Leu or Gln) and therefore they were termed LPXRFamide (X=L or Q) peptides. Their precursors encode for two to four peptides in vertebrates. In fish three forms have been identified, LPXRFa-1, LPXRFa-2, and LPXRFa-3 (Sawada et al. 2002b, Zhang et al. 2010, Shahjahan et al. 2011).

In fish, the physiological action of GnIH system on gonadotrophin release is not clear. To date, contradictory findings have been reported. The three goldfish LPXRFa increased the LH, FSH and GH release, whereas they did not affect the secretion of prolactin and somatolactin from pituitary cells of sockeye salmon (Amano et al. 2006). However, intraperitoneal injections of zebrafish LPXRFa-3 in goldfish decreased plasma LH levels (Zhang et al. 2010). In contrast, goldfish LPXRFa-1 stimulated the LHβ and FSHβ gene expression in cultured grass puffer pituitary (Shahjahan et al. 2011).

The localization of fish GnIH cells has been addressed in goldfish (Sawada et al. 2002b). GnIH cells were mostly localized in the posterior periventricular nucleus of the hypothalamus and in the terminal nerve region and the GnIH fibers reached the lateral tuberal nucleus, the ventral telencephalon, the optic tectum and the pituitary gland (Sawada et al. 2002b). Taking into account the GnIH expression in the terminal nerve, the potential structural and functional relationship between GnRH3 and GnIH, as suggested in zebrafish (Smith et al. 2012) is not surprising. Moreover, GnIH functions are dependent of the reproductive stage of fish and they appear modulated by photoperiod, via the melatonin action (Shahjahan et al. 2011, Moussavi et al. 2012).

In the European sea bass, the characterization and analysis of GnIH variants is currently being approached in our laboratory. Ongoing investigations addressing the incidence and role of GnIH system could represent an important step forward in the comprehension of the complex mechanisms modulating the control of the pineal-brain-pituitary-gonad axis in this species. Notably, it would be very interesting to understand the role of the GnIH system in a species like the European sea bass, where dopamine does not seem to have an evident inhibitory role on the reproductive processes.

Sex Steroids Feedback

Until now, in this chapter, we have reviewed the mechanisms underlining the control that the brain and pituitary gland exert on gonads. But there is also a critical communication from the other direction, i.e., from gonads up to the brain and pituitary. This bidirectional communication is crucial since it allows the coordinated activity of the different components of the pineal-brain-pituitary-gonad axis at all steps of the life cycle and, thus, assures the elaboration of the adequate response at the right moment. Sexual steroids play a key role in referring the sexual status to the brain and pituitary. They are produced by the gonads and are able to modulate the central expression of neuropeptides and neurotransmitters and also that of their cognate receptors.

The specific effects of sexual steroids are likely to vary between species and reproductive stages and their precise mode of action is not completely understood (Prat et al. 1990, Rocha et al. 2009). Nevertheless, it is known that steroids can act directly on gonadotrophin cells at pituitary level (short or direct regulatory feedback) or via neural pathways modulating pituitary functions (long or indirect regulatory feedback). The activation of these two different loops depends on the phase of the development and of the reproductive cycle (Blazquez et al. 1998, Yaron et al. 2003). In general, both positive and negative effects have been observed on the synthesis and release of LH and FSH in teleosts (Crim and Evans 1983, Breton et al. 1997, Trudeau 1997, Saligaut et al. 1998, Huggard-Nelson et al. 2002, Aroua et al. 2007). For instance, in the European sea bass the exogenous estrogen implantation inhibits FSHβ but stimulates LHβ expression (Mateos et al. 2003).

The actions of sex steroids are mainly mediated by the binding to specific intra-cellular receptors that act as ligand-dependent transcription factors, to regulate the transcription rate of target genes (Beato et al. 1995). To date, one androgen receptor (AR) and three estrogen receptor (ER) subtypes have been cloned in the European sea bass (Halm et al. 2004, Blazquez and Piferrer 2005). Within the reproductive axis, AR was mostly detected by RT-PCR at the level of the brain and gonads, suggesting that androgens can directly target the central nervous systems and also act locally in the ovary and testis. In sea bass, AR is proposed to control the sex differentiation since sex-related differences in AR expression profiles were reported between males and females at the time of gonadal differentiation (Blazquez and Piferrer 2005).

The estrogen receptors localization has been reported in the brain and pituitary of this species at mRNA levels by *in situ* hybridization studies (Muriach et al. 2008a, Muriach et al. 2008b). The estrogen receptor alpha (ERα or Esr1) is widely distributed within the sea bass brain, including regions like the telencephalon, preoptic area, thalamus, hypothalamus, mesencephalic tectum and tegmentum and rhombencephalon (Muriach et al. 2008b). Also the estrogen receptor beta2 (ERβ2 or Esr2a) presents a widespread distribution and mRNAs were detected within the ventral telencephalon, preoptic area, hypothalamus, thalamus, posterior tubercle, mesencephalon and rhombencephalon (Muriach et al. 2008a). On the contrary, the estrogen receptor beta1 (ERβ1 or Esr2b) transcripts distribution was restricted to the preoptic area and tuberal hypothalamus (Muriach et al. 2008a). Thus, all the ERs were profusely expressed in the main neuroendocrine areas such as the preoptic region and the hypothalamus (Kah et al. 1993). These results suggest that ERs in sea bass brain mediate steroid actions on hypophysiotrophic neurons by using the long or indirect regulatory feedback pathway to modulate the pituitary activity. However, a direct effect of estrogens on gonadotrophin synthesis is not excluded since expression of all three ERs was detected in FSH and LH cells (Muriach et al. 2008a, Muriach et al. 2008b). As mentioned above, GnRH neurons of sea bass are placed in the telencephalon and preoptic area, and most precisely in the ventral area of the ventral telencephalon as well as in the parvocellular and anteroventral preoptic nuclei (Gonzalez-Martinez et al. 2002a). In addition, GnRH receptors were evidenced in the anterior periventricular nucleus and magnocellular preoptic nucleus (Gonzalez-Martinez et al. 2004a). All these areas show ERs transcripts suggesting that estrogens could modulate GnRH synthesis via these estrogen receptors (Muriach et al. 2008a, Muriach et al. 2008b). Even if an estrogenic modulation of GnRH synthesis has been reported in several piscine species (Parhar et al. 2000, Okuzawa et al. 2002, Vetillard et al. 2006), GnRH cells appear to lack ER in fish (Navas et al. 1995). A possible explanation could be the existence of interneurons able to bind estrogens and acting upstream to the GnRH neurons. Kisspeptins are proposed as candidates to this role since in sea bass some preoptic GnRH1 neurons express kiss receptors and hypothalamic kiss cells contains ERs transcripts (Escobar et al. 2013). In addition, ERs are widely expressed in the paraventricular organ, close to the opening of the lateral recess (Muriach et al. 2008a, Muriach et al. 2008b), in a region where most kiss2-expressing cells were placed (Escobar et al. 2013). Nevertheless, the number of GnRH cells expressing kiss receptors is really small, suggesting that further interneurons and factors could act in the connection of GnRH and kiss systems with steroids in this species.

In sea bass brain the ERs expression sites are not restricted to hypophysiotrophic areas. Indeed, ERs were also detected in areas containing neuronal pathways associated to motor activity, sensory perception and food intake (Muriach et al. 2008b). Recently, it has been reported that sea basses treated with 17-beta estradiol or testosterone have decreased self-feeding activity, feeding efficiency and growth rate. In contrast, no effect was observed when fish receive a dose of 11-keto androstenedione, a precursor of the main fish androgen (11-keto testosterone). The authors of this work concluded that the inhibitory effect of testosterone on food intake could be mediated by its aromatization to estradiol, whereas the growth processes are mediated by androgens per se (Leal et al. 2009).

The complexity of the sex steroid regulation of reproduction is increased by the fact that the brain of fish shows high capacity to convert aromatizable androgens into estrogens (Pasmanik and Callard 1988, Callard et al. 1990, Zohar et al. 2010). Therefore, many effects of aromatizable androgens, i.e., testosterone, can be mediated by ERs in fish. In addition, in contrast to mammals, two aromatase genes were isolated in sea bass, and other fish species, corresponding to the *cyp19a1b* gene, more specific of the brain, and the *cyp19a1a* gene, whose expression is restricted to gonads (Blazquez and Piferrer 2004). The *cyp19a1b* expression has been analyzed in the brain of juvenile sea bass. At the time of gonadal sexual differentiation, females showed higher expression levels than males. Thereafter (after 200 days after fertilization), male expression starts to overpass the female one (Blazquez and Piferrer 2004). These sex-related differences in the expression profile are in agreement with the enzymatic activity measured in adults (Gonzalez and Piferrer 2003). These observations may suggest a role for the brain aromatase in sea bass sex differentiation. However, due to the continuous growth of the teleost brain throughout life, a role of brain aromatase in neurogenesis cannot be discarded. In fact, fish aromatase activity is mainly localized in the periventricular forebrain, exclusively in radial glial cells, and not in neurons as in the case of mammals (Forlano et al. 2001, Menuet et al. 2003, Pellegrini et al. 2005, Pellegrini et al. 2007). Recently, it has been elegantly shown that these cyp19a1b expressing radial glial cells represent progenitor cells in the brain of adult fish (Pellegrini et al. 2007).

Others Factors Involved in Sea Bass Reproduction

Neuropeptide Y (NPY)

Neuropeptide Y (NPY) belongs to a 36-amino acid pancreatic peptide family which includes NPY, gut endocrine peptide YY (PYY), pancreatic polypeptide (PP) and fish pancreatic peptide Y (Cerdá-Reverter and Larhammar 2000). The name NPY reflects the fact that its sequence begins and ends with a tyrosine residue, whose abbreviation is Y. In contrast to mammals, in sea bass as well as in other bony fishes, all three NPY-family peptides are expressed in the brain (Cerdá-Reverter et al. 2000a,b). These findings add unexpected complexity to the physiological functions of this family of neuropeptides. It is an intriguing possibility that this complexity is also reflected by an increase in the number of receptor subtypes, as it occurs in zebrafish (Ringvall et al. 1997).

The distribution of NPY-secreting neurons has been identified in the brain of teleost fishes, including the European sea bass, by using immunohistochemical and *in situ* hybridization techniques. These cells are located in the olfactory placodes, the olfactory bulbs, the central portion of the dorsal telencephalon, the ventral telencephalon, the caudal preoptic area, the thalamus, the optic tectum, the entopeduncular nucleus and the locus coeruleus (Kah et al. 1993, Cerdá-Reverter et al. 2000a). These two last cell masses seem to represent the origin of NPY fibers that reach the pituitary and innervate the gonadotrophic cells.

In fish and mammals, the NPY is strongly involved in the control of feeding (Kaiyala et al. 1995, Peyon et al. 2003, Zohar et al. 2010). Moreover, this neuropeptide was also shown to participate in the regulation of gonadotrophin release and in the stimulation of growth hormone. Actually, NPY is considered the putative factor linking growth, feeding and reproduction, at least in fish, where nutritional status could be involved in the modulation of both somatotrophic and gonadotrophic axis (Peng et al. 1993a, Peng et al. 1993b, Cerdá-Reverter et al. 1999, Peyon et al. 2001). In fact, the European sea bass, as well as other fish species, reduces the food intake during gonadal development (October–December) and it reaches minimal values during the spawning period (December-February) (Zanuy and Carrillo 1985). Further evidences reported in this species showed that NPY stimulate LH secretion and this NPY-induced LH increase was modulated by the energetic status (Cerdá-Reverter et al. 1999). Indeed, the *in vivo* NPY effect on LH stimulation was much evident in fasted animals. In contrast, positive energetic status suppressed the ability of NPY to stimulate LH secretion (Cerdá-Reverter et al. 1999). *In vitro* studies reported similar evidences since pituitary cells from fasted juvenile sea bass incubated in restricted media (sea bass Ringer) were more responsive to NPY compared with fasted cells incubated in L-15 or with pituitary cells from fed animals. In this work, the authors also showed that NPY is capable of modifying glucose metabolism (Cerdá-Reverter et al. 1999).

Subsequent studies also found a stimulatory effect of NPY on LH release from dispersed pituitary cells of late prepubertal and adult male fish, supporting the role of NPY in gonadotrophin regulation at the pituitary level in sea bass (Peyon et al. 2001). Moreover, it has been proposed that NPY mediates some of the effects of leptin in the control of feeding behavior and reproductive function. In fact, porcine NPY treatment on sea bass pituitary cells alone was weakly effective on basal LH release, although it increased LH and somatolactin release induced by leptin in late prepuberty but not in early postpuberty (Peyon et al. 2001, Peyon et al. 2003). Thus, the effect of co-treatment with leptin and NPY appeared dependent on the stage of natural reproductive development (Peyon et al. 2001). Previous data in mammals suggested that sex steroids regulate the effects of NPY on gonadotrophin secretion from the anterior pituitary, in addition to modulating NPY-induced GnRH release from the hypothalamus (McDonald and Koenig 1993). In sea bass, what is known is that the action of NPY on LH release varies with the seasonal reproductive cycle. Furthermore, regions containing NPY expressing cells, like the periventricular pretectal nucleus (PPd) and paracommissural nucleus (NP) of the synencephalon and the periventricular gray zone (PGZ) of the optic tectum, exhibit ERs mRNA (Muriach et al. 2008b). Taken together, these results strongly suggest that the orexigenic neuropeptide Y (NPY) is one of the factors implicated in the hypothalamic regulation of reproduction and energy homeostasis in sea bass, as reported in other fish and vertebrates.

γ-aminobutyric acid (GABA)

GABA is considered as a major inhibitory neurotransmitter in the vertebrate central nervous system. It is largely synthesized from the precursor glutamate in a single step reaction controlled by glutamic acid decarboxylase (GAD65 and GAD67) and is degraded by the enzyme GABA transaminase (GABA-T) into succinic semialdehyde (Popesku et al. 2008). As a further evidence of the teleost genome duplication, teleost fish exhibit a novel GAD isoform, termed GAD3 (Lariviere et al. 2002).

Several reports showed that GABA is strongly involved in the LH release in goldfish brain, via the stimulation of GnRH secretion (Kah et al. 1992, Sloley et al. 1992) and also by inhibiting dopaminergic neurons in the telencephalic, preoptic and hypothalamic regions (Trudeau et al. 2000b). In addition, GABA projections originating from the preoptic region or from the lateral tuberal nucleus in the hypothalamus reach the proximity of gonadotrophic cells in the pituitary of goldfish and rainbow trout (Kah et al. 1987, Mañanos et al. 1999). In rainbow trout, GABA was also shown to stimulate LH and FSH releases both *in vivo* and *in vitro* (Mañanos et al. 1999). Further evidences support a direct action of GABA on pituitary cells in some fish species (Trudeau et al. 2000b). Thus, the administration of GABA to pituitary slices caused a dose-related increase of GnRH release at the level of the pituitary.

GABA actions of the reproductive axis appear to be dependent on factors like sex or sex steroid levels. The existence of season-related sex differences in the effects of sex steroids on GABA synthesis (Bosma et al. 2001, Lariviere et al. 2005) has also been demonstrated. Moreover, an increase in water temperature is associated with stimulation of GABA synthesis, LH release and spawning (Fraser et al. 2002). Therefore, the GABAergic system may integrate both external environmental and internal hormonal feedback signals to modulate LH release (Popesku et al. 2008). To date, GABA actions on the reproductive axis have been extensively investigated in goldfish and, to a lesser extent, in rainbow trout (Trudeau et al. 2000b, Popesku et al. 2008), but there appear to exist species differences. Indeed, while the stimulatory effects of GABA on LH were evidenced in at least four fish species, convincing data for the inhibitory effects of GABA on LH release was only observed in one fish species (Trudeau et al. 2000a). In the European sea bass, the GABA specific effects and roles on the brain and the pituitary functions are not defined yet. An isolated study performed in this species reports the description of the GABA concentrations in different cerebral areas and showed the highest levels of GABA in the telencephalic hemispheres, mean values in the diencephalon and the optic lobes and the lowest ones in the mesencephalon (Di Summa et al. 1993). Further analysis is required to determine the nature of the GABA effects on sea bass reproductive axis.

Leptin

Leptin is a protein encoded by the obese gene (ob), which in mammals is mainly produced by the adipocites, but also by other tissues including brain and gastric epithelium (Harvey and Ashford 2003). Leptin concentration in plasma is in proportion with body adiposity. It has been described as an "adipostat", a humoral signal carrying information about energy reserves and, therefore, serving important functions in the regulation of body weight and metabolism (Zhang et al. 1994, Campfield et al. 1995, Halaas et al. 1995). Indeed, the administration of leptin decreases food intake and causes weight loss in birds and mammals (Neary et al. 2004). Further evidences suggest that leptin also plays an important role in neuroendocrine signalling and reproduction (Auwerx and Staels 1998). In mammals, leptin receptors have been identified in the pituitary gland and leptin administration increases gonadotrophin and growth hormone release (Yu et al. 1997, Mizuno et al. 1999, Sone and Osamura 2001). Altogether these findings indicate that leptin could be a link between the regulation of energy balance and the control of reproduction (Magni et al. 2000).

Up to two leptin-like hormones have been identified in several fish species (Johnson et al. 2000, Yaghoubian et al. 2001, Mustonen et al. 2002, Nieminen et al. 2003). Moreover, fish leptins appear to be mainly expressed in the liver, and not in adipocytes like in mammals (Zhang et al. 1994, Montague et al. 1997). However, leptin expression has also been reported in different peripheral tissues, including intestine, kidney, ovary, muscle and adipose tissue (Murashita et al. 2008, Kurokawa and Murashita 2009). In fish, leptin actions appear to vary among species. Leptin treatments neither affect food intake nor body weight of salmon (Baker et al. 2000), catfish (Silverstein and Plisetskaya 2000) and green sunfish (Londraville and Duvall 2002), but in goldfish, both peripheral and central injections of mammalian leptin decrease food intake (Volkoff et al. 2003).

In the European sea bass, leptin seems to act directly at the pituitary level (Peyon et al. 2001, Peyon et al. 2003). Leptin treatment on dispersed pituitary cell cultures had a potent effect on LH but also on somatolactin (SL) release (Peyon et al. 2001, Peyon et al. 2003). The LH response was independent of the GnRH1 stimulation, suggesting that leptin and GnRH1 stimulations of LH secretion are controlled by different mechanisms (Peyon et al. 2001). Moreover, the magnitude of LH and SL responses to leptin administration depends on the stage of sexual development of sea bass, showing higher response at the prepubertal and first spermiating stages (Peyon et al. 2001). These results suggest a change in the sensitivity to leptin of both SL- and gonadotrophin-secreting cells, during sexual development of sea bass. Further data indicate that leptin would also act centrally to stimulate LH release, but only during late juvenile development. Therefore, leptin most likely plays a stimulatory role in late juvenile LH secretion, but does not drive the GnRH induction of LH release leading to sexual maturity (Dearth et al. 2000). Taken together, these observations suggest that leptin plays a role in the neuroendocrine regulation of pituitary hormone release in sea bass, acting at the onset of puberty.

Leptin actions are mediated by the leptin or obesity receptor, which is a member of class I cytokine receptor family (Gorska et al. 2010). To date, just one leptin receptor type has been identified in fish, although a very recent analysis of European eel genome reported the presence of two different leptin receptor genes (Lafont et al. 2012). In sea bass, one leptin receptor has been recently isolated (Gomez, Felip, Zanuy unpublished). Leptin receptor-expressing cells in sea bass brain were mostly identified at the level of the preoptic area, the habenula, the mediobasal hypothalamus and along the later recess (Escobar, Servili, Felip, Zanuy, Kah unpublished).

Interestingly, most of these regions exhibiting leptin receptor mRNAs also express kisspeptins. Notably, the region of the mediobasal hypothalamus in breeding sea bass shows kiss1-expressing cells that could contain leptin receptor mRNA, but this hypothesis needs to be confirmed by double *in situ* hybridation experiments. In this case, leptin would be the ideal candidate to link appetite, metabolic rate and energy stores with the central systems controlling reproduction in this species. In several mammalian species as rat, mouse and sheep, compelling evidences show that low leptin levels are associated with weak hypothalamic KISS1 expression, whereas the administration of leptin in animals with negative energy balance (exhibiting defective levels of endogenous leptin) increases KISS1 mRNA expression in the hypothalamus. Altogether, these observations suggest a strict connection between kisspeptin system and leptin in fish and mammals.

Serotonin and melatonin

Serotonin (5-hydroxy-tryptamine or 5-HT) is a neurotransmitter produced in the brain by neurons that store it in vesicles on the presynaptic membrane (Halford and Blundell 2000). Peripheral serotonin is synthesized by enterochromaffin cells in gut mucous (Cooke 1986) in response to stimulus like gastric distension, hyperosmotic solutions or presence of glucose (Hubel 1985). In fish as in mammals, serotonin is shown to be involved in several functions such as feeding behavior and body weight (Khan and Thomas 1993). Moreover, this neurotransmitter has been implicated in the neuroendocrine modulation of the reproductive process by acting across the brain-pituitary-gonad axis in fish. In fact, serotonin stimulates the release of gonadotrophins from the pituitary gland of goldfish (Somoza et al. 1988) and Atlantic croaker (Khan and Thomas 1992). Furthermore, serotonin is able to stimulate the GnRH release from the anterior hypothalamus in goldfish (Yu et al. 1991) and in seabream (Senthilkumaran et al. 2001).

The neuroanatomical localization of serotonin-immunoreactive cells has been investigated in several fish species including the European sea bass (Batten et al. 1993). The main 5HT-immunoreactive cells population was located at the level of the raphe nuclei. This serotoninergic system is conserved in vertebrates. The projections from the raphe nuclei densely innerve forebrain areas, like the preoptic region and the inferior lobe (Batten et al. 1993). Very recent data suggest a role for the kiss1 neurons placed in the habenular region on the action of the serotoninergic population in the raphe of zebrafish (Ogawa et al. 2012a). In fact, kiss1 administration increased the expression of genes (*pet1* and *slc6a4a*) responsible for 5-HT release, but there was no effect on the mRNA levels of *tph2*, which is responsible for 5-HT synthesis (Ogawa et al. 2012a). The kiss1 action would be indirect via interneurons since raphe nuclei do not exhibit kiss1r-expressing cells.

No information is available concerning the possible existence of the kiss1 modulation of serotonin system in the raphe of sea bass. Nevertheless, both the habenular kiss1 and raphe serotoninergic systems are conserved in both species.

Further 5-HT-ir cells were observed in the dorsolateral nucleus and in the pretectal area at the boundary of the thalamus with the mesencephalic tegmentum, as well as in the pituitary gland, mainly restricted to the *proximal pars distalis*. In the paraventricular organ of sea bass a conspicuous 5-HT-ir population was also observed (Batten et al. 1993). In this region is located the fish specific population of serotoninergic or dopaminergic cerebrospinal fluid-contacting neurons (Kah et al. 1978, Geffard et al. 1982, Kah and Chambolle 1983). Although the function of this prominent seronotinergic cell mass is not defined yet, very recent observations suggested a role in neurogenesis, at least in zebrafish (Pérez et al. 2012).

Serotonin is a precursor in the chemical pathway producing melatonin, the main pineal hormone in fish and other vertebrates (Falcon et al. 2010). In fact, melatonin is synthesized from tryptophan taken up by the pineal cells. Two enzymatic steps (by the action of tryptophan hydroxylase and aromatic amino acid decarboxylase) convert tryptophan in serotonin. Two other enzymatic reactions, involving arylalkylamine N-acetyltransferase and hydroxyindole-O-methyltransferase, transform serotonin to melatonin (Falcon et al. 2007, Falcon et al. 2010). Melatonin (N-acetyl-5-methoxytryptamine) is an indolamine mainly synthesized by the pineal organ that is implicated in the synchronization of several rhythmic physiological processes to the environmental cues (Zachmann et al. 1992, Cassone 1998, Malpaux et al. 2001, Falcon et al. 2010). Both serotonin and melatonin present daily variation but while serotonin levels are high during the day and low at night, melatonin levels exhibit a reverse pattern, i.e., elevated levels at night and reduced levels during the day (Bromage et al. 1990, Falcon 1999). In non-mammalian vertebrates, melatonin acts both as a clock, since the duration of the plasma melatonin rise corresponds to the length of the night, and as a calendar, because the seasonal changes in the length of the days and temperature modulates the duration and amplitude of the nocturnal melatonin increase, respectively (Molina-Borja et al. 1994, Sánchez-Vázquez et al. 1997, Roy et al. 2001).

Many evidences show that melatonin exerts a role in fish reproduction, acting at all the levels of the reproductive axis. Data collected in fish show that melatonin effects on the reproductive axis can be stimulatory (Carrillo et al. 1995, Khan and Thomas 1996, Amano et al. 2000) and inhibitory (Amano et al. 2004, Ghosh and Nath 2005) depending on the dose of the melatonin treatment, age and reproductive state of animals. Melatonin has been shown to activate the brain dopaminergic system in eel (Sebert et al. 2008). In the Atlantic croaker, melatonin administration causes an increase in LH levels by stimulating the preoptic area and also by acting on the pituitary gland (Khan and Thomas 1996). Moreover, pituitary has been identified as a melatonin target tissue because of the presence of melatonin receptor mRNAs in different species (Gaildrat and Falcon 1999, 2000, Falcon et al. 2003, Sauzet et al. 2008, Confente et al. 2010). Melatonin receptors or melatonin binding sites have also been revealed in fish gonads (Molina-Borja et al. 1994, Chattoraj et al. 2009).

Also, in the European sea bass, many evidences indicate that melatonin can act across the reproductive axis. In fact, daily and seasonal variations in melatonin release (Iigo and Aida 1995, Sánchez-Vázquez et al. 1997, Garcia-Allegue et al. 2001), as well as in melatonin binding sites (Bayarri et al. 2004a, Bayarri et al. 2010a) and in reproductive hormone levels (Prat et al. 1990, Mañanos et al. 1997, Rodriguez et al. 2000, Moles et al. 2007) have been reported. In addition, melatonin receptors have been detected in all the components of the sea bass reproductive axis: in the neuroendocrine brain regions, in the pituitary gland and in gonads (Sauzet et al. 2008, Herrera-Perez et al. 2010). Moreover, it has been observed that changes in photoperiod conditions interfere in sea bass reproductive performance and the onset of puberty (Bayarri et al. 2004b, Bayarri et al. 2009, Bayarri et al. 2010b).

As stated above, our recent findings revealed the existence of day-night variations in the brain expression of two GnRH isoforms (GnRH1 and GnRH3) as well as of two GnRH receptors (dlGnRHR-II-1c and dlGnRHR-II-2a), which exhibit higher mRNA levels at midday in relation to dusk and mid-dark periods (Servili et al. 2013). We have also noted an inhibitory effect of the melatonin injection on the nocturnal brain expression of GnRH1, GnRH3, and the GnRH receptor subtypes dlGnRHR-II-1c, dlGnRHR-II-2a and dlGnRHR-II-2b in sea bass (Servili et al. 2013). Therefore, we have revealed that melatonin could mediate a central effect on the reproductive axis of sea bass by acting on the GnRH systems that control the gonadotrophin release. The presence of melatonin receptors described in neuroendocrine brain areas of the sea bass brain (Herrera-Perez et al. 2010) reinforces this evidence. Nevertheless, a direct effect of the pineal hormone on GnRH cells is unlikely because GnRH and melatonin receptor-expressing cells do not appear to be co-localized in the same brain areas. With the exception of dlGnRHR-II-2b, the inhibitory effect of melatonin was evident on GnRH forms and GnRH receptors that exhibited significant day-night fluctuations in their expression, suggesting that exogenous melatonin could be reinforcing physiological mechanisms already established. Thus, it appears that melatonin injection is mimicking and enhancing physiological inhibitory actions of the endogenous melatonin on particular *GnRH* and *GnRH receptor* genes. In the case of the receptor dlGnRHR-II-2b, the exogenous melatonin administration inhibits its brain expression levels at midnight, even thought this receptor subtype does not show apparent physiological day-night variations. As it was illustrated above, GnRH2 is the ligand showing the highest preference for dlGnRHR-II-2b receptor (Kah et al. 2007, Servili et al. 2010) and it does not exhibit significant variations along the day-night cycle. Nevertheless, this receptor is the most expressed in the pineal gland, where GnRH2 acts as a melatonin-releasing factor (Servili et al. 2010). Thus, this inhibitory effect of melatonin on dlGnRHR-II-2b expression could be a part of a regulatory feedback mechanism.

The results obtained in our study of sea bass suggest that melatonin, through its effects on hypophysiotrophic GnRH systems, could act as a synchronizer of the hormonal rhythms at precise times of the day and, presumably, of the year. Therefore, these interactions between melatoninergic and GnRH systems could represent a substrate of photoperiod effects on reproductive and other rhythmic physiological events in the European sea bass.

Conclusion

The European sea bass is a perciform fish highly relevant for European aquaculture. It is a seasonal breeder showing a highly photoperiodic annual reproductive cycle. The neuroendocrine mechanisms responsible for driving the reproductive events behave across the pineal-brain-pituitary-gonad axis and consist of several neurohormones and neurotransmitters interacting with each other at several levels of this axis.

The extreme complexity of this central neuroendocrine control of reproduction is further increased by the fact that sea bass, as a result of the teleost genome third round duplication event, exhibits multiple isoforms of the homologous mammalian genes that play in the regulation of reproduction. After duplication, paralog genes may evolve in different ways. One of them can lose the ability to express any RNA or protein, becoming a pseudogene. Another possibility is that the two paralogs act in a complementary way to exert the original function of the ancestral gene. Alternatively, one gene can evolve and develop a completely new role.

For instance, the European sea bass exhibits three *GnRH* variants and five different *GnRH receptor* genes. The critical role of GnRH1 in stimulating the synthesis and release of gonadotrophin is very well defined in all classes of vertebrates. However, the specific functions of GnRH2 and GnRH3 forms are still not completely understood. Many evidences show these forms would act as neuromodulators of several neural systems in fish. Thus, GnRH3 is likely involved in the modulation of sensory, odorant and motivational systems while it is suggested that GnRH2 controls sex behavior and sensory-motor activity. Our results, obtained in sea bass studies, have demonstrated that GnRH1 and GnRH3 represent the hypophysiotrophic forms controlling gonadotrophin secretion in this species. Moreover, we have shown that GnRH2 and GnRH3 could be involved in the modulation of the activity of the two photoreceptive pacemaker structures, i.e., the pineal organ and the retina, respectively.

Also the important role of kisspeptins in stimulating gonadotrophin release has been stated in sea bass. Nevertheless, the precise action mechanisms of kisspeptin systems and whether they would act directly or indirectly on preoptic GnRH1 system remain to be clarified. The hypothesis of the mediation of interneurons that would transfer kiss inputs to GnRH neurons is plausible in sea bass, as in fish in general, and nNOS cells could be possible candidates for this role.

We have also discussed the possibility of an involvement of kiss genes in diverse functions other than the stimulation of pituitary secretion. They could be responsible for connecting information about the energy status of the fish with the reproductive axis, since putative co-expression between kiss system and NPY- or leptin receptor-expressing cells was observed in sea bass. In fact, gonadal maturation and nutritional status are strictly linked in sea bass. It is known that sea bass decreases the amount of food intake during the breeding period. Moreover, the interaction of kisspeptins with the somatotrophic axis has also been suggested, via the involvement of somatolactin cells that express kiss receptor.

Due to the localization of kiss1 neuron and kiss1r-expressing cells into the habenular nuclei, an interaction of kisspeptin and pineal melatoninergic systems is conceivable. Photoperiod condition is certainly playing a role in the modulation of the reproductive axis in sea bass. However, the effect of melatonin on sea bass kiss systems has not been elucidated yet. On the other hand, we have observed the existence of a particular relationship between melatonin and GnRH systems in sea bass (Fig. 5). We have demonstrated a role of GnRH2 system on the stimulation of melatonin release. In addition, we have reported the effects of melatonin on the reproductive axis at the central level, by inhibiting the expression of GnRH1 and GnRH3 (Fig. 5). Taking into account that GnRH-1 and GnRH-3 represent the hypophysiotrophic GnRH forms in sea bass our results suggest that these melatonin effects at the brain level could also have a functional correlate in the pituitary gland and in gonadotrophin secretion. Anyway, whether changes in GnRH mRNA levels are correlated with changes in biologically active peptides remains to be revealed. In the light of the discovery of kisspeptin system in fish and its putative role on GnRH and reproduction, it would be interesting to investigate whether kisspeptins are also functionally related with melatonin signal and seasonal processes in sea bass.

To our knowledge, in the European sea bass there are no clear evidences of an inhibitory action on the reproductive axis. So far, no dopamine

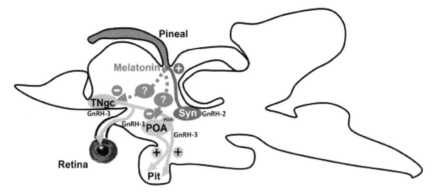

Figure 5. Schematic representation of the interactions between GnRH systems and photoreceptor organs in the European sea bass. Preoptic (POA) GnRH1 cells (light gray circle) massively project to the pituitary gland. GnRH2 cells (black oval) located in the synencephalon (Syn) and GnRH3 neurons (dark gray diamond) of the terminal nerve (TN) are connected to the pineal gland and retina, respectively. Moreover, GnRH2 stimulates (+) the pineal melatonin secretion, which in turns, inhibits (–) brain expression levels of GnRH1 and GnRH3 forms (gray dotted arrows). Cc, corpus of the cerebellum; MO, medulla oblongata; OB, olfactory bulb; Pit, pituitary gland; SC, spinal cord; Tel, telencephalon.

Color image of this figure appears in the color plate section at the end of the book.

inhibition has been reported in this species. Possibly, the main inhibition of reproductive processes is exerted by the GnIH system. Currently, several laboratories are addressing investigations to characterize the GnIH system in fish, including the sea bass (in Dr. Muñoz-Cueto laboratory), in order to elucidate its effect at the central level (on GnRH cells), as well as at the pituitary level (on gonadotrophins, somatolactin and also prolactin release). Also, the possible interaction of GnIH and kisspeptin is worthy of investigation.

Once the mechanisms controlling the pineal-brain-pituitary-gonad axis are completely understood, it will be much easier to define the feedback actions of sex steroids at all the levels and to manage the reproductive process. A full comprehension of the reproductive events as well as the mechanism underlying the onset of puberty would be interesting in light of the wide range of the potential applications in sea bass farming. Notably, solving the problem of precocious puberty of the sea bass male under farming conditions would remarkably increase the total fish production; sea bass research, therefore, must focus its efforts in the near future on these aspects.

Acknowledgements

We thank the staff from the "Planta de Cultivos Marinos" (University of Cádiz) for their support in maintaining the experimental animals used in these studies. We thank Dr. Mairi Cowan for the critical reading of the manuscript. Grant sponsors: Regional Government of Andalusia (Junta de Andalucía, Excellence Project no. P10-AGR-05916); Spanish Ministry of Science and Innovation (Grant no. AGL2001-0593-C03-02).

References

Amano, M., M. Iigo, K. Ikuta, S. Kitamura, H. Yamada and K. Yamamori. 2000. Roles of melatonin in gonadal maturation of underyearling precocious male masu salmon. Gen. Comp. Endocrinol. 120: 190–7.

Amano, M., M. Iigo, K. Ikuta, S. Kitamura, K. Okuzawa, H. Yamada and K. Yamamori. 2004. Disturbance of plasma melatonin profile by high dose melatonin administration inhibits testicular maturation of precocious male masu salmon. Zoolog. Sci. 21: 79–85.

Amano, M., S. Moriyama, M. Iigo, S. Kitamura, N. Amiya, K. Yamamori, K. Ukena and K. Tsutsui. 2006. Novel fish hypothalamic neuropeptides stimulate the release of gonadotrophins and growth hormone from the pituitary of sockeye salmon. J. Endocrinol. 188: 417–23.

Amo, R., H. Aizawa, M. Takahoko, M. Kobayashi, R. Takahashi, T. Aoki and H. Okamoto. 2010. Identification of the zebrafish ventral habenula as a homolog of the mammalian Lateral Habenula. J. Neurosci. 30: 1566–74.

Aroua, S., F.A. Weltzien, N. Le Belle and S. Dufour. 2007. Development of Real-Time Rt-Pcr Assays for Eel Gonadotropins and their Application to the Comparison of *in vivo* and *in vitro* Effects of Sex Steroids. Gen. Comp. Endocrinol. 153: 333–43.

Auwerx, J. and B. Staels. 1998. Leptin. Lancet 351: 737–42.

Baker, D.M., D.A. Larsen, P. Swanson and W.W. Dickhoff. 2000. Long-term peripheral treatment of immature coho salmon (Oncorhynchus Kisutch) with human leptin has no clear physiologic effect. Gen. Comp. Endocrinol. 118: 134–8.

Barnabé, G. and R. Billard. 1984. L'aquaculture du bar et des sparidés. INRA Publication. Isbn: 2-85340-600-8 – 1984. Paris, France.

Batten, T.F., P.A. Berry, A. Maqbool, L. Moons and F. Vandesande. 1993. Immunolocalization of Catecholamine Enzymes, Serotonin, Dopamine and L-Dopa in the Brain of *Dicentrarchus Labrax* (Teleostei). Brain Res. Bull. 31: 233–52.

Bayarri, M.J., R. Garcia-Allegue, J.A. Muñoz-Cueto, J.A. Madrid, M. Tabata, F.J. Sanchez-Vazquez and M. Iigo. 2004a. Melatonin Binding Sites in the Brain of European Sea Bass (*Dicentrarchus Labrax*). Zoolog. Sci. 21: 427–34.

Bayarri, M.J., L. Rodriguez, S. Zanuy, J.A. Madrid, F.J. Sanchez-Vazquez, H. Kagawa, K. Okuzawa and M. Carrillo. 2004b. Effect of Photoperiod Manipulation on the Daily Rhythms of Melatonin and Reproductive Hormones in Caged European Sea Bass (*Dicentrarchus Labrax*). Gen. Comp. Endocrinol. 136: 72–81.

Bayarri, M.J., S. Zanuy, O. Yilmaz and M. Carrillo. 2009. Effects of Continuous Light on the Reproductive System of European Sea Bass Gauged by Alterations of Circadian Variations During Their First Reproductive Cycle. Chronobiol. Int. 26: 184–99.

Bayarri, M.J., J. Falcon, S. Zanuy and M. Carrillo. 2010a. Continuous Light and Melatonin: Daily and Seasonal Variations of Brain Binding Sites and Plasma Concentration During the First Reproductive Cycle of Sea Bass. Gen. Comp. Endocrinol. 169: 58–64.

Bayarri, S., M.J. Gracia, R. Lazaro, C. Pe Rez-Arquillue, M. Barberan and A. Herrera. 2010b. Determination of the Viability of Toxoplasma Gondii in Cured Ham Using Bioassay: Influence of Technological Processing and Food Safety Implications. J. Food Prot. 73: 2239–43.

Beato, M., P. Herrlich and G. Schutz. 1995. Steroid Hormone Receptors: Many Actors in Search of a Plot. Cell 83: 851–7.

Bentley, G.E., L.J. Kriegsfeld, T. Osugi, K. Ukena, S. O'brien, N. Perfito, I.T. Moore, K. Tsutsui and J.C. Wingfield. 2006. Interactions of Gonadotropin-Releasing Hormone (Gnrh) and

Gonadotropin-Inhibitory Hormone (Gnih) in Birds and Mammals. J. Exp. Zool. A. Comp. Exp. Biol. 305: 807–14.

Bentley, G.E., T. Ubuka, N.L. Mcguire, V.S. Chowdhury, Y. Morita, T. Yano, I. Hasunuma, M. Binns, J.C. Wingfield and K. Tsutsui. 2008. Gonadotropin-Inhibitory Hormone and Its Receptor in the Avian Reproductive System. Gen. Comp. Endocrinol. 156: 34–43.

Bianco, I.H. and S.W. Wilson. 2009. The Habenular Nuclei: A Conserved Asymmetric Relay Station in the Vertebrate Brain. Philos. Trans. R. Soc. Lond. B. Biol. Sci. 364: 1005–20.

Biran, J., S. Ben-Dor and B. Levavi-Sivan. 2008. Molecular Identification and Functional Characterization of the Kisspeptin/Kisspeptin Receptor System in Lower Vertebrates. Biol. Reprod. 79: 776–86.

Biran, J., O. Palevitch, S. Ben-Dor and B. Levavi-Sivan. 2012. Neurokinin Bs and Neurokinin B Receptors in Zebrafish-Potential Role in Controlling Fish Reproduction. Proc. Natl. Acad. Sci. USA 109: 10269–74.

Blazquez, M. and F. Piferrer. 2004. Cloning, Sequence Analysis, Tissue Distribution, and Sex-Specific Expression of the Neural Form of P450 Aromatase in Juvenile Sea Bass (*Dicentrarchus Labrax*). Mol. Cell Endocrinol. 219: 83–94.

Blazquez, M. and F. Piferrer. 2005. Sea Bass (*Dicentrarchus Labrax*) Androgen Receptor: Cdna Cloning, Tissue-Specific Expression, and Mrna Levels During Early Development and Sex Differentiation. Mol. Cell Endocrinol. 237: 37–48.

Blazquez, M., P.T. Bosma, E.J. Fraser, K.J. Van Look and V.L. Trudeau. 1998. Fish as Models for the Neuroendocrine Regulation of Reproduction and Growth. Comp. Biochem. Physiol. C Pharmacol. Toxicol. Endocrinol. 119: 345–64.

Bosma, P.T., M. Blazquez, E.J. Fraser, R.W. Schulz, K. Docherty and V.L. Trudeau. 2001. Sex Steroid Regulation of Glutamate Decarboxylase Mrna Expression in Goldfish Brain Is Sexually Dimorphic. J. Neurochem. 76: 945–56.

Breton, B., B. Jalabert, R. Billard and C. Weil. 1971. *In vitro* Stimulation of the Release of Pituitary Gonadotropic Hormone by a Hypothalamic Factor in the Carp *Cyprinus Carpio* L. C. R. Acad. Sci. Hebd. Seances Acad. Sci. D. 273: 2591–4.

Breton, B., C. Weil, B. Jalabert and R. Billard. 1972. Reciprocal Activity of Hypothalamic Factors of Rams (Ovis Aries) and of Teleostean Fishes on the Secretion *in vitro* of Gonadotropic Hormones C-Hg and Lh, Respectively, by Carp and Ram Hypophyses. C. R. Acad. Sci. Hebd. Seances Acad. Sci. D. 274: 2530–3.

Breton, B., E. Sambroni, M. Govoroun and C. Weil. 1997. Effects of Steroids on Gth I and Gth Ii Secretion and Pituitary Concentration in the Immature Rainbow Trout Oncorhynchus Mykiss. C. R. Acad. Sci. III. 320: 783–9.

Bromage, N., J. Duston, C. Randall, A. Brook, M. Thrush, M. Carrillo and S. Zanuy. 1990. Photoperiodic Control of Teleost Reproduction. Prog. Clin. Biol. Res. 342: 620–6.

Callard, G., B. Schlinger, M. Pasmanik and K. Corina. 1990. Aromatization and Estrogen Action in Brain. Prog. Clin. Biol. Res. 342: 105–11.

Campfield, L.A., F.J. Smith, Y. Guisez, R. Devos and P. Burn. 1995. Recombinant Mouse Ob Protein: Evidence for a Peripheral Signal Linking Adiposity and Central Neural Networks. Science 269: 546–9.

Cardinaud, B., J.M. Gilbert, F. Liu, K.S. Sugamori, J.D. Vincent, H.B. Niznik and P. Vernier. 1998. Evolution and Origin of the Diversity of Dopamine Receptors in Vertebrates. Adv. Pharmacol. 42: 936–40.

Carrillo, M., S. Zanuy, F. Prat, J. Cerdá, J. Ramos, E. Mañanos and N.R. Bromage. 1995. Sea bass. pp. 138–168. *In*: Broodstock Management and Egg and Larval Quality. Blackwell, Oxford, UK.

Carrillo, M., S. Zanuy, A. Felip, M.J. Bayarri, G. Moles and A. Gomez. 2009. Hormonal and Environmental Control of Puberty in Perciform Fish: The Case of Sea Bass. Ann. N. Y. Acad. Sci. 1163: 49–59.

Cassone, V.M. 1998. Melatonin's Role in Vertebrate Circadian Rhythms. Chronobiol. Int. 15: 457–73.

Cerdá-Reverter, J.M. and D. Larhammar. 2000. Neuropeptide Y Family of Peptides: Structure, Anatomical Expression, Function, and Molecular Evolution. Biochem. Cell Biol. 78: 371–92.

Cerdá-Reverter, J.M., L.A. Sorbera, M. Carrillo and S. Zanuy. 1999. Energetic Dependence of Npy-Induced Lh Secretion in a Teleost Fish (*Dicentrarchus Labrax*). Am. J. Physiol. 277: R1627–34.

Cerdá-Reverter, J.M., I. Anglade, G. Martinez-Rodriguez, D. Mazurais, J.A. Muñoz-Cueto, M. Carrillo, O. Kah and S. Zanuy. 2000a. Characterization of Neuropeptide Y Expression in the Brain of a Perciform Fish, the Sea Bass (*Dicentrarchus Labrax*). J. Chem. Neuroanat. 19: 197–210.

Cerdá-Reverter, J.M., G. Martinez-Rodriguez, I. Anglade, O. Kah and S. Zanuy. 2000b. Peptide Yy (Pyy) and Fish Pancreatic Peptide Y (Py) Expression in the Brain of the Sea Bass (*Dicentrarchus Labrax*) as Revealed by *in situ* Hybridization. J. Comp. Neurol. 426: 197–208.

Cerdá-Reverter, J.M., S. Zanuy and J.A. Muñoz-Cueto. 2001a. Cytoarchitectonic Study of the Brain of a Perciform Species, the Sea Bass (*Dicentrarchus Labrax*). I. The Telencephalon. J. Morphol. 247: 217–28.

Cerdá-Reverter, J.M., S. Zanuy and J.A. Muñoz-Cueto. 2001b. Cytoarchitectonic Study of the Brain of a Perciform Species, the Sea Bass (*Dicentrarchus Labrax*). Ii. The Diencephalon. J. Morphol. 247: 229–51.

Cerdá-Reverter, J.M., B. Muriach, S. Zanuy and J.A. Muñoz-Cueto. 2008. A Cytoarchitectonic Study of the Brain of a Perciform Species, the Sea Bass (*Dicentrarchus Labrax*): The Midbrain and Hindbrain. Acta Histochem. 110: 433–50.

Chang, J.P. and R.E. Peter. 1983. Effects of Dopamine on Gonadotropin Release in Female Goldfish, Carassius Auratus. Neuroendocrinology 36: 351–7.

Chartrel, N., C. Dujardin, J. Leprince, L. Desrues, M.C. Tonon, E. Cellier, P. Cosette, T. Jouenne, G. Simonnet and H. Vaudry. 2002. Isolation, Characterization, and Distribution of a Novel Neuropeptide, Rana Rfamide (R-Rfa), in the Brain of the European Green Frog Rana Esculenta. J. Comp. Neurol. 448: 111–27.

Chattoraj, A., M. Seth and S.K. Maitra. 2009. Localization and Dynamics of Mel(1a) Melatonin Receptor in the Ovary of Carp Catla Catla in Relation to Serum Melatonin Levels. Comp. Biochem. Physiol. A. Mol. Integr. Physiol. 152: 327–33.

Chistiakov, D.A., B. Hellemans and F.A. Volckaert. 2007. Review on the Immunology of European Sea Bass Dicentrarchus Labrax. Vet. Immunol. Immunopathol. 117: 1–16.

Clements, M.K., T.P. Mcdonald, R. Wang, G. Xie, B.F. O'dowd, S.R. George, C.P. Austin and Q. Liu. 2001. Fmrfamide-Related Neuropeptides Are Agonists of the Orphan G-Protein-Coupled Receptor Gpr54. Biochem. Biophys. Res. Commun. 284: 1189–93.

Colledge, W.H. 2008. Gpr54 and Kisspeptins. Results Probl. Cell Differ. 46: 117–43.

Confente, F., M.C. Rendón, L. Besseau, J. Falcón and J.A. Muñoz-Cueto. 2010. Melatonin Receptors in a Pleuronectiform Species, *Solea senegalensis*: Cloning, Tissue Expression, Day-Night and Seasonal Variations. Gen. Comp. Endocrinol. 167: 202–14.

Cooke, H.J. 1986. Neurobiology of the Intestinal Mucosa. Gastroenterology 90: 1057–81.

Crim, L.W. and D.M. Evans. 1983. Influence of Testosterone and/or Luteinizing Hormone Releasing Hormone Analogue on Precocious Sexual Development in the Juvenile Rainbow Trout. Biol. Reprod. 29: 137–42.

De Roux, N., E. Genin, J.C. Carel, F. Matsuda, J.L. Chaussain and E. Milgrom. 2003. Hypogonadotropic Hypogonadism Due to Loss of Function of the Kiss1-Derived Peptide Receptor Gpr54. Proc. Natl. Acad. Sci. USA 100: 10972–6.

Dearth, R.K., J.K. Hiney and W.L. Dees. 2000. Leptin Acts Centrally to Induce the Prepubertal Secretion of Luteinizing Hormone in the Female Rat. Peptides 21: 387–92.

Di Summa, A., F. Abbate, G.P. Germana, F. Naccari, L. Dhaskali and G.F. Passantino. 1993. Gaba Concentrations in Different Cerebral Areas of Bass (*Dicentrarchus Labrax*). Boll. Soc. Ital. Biol. Sper. 69: 711–6.

Douglas, R.H., H.J. Wagner, M. Zaunreiter, U.D. Behrens and M.B. Djamgoz. 1992. The Effect of Dopamine Depletion on Light-Evoked and Circadian Retinomotor Movements in the Teleost Retina. Vis. Neurosci. 9: 335–43.

Dufour, S., F.A. Weltzien, M.E. Sebert, N. Le Belle, B. Vidal, P. Vernier and C. Pasqualini. 2005. Dopaminergic Inhibition of Reproduction in Teleost Fishes: Ecophysiological and Evolutionary Implications. Ann. N. Y. Acad. Sci. 1040: 9–21.

Dufour, S., M.E. Sebert, F.A. Weltzien, K. Rousseau and C. Pasqualini. 2010. Neuroendocrine Control by Dopamine of Teleost Reproduction. J. Fish Biol. 76: 129–60.

Dungan, H.M., D.K. Clifton and R.A. Steiner. 2006. Minireview: Kisspeptin Neurons as Central Processors in the Regulation of Gonadotropin-Releasing Hormone Secretion. Endocrinology 147: 1154–8.

Eisthen, H.L., R.J. Delay, C.R. Wirsig-Wiechmann and V.E. Dionne. 2000. Neuromodulatory Effects of Gonadotropin Releasing Hormone on Olfactory Receptor Neurons. J. Neurosci. 20: 3947–55.

Ekstrom, P. and H. Meissl. 2003. Evolution of Photosensory Pineal Organs in New Light: The Fate of Neuroendocrine Photoreceptors. Philos. Trans. R. Soc. Lond. B. Biol. Sci. 358: 1679–700.

Escobar, S., F. Espigares Puerto, A. Felip, S. Zanuy, M.M. Gueguen, M. Carrillo, O. Kah and A. Servili. 2012. Kisspeptin/GPR54 expression in the brain of the European sea bass. Abstract book of the 7th International Symposium of Fish Endocrinology. Buenos Aires, Argentina. pp. 152.

Escobar, S., A. Felip, M.M. Gueguen, S. Zanuy, M. Carrillo, O. Kah and A. Servili. 2013. Expression of Kisspeptins in the Brain and Pituitary of the European Sea Bass (*Dicentrarchus Labrax*). J. Comp. Neurol. 2013 Mar 1; 521(4): 933–48.

Falcon, J. 1999. Cellular Circadian Clocks in the Pineal. Prog. Neurobiol. 58: 121–62.

Falcon, J., L. Besseau, D. Fazzari, J. Attia, P. Gaildrat, M. Beauchaud and G. Boeuf. 2003. Melatonin Modulates Secretion of Growth Hormone and Prolactin by Trout Pituitary Glands and Cells in Culture. Endocrinology 144: 4648–58.

Falcon, J., L. Besseau, S. Sauzet and G. Boeuf. 2007. Melatonin Effects on the Hypothalamo-Pituitary Axis in Fish. Trends Endocrinol. Metab. 18: 81–8.

Falcon, J., L. Besseau, M. Fuentes, S. Sauzet, E. Magnanou and G. Boeuf. 2009. Structural and Functional Evolution of the Pineal Melatonin System in Vertebrates. Ann. N. Y. Acad. Sci. 1163: 101–11.

Falcon, J., H. Migaud, J.A. Muñoz-Cueto and M. Carrillo. 2010. Current Knowledge on the Melatonin System in Teleost Fish. Gen. Comp. Endocrinol. 165: 469–82.

FAO. © 2009–2012. Cultured Aquatic Species Information Programme *Dicentrarchus Labrax*. Cultured Aquatic Species Fact Sheets. *In*: Département des pêches et de l'aquaculture de la FAO [http://www.fao.org/fishery/culturedspecies/Dicentrarchus_labrax/fr]. Rome.

Felip, A., S. Zanuy, R. Pineda, L. Pinilla, M. Carrillo, M. Tena-Sempere and A. Gomez. 2009. Evidence for Two Distinct Kiss Genes in Non-Placental Vertebrates That Encode Kisspeptins with Different Gonadotropin-Releasing Activities in Fish and Mammals. Mol. Cell Endocrinol. 312: 61–71.

Fernald, R.D. and R.B. White. 1999. Gonadotropin-Releasing Hormone Genes: Phylogeny, Structure, and Functions. Front Neuroendocrinol. 20: 224–40.

Forlano, P.M., D.L. Deitcher, D.A. Myers and A.H. Bass. 2001. Anatomical Distribution and Cellular Basis for High Levels of Aromatase Activity in the Brain of Teleost Fish: Aromatase Enzyme and Mrna Expression Identify Glia as Source. J. Neurosci. 21: 8943–55.

Fraser, E.J., P.T. Bosma, V.L. Trudeau and K. Docherty. 2002. The Effect of Water Temperature on the Gabaergic and Reproductive Systems in Female and Male Goldfish (Carassius Auratus). Gen. Comp. Endocrinol. 125: 163–75.

Fujita, I., P.W. Sorensen, N.E. Stacey and T.J. Hara. 1991. The Olfactory System, Not the Terminal Nerve, Functions as the Primary Chemosensory Pathway Mediating Responses to Sex Pheromones in Male Goldfish. Brain Behav. Evol. 38: 313–21.

Fukusumi, S., Y. Habata, H. Yoshida, N. Iijima, Y. Kawamata, M. Hosoya, R. Fujii, S. Hinuma, C. Kitada, Y. Shintani, M. Suenaga, H. Onda, O. Nishimura, M. Tanaka, Y. Ibata and M. Fujino. 2001. Characteristics and Distribution of Endogenous Rfamide-Related Peptide-1. Biochim. Biophys. Acta 1540: 221–32.

Gaildrat, P. and J. Falcon. 1999. Expression of Melatonin Receptors and 2-[125i]Iodomelatonin Binding Sites in the Pituitary of a Teleost Fish. Adv. Exp. Med. Biol. 460: 61–72.

Gaildrat, P. and J. Falcon. 2000. Melatonin Receptors in the Pituitary of a Teleost Fish: Mrna Expression, 2-[(125)I]Iodomelatonin Binding and Cyclic Amp Response. Neuroendocrinology 72: 57–66.

Garcia-Allegue, R., J.A. Madrid and F.J. Sanchez-Vazquez. 2001. Melatonin Rhythms in European Sea Bass Plasma and Eye: Influence of Seasonal Photoperiod and Water Temperature. J. Pineal. Res. 31: 68–75.

Geffard, M., O. Kah, P. Chambolle, M. Le Moal and M. Delaage. 1982. 1st Immunocytochemical Application of an Anti-Dopamine Antibody in the Study of the Central Nervous System. C. R. Seances Acad. Sci. III. 295: 797–802.

Ghosh, J. and P. Nath. 2005. Seasonal Effects of Melatonin on Ovary and Plasma Gonadotropin and Vitellogenin Levels in Intact and Pinealectomized Catfish, Clarias Batrachus (Linn). Indian J. Exp. Biol. 43: 224–32.

Gonzalez-Martinez, D., T. Madigou, N. Zmora, I. Anglade, S. Zanuy, Y. Zohar, A. Elizur, J.A. Muñoz-Cueto and O. Kah. 2001. Differential Expression of Three Different Prepro-Gnrh (Gonadotrophin-Releasing Hormone) Messengers in the Brain of the European Sea Bass (*Dicentrarchus Labrax*). J. Comp. Neurol. 429: 144–55.

Gonzalez-Martinez, D., N. Zmora, E. Mañanos, D. Saligaut, S. Zanuy, Y. Zohar, A. Elizur, O. Kah and J.A. Muñoz-Cueto. 2002a. Immunohistochemical Localization of Three Different Prepro-Gnrhs in the Brain and Pituitary of the European Sea Bass (*Dicentrarchus Labrax*) Using Antibodies to the Corresponding Gnrh-Associated Peptides. J. Comp. Neurol. 446: 95–113.

Gonzalez-Martinez, D., N. Zmora, S. Zanuy, C. Sarasquete, A. Elizur, O. Kah and J.A. Muñoz-Cueto. 2002b. Developmental Expression of Three Different Prepro-Gnrh (Gonadotrophin-Releasing Hormone) Messengers in the Brain of the European Sea Bass (*Dicentrarchus Labrax*). J. Chem. Neuroanat. 23: 255–67.

Gonzalez-Martinez, D., T. Madigou, E. Mañanos, J.M. Cerda-Reverter, S. Zanuy, O. Kah and J.A. Muñoz-Cueto. 2004a. Cloning and Expression of Gonadotropin-Releasing Hormone Receptor in the Brain and Pituitary of the European Sea Bass: An *in situ* Hybridization Study. Biol. Reprod. 70: 1380–91.

Gonzalez-Martinez, D., N. Zmora, D. Saligaut, S. Zanuy, A. Elizur, O. Kah and J.A. Muñoz-Cueto. 2004b. New Insights in Developmental Origins of Different Gnrh (Gonadotrophin-Releasing Hormone) Systems in Perciform Fish: An Immunohistochemical Study in the European Sea Bass (*Dicentrarchus Labrax*). J. Chem. Neuroanat. 28: 1–15.

Gonzalez, A. and F. Piferrer. 2003. Aromatase Activity in the European Sea Bass (*Dicentrarchus Labrax* L.) Brain. Distribution and Changes in Relation to Age, Sex, and the Annual Reproductive Cycle. Gen. Comp. Endocrinol. 132: 223–30.

Gorska, E., K. Popko, A. Stelmaszczyk-Emmel, O. Ciepiela, A. Kucharska and M. Wasik. 2010. Leptin Receptors. Eur. J. Med. Res. 15: 50–4.

Grens, K.E., A.K. Greenwood and R.D. Fernald. 2005. Two Visual Processing Pathways Are Targeted by Gonadotropin-Releasing Hormone in the Retina. Brain Behav. Evol. 66: 1–9.

Guglielmotti, V. and L. Cristino. 2006. The Interplay between the Pineal Complex and the Habenular Nuclei in Lower Vertebrates in the Context of the Evolution of Cerebral Asymmetry. Brain Res. Bull. 69: 475–88.

Guilgur, L.G., N.P. Moncaut, A.V. Canario and G.M. Somoza. 2006. Evolution of Gnrh Ligands and Receptors in Gnathostomata. Comp. Biochem. Physiol. A. Mol. Integr. Physiol. 144: 272–83.

Halaas, J.L., K.S. Gajiwala, M. Maffei, S.L. Cohen, B.T. Chait, D. Rabinowitz, R.L. Lallone, S.K. Burley and J.M. Friedman. 1995. Weight-Reducing Effects of the Plasma Protein Encoded by the Obese Gene. Science 269: 543–6.

Halford, J.C. and J.E. Blundell. 2000. Separate Systems for Serotonin and Leptin in Appetite Control. Ann. Med. 32: 222–32.

Halm, S., G. Martinez-Rodriguez, L. Rodriguez, F. Prat, C.C. Mylonas, M. Carrillo and S. Zanuy. 2004. Cloning, Characterisation, and Expression of Three Oestrogen Receptors (Eralpha, Erbeta1 and Erbeta2) in the European Sea Bass, *Dicentrarchus Labrax*. Mol. Cell Endocrinol. 223: 63–75.

Hanchate, N.K., J. Parkash, N. Bellefontaine, D. Mazur, W.H. Colledge, X. D'anglemont De Tassigny and V. Prevot. 2012. Kisspeptin-Gpr54 Signaling in Mouse No-Synthesizing Neurons Participates in the Hypothalamic Control of Ovulation. J. Neurosci. 32: 932–45.

Harvey, J. and M.L. Ashford. 2003. Leptin in the Cns: Much More Than a Satiety Signal. Neuropharmacology 44: 845–54.

Hendricks, M. and S. Jesuthasan. 2007. Asymmetric Innervation of the Habenula in Zebrafish. J. Comp. Neurol. 502: 611–9.

Herrera-Pérez, P., M. Del Carmen Rendon, L. Besseau, S. Sauzet, J. Falcon and J.A. Muñoz-Cueto. 2010. Melatonin Receptors in the Brain of the European Sea Bass: An *in situ* Hybridization and Autoradiographic Study. J. Comp. Neurol. 518: 3495–511.

Holland, M.C., S. Hassin and Y. Zohar. 1998. Effects of Long-Term Testosterone, Gonadotropin-Releasing Hormone Agonist, and Pimozide Treatments on Gonadotropin Ii Levels and Ovarian Development in Juvenile Female Striped Bass (Morone Saxatilis). Biol. Reprod. 59: 1153–62.

Hubel, K.A. 1985. Intestinal Nerves and Ion Transport: Stimuli, Reflexes, and Responses. Am. J. Physiol. 248: G261–71.

Huggard-Nelson, D.L., P.S. Nathwani, A. Kermouni and H.R. Habibi. 2002. Molecular Characterization of Lh-Beta and Fsh-Beta Subunits and Their Regulation by Estrogen in the Goldfish Pituitary. Mol. Cell Endocrinol. 188: 171–93.

Iigo, M. and K. Aida. 1995. Effects of Season, Temperature, and Photoperiod on Plasma Melatonin Rhythms in the Goldfish, Carassius Auratus. J. Pineal Res. 18: 62–8.

Iigo, M., F.J. Sanchez-Vazquez, J.A. Madrid, S. Zamora and M. Tabata. 1997. Unusual Responses to Light and Darkness of Ocular Melatonin in European Sea Bass. Neuroreport 8: 1631–5.

Johnson, R.M., T.M. Johnson and R.L. Londraville. 2000. Evidence for Leptin Expression in Fishes. J. Exp. Zool. 286: 718–24.

Kah, O. and P. Chambolle. 1983. Serotonin in the Brain of the Goldfish, Carassius Auratus. An Immunocytochemical Study. Cell Tissue Res. 234: 319–33.

Kah, O. and S. Dufour. 2010. Conserved and divergent features of reproductive neuroendocrinology in fish. pp. 15–42. *In*: D.O. Norris and K.H. Lopez (eds.). Hormones and Reproduction of Vertebrates. Vol. 1 : Fishes. Elsevier, San Diego, USA.

Kah, O., P. Chambolle and M. Olivereau. 1978. Aminergic Innervation of the Hypothalamo-Hypophyseal System in *Gambusia* Sp. (Teleost, Poecilidae) Studied by Two Fluorescence Techniques. C. R. Acad. Sci. Hebd. Seances Acad. Sci. D. 286: 705–8.

Kah, O., B. Breton, J.G. Dulka, J. Nunez-Rodriguez, R.E. Peter, A. Corrigan, J.E. Rivier and W.W. Vale. 1986. A Reinvestigation of the Gn-Rh (Gonadotrophin-Releasing Hormone) Systems in the Goldfish Brain Using Antibodies to Salmon Gn-Rh. Cell Tissue Res. 244: 327–37.

Kah, O., P. Dubourg, M.G. Martinoli, M. Rabhi, F. Gonnet, M. Geffard and A. Calas. 1987. Central Gabaergic Innervation of the Pituitary in Goldfish: A Radioautographic and Immunocytochemical Study at the Electron Microscope Level. Gen. Comp. Endocrinol. 67: 324–32.

Kah, O., V.L. Trudeau, B.D. Sloley, J.P. Chang, P. Dubourg, K.L. Yu and R.E. Peter. 1992. Influence of Gaba on Gonadotrophin Release in the Goldfish. Neuroendocrinology 55: 396–404.

Kah, O., I. Anglade, E. Leprêtre, P. Dubourg and D. De Monbrison. 1993. The Reproductive Brain in Fish. Fish Physiol. and Biochem. 11: 85–98.

Kah, O., C. Lethimonier, G. Somoza, L.G. Guilgur, C. Vaillant and J.J. Lareyre. 2007. Gnrh and Gnrh Receptors in Metazoa: A Historical, Comparative, and Evolutive Perspective. Gen. Comp. Endocrinol. 153: 346–64.

Kaiyala, K.J., S.C. Woods and M.W. Schwartz. 1995. New Model for the Regulation of Energy Balance and Adiposity by the Central Nervous System. Am. J. Clin. Nutr. 62: 1123S–1134S.

Kanda, S., Y. Akazome, T. Matsunaga, N. Yamamoto, S. Yamada, H. Tsukamura, K. Maeda and Y. Oka. 2008. Identification of Kiss-1 Product Kisspeptin and Steroid-Sensitive Sexually Dimorphic Kisspeptin Neurons in Medaka (Oryzias Latipes). Endocrinology 149: 2467–76.

Kanda, S., T. Karigo and Y. Oka. 2012. Steroid Sensitive Kiss2 Neurones in the Goldfish: Evolutionary Insights into the Duplicate Kisspeptin Gene-Expressing Neurones. J. Neuroendocrinol. 24: 897–906.

Kaslin, J. and P. Panula. 2001. Comparative Anatomy of the Histaminergic and Other Aminergic Systems in Zebrafish (Danio Rerio). J. Comp. Neurol. 440: 342–77.

Kauffman, A.S. 2009. Sexual Differentiation and the Kiss1 System: Hormonal and Developmental Considerations. Peptides 30: 83–93.

Kebabian, J.W. and D.B. Calne. 1979. Multiple Receptors for Dopamine. Nature 277: 93–6.

Khan, I.A. and P. Thomas. 1992. Stimulatory Effects of Serotonin on Maturational Gonadotropin Release in the Atlantic Croaker, Micropogonias Undulatus. Gen. Comp. Endocrinol. 88: 388–96.

Khan, I.A. and P. Thomas. 1993. Immunocytochemical Localization of Serotonin and Gonadotropin-Releasing Hormone in the Brain and Pituitary Gland of the Atlantic Croaker Micropogonias Undulatus. Gen. Comp. Endocrinol. 91: 167–80.

Khan, I.A. and P. Thomas. 1996. Melatonin Influences Gonadotropin Ii Secretion in the Atlantic Croaker (Micropogonias Undulatus). Gen. Comp. Endocrinol. 104: 231–42.

Kirsch, M., M.B.A. Djamgoz and H.J. Wagner. 1990. Correlation of Spinule Dynamics and Plasticity of the Horizontal Cell Spectral Response in Cyprinid Fish Retina. Tissue Res. 260: 123–130.

Kitahashi, T., S. Ogawa and I.S. Parhar. 2009. Cloning and Expression of Kiss2 in the Zebrafish and Medaka. Endocrinology 150: 821–31.

Kobayashi, M., M. Amano, M.H. Kim, K. Furukawa, Y. Hasegawa and K. Aida. 1994. Gonadotropin-Releasing Hormones of Terminal Nerve Origin Are Not Essential to Ovarian Development and Ovulation in Goldfish. Gen. Comp. Endocrinol. 95: 192–200.

Koda, A., K. Ukena, H. Teranishi, S. Ohta, K. Yamamoto, S. Kikuyama and K. Tsutsui. 2002. A Novel Amphibian Hypothalamic Neuropeptide: Isolation, Localization, and Biological Activity. Endocrinology 143: 411–9.

Kotani, M., M. Detheux, A. Vandenbogaerde, D. Communi, J.M. Vanderwinden, E. Le Poul, S. Brezillon, R. Tyldesley, N. Suarez-Huerta, F. Vandeput, C. Blanpain, S.N. Schiffmann, G. Vassart and M. Parmentier. 2001. The Metastasis Suppressor Gene Kiss-1 Encodes Kisspeptins, the Natural Ligands of the Orphan G Protein-Coupled Receptor Gpr54. J. Biol. Chem. 276: 34631–6.

Kumakura, N., K. Okuzawa, K. Gen and H. Kagawa. 2003. Effects of Gonadotropin-Releasing Hormone Agonist and Dopamine Antagonist on Hypothalamus-Pituitary-Gonadal Axis of Pre-Pubertal Female Red Seabream (Pagrus Major). Gen. Comp. Endocrinol. 131: 264–73.

Kurokawa, T. and K. Murashita. 2009. Genomic Characterization of Multiple Leptin Genes and a Leptin Receptor Gene in the Japanese Medaka, Oryzias Latipes. Gen. Comp. Endocrinol. 161: 229–37.

Lafont, A.G., M. Morini, J. Pasquier, K. Rousseau and S. Dufour. 2012. Duplicated leptin/leptin receptor system in a basal teleost, the European eel, Abstract book of the 7th International Symposium of Fish Endocrinology. Buenos Aires, Argentina. pp. 211.

Lariviere, K., L. Maceachern, V. Greco, G. Majchrzak, S. Chiu, G. Drouin and V.L. Trudeau. 2002. Gad(65) and Gad(67) Isoforms of the Glutamic Acid Decarboxylase Gene Originated before the Divergence of Cartilaginous Fishes. Mol. Biol. Evol. 19: 2325–9.

Lariviere, K., M. Samia, A. Lister, G. Van Der Kraak and V.L. Trudeau. 2005. Sex Steroid Regulation of Brain Glutamic Acid Decarboxylase (Gad) Mrna Is Season-Dependent and Sexually Dimorphic in the Goldfish Carassius Auratus. Brain Res. Mol. Brain Res. 141: 1–9.

Leal, E., E. Sanchez, B. Muriach and J.M. Cerda-Reverter. 2009. Sex Steroid-Induced Inhibition of Food Intake in Sea Bass (*Dicentrarchus Labrax*). J. Comp. Physiol. B. 179: 77–86.

Lee, Y.R., K. Tsunekawa, M.J. Moon, H.N. Um, J.I. Hwang, T. Osugi, N. Otaki, Y. Sunakawa, K. Kim, H. Vaudry, H.B. Kwon, J.Y. Seong and K. Tsutsui. 2009. Molecular Evolution of Multiple Forms of Kisspeptins and Gpr54 Receptors in Vertebrates. Endocrinology 150: 2837–46.

Lethimonier, C., T. Madigou, J.A. Muñoz-Cueto, J.J. Lareyre and O. Kah. 2004. Evolutionary Aspects of Gnrhs, Gnrh Neuronal Systems and Gnrh Receptors in Teleost Fish. Gen. Comp. Endocrinol. 135: 1–16.

Li, S., Y. Zhang, Y. Liu, X. Huang, W. Huang, D. Lu, P. Zhu, Y. Shi, C.H. Cheng, X. Liu and H. Lin. 2009. Structural and Functional Multiplicity of the Kisspeptin/Gpr54 System in Goldfish (Carassius Auratus). J. Endocrinol. 201: 407–18.

Londraville, R.L. and C.S. Duvall. 2002. Murine Leptin Injections Increase Intracellular Fatty Acid-Binding Protein in Green Sunfish (Lepomis Cyanellus). Gen. Comp. Endocrinol. 129: 56–62.

Magni, P., M. Motta and L. Martini. 2000. Leptin: A Possible Link between Food Intake, Energy Expenditure, and Reproductive Function. Regul. Pept. 92: 51–6.

Malpaux, B., M. Migaud, H. Tricoire and P. Chemineau. 2001. Biology of Mammalian Photoperiodism and the Critical Role of the Pineal Gland and Melatonin. J. Biol. Rhythms. 16: 336–47.

Mañanos, E.L., S. Zanuy and M. Carrillo. 1997. Photoperiodic Manipulations of the Reproductive Cycle of Sea Bass (*Dicentrarchus Labrax*) and Their Effects on Gonadal Development, and Plasma 17ß-Estradiol and Vitellogenin Levels. Fish Physiol. and Biochem. 16: 211–222.

Mañanos, E.L., I. Anglade, J. Chyb, C. Saligaut, B. Breton and O. Kah. 1999. Involvement of Gamma-Aminobutyric Acid in the Control of Gth-1 and Gth-2 Secretion in Male and Female Rainbow Trout. Neuroendocrinology 69: 269–80.

Maruska, K.P. and T.C. Tricas. 2007. Gonadotropin-Releasing Hormone and Receptor Distributions in the Visual Processing Regions of Four Coral Reef Fishes. Brain Behav. Evol. 70: 40–56.

Mateos, J., E. Mañanos, G. Martinez-Rodriguez, M. Carrillo, B. Querat and S. Zanuy. 2003. Molecular Characterization of Sea Bass Gonadotropin Subunits (Alpha, Fshbeta, and Lhbeta) and Their Expression During the Reproductive Cycle. Gen. Comp. Endocrinol. 133: 216–32.

McDonald, J.K. and J.I. Koenig. Neuropeptide Y actions on reproductive and endocrine functions. pp 419–456. *In*: W.F. Colmers and C. Wahlestedt (eds.). 1993. Neuropeptide Y and Related Peptides. Humana, Totowa, Newyork, USA.

Menuet, A., I. Anglade, R. Le Guevel, E. Pellegrini, F. Pakdel and O. Kah. 2003. Distribution of Aromatase Mrna and Protein in the Brain and Pituitary of Female Rainbow Trout: Comparison with Estrogen Receptor Alpha. J. Comp. Neurol. 462: 180–93.

Migaud, H., A. Davie, C.C. Martinez Chavez and S. Al-Khamees. 2007. Evidence for Differential Photic Regulation of Pineal Melatonin Synthesis in Teleosts. J. Pineal Res. 43: 327–35.

Mitani, Y., S. Kanda, Y. Akazome, B. Zempo and Y. Oka. 2010. Hypothalamic Kiss1 but Not Kiss2 Neurons Are Involved in Estrogen Feedback in Medaka (Oryzias Latipes). Endocrinology 151: 1751–9.

Mizuno, I., Y. Okimura, Y. Takahashi, H. Kaji, H. Abe and K. Chihara. 1999. Leptin Stimulates Basal and Ghrh-Induced Gh Release from Cultured Rat Anterior Pituitary Cells *in vitro*. Kobe J. Med. Sci. 45: 221–7.

Mohamed, J.S., P. Thomas and I.A. Khan. 2005. Isolation, Cloning, and Expression of Three Prepro-Gnrh Mrnas in Atlantic Croaker Brain and Pituitary. J. Comp. Neurol. 488: 384–95.

Moles, G., M. Carrillo, E. Mañanos, C.C. Mylonas and S. Zanuy. 2007. Temporal Profile of Brain and Pituitary Gnrhs, Gnrh-R and Gonadotropin Mrna Expression and Content During Early Development in European Sea Bass (*Dicentrarchus Labrax* L.). Gen. Comp. Endocrinol. 150: 75–86.

Moles, G., A. Gomez, M. Carrillo and S. Zanuy. 2012. Development of a Homologous Enzyme-Linked Immunosorbent Assay for European Sea Bass Fsh. Reproductive Cycle Plasma Levels in Both Sexes and in Yearling Precocious and Non-Precocious Males. Gen. Comp. Endocrinol. 176: 70–8.

Molina-Borja, M., J. Falcon, E. Urquiola and S. Oaknin. 1994. Characterization of 2-125i-Iodomelatonin Binding Sites in the Brain, Intestine and Gonad of the Gilthead Sea Bream (Sparus Aurata). Pflügers Arch. 427 (Suppl.1) R5.

Moncaut, N., G. Somoza, D.M. Power and A.V. Canario. 2005. Five Gonadotrophin-Releasing Hormone Receptors in a Teleost Fish: Isolation, Tissue Distribution and Phylogenetic Relationships. J. Mol. Endocrinol. 34: 767–79.

Montague, C.T., I.S. Farooqi, J.P. Whitehead, M.A. Soos, H. Rau, N.J. Wareham, C.P. Sewter, J.E. Digby, S.N. Mohammed, J.A. Hurst, C.H. Cheetham, A.R. Earley, A.H. Barnett, J.B. Prins and S. O'rahilly. 1997. Congenital Leptin Deficiency Is Associated with Severe Early-Onset Obesity in Humans. Nature 387: 903–8.

Morgan, K. and R.P. Millar. 2004. Evolution of Gnrh Ligand Precursors and Gnrh Receptors in Protochordate and Vertebrate Species. Gen. Comp. Endocrinol. 139: 191–7.

Moussavi, M., M. Wlasichuk, J.P. Chang and H.R. Habibi. 2012. Gonadotropin inhibitory hormone mediated control of gonadotrophin production in goldfish. Abstract book of the 7th International Symposium of Fish Endocrinology. Buenos Aires, Argentina. pp. 151.

Muir, A.I., L. Chamberlain, N.A. Elshourbagy, D. Michalovich, D.J. Moore, A. Calamari, P.G. Szekeres, H.M. Sarau, J.K. Chambers, P. Murdock, K. Steplewski, U. Shabon, J.E. Miller, S.E. Middleton, J.G. Darker, C.G. Larminie, S. Wilson, D.J. Bergsma, P. Emson, R. Faull, K.L. Philpott and D.C. Harrison. 2001. Axor12, a Novel Human G Protein-Coupled Receptor, Activated by the Peptide Kiss-1. J Biol. Chem. 276: 28969–75.

Murashita, K., S. Uji, T. Yamamoto, I. Ronnestad and T. Kurokawa. 2008. Production of Recombinant Leptin and Its Effects on Food Intake in Rainbow Trout (Oncorhynchus Mykiss). Comp. Biochem. Physiol. B. Biochem. Mol. Biol. 150: 377–84.

Muriach, B., M. Carrillo, S. Zanuy and J.M. Cerda-Reverter. 2008a. Distribution of Estrogen Receptor 2 mRNAs (Esr2a and Esr2b) in the Brain and Pituitary of the Sea Bass (*Dicentrarchus Labrax*). Brain Res. 1210: 126–41.

Muriach, B., J.M. Cerda-Reverter, A. Gomez, S. Zanuy and M. Carrillo. 2008b. Molecular Characterization and Central Distribution of the Estradiol Receptor Alpha (ERalpha) in the Sea Bass (*Dicentrarchus Labrax*). J. Chem. Neuroanat. 35: 33–48.

Muske, L.E. 1993. Evolution of Gonadotropin-Releasing Hormone (GnRH) Neuronal Systems. Brain Behav. Evol. 42: 215–30.

Muske, L.E. and F.L. Moore. 1994. Antibodies against Different Forms of GnRH Distinguish Different Populations of Cells and Axonal Pathways in a Urodele Amphibian, Taricha Granulosa. J. Comp. Neurol. 345: 139–47.

Mustonen, A.M., P. Nieminen and H. Hyvarinen. 2002. Leptin, Ghrelin, and Energy Metabolism of the Spawning Burbot (*Lota Lota* L.). J. Exp. Zool. 293: 119–26.

Navarro, V.M. and M. Tena-Sempere. 2011. Neuroendocrine Control by Kisspeptins: Role in Metabolic Regulation of Fertility. Nat. Rev. Endocrinol. 8: 40–53.

Navas, J.M., I. Anglade, T. Bailhache, F. Pakdel, B. Breton, P. Jego and O. Kah. 1995. Do Gonadotrophin-Releasing Hormone Neurons Express Estrogen Receptors in the Rainbow Trout? A Double Immunohistochemical Study. J. Comp. Neurol. 363: 461–74.

Neary, N.M., A.P. Goldstone and S.R. Bloom. 2004. Appetite Regulation: From the Gut to the Hypothalamus. Clin. Endocrinol. (Oxf). 60: 153–60.

Nelson, J.S. 2006. Fishes of the World, 4th ed. Wiley, Hoboken, New York.

Nieminen, P., A.M. Mustonen and H. Hyvarinen. 2003. Fasting Reduces Plasma Leptin-and Ghrelin-Immunoreactive Peptide Concentrations of the Burbot (Lota Lota) at 2 Degrees C but Not at 10 Degrees C. Zoolog. Sci. 20: 1109–15.

Oakley, A.E., D.K. Clifton and R.A. Steiner. 2009. Kisspeptin Signaling in the Brain. Endocr. Rev. 30: 713–43.

Ogawa, S., K.W. Ng, P.N. Ramadasan, F.M. Nathan and I.S. Parhar. 2012a. Habenular Kiss1 Neurons Modulate the Serotonergic System in the Brain of Zebrafish. Endocrinology 153: 2398–407.

Ogawa, S., P.N. Ramadasan, M. Goschorska, A. Anantharajah, K.W. Ng and I.S. Parhar. 2012b. Cloning and Expression of Tachykinins and Their Association with Kisspeptins in the Brains of Zebrafish. J. Comp. Neurol. 520: 2991–3012.

Oka, Y. and M. Ichikawa. 1990. Gonadotropin-Releasing Hormone (Gnrh) Immunoreactive System in the Brain of the Dwarf Gourami (Colisa Lalia) as Revealed by Light Microscopic Immunocytochemistry Using a Monoclonal Antibody to Common Amino Acid Sequence of Gnrh. J. Comp. Neurol. 300: 511–22.

Okubo, K., F. Sakai, E.L. Lau, G. Yoshizaki, Y. Takeuchi, K. Naruse, K. Aida and Y. Nagahama. 2006. Forebrain Gonadotropin-Releasing Hormone Neuronal Development: Insights from Transgenic Medaka and the Relevance to X-Linked Kallmann Syndrome. Endocrinology 147: 1076–84.

Okuzawa, K., N. Kumakura, A. Mori, K. Gen, S. Yamaguchi and H. Kagawa. 2002. Regulation of Gnrh and Its Receptor in a Teleost, Red Seabream. Prog. Brain Res. 141: 95–110.

Olivereau, M. and J.N. Ball. 1964. Contribution to the Histophysiology of the Pituitary Gland of Teleosts, Particularly Those of the Poecilia Species. Gen. Comp. Endocrinol. 47: 523–32.

Onuma, T.A. and C. Duan. 2012. Duplicated Kiss1 Receptor Genes in Zebrafish: Distinct Gene Expression Patterns, Different Ligand Selectivity, and a Novel Nuclear Isoform with Transactivating Activity. FASEB J. 26: 2941–50.

Pandolfi, M., J.A. Muñoz Cueto, F.L. Lo Nostro, J.L. Downs, D.A. Paz, M.C. Maggese and H.F. Urbanski. 2005. Gnrh Systems of Cichlasoma Dimerus (Perciformes, Cichlidae) Revisited: A Localization Study with Antibodies and Riboprobes to Gnrh-Associated Peptides. Cell Tissue Res. 321: 219–32.

Parhar, I.S., T. Soga and Y. Sakuma. 2000. Thyroid Hormone and Estrogen Regulate Brain Region-Specific Messenger Ribonucleic Acids Encoding Three Gonadotropin-Releasing Hormone Genes in Sexually Immature Male Fish, Oreochromis Niloticus. Endocrinology 141: 1618–26.

Pasmanik, M. and G.V. Callard. 1988. A High Abundance Androgen Receptor in Goldfish Brain: Characteristics and Seasonal Changes. Endocrinology 123: 1162–71.

Pasquier, J., A.G. Lafont, J. Leprince, H. Vaudry, K. Rousseau and S. Dufour. 2011. First Evidence for a Direct Inhibitory Effect of Kisspeptins on Lh Expression in the Eel, Anguilla Anguilla. Gen. Comp. Endocrinol. 173: 216–25.

Pellegrini, E., A. Menuet, C. Lethimonier, F. Adrio, M.M. Gueguen, C. Tascon, I. Anglade, F. Pakdel and O. Kah. 2005. Relationships between Aromatase and Estrogen Receptors in the Brain of Teleost Fish. Gen. Comp. Endocrinol. 142: 60–6.

Pellegrini, E., K. Mouriec, I. Anglade, A. Menuet, Y. Le Page, M.M. Gueguen, M.H. Marmignon, F. Brion, F. Pakdel and O. Kah. 2007. Identification of Aromatase-Positive Radial Glial Cells as Progenitor Cells in the Ventricular Layer of the Forebrain in Zebrafish. J. Comp. Neurol. 501: 150–67.

Peng, C., J.P. Chang, K.L. Yu, A.O. Wong, F. Van Goor, R.E. Peter and J.E. Rivier. 1993a. Neuropeptide-Y Stimulates Growth Hormone and Gonadotropin-Ii Secretion in the Goldfish Pituitary: Involvement of Both Presynaptic and Pituitary Cell Actions. Endocrinology 132: 1820–9.

Peng, C., S. Humphries, R.E. Peter, J.E. Rivier, A.G. Blomqvist and D. Larhammar. 1993b. Actions of Goldfish Neuropeptide Y on the Secretion of Growth Hormone and Gonadotropin-Ii in Female Goldfish. Gen. Comp. Endocrinol. 90: 306–17.

Pérez, M.R., E. Pellegrini, M.M. Gueguen, C. Vaillant, G.M. Somoza and O. Kah. 2012. Serotonin modulates proliferation of radial glial cells in the hypothalamus of adult zebrafish (Danio rerio). Abstract book of the 7th International Symposium of Fish Endocrinology. Buenos Aires, Argentina. pp. 156.

Peter, R.E. and C.R. Paulencu. 1980. Involvement of the Preoptic Region in Gonadotropin Release-Inhibition in Goldfish, Carassius Auratus. Neuroendocrinology 31: 133–41.

Peter, R.E., J.P. Chang, C.S. Nahorniak, R.J. Omeljaniuk, M. Sokolowska, S.H. Shih and R. Billard. 1986. Interactions of Catecholamines and Gnrh in Regulation of Gonadotropin Secretion in Teleost Fish. Recent Prog. Horm. Res. 42: 513–48.

Peter, R.E., V. Trudeau, B.D. Slolely, C. Peng and C.S. Nahorniak. 1991. Actions of catecholamines, peptides and sex steroids in regulation of gonadotrophin-II in the goldfish. pp. 30–40. *In*: A.P. Scott, J.P. Sumpter, D.E. Kime and M.S. Rolfe (eds.). Reproductive Physiology of Fish. Department of Biological Sciences, University of Sheffield, Sheffield.

Peyon, P., S. Zanuy and M. Carrillo. 2001. Action of Leptin on *in vitro* Luteinizing Hormone Release in the European Sea Bass (*Dicentrarchus Labrax*). Biol. Reprod. 65: 1573–8.

Peyon, P., S. Vega-Rubin De Celis, P. Gomez-Requeni, S. Zanuy, J. Perez-Sanchez and M. Carrillo. 2003. *In vitro* Effect of Leptin on Somatolactin Release in the European Sea Bass (*Dicentrarchus Labrax*): Dependence on the Reproductive Status and Interaction with Npy and Gnrh. Gen. Comp. Endocrinol. 132: 284–92.

Piferrer, F., M. Blazquez, L. Navarro and A. Gonzalez. 2005. Genetic, Endocrine, and Environmental Components of Sex Determination and Differentiation in the European Sea Bass (*Dicentrarchus Labrax* L.). Gen. Comp. Endocrinol. 142: 102–10.

Pineda, R., E. Aguilar, L. Pinilla and M. Tena-Sempere. 2010. Physiological Roles of the Kisspeptin/Gpr54 System in the Neuroendocrine Control of Reproduction. Prog. Brain Res. 181: 55–77.

Popesku, J.T., C.J. Martyniuk, J. Mennigen, H. Xiong, D. Zhang, X. Xia, A.R. Cossins and V.L. Trudeau. 2008. The Goldfish (Carassius Auratus) as a Model for Neuroendocrine Signaling. Mol. Cell Endocrinol. 293: 43–56.

Prat, F., S. Zanuy, M. Carrillo, A. De Mones and A. Fostier. 1990. Seasonal Changes in Plasma Levels of Gonadal Steroids of Sea Bass, *Dicentrarchus Labrax* L. Gen. Comp. Endocrinol. 78: 361–73.

Prat, F., S. Zanuy and M. Carrillo. 2001. Effect of Gonadotropin-Releasing Hormone Analogue Gnrha/and Pimozide on Plasma Levels of Sex Steroids and Ovarian Development in Sea Bass *Dicentrarchus Labrax* L. Aquaculture 198: 325–338.

Ringvall, M., M.M. Berglund and D. Larhammar. 1997. Multiplicity of Neuropeptide Y Receptors: Cloning of a Third Distinct Subtype in the Zebrafish. Biochem. Biophys. Res. Commun. 241: 749–55.

Roa, J., J.M. Castellano, V.M. Navarro, D.J. Handelsman, L. Pinilla and M. Tena-Sempere. 2009. Kisspeptins and the Control of Gonadotropin Secretion in Male and Female Rodents. Peptides 30: 57–66.

Rocha, A., S. Zanuy, M. Carrillo and A. Gomez. 2009. Seasonal Changes in Gonadal Expression of Gonadotropin Receptors, Steroidogenic Acute Regulatory Protein and Steroidogenic Enzymes in the European Sea Bass. Gen. Comp. Endocrinol. 162: 265–75.

Rodriguez, L., M. Carrillo, L.A. Sorbera, M.A. Soubrier, E. Mañanos, M.C. Holland, Y. Zohar and S. Zanuy. 2000. Pituitary Levels of Three Forms of Gnrh in the Male European Sea Bass (*Dicentrarchus Labrax* L.) During Sex Differentiation and First Spawning Season. Gen. Comp. Endocrinol. 120: 67–74.

Rodriguez, L., M. Carrillo, L.A. Sorbera, Y. Zohar and S. Zanuy. 2004. Effects of Photoperiod on Pituitary Levels of Three Forms of Gnrh and Reproductive Hormones in the Male European Sea Bass (*Dicentrarchus Labrax* L.) During Testicular Differentiation and First Testicular Recrudescence. Gen. Comp. Endocrinol. 136: 37–48.

Roy, D., N.L. Angelini, H. Fujieda, G.M. Brown and D.D. Belsham. 2001. Cyclical Regulation of Gnrh Gene Expression in Gt1-7 Gnrh-Secreting Neurons by Melatonin. Endocrinology 142: 4711–20.

Saligaut, C., B. Linard, E.L. Mañanos, O. Kah, B. Breton and M. Govoroun. 1998. Release of Pituitary Gonadotrophins Gth I and Gth Ii in the Rainbow Trout (Oncorhynchus Mykiss): Modulation by Estradiol and Catecholamines. Gen. Comp. Endocrinol. 109: 302–9.

Sánchez-Vázquez, F.J., J.A. Madrid and S. Zamora. 1995. Circadian Rhythms of Feeding Activity in Sea Bass, *Dicentrarchus Labrax* L.: Dual Phasing Capacity of Diel Demand-Feeding Pattern. J. Biol. Rhythms 10: 256–66.

Sánchez-Vázquez, F.J., M. Iigo, J.A. Madrid, S. Zamora and M. Tabata. 1997. Daily Cycles in Plasma and Ocular Melatonin in Demand-Fed Sea Bass, *Dicentrarchus Labrax* L. J. Comp. Physiol. B. 167: 409–415.

Sauzet, S., L. Besseau, P. Herrera Perez, D. Coves, B. Chatain, E. Peyric, G. Boeuf, J.A. Muñoz-Cueto and J. Falcon. 2008. Cloning and Retinal Expression of Melatonin Receptors in the European Sea Bass, *Dicentrarchus Labrax*. Gen. Comp. Endocrinol. 157: 186–95.

Sawada, K., K. Ukena, S. Kikuyama and K. Tsutsui. 2002a. Identification of a Cdna Encoding a Novel Amphibian Growth Hormone-Releasing Peptide and Localization of Its Transcript. J. Endocrinol. 174: 395–402.

Sawada, K., K. Ukena, H. Satake, E. Iwakoshi, H. Minakata and K. Tsutsui. 2002b. Novel Fish Hypothalamic Neuropeptide. Eur. J. Biochem. 269: 6000–8.

Scapigliati, G., N. Romano, F. Buonocore, S. Picchietti, M.R. Baldassini, D. Prugnoli, A. Galice, S. Meloni, C.J. Secombes, M. Mazzini and L. Abelli. 2002. The Immune System of Sea Bass, *Dicentrarchus Labrax*, Reared in Aquaculture. Dev. Comp. Immunol. 26: 151–60.

Sebert, M.E., C. Legros, F.A. Weltzien, B. Malpaux, P. Chemineau and S. Dufour. 2008. Melatonin Activates Brain Dopaminergic Systems in the Eel with an Inhibitory Impact on Reproductive Function. J. Neuroendocrinol. 20: 917–29.

Seminara, S.B., S. Messager, E.E. Chatzidaki, R.R. Thresher, J.S. Acierno, Jr., J.K. Shagoury, Y. Bo-Abbas, W. Kuohung, K.M. Schwinof, A.G. Hendrick, D. Zahn, J. Dixon, U.B. Kaiser, S.A. Slaugenhaupt, J.F. Gusella, S. O'rahilly, M.B. Carlton, W.F. Crowley, Jr., S.A. Aparicio and W.H. Colledge. 2003. The Gpr54 Gene as a Regulator of Puberty. N. Engl. J. Med. 349: 1614–27.

Senthilkumaran, B., K. Okuzawa, K. Gen and H. Kagawa. 2001. Effects of Serotonin, Gaba and Neuropeptide Y on Seabream Gonadotropin Releasing Hormone Release *in vitro* from Preoptic-Anterior Hypothalamus and Pituitary of Red Seabream, Pagrus Major. J. Neuroendocrinol. 13: 395–400.

Servili, A., C. Lethimonier, J.J. Lareyre, J.F. Lopez-Olmeda, F.J. Sanchez-Vazquez, O. Kah and J.A. Muñoz-Cueto. 2010. The Highly Conserved Gonadotropin-Releasing Hormone-2 Form Acts as a Melatonin-Releasing Factor in the Pineal of a Teleost Fish, the European Sea Bass *Dicentrarchus Labrax*. Endocrinology 151: 2265–75.

Servili, A., P. Herrera-Pérez, J. Yáñez and J.A. Muñoz-Cueto. 2011a. Afferent and Efferent Connections of the Pineal Organ in the European Sea Bass *Dicentrarchus Labrax*: a Carbocyanine Dye Tract-Tracing Study. Brain Behav. Evol. 78: 272–285.

Servili, A., Y. Le Page, J. Leprince, A. Caraty, S. Escobar, I.S. Parhar, J.Y. Seong, H. Vaudry and O. Kah. 2011b. Organization of Two Independent Kisspeptin Systems Derived from Evolutionary-Ancient Kiss Genes in the Brain of Zebrafish. Endocrinology 152: 1527–40.

Servili, A., P. Herrera-Perez, O. Kah and J.A. Muñoz-Cueto. 2012. The Retina Is a Target for GnRH3 System in the European Sea Bass, *Dicentrarchus Labrax*. Gen. Comp. Endocrinol. 175: 398–406.

Servili, A., P. Herrera-Pérez, M.C. Rendón and J.A. Muñoz-Cueto. 2013. Melatonin Inhibits GnRH-1, GnRH-3 and GnRH Receptor Expression in the Brain of the European Sea Bass, *Dicentrarchus Labrax*. Int. J. Mol. Sci. 14: 7603–16.

Shahjahan, M., T. Ikegami, T. Osugi, K. Ukena, H. Doi, A. Hattori, K. Tsutsui and H. Ando. 2011. Synchronised Expressions of Lpxrfamide Peptide and Its Receptor Genes: Seasonal, Diurnal and Circadian Changes During Spawning Period in Grass Puffer. J. Neuroendocrinol. 23: 39–51.

Sherwood, N.M., D.B. Parker, J.E. McRoy and D.W. Leischeid. 1994. Molecular evolution of growth hormone-releasing hormone and gonadotropin-releasing hormone. pp. 3–66.

In: N.M. Sherwood and C.L. Hew (eds.). Fish Physiology, vol 13. Academic Press, San Diego, CA.

Silverstein, J.T. and E.M. Plisetskaya. 2000. The Effects of Npy and Insulin on Food Intake Regulation in Fish. Am. Zool. 40: 296–308.

Sloley, B.D., O. Kah, V.L. Trudeau, J.G. Dulka and R.E. Peter. 1992. Amino Acid Neurotransmitters and Dopamine in Brain and Pituitary of the Goldfish: Involvement in the Regulation of Gonadotropin Secretion. J. Neurochem. 58: 2254–62.

Smith, O., W. Xia, N. Zmora and Y. Zohar. 2012. Localization of Gonadotropin-Inhibitory Hormone (GnIH) neurons in the brain of zebrafish (Danio rerio). Abstract book of the 7th International Symposium of Fish Endocrinology. Buenos Aires, Argentina. pp. 150.

Somoza, G.M., K.L. Yu and R.E. Peter. 1988. Serotonin Stimulates Gonadotropin Release in Female and Male Goldfish, *Carassius Auratus* L. Gen. Comp. Endocrinol. 72: 374–82.

Sone, M. and R.Y. Osamura. 2001. Leptin and the Pituitary. Pituitary 4: 15–23.

Sower, S.A. 1997. Evolution of GnRH in fish of ancient origins. pp. 486. *In*: I.S. Parhar and Y. Sakuma (eds.). GnRH Neurons: Gene to Behavior. Brain Shuppan Publishers, Tokyo.

Tena-Sempere, M. 2010. Roles of Kisspeptins in the Control of Hypothalamic-Gonadotropic Function: Focus on Sexual Differentiation and Puberty Onset. Endocr. Dev. 17: 52–62.

Tena-Sempere, M., A. Felip, A. Gomez, S. Zanuy and M. Carrillo. 2012. Comparative Insights of the Kisspeptin/Kisspeptin Receptor System: Lessons from Non-Mammalian Vertebrates. Gen. Comp. Endocrinol. 175: 234–43.

Trudeau, V.L. 1997. Neuroendocrine Regulation of Gonadotrophin Ii Release and Gonadal Growth in the Goldfish, Carassius Auratus. Rev. Reprod. 2: 55–68.

Trudeau, V.L., O. Kah, J.P. Chang, B.D. Sloley, P. Dubourg, E.J. Fraser and R.E. Peter. 2000a. The Inhibitory Effects of (Gamma)-Aminobutyric Acid (Gaba) on Growth Hormone Secretion in the Goldfish Are Modulated by Sex Steroids. J. Exp. Biol. 203: 1477–85.

Trudeau, V.L., D. Spanswick, E.J. Fraser, K. Lariviere, D. Crump, S. Chiu, M. Macmillan and R.W. Schulz. 2000b. The Role of Amino Acid Neurotransmitters in the Regulation of Pituitary Gonadotropin Release in Fish. Biochem. Cell Biol. 78: 241–59.

Tsutsui, K. and K. Ukena. 2006. Hypothalamic Lpxrf-Amide Peptides in Vertebrates: Identification, Localization and Hypophysiotropic Activity. Peptides 27: 1121–9.

Tsutsui, K., E. Saigoh, K. Ukena, H. Teranishi, Y. Fujisawa, M. Kikuchi, S. Ishii and P.J. Sharp. 2000. A Novel Avian Hypothalamic Peptide Inhibiting Gonadotropin Release. Biochem. Biophys. Res. Commun. 275: 661–7.

Tsutsui, K., G.E. Bentley, T. Ubuka, E. Saigoh, H. Yin, T. Osugi, K. Inoue, V.S. Chowdhury, K. Ukena, N. Ciccone, P.J. Sharp and J.C. Wingfield. 2007. The General and Comparative Biology of Gonadotropin-Inhibitory Hormone (Gnih). Gen. Comp. Endocrinol. 153: 365–70.

Ukena, K., E. Iwakoshi, H. Minakata and K. Tsutsui. 2002. A Novel Rat Hypothalamic Rfamide-Related Peptide Identified by Immunoaffinity Chromatography and Mass Spectrometry. FEBS Lett. 512: 255–8.

Ukena, K., A. Koda, K. Yamamoto, T. Kobayashi, E. Iwakoshi-Ukena, H. Minakata, S. Kikuyama and K. Tsutsui. 2003. Novel Neuropeptides Related to Frog Growth Hormone-Releasing Peptide: Isolation, Sequence, and Functional Analysis. Endocrinology 144: 3879–84.

Um, H.N., J.M. Han, J.I. Hwang, S.I. Hong, H. Vaudry and J.Y. Seong. 2010. Molecular Coevolution of Kisspeptins and Their Receptors from Fish to Mammals. Ann. N. Y Acad. Sci. 1200: 67–74.

Umino, O. and J.E. Dowling. 1991. Dopamine Release from Interplexiform Cells in the Retina: Effects of Gnrh, Fmrfamide, Bicuculline, and Enkephalin on Horizontal Cell Activity. J. Neurosci. 11: 3034–46.

Vetillard, A., F. Ferriere, P. Jego and T. Bailhache. 2006. Regulation of Salmon Gonadotrophin-Releasing Hormone Gene Expression by Sex Steroids in Rainbow Trout Brain. J. Neuroendocrinol. 18: 445–53.

Vickers, E.D., F. Laberge, B.A. Adams, T.J. Hara and N.M. Sherwood. 2004. Cloning and Localization of Three Forms of Gonadotropin-Releasing Hormone, Including the Novel Whitefish Form, in a Salmonid, Coregonus Clupeaformis. Biol. Reprod. 70: 1136–46.

Volkoff, H., A.J. Eykelbosh and R.E. Peter. 2003. Role of Leptin in the Control of Feeding of Goldfish Carassius Auratus: Interactions with Cholecystokinin, Neuropeptide Y and Orexin a, and Modulation by Fasting. Brain Res. 972: 90–109.

Wagner, H.J. 1980. Light-Dependent Plasticity of the Morphology of Horizontal Cell Terminals in Cone Pedicles of Fish Retinas. J. Neurocytol. 9: 573–90.

Wagner, H.J. and U.D. Behrens. 1993. Microanatomy of the Dopaminergic System in the Rainbow Trout Retina. Vision Res. 33: 1345–58.

Wayne, N.L., K. Kuwahara, K. Aida, Y. Nagahama and K. Okubo. 2005. Whole-Cell Electrophysiology of Gonadotropin-Releasing Hormone Neurons That Express Green Fluorescent Protein in the Terminal Nerve of Transgenic Medaka (Oryzias Latipes). Biol. Reprod. 73: 1228–34.

White, S.A., T.L. Kasten, C.T. Bond, J.P. Adelman and R.D. Fernald. 1995. Three Gonadotropin-Releasing Hormone Genes in One Organism Suggest Novel Roles for an Ancient Peptide. Proc. Natl. Acad. Sci. USA 92: 8363–7.

Yaghoubian, S., M.F. Filosa and J.H. Youson. 2001. Proteins Immunoreactive with Antibody against a Human Leptin Fragment Are Found in Serum and Tissues of the Sea Lamprey, *Petromyzon Marinus* L. Comp. Biochem. Physiol. B. Biochem. Mol. Biol. 129: 777–85.

Yamamoto, N., Y. Oka and S. Kawashima. 1997. Lesions of Gonadotropin-Releasing Hormone-Immunoreactive Terminal Nerve Cells: Effects on the Reproductive Behavior of Male Dwarf Gouramis. Neuroendocrinology 65: 403–12.

Yaron, Z., G. Gur, P. Melamed, H. Rosenfeld, A. Elizur and B. Levavi-Sivan. 2003. Regulation of Fish Gonadotropins. Int. Rev. Cytol. 225: 131–85.

Yu, K.L., P.M. Rosenblum and R.E. Peter. 1991. *In vitro* Release of Gonadotropin-Releasing Hormone from the Brain Preoptic-Anterior Hypothalamic Region and Pituitary of Female Goldfish. Gen. Comp. Endocrinol. 81: 256–67.

Yu, W.H., M. Kimura, A. Walczewska, S. Karanth and S.M. Mccann. 1997. Role of Leptin in Hypothalamic-Pituitary Function. Proc. Natl. Acad. Sci. USA 94: 1023–8.

Zachmann, A., J. Falcon, S.C. Knijff, V. Bolliet and M.A. Ali. 1992. Effects of Photoperiod and Temperature on Rhythmic Melatonin Secretion from the Pineal Organ of the White Sucker (Catostomus Commersoni) *in vitro*. Gen. Comp. Endocrinol. 86: 26–33.

Zanuy, S. and M. Carrillo. 1985. Annual Cycles of Growth, Feeding Rate, Gross Conversion Efficiency and Hematocrit Levels of Sea Bass (*Dicentrarchus Labrax* L.) Adapted to Different Osmotic Media. Aquaculture 44: 11–25.

Zhang, Y., R. Proenca, M. Maffei, M. Barone, L. Leopold and J.M. Friedman. 1994. Positional Cloning of the Mouse Obese Gene and Its Human Homologue. Nature 372: 425–32.

Zhang, Y., S. Li, Y. Liu, D. Lu, H. Chen, X. Huang, X. Liu, Z. Meng, H. Lin and C.H. Cheng. 2010. Structural Diversity of the Gnih/Gnih Receptor System in Teleost: Its Involvement in Early Development and the Negative Control of Lh Release. Peptides 31: 1034–43.

Zmora, N., D. Gonzalez-Martinez, J.A. Muñoz-Cueto, T. Madigou, E. Mañanos-Sanchez, S.Z. Doste, Y. Zohar, O. Kah and A. Elizur. 2002. The Gnrh System in the European Sea Bass (*Dicentrarchus Labrax*). J. Endocrinol. 172: 105–16.

Zmora, N., J. Stubblefield, Z. Zulperi, J. Biran, B. Levavi-Sivan, J.A. Muñoz-Cueto and Y. Zohar. 2012. Differential and Gonad Stage-Dependent Roles of Kisspeptin1 and Kisspeptin2 in Reproduction in the Modern Teleosts, Morone Species. Biol. Reprod. 86: 177.

Zohar, Y., M. Harel, S. Hassin and A. Tandler. 1995. Broodstock management and manipulation of spawning in the gilthead seabream, Sparus aurata. pp. 94–114. *In*: N.R. Bromage and R.J. Roberts (eds.). Broodstock Management and Egg and Larval Quality. Blackwell Press, London.

Zohar, Y., J.A. Muñoz-Cueto, A. Elizur and O. Kah. 2010. Neuroendocrinology of Reproduction in Teleost Fish. Gen. Comp. Endocrinol. 165: 438–55.

Zucker, C.L. and J.E. Dowling. 1987. Centrifugal Fibres Synapse on Dopaminergic Interplexiform Cells in the Teleost Retina. Nature 330: 166–8.

European Sea Bass Larval Culture

Enric Gisbert,[1,] Ignacio Fernández,[2] Natalia Villamizar,[3]
Maria J. Darias,[4] Jose L. Zambonino-Infante[5]
and Alicia Estévez[1]*

Introduction

Marine fish larval culture is one of the main bottlenecks in the aquaculture industry. At hatching, most of the organs of marine fish larvae are not fully developed, and developmental processes initiated during embryogenesis continue during larval development in order to ensure the proper ontogenesis of organs and associated physiological functions, essential for the survival of individuals. Consequently, during the first weeks of life, marine fish larvae undergo significant morphological and physiological modifications to acquire all the adult features by the end of the larval period. This phase is certainly the most critical in the life cycle of fish, not only in natural environments, but also under controlled conditions. In aquaculture, in spite of the many advances made to control environmental cues, this stage of production can still suffer from substantial mortality or

[1] IRTA, Centre de Sant Carles de la Ràpita (IRTA-SCR), Unitat de Cultius Experimentals, Crta. del Poble Nou s/n, 43540 Sant Carles de la Ràpita, Spain.
[2] Centro de Ciências do Mar (CCMAR/CIMAR-LA), Universidade do Algarve, Campus de Gambelas, 8005-139 Faro, Portugal.
[3] Department of Physiology, Faculty of Biology University of Murcia, 30100-Murcia, Spain.
[4] IRD, UMR 226 ISE-M, Université Montpellier II, 34095 Montpellier Cedex 5, France.
[5] Ifremer, Department of Functional Physiology of Marine Organisms, Fish Nutrition Unit, BP. 70, 29280 Plouzané, France.
*Corresponding author: enric.gisbert@irta.cat

impaired larval quality, influencing the overall performance of the rearing production, whose final objective is the obtention of enough fry of good quantity and quality for on-growing purposes.

The early life stages of fish have stage-specific environmental and nutritional requirements. Information about the morphological development and growth patterns of young fish is important for fisheries management and aquaculture. Recognition of normal patterns and detection of developmental defects can be used to improve larval rearing techniques through the modification of environmental parameters and feeding practices. As there are interspecific variations in the timing of organ formation, development and functionality, it is necessary to conduct studies of organogenesis for individual species. This approach can also be used to compare specimens from different egg batches and to estimate juvenile quality and evaluate their suitability for re-stocking or further rearing. Length and mass are the most widely used markers although they do not show any strict relationship with development, whereas morphological and functional markers might be more useful indicators for the development of optimal rearing techniques. Developments in rearing technology and start-feeding practices have benefitted significantly from the above-mentioned studies on the developmental biology and ecology of fish eggs and larvae. Thus, this chapter will firstly focus on the embryonic and larval development of European sea bass with a comprehensive analysis of the ontogenetic changes during the early life stages of fish that help identifying limiting factors during their rearing, optimizing the rearing technology, and determining the appropriate time for weaning and synchronizing feeding practices with the developmental stage of the fish. Secondly, a description of the current larval production systems at an industrial and laboratory scale will be presented with special emphasis on the optimal rearing factors affecting fry production; this chapter ends by reviewing how some of the most important abiotic (light and temperature) and biotic (nutrition) factors affect larval performance.

Embryonic and Larval Development

Fertilized eggs of European sea bass are pelagic, spherical and translucent with mean diameters ranging between 1.07 mm and 1.32 mm (Barnabé 1976, Devauchelle and Coves 1988, Saka et al. 2001). They contain one to five oil globules that represent around 2%–3% of the total volume. The incubation period (IP, hours) depends on temperature (T, °C), obeying to the following equation when temperatures range from 13°C to 22°C: IP = 414.46–119.73 Ln T (r = 0.93) (Devauchelle and Coves 1988).

The sequence of appearance of embryonic structures remains constant during development. Nevertheless, embryos of the same age can be at different degrees of development depending on the environmental parameters, egg size and quality. Thus, it is of practical significance to divide the embryonic and larval development into different stages that consider morphologic characters and their timing of differentiation. Classification into developmental stages is a more accurate method to determine and standardize larval development than other criteria frequently used, such as the age from hatching (dph) or larval size, because it is independent of the rearing conditions and the water-rearing temperatures. This information may be used for evaluating and comparing the quality of embryos and the environmental conditions at which they are incubated and reared.

According to Saka et al. (2001), the **embryonic development** of European sea bass at 15°C is summarized as follows (Fig. 1): after 1.15 hpf (hours post fertilization) embryo is at two-cell (blastomer) stage, the second division (4 blastomers) occurs at 1.50 hpf. The 8-cell stage appears after 2.40 hpf. The subsequent symmetrical divisions result in smaller blastomers, and the morula and blastula stages are observed at 5.05 and 9.30 hpf, respectively. Gastrulation starts at 13.40 hpf and continues until 28.35 h. At 23.35 hpf, the germ disc covers half of the embryo, whereas it increases to 3/4 at 26.50 hpf and then, the blastopore closes (27.20 hpf). At 38.50 hpf, between five and six pairs of somites are visible in the developing embryo and somitogenesis proceeds in anterior-caudal direction. At 42 hpf, the kupffer apparatus is visible, while at 42.40 hpf, the heart starts to differentiate, while the embryonic pigmentation begins to develop at 43.50 hpf. During embryogenesis, the embryo progressively grows and it occupies 2/3 and 3/4 of the egg volume at 47.20 and 51.30 hpf, respectively. The optic cup is clearly visible at 58.50 hpf, whereas the primordial fin-fold starts to form at 60.30 hpf. The first heart beats start at 65.30 hpf. Then, the number of somites increases and pigmented melanophores appear in the cephalic region. At the stage of 25 somites, the pharynx is wide and the gill slits are opened laterally into the perivitelline space, whereas at the 28-somites stage the gut is formed and the anus opens. Hatching takes place between 84.25 and 87.10 hpf, when all embryos emerge from the egg.

After hatching, **larval development** may be divided into different stages according to the use of endogenous (yolk sac) and exogenous nutrients (Fig. 2), as well as the development of the main organs and systems. Although European sea bass is one of the most studied marine aquaculture species, there is no study that considers the morphological and anatomical development of this species under a holistic approach; thus, the following information is summarized from different studies on different organs and systems in developing larvae, and it contains data from both morphological and physiological points of view.

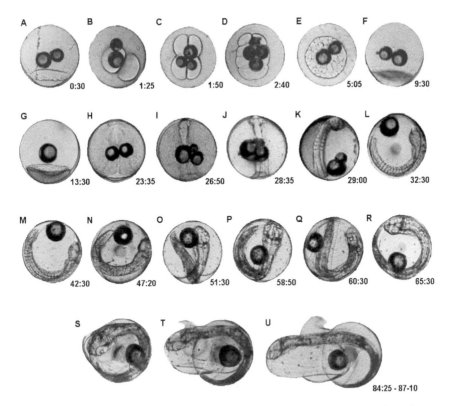

Figure 1. Summary of the main embryonic developmental stages of European sea bass shown chronologically (hours: minutes). Eggs (1.16 ± 0.004 mm diameter) were at 15.0 ± 0.4°C and their hatching rate ranged between 84.8% to 89.4%. A) Germinal cytoplasm separated from the chorion by the periviteline space; B) fist cell division stage (2 blastomers); C) second cell division stage (4 balstomers); D) third cell division stage (8 blastomers). The symmetry and shape of blastomers at this stage is used as a visual criterion to assess the quality of the spawning [see review in Kjørsvik et al. (1990)]; E) morula stage; F) blastula stage; G) onset of the gastrulation stage; H-I) gastrulation proceeds and the germinal disc covers half and three-quarter parts of the egg, respectively; J) the germinal disc covers the totality of the egg; K) onset of embryo's segmentation in the middle part of the body; L) the segmentation of the embryo proceeds towards the caudal region; M) formation of the Kupffer apparatus and heart; N-O) the embryo occupies two-thirds and three-quarter parts of the egg, respectively; P) formation of the optic cup; Q) formation of the primordial fin-fold; R) onset of heart beating; S-U) different stages of hatching. Authors would like to highlight that this information is merely descriptive and may substantially change according to egg size and egg incubation conditions. Data were redrawn from Saka et al. (2001).

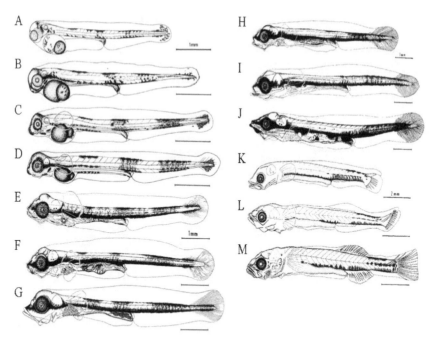

Figure 2. European sea bass larval development at 15°C (Barnabé 1976). A) 1 day post hatch (dph.), the oil globule orients the larvae in hyponeustomic position, the ventral part upwards or laterally. B) 3 dph, the larva is enlarged and yolk reserves decrease. C) 5 dph, pectoral fins and eye pigmentation start to develop. Larvae are immobile and their heads face the bottom of the tank. D) 7 dph, the mouth opens, the eyes are pigmented and larvae are oriented almost horizontally. Larvae show periods of active swimming alternated with periods of immobility in the water column. E) 12 dph, the yolk sac reserves are exhausted, with the exception of vestiges of the oil globule, and active feeding is well established (endoexotrophic period). F) 15 dph, the liver is clearly visible; larvae are well pigmented and feed actively. The rays of the caudal fin start to form. G) 17 dph, the yolk sac is completely depleted and swimming activity is almost constant. The skin is heavily pigmented along the vertebral axis. H) 20 dph, vertebrae are visible along the vertebral column. I) 25 dph, the swim bladder inflates and is clearly visible. J) 28 dph, the rays of the caudal fins are formed. K) 32 dph, the anal fin begins to form. L) 35 dph, larvae present 17, 13 and 12 rays in the caudal, anal and dorsal fins, respectively, and their swimming capacities are improved. M) 40 dph, the caudal fin becomes homocercal and the primordial fin fold is almost reduced to the post anal region. The pelvic fins appear 2–3 days later. The main structures, with the exception of the first dorsal fin, are acquired at this stage. The scales appear at 75–80 dph (not shown).

Stage I: Yolk sac larvae or the endotrophic period

At hatching (ca. 93 hpf and 3.5 mm total length (TL) at 18–19°C), European sea bass larvae have a large yolk sac located beneath the digestive tract that extends along half of the total body length (García-Hernández et al. 2001). The yolk sac contains two kinds of nutrient reserves, the external vitellus and the internal oil globule surrounded and separated from each other by the periblast, a syncytial envelope (Diaz et al. 2002). At this stage, the anterior part of the gut (pharynx and first gill slit), the otic vesicle and the pancreatic and hepatic primordial, located dorsally and ventrally to the anterior midgut, respectively, are visible (Tan-Tue 1976, Beccaria et al. 1991, García-Hernández and Agulleiro 1992, Sucré et al. 2009). The gut consists of a straight tube, limited by a single cellular layer, with a smooth lumen that curves in the most caudal zone (García-Hernández et al. 2001).

The structures necessary to articulate the mouth for feeding, involving prey capture and ingestion, the maxillary, Meckel's cartilage, hyosymplectic and quadrate, and the gill arches for breathing (hyoid and gill arches) start to form at ca. 4 mm TL (Gluckmann et al. 1999). The operculum appears as a bud (Sucré et al. 2009), the *cleithrum* is already ossified (Marino et al. 1993) and gives the support for the pectoral fins that begin to develop and are the main way of movement during the early life history. The kidney is present as an aglomerular pronephros located along each side of the digestive tract that connects in the posterior part to the urinary bladder (Nebel et al. 2005). At 2 dph (days post hatch), the dorsal side of the urinary bladder is differentiated into a transport epithelium, showing a strong capacity of osmoregulation (Nebel et al. 2005). At 3 dph, the gall bladder is visible between the liver and the pancreas, ventrolaterally to the posterior region of the foregut. The exocrine pancreas consists of tubules connected to a common epithelial duct that opens to the ventral side of this region of the gut (Beccaria et al. 1991). More information about the ontogenesis of the pancreas may be found in Tan-Tue (1976), Beccaria et al. (1990), García and Agulleiro (1992) and García-Hernández et al. (1994).

At the end of Stage I (5 dph, 5 mm TL at 15°C), the pharyngeal cavity extends anteriorly to form the mouth and gill arches and opercula are also well differentiated (García-Hernández et al. 2001, Sucré et al. 2009). The digestive tract is lined by a simple squamous epithelium, except in the most caudal region, where the epithelium is pseudostratified and lined by ciliated cells (García-Hernández et al. 2001). A large number of flat chloride cells are visible on the surface along the entire body, including the fins, and their function is to maintain the hydromineral balance of the organism before osmoregulatory organs are fully differentiated and functional (Varsamos et al. 2001, 2002). The low ionic regulation observed in the region of the future stomach (Giffard-Mena et al. 2006) and the presence of lamellar structures associated with mitochondria in the epithelial cells of the gut (García-Hernández et al. 2001) suggest that the digestive tract is involved in osmoregulation before the digestive function. During this stage and the transition to exogenous feeding (endoexotrophic period), the periblast uses yolk reserves and a large amount of oil globule as a source of energy (Diaz et al. 2002). Lipids contained in yolk are cleaved into lipoproteins that are released into the general blood circulation system through the hepatic sinusoid (Diaz et al. 2002). Signs of lipid digestion are also seen within the epithelial cells of the gut (Deplano et al. 1991, García-Hernández et al. 2001), whereas there are no evidences for protein absorption in the posterior region of the digestive tract before first feeding (García-Hernández et al., 2001). The ultrastructural characteristics of the enterocytes are essentially the same as in adult fish and the existence of functional lipid absorption structures, such as well developed endoplasmatic reticulum and Golgi apparatus at the time of first-feeding has been noted. These facts suggest that the enterocytes are not only cytologically differentiated but also physiologically functional during the transition to exogenous feeding.

Stage II: Mouth opening

The mouth opening occurs in larvae measuring 5 mm TL at ca. 5 dph (15°C) and the yolk sac progressively decreases in size, until its complete reabsorption during this phase. The swim bladder undergoes the primary inflation (Chatain 1986). The sensory organs develop and foraging behaviour is improved due to the development of three pairs of neuromasts on the head and seven on the flanks, aligned behind de pectoral fins, which also confer positive rheotaxis to the larvae (Diaz et al. 2003). Regarding the digestive system, the liver and pancreas increase in size and volume, and zymogen granules (precursors of pancreatic digestive enzymes) increase in the exocrine pancreatic tissue (Beccaria et al. 1991). Such increase in zymogen granules between 3 dph and 5 dph coincides with an increase in trypsin activity levels (Zambonino-Infante and Cahu 1994). Synthesis of the pancreatic enzymes is genetically programmed until this stage; however, larvae acquire the ability to modulate their digestive capacities according to the composition of the diet after this phase (Beccaria et al. 1991, Zambonino-Infante et al. 1994). At the end of Stage II, four differentiated regions of the digestive tract may be identified: the oesophagus, gastric region, intestine and rectum (Tan-Tue 1976, García-Hernández et al. 2001). Goblet cells producing mucous substances appear in the oesophagus and in both intestine and rectum, together with an incipient valvular structure that divides the intestine from the rectum (Tan Tue 1976, García-Hernández et al. 2001). The pronephric kidney becomes functional at this stage. It is composed of a pronephric renal corpuscle with a single large glomerulus located between the notochord and the digestive tract (Nebel et al. 2005). Two different sections are discernible: the anterior long urinary tubules and the short urinary ducts. During larval development, the pronephric tubules increase in diameter and length, and become folded and differentiated

(Nebel et al. 2005). Chloride cells increase in the gill arches and later will exclusively be located in the developing gill filament (Varsamos et al. 2002).

At 5.4 mm TL, the notochord starts its segmentation. The second neural process and those near the caudal fin start to be visible and continue to develop into *caudad* and *cephalad* direction, respectively. The haemal processes of vertebral bodies are also distinguishable at that time. The development of cartilaginous parapophyses takes place *cephalad*. At 6.6 mm TL, the neural and haemal processes from each side of the notochord join together to form the neural and haemal arches, respectively (M.J. Darias, personal communication). The flexion of the notochord occurs at 6.3 mm SL, together with the appearance of the caudal fin hypuralia, and is completed at 8.5 mm SL (Marino et al. 1993). Splanchnocranial bones are the first skeletal structures to mineralize (Gluckmann et al. 1999, M.J. Darias, personal communication).

Stage III: Complete yolk sac resorption and swim bladder final differentiation

During this period, the swim bladder undergoes the second stage of development (ca. 11 mm TL) that is characterized by the apparition of a new air bubble that merges with the first one, resulting in a larger hydrostatic organ (Chatain 1986). Neuromasts dramatically increase in number and get aligned to conform the future position of the sensory canals. At 15 mm TL, larvae have 20 neuromasts on each side of the head, around the eye and the nasal cavity and on the operculum. The future lateral line includes 35 neuromasts, some of which are aligned dorso-ventrally near the caudal fin. At the end of this stage, larvae begin to display gregarious behaviour coinciding with the development of the lateral line and eyes (Diaz et al. 2003). Chloride cells attain their final stage of differentiation and their activity increases (Varsamos et al. 2002). At this point, there is a transition from the skin to the gills and kidney, and oesophagus to a lesser extent, as osmoregulatory organs (Varsamos et al. 2002, Giffard-Mena et al. 2006).

Regarding the digestive system, the oesophagus continues to develop and their mucosal folds become deeper, as well as the number of goblet cells. The gastric region presents longitudinal and transversal mucosa folds and the pyloric caeca starts to form. The mucosa folds of the anterior gut increase in length and number during this stage, as well as the number of goblet cells containing and secreting neutral mucosubstances. The connective tissue and the muscle layer enter at the junction of the stomach and the intestine to form the pyloric caeca (García-Hernández et al. 2001). The pancreas becomes more diffuse and increases in size. By 14.5 mm TL, zymogen granules are homogeneous and occupy most of the pancreocyte (Beccaria et al. 1991). This change in structure has been related to variations in enzymatic activity, especially in the amount of trypsin produced (Alliot et al. 1977). The activity of pancreatic digestive enzymes, trypsin and amylase, follows the expression profile of their coding genes, trypsinogen and amylase (Darias et al. 2008). Trypsin activity increases from 16 dph to 23 dph followed by a sharp decrease (19°C, Zambonino-Infante and Cahu 1994), while the activity of amylase and brush border membrane enzymes remains constant until 23 dph. From that day onwards, the activity of these enzymes is lower. At this developmental stage, a more efficient digestive system begins to be established, where the basic protein digestion in the intestinal lumen is replaced by a more efficient acidic digestion in the stomach with the progressive abundance of other more effective protein digestive enzymes such as gastricsin (Darias et al. 2008). Because protein is one of the major components of fish larval diet, the activity levels of pancreatic proteolytic enzymes, i.e., trypsin and chymotrypsin, are well suited as indicators of the nutritional status of the organism. Thus, secretion rate of trypsin has been related to feed intake and stomach replenishment, and starvation or reduced food intake normally results in a decrease in enzyme activity. Nevertheless, other alkaline proteases, like chymotrypsin, may be an even better indicator of nutritional status during larval stages in some species, whereas Cara et al. (2007) demonstrated that both trypsin and chymotrypsin allow an early discrimination between batches of European sea bass larvae that will show differences in survival rates. On the other hand, comparatively high activity of both enzymes was linked to a sub-optimal feeding status, possibly indicating a short-term compensation effect of both enzymes, and particularly of chymotrypsin, to decreased nutrient availability.

The growth in size and number of muscular fibres takes place continuously during larval development by hypertrophy and hyperplasia that are dependent on environmental conditions (Ramírez-Zarzosa et al. 1995, Lopez-Albors et al. 1998). The fibres of the red muscle (superficial layer) increase in number and size from 25 dph until they attain the properties of the adult red fibres (Scapolo et al. 1988, Ramírez-Zarzoza et al. 1995). The growth of the white muscle (deep layer) occurs by the generation of new fibres from the presumptive myoblasts existing inside the myotome at hatching (Veggetti et al. 1990). A new stratum of fibres is continuously added in the most superficial zones of the white muscle and may be the origin of the pink muscle or intermediate layer (Lopez-Albors et al. 1998). Regarding the skeleton, fin development does not occur in a continuous way but through pauses and periods alternated with rapid changes. In particular, seven cartilaginous anal fin buds appear corresponding to the median region of the anal fin, improving larval swimming performance

(Marino et al. 1993). At 11.5 mm SL (standard length), the first eight pterygophores appear in the posterior region of the future second dorsal fin, and develop *rostrad*. At 14.4 mm SL, the first dorsal fin separates from the second one. Ossification proceeds *caudad*. The most anterior dorsal pterygophore originates from two pieces of cartilage, contrary to the anal one, which is formed from one single piece of cartilage (Marino et al. 1993). The final cartilaginous development of the vertical fins precedes that of paired fins. The components of the pelvic fin are the last ones to appear. The *basipterygium* is visible first at 11.30 mm SL and the development of pelvic fin rays occurs fast, whereas fin rays were completed at 14 mm SL. Besides, ossification of fin supports occurs simultaneously from 21.7 mm SL (Marino et al. 1993). Regarding the axial skeleton, the four paraphophyses are visible at 8.3 mm TL, the first one corresponding to the eighth vertebra. Intramembranous ossification of the vertebrae begins at the *centra* and ocurrs *caudad* (M.J. Darias, personal communication). At 12.8 mm TL, the vertebral column is more developed and shows higher number of ossified vertebral *centra*. Besides, neural and haemal arches start to ossify. At 15.5 mm TL, the basioccypital articulatory process, that connects the skull to the vertebral column by the first pleural vertebrae, vertebral *centra* and neural and haemal arches are completely ossified. At this time, dorsal and pleural ribs are completely formed. The number of vertebrae varies between 24 and 26, 25 being the most frequent number. The pleural vertebrae are composed of neural arches and spines dorsally and parapophyses ventrally from the eighth to the eleventh vertebrae. The caudal vertebrae are provided of neural arches and spines dorsally and haemal arches and spines ventrally. The neural and haemal spines of preurals 1 and 2 are modified to support the caudal fin rays (M.J. Darias, personal communication). The degree of bone mineralization during the larval development is well correlated with the exponential pattern of osteocalcin gene expression, which is involved in differentiation and mineralization of osteoblasts, the bone forming cells. From 8.3 to 15.5 mm TL, the bones mineralize notably, nearly concerning the whole larval skeleton (skull, vertebral column and caudal complex) (Darias et al. 2010a).

Stage IV: Development of gastric glands, end of the larval development

This stage encloses the end of the larval development and the beginning of the juvenile phase. A significant increase of the osmoregulatory capacity occurs at this stage, as the organism has the capacity to maintain constant plasma osmolarity (Varsamos et al. 2001, 2002, Nebel et al. 2005). From the juvenile stage onwards, the structure of the gills and the distribution of chloride cells are those observed in young adults (Varsamos et al. 2002). In juveniles, the amount of hematopoietic tissue increases in the head kidney, whereas the pronephric tubules present at earlier stages degenerate. The opisthonephric tubules differentiate and become more numerous and longer in the median part of the urinary system (Nebel et al. 2005). Juveniles form isotonic urine in seawater and hypotonic urine in diluted brackish water. The ability to produce diluted urine at low salinity increases in pre-adults (Nebel et al. 2005).

Concerning the digestive system, the oesophagus and rectum present minor changes during this phase. The intestine folds form two intestinal loops and increase in length in the anterior zone. The goblet cells become more numerous and increase towards the caudal zone. Four or five pyloric caeca are visible and increase in diameter progressively. The first goblet cells containing neutral mucosubstances appear in this region. The development of the digestive tract is completed during this stage with the formation of the gastric glands. The anterior gastric region develops caudally and surpasses the posterior one, producing a blind sac, dorsal to the intestine. The lining epithelium folds transversally and forms the gastric pits, where tubular gastric glands develop at around 55 dph and proliferate fast. Differentiated gastric glands of the anterior region present a wide lumen limited by cubic cells with large, irregular apical cell processes, whereas no glands differentiate in the posterior gastric region. The completely differentiated stomach consists of a glandular, descending branch and a non-glandular ascendant or pyloric one (caecal type) (García-Hernández et al. 2001). The decrease in expression level of trypsin precursors, combined with the most intense gastricsin (pepsinogen C) expression level at 43 dph, illustrates the importance of acid digestion in sea bass (Darias et al. 2008). A fully developed stomach together with an acid digestion is considered as the end of the transition from the larval to the juvenile stage and the beginning of adult-type feeding characteristics in fish (Govoni et al. 1986).

Juveniles can swim in faster water currents due to their improved swimming performance. The lateral musculature is characterized by a myotomal organization, in which two main fibre types are grouped into three muscle layers, the superficial (red muscle), the intermediate (pink muscle) and the deep (white muscle) layers. New intermediate muscle fibres are produced from myosatellite cells (Ramírez-Zarzosa et al. 1995), and at the transition zones, intermediate fibres progressively transform and incorporate into the red or white muscles (López-Albors et al. 1998). The lateral line system develops in concomitance, forming canals that respond to hydrodynamic biological stimuli, even in the presence of background noise, through mechanical filtering properties (Diaz et al. 2003).

Larval Culture Systems

A bit of history

The first attempts to cultivate sea bass larvae under intensive conditions started in France in the seventies (Barnabé 1974, 1976, Girin 1976, 1979, Barahona-Fernandes 1978). The experiments carried out from 1976 to 1978 were used to define the optimal environmental conditions (e.g., temperature, aeration, flow rate, salinity and light conditions) needed for larval rearing, as well as for establishing the most convenient feeding protocols using enriched live preys (rotifers and *Artemia* nauplii and metanauplii). Although initially, microalgae were added at experimental level to the tanks to assure the nutritional quality of the live prey, later, in 1982–1983, when the "green water" larval rearing technique was tested in larger water volumes (1,000 to 20,000 litres), two main problems arose: i) a high percentage of fish with skeletal deformations, mostly abnormal mandibular arches (Barahona-Fernandes 1982) and animals with severe lordosis derived from the absence of a functional swim bladder (Chatain 1986, 1987), and ii) the abnormal whirling swimming behaviour of some larvae with loss of appetite and emission of white faeces. Thus, in the eighties most of the effort was centred in improving rearing techniques in order to increase the survival and quality of the juveniles produced.

In contrast to other marine fish species, the green water technique was abandoned and continuous water renewal was used from the beginning of the larval rearing period in order to maintain and improve water quality. In addition, artificial and high light intensity regimes were also discarded by the industry and replaced by a more natural lighting, using lower intensity and a photoperiod regime with only 9 h to 12 h of light per day (9–12hL:15-12hD); the colour of the tanks was also changed to black walls (Ronzani-Cerqueira 1986). With these changes in the rearing protocol, a substantial improvement in the frequency of juveniles with normally inflated swim bladders was obtained, as well as an important increase in the final survival, from 15% using the green water technique to 35%–50% with this new protocol (see reviews in Coves 1985, Coves et al. 1991). This information generated was then transferred to industrial European sea bass hatcheries where different rearing techniques (intensive, semi-intensive, mesocosm) have been applied since the late eighties.

Larval rearing methods

Intensive systems

A method known as the "French Technique" was developed by the Equipe Merea-IFREMER during the eighties (Coves 1985) and has been used, with modifications, for most of the 90 European marine commercial hatcheries to produce approximately 1–5 million fry per year for their own use and also by the 10 companies that produce more than 10 million juveniles per year for sale to external customers (Shields 2001). According to FAO, total production of European sea bass juveniles in the Mediterranean countries reached 53 million in 2009 (FAO 2010). Some hatcheries incubate the eggs directly into the larval rearing tanks or into floating containers inside the tank (4,000–7,000 eggs/l in each container) with a water renewal of 5%–10%/h, to assure a good microbial environment, at 14°C–15°C temperature for 110h, with a final hatching rate between 80% and 90%. In other hatcheries, egg incubation takes place in small volume tanks (100–500 litres) and newly hatched larvae are stocked in large volume rearing tanks at a density of 100–150 larvae/l. The general rearing conditions used by most hatchery managers are summarized in Table 1. These rearing conditions can be applied using small and/or large volume, cylindro-conical tanks, ranging from 1,000 litres to 20,000 litres (Estévez and Planas 1987), always assuring good water quality (UV filtered seawater) and renewal.

Swim bladder inflation is essential for functional buoyancy control, swimming ability and feeding success of most marine cultured larvae. Failure to inflate the swim bladder has been considered one of the main obstacles in the commercial rearing of several species. Fish lacking a

Table 1. Culture conditions for European sea bass larval rearing (after Coves et al. 1991, Barnabé 1994, Moretti et al. 1999, Büke 2002). SWB= Swim bladder, dph = Days post hatching.

Factor	Range of values
Temperature (°C)	15°C–17°C from hatching until SWB inflation (0.5°C/day increase) 17°C–20°C afterwards
Salinity (ppt)	25–26 ppt from mouth opening to 17 dph
Light Intensity	Incubation in total darkness 20–100 lux from hatching until 13 dph 500 lux 17 dph onwards
Photoperiod	0hL:24hD the first 10 days of culture or 8hL:16hD from 0 to 16 dph 16hL:8hD 17 dph onwards
Water Renewal (%/h)	10% 0–10 dph, 20% 10–20 dph, 30%–50% 20–40 dph
Oxygen (mg/l)	5–7 mg/l, aeration only after SWB inflation, or gentle aeration (0.2–0.6 l/min) to allow the larvae to take air from the water surface of the tank
Surface skimmers	During swim bladder inflation from 4–20 dph

functional swim bladder have been reported to show high mortality (Chatain 1986), increased metabolic rate (Marty et al. 1995), delayed growth and severe skeletal deformities affecting their quality and compromising their commercialization (Chatain 1994, Kitajima et al. 1994, Trotter et al. 2001, Boglione et al. 2013). Light and aeration have been proved as the most important factors affecting swim bladder inflation, thus gentle aeration (or no aeration at the beginning of larval rearing) is recommended to allow the larvae to take gulps of air from the surface during the first 10 days of culture (primary inflation). After this period, the pneumatic duct that connects the swim bladder with the digestive tract degenerates (Woolley and Qin 2010) and the swim bladder continues its development with a second period of expansion and the formation of a second gas bubble that merges with the one already present (Chatain 1986). Photoperiod and light intensity are other clues for changes in swim bladder volume; thus, before and during the initial swim bladder inflation, a dark or low intensity light phase is necessary due to the nocturnal swim-up behaviour of the larvae to gulp air from the surface. Later on, light intensity, as well as the number of hours of light is progressively increased to allow the larvae to feed on live prey. Recent research regarding not only light intensity, but its spectrum has shown that under the light conditions that best approached those of the natural aquatic environment, that is using blue light, and under a 12hL:12hD photoperiod, the performance of European sea bass larvae, in terms of growth, development, and swim bladder inflation, was significantly improved (Villamizar et al. 2009). To allow a clean surface for the larvae to gulp the air, apart from the absence of light it is necessary to use surface skimmers that remove any floating debris and/or oil derived from the enriched live prey, by blowing air at low pressure tangentially to the water surface (Moretti et al. 1999).

Another factor with a high influence on European sea bass larval growth is salinity; thus, larvae can be reared from hatching until metamorphosis (day 45) at low salinity (25–26 ppt) with a positive effect on growth performance due to the high tolerance to low salinity of this species at early stages of development (Varsamos et al. 2001). Low salinity induces a decrease in the energy expenditure and consequently, an increase in larval survival and growth rates, as well as an improved efficiency in the process of swim bladder inflation (Saillant et al. 2003). Although some hatcheries apply this low salinity method for European sea bass larval rearing, others do not apply it due to the production costs, since rearing larvae in low salinity environments implies having a double water line system, one line for marine and another for fresh water, as well as a mixing reservoir.

There almost exists one feeding protocol for European sea bass larvae for each known hatchery, since hatchery managers tend to adapt the procedures according to their expertise, available facilities and previous results. Whatever the protocol chosen, the larval rearing of European sea bass is a typical intensive rearing technology ideally involving complete control over the environmental parameters and fish population. Two different protocols are acknowledged. The first takes place in a lighted environment using rotifers as first feeding similar to gilthead sea bream larval feeding protocols, whereas the second protocol, frequently referred to as the "French technique", is characterised by a dark environment during the first days after hatching, and by the use of small newly hatched brine shrimp nauplii as first food.

A standard feeding protocol for European sea bass using lighted environment during the first developmental stages (Moretti et al. 1999) may be summarized as follows: larvae aged from four to seven days receive a daily amount of 20 million rotifers, 2 million small size *Artemia* nauplii (*Artemia* quality grade: AF or BE) and 40 litres of mature algal culture, the latter one decreasing gradually till day 23. In some cases, when no microalgae are added to the rearing tanks, rotifers are previously enriched with tailor-made or commercial enrichment products. Then, from day 8 to 12 the amount of rotifers is progressively increased to 25 million and *Artemia* nauplii to 3 million. As larval growth significantly increased during this period, the amount of rotifers distributed between 13 to 16 days starts to decrease to 15 million, whereas the *Artemia* nauplii ration is increased to 4 million. From day 17, in order to begin getting larvae accustomed to inert feeds and facilitate weaning, a small quantity of a microdiet (80–200 µm) is distributed together with 10 million rotifers for the smaller part of the fish population with 6 million of *Artemia* nauplii (quality grade: AF or BE) with 2 million of *Artemia* nauplii (quality grade: EG or RH) for the larger fish. The distribution of two types of *Artemia* with different naupliar sizes guarantees the proper feeding of smaller and larger fish, and prevents them from starvation, since fish population inside the larval rearing tanks is not uniform. Enriched *Artemia* metanauplii are offered from day 20 at a ration of 14 million, whereas rotifers decrease to 5 million and inert feed is gradually increased to 10 g. At 24 days, no more microalgae and rotifers are added to the larval rearing tanks, and the quantity of *Artemia* nauplii (quality grade EG or RH) is increased to 16 million and the quantity of inert feed to 10g–15 g. Between 28 and 34 days, the amount of enriched *Artemia* metanauplii is increased to 20 million, as well as the amount of inert feed (150–300 µm) that is increased to 20 g. From day 40, when metamorphosis from post-larval to juvenile (fry stage) is almost completed, the enriched *Artemia* metanauplii ration is decreased (down to 16 million) and the 150–300 µm inert feed is increased to 20 g. From now on fish will be ready to move to the weaning sector.

The "French larval rearing technique" is based on the principle that at the end of the 8–10 days darkness/low light intensity period (depending on the facility and rearing conditions), larvae can be fed either directly, enriched *Artemia* metanauplii from 8 dph–10 dph onwards (Coves et al. 1991, Saillant et al. 2003) or enriched rotifers (*Brachionus plicatilis*) for only 2–3 days before *Artemia* feeding, or from the mouth opening stage to 20 dph (Büke 2002) together with *Artemia* nauplii, in order to reduce live prey production costs (rotifer production and maintenance is cheaper than *Artemia*). Feeding with enriched *Artemia* metanauplii at increasing densities and number of doses continues until weaning that usually occurs at day 42–45 when the larvae reach a weight of 40–50 mg (Büke, 2002). Under the above-mentioned intensive larval rearing conditions, growth in weight can reach from 8.5–8.7%/day at 30 dph (Hatziathanasiou et al. 2002) to 4.0%–4.5%/day at 45 dph (Coves et al. 1991), and survival rate may average 50% (35%–65%) with 80%–100% of 45 dph fry with normally inflated swim bladder (Coves et al. 1991, Büke 2002).

The replacement of live prey, rotifers and *Artemia*, by compound diets has been one of the main objectives of fish larval rearing for the last three decades. Thus, the development of high-quality artificial microparticulate diets may potentially ameliorate water quality and overcome some disease problems, as well as reduce the high cost of live feed production, since rotifers and *Artemia* production and their enrichment procedures require considerable space, manpower and labour. In contrast, microdiets have a high and constant nutritional value, they are easier to maintain and have lower production costs. These advantages have significant implications for the future sustainability of marine fish larvae production. Although the formulation and manufacturing of microdiets have been improved during the last years and several commercial microdiets exist in the market (Holt et al. 2011), artificial diets still led to poor larval performance compared to live preys and their successful replacement has only been fully or partially achieved in a very limited number of marine fish species (Holt et al. 2011), including European sea bass (Zambonino-Infante et al. 1997, Cahu and Zambonino-Infante 2001, Villeneuve et al. 2005, among other publications from the same research group). In this sense, larvae of this species can also be reared from mouth opening onwards using only microdiets, although this technique has only been applied experimentally for nutritional studies and hatchery managers prefer to use live prey to assure a high survival rate.

Regardless of the feeding protocol chosen, live preys are usually distributed in larval rearing tanks by hand into the areas of the tank surface where larval density is lower. Feed is distributed three times per day, starting as soon as the lights have been switched on in the morning until four hours before the artificial sunset, in late evening. A quick distribution of the first ration in the morning is recommended to stop the forced starvation, which takes place during darkness, as larvae are visual feeders and do not feed under darkness. As enriched live prey tend to lose their nutritional

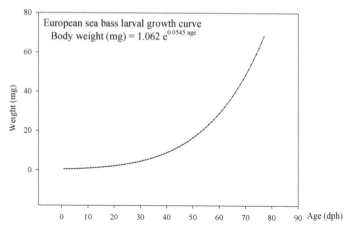

Figure 3. Growth of European sea bass larvae reared under intensive culture conditions until 80 days post hatching and under the environmental parameters shown in Table 1 (C. Aguilera, pers. com. 2012).

value over time, especially with regard to their content in essential highly polyunsaturated fatty acids (HUFAs) (Navarro et al. 1999), outlet removable screen meshes are increased to 500 µm overnight in order to remove uneaten live prey. During light hours and if water renewal is needed during feeding, the outlet screen mesh is set up between 100 µm and 250 µm in order to keep enriched rotifers and *Artemia* in the tank.

Using the "French technique" for European sea bass larvae, weaning starts when larvae are 42–45 days old, usually after discarding those without functional swim bladders. For this purpose, larvae are anesthetized with Tricaine methanesulfonate (MS 222, 0.07 g/l) and introduced in a container filled with high salinity water (60 ppt). Juveniles with normal swim bladder float whereas those without it sink to the bottom (Coves et al. 1991). After collecting normally developed (inflated swim bladder) post-larvae, they are transferred to on-growing tanks at a density that can go from 4 fry/l to 1.5–2 fry/l (Coves et al. 1991, Büke 2002, respectively). Although higher densities (up to 20 post-larvae/l) have been reported, low density is recommended to reduce the incidence of cannibalism (Hatziathanasiou et al. 2002). Weaning conditions differ from those used for larval culture, temperature is 18–22°C, salinity is increased to ambient 36–40 ppt, water renewal ranges between 50% and 100% total volume h⁻¹ and photoperiod increased to 14hL:10hD (Büke 2002) or natural (Coves et al. 1991). From day 42–45 until day 57 enriched *Artemia* metanauplii are given in decreasing quantities together with dry feed. Some mortality from 20–25% (Büke 2002) to 40% (Coves et al. 1991) is recorded during this adaptive period, mostly due to lack of adaptation of some animals to the new dry diet, although the above-mentioned losses depend on the initial larval weight and the co-feeding strategy used. Early weaning strategies have also been used; thus, Cahu and Zambonino-Infante (1994) obtained similar survival rates with lower final growth when larvae are weaned at day 20 compared to larvae fed on *Artemia*. If the dry feed contains the adequate amounts of lipids, larvae can be weaned even earlier (day 14) with good results in survival (48%) and growth performance (Zambonino-Infante and Cahu 1999). Commercially available microdiets of less than 100 µm are recently being used as a replacement of *Artemia* on larvae from 13–15 dph with encouraging results, especially considering the erratic quality of *Artemia* and its high cost and labour demanding production.

The mesocosm technique

Mesocosm is an indoor or semi-outdoor hatchery technique that has been applied successfully with European sea bass larvae (Nehr 1996, Divanach and Kentouri 2000). Although the method is not currently used in commercial hatcheries, it is notable for the high quality of the juveniles produced and recommended for small and resource limited producers (Shields 2001). Larvae are reared at relatively low densities (2–8 larvae/l) in relatively large (30–100 m^3), deep (1.5–2.5 m) tanks using long photoperiod, 15–21°C water temperature, and providing the tanks with surface skimmers to clean the oily film from water surface. Prior to introducing the larvae, a plankton bloom is created by adding *Chlorella* sp. or *Tetraselmis suecica* to the rearing water and rotifers, if used. Thus, when the larvae are introduced in the tank, they can feed on the prey as soon as they start exogenous feeding. To maintain water quality and prey density, newly enriched rotifers and *Artemia* are daily added to the rearing tank to adjust the desired densities. Using a mesocosm technique, in this case a 40 m^3 cylindro-conical enclosure fitted to a sea cage and using natural phytoplankton and enriched *Artemia* as live prey, Nehr (1996) obtained 82 dph larvae with a mean body weight of 1.8 g and without any skeletal deformation.

Several studies have compared fish larval performance when reared under intensive and mesocosm rearing systems. Distinction between systems takes into account factors concerning mainly larval stocking density, tank or enclosure size, water supply, prey source or feed contribution. According to Zouiten et al. (2011), European sea bass larvae reared in mesocosm showed a better growth performance than those reared in intensive systems, a more advanced level of maturation of their digestive systems, as well as a lesser incidence of skeletal deformities. As fish larvae ossification is size dependent (Vagner et al. 2007), fish reared under mesocosm conditions generally show a more advanced skeletal development that is suspected to contribute to ameliorate larvae nutritional status by improving their predation efficiency. In this sense, fish reared under intensive rearing systems present a behavioural delay when compared to those reared under mesocosm, and this can be justified by the fact that mesocosm conditions are approaching those found in the field, whereas the intensive ones comprise a more stressful environment for cultivated fish. In this sense, fish larvae in mesocosm rearing systems are subjected to lesser competition for food, which in combination with the larger space availability (tank volume), potentially may enhance their welfare (Kristiansen 2009).

Gnotobiotic culture for experimental purposes

Due to the risks related to antibiotic use in aquaculture (antibiotic resistance represents a threat for public health), the use of pre- and probiotics has been considered as another approach in health management practices of cultured fish. The presence of normal bacterial flora in larval fish makes it impossible to define the exact role of probiotics, thus a germ-free or so-called gnotobiotic culture for European sea bass larvae has been recently developed (Rekecki et al. 2009). Axenic rearing conditions needed for gnotobiotic larval rearing start with the disinfection of eggs. Embryos are disinfected for 3 min with 200 mg/l glutaraldehyde, incubated for three days in filtered sea water with 10 m/l ampicillin and 10 mg/l rifampicin. Newly hatched larvae are then transferred in vials filled with autoclaved sea water and 10 mg/l rifampicin. Eggs and larvae are kept in a temperature controlled room at 16°C in a constant dim light at 37 ppt salinity. Survival is relatively high (from 100% at 10 dph to 43% at 15 dph) and growth is even higher than non germ-free larvae showing that these germ-free larvae have a more developed gastrointestinal tract. This gnotobiotically produced European sea bass and also Atlantic cod larvae (Forberg et al. 2011) opens the door for more controlled studies on host-microbe interaction, host-response to bacteria or the use of pre- and probiotics.

Effects of Biotic and Abiotic Factors on European Sea Bass Larval Rearing and Performance

Larval stage is a very sensitive period influenced by many factors: abiotic (e.g., temperature, salinity, pH, CO_2 and O_2 water concentration, light intensity and photoperiod, radiation, flow rate and tank shape, volume and color) (Doroshev and Aronovich 1974, Bengtsson et al. 1988, Divanach et al. 1997, Divanach and Kentouri 2000, Villamizar et al. 2011, among others); biotic (e.g., stocking density, handling, parasites, pathological infections) (Pommeranz 1974, Lom et al. 1991, Boglione et al. 2013, among others); physiological (e.g., stress, infectious diseases); xenobiotic (e.g., pollutants, algaecides, insecticides, heavy metals, pesticides); genetic (e.g., hybridization, inbreeding) (Daoulas et al. 1991, Madsen et al. 2001, Sadler et al. 2001, Finn 2007, among others) and nutritional (deficiencies or excesses in macro- and micronutrients) (Cahu et al. 2003, Boglione et al. 2013, among others). Among the above-mentioned parameters, not all of them have received the same attention from a research point of view; however, in this section a special emphasis is going to be devoted to the effects of light regimes and intensity, temperature, as well as nutrition on European sea bass larvae under different rearing systems, since these are some of the main critical factors affecting fish larval performance.

Effects of environmental parameters and cycles on larvae

The circadian system in fish comprises multioscillatory mechanisms entrained by light, temperature and food (López-Olmeda and Sánchez-Vázquez 2011). However, the signalling pathway that couples the environmental signals with the ontogeny and survival is yet unclear and currently most of the research performed is based on the response of fish larvae to constant environmental conditions; therefore, the natural photo- and thermo-cycles are rarely considered. Regarding constant conditions, there is strong evidence of the early effect of light and temperature on fish larvae as the exposure to these entraining signals over the first days of life seems to be necessary for the initiation of robust behavioural rhythmicity (Hurd and Cahill 2002). Indeed, recent research has suggested the existence of rhythmic processes occurring at different stages of embryogenesis, indicating the presence of a time-control mechanism in embryos and even more, a high degree of synchronism among them that allows researchers to predict the temporal formation of structures (i.e., somites) (Gorodilov 2010). Although the early development of fish has been classically considered a succession of developmental stages which are mainly controlled by temperature, light also plays a significant role as fish seem to be able to detect light as early as five hours post fertilization (Tamai et al. 2004). However, more important are the daily changes in lighting and temperature conditions (i.e., sunrise, sunset) which represent an important time cue for the optimal temporal distribution of activities and physiological processes of the organism, such as cell proliferation and the transcription of genes involved in the circadian clock and DNA repair (Weger et al. 2011). Regarding European sea bass ontogeny, research under those conditions that mimic the daily oscillations of the natural environment remain poorly explored and at present, most of the investigation still focuses on the development of embryo and larvae under constant conditions (Saka et al. 2001, Pavlidis et al. 2011, Cucchi et al. 2012).

Important physiological processes involved in early larval development and growth respond to light and dark cycles by mitogenic stimulation, which is timed to occur at the end of the day or during the night in a remarkably great variety of species (Nikaido and Johnson 2000). Although light signals have been suggested as necessary cues for the rhythmicity of cell proliferation and consequently, normal growth and development (Dekens et al. 2003), the intensive "French technique" rearing protocol for European sea bass is characterized by the application of a dark period (constant darkness, DD) during the first days after hatching (Coves et al. 1991). Experimentally, European sea bass larvae raised under DD delayed the consumption of the endogenous reserves (yolk sac) by decreasing larval metabolism and consequently, decreasing body growth. This effect on growth was observed at a very early stage during endogenous nutrition, suggesting that light affected larval development before active feeding had started, therefore before food conversion played a key role. During the exogenous feeding, larvae reared under DD had very low foraging activity and died by 18 days post hatching (dph) (Villamizar et al. 2009). As a visual feeder, European sea bass larvae relies on ambient light for prey detection and this is even more critical during the early stages due to their short visual range that increases as larvae develop (Huse and Fiksen 2010). When European sea bass larvae were reared under constant light conditions (LL), larval growth was promoted and teeth and fins showed faster development (Villamizar et al. 2009). However, the above-mentioned larvae showed a higher incidence of skeletal malformations and swim bladder inflation anomalies (Villamizar et al. 2009). A strong correlation between abnormal development of the swim bladder and the apparition of vertebral deformities such as lordosis and kyphosis has been reported (Koumoundouros et al. 2002), which in turn may increase fry mortality rate. Another consequence of swim bladder anomalies or the lack of it is the poor growth performance resulting from the high energy cost for the larvae to maintain their buoyancy (Chatain and Ounais-Guschemann 1991).

Continuous illumination or dark conditions are evaluated by the hatchery managers as a balance of benefits and losses in order to obtain the best outcome. However, this is often a complicated issue especially when the natural physiology and behaviour of the species is not taken into account. Many animals including fish, exhibit an approximately 24-h (dial) cycle in their activities which mark several, if not all aspects of the fish, from the secretion of growth hormones to their locomotor and feeding activities. Several studies have focused on the light requirements for the culture of different marine species and the contrasting results found among species and light regimes suggest that the response to a particular light regime is a species-specific trait (Downing and Litvak 2001, Saka et al. 2001, Monk et al. 2006, Blanco-Vives et al. 2010). Regarding European sea bass, Roncarati et al. (2001) compared fish performance under intensive (DD, clear water, high stocking density) and extensive (natural photoperiod, mesocosm, low density) rearing conditions and they did not find differences regarding larval specific growth rate, feed conversion ratio and body weight. In addition, the above mentioned authors found that larvae reared under a mesocosm system had higher survival and better body shape. These results are in agreement with those obtained in other studies where European sea bass reared under natural conditions grew better when compared with intensively cultured groups (Melotti et al. 1993, 1994, Zouiten et al. 2011). Apart from the specificity or the response of each fish species to the photoperiod, some studies have found that there may also be diverse light requirements for different populations of the same species as reported for two different populations of Atlantic cod, where the best results were obtained when the natural specific photoperiod was applied to each population (Puvanendran and Brown 1998, Van der Meeren and Jorstad 2001). Based on the above mentioned results and many others, there is enough information that suggests that under artificial conditions, fish larvae performed better under the photoperiod that best resembles their natural environment.

In addition to the photoperiod, **light spectrum** also plays an important role on fish larval development and performance. In this sense, the response to light is the result of long term adaptations of the physiology and behaviour of fish to their surrounding environment. This response to light comprises two systems, the retinal mechanism for visual processing and the extraretinal mechanism with non-visual photoreceptors (Foster et al. 2007). Although some efforts have been made in order to clarify the way fish detect process and respond to light stimuli, the complexity of the system is far from being fully understood, especially with regards to the larval stage. Recent research has suggested that the response to light is species specific with a strong relationship between behavioural responses to artificial light and the behaviour under natural light conditions, as they can be related to both phylogenetic and ecological factors (Marchesan et al. 2005). The understanding of the fish's natural environment is mandatory in order to elucidate the complete mechanism behind the photoresponse behaviour, as well as the ontogeny of their photoreceptor cells.

European sea bass is a species adapted to brightness variations and dim light conditions as its retina has a well-developed pigmentary epithelium, abundant rods and large cones, whose mosaic layout (single and twin cones) contributes to the perception of movements, and hence the capture of fast moving prey (Mani-Ponset et al. 1993). A study on retinal organogenesis during larval stages showed that light detection may appear with photoreceptor differentiation at 3 dph, whereas after 5 dph larval vision progressively improves as the retinal structure starts to resemble the adult retina, although in a simpler way. This stage coincides with the start of the exogenous feeding when larvae is only able to detect their prey at 1 mm distance, but rapidly develops and increases the distance to 3 mm by day 13 and to 5–6 mm by day 20. In this period, light plays a crucial role, since the retina of larvae is formed only by cones, limiting their vision to day sight (Mani-Ponset et al. 1993). Although visual photoreception is well studied in European sea bass, there is no information whether this species has an extraretinal photoreception system as it has been described in other teleost species (Whitmore et al. 2000, Pando et al. 2001).

Conclusive studies about the spectral sensitivity of European sea bass larvae and fish in general are scarce, perhaps due to the lack of standardisation of experimental components such as the units of light measurement (i.e., lux, watts, photon flux), the light sources of very different spectral compositions (i.e., halogen, tungsten, light emitting diodes LED), the rearing system (i.e., tank dimensions) and husbandry protocols (green vs. clear water). In a recent study, European sea bass embryos and larvae were reared under three pure light spectra using light emitting diodes (LED) [blue (463 nm), red (685 nm) or white (367–700 nm)] under a 12L:12D photoperiod and normalized light intensity of 0.5 W/m^2 (Villamizar et al. 2009). The above mentioned authors found that larvae reared in the blue light grew better, developed faster (fins and swim bladder) and fed more actively than larvae from the rest of the experimental groups. This may be explained by the possibility that the larval visual system might be predisposed to better perform under light spectral conditions (short wavelengths) most frequently encountered in natural environments. In addition, blue light in larval rearing tanks increases the contrast of live prey in the water column, enhancing larval feeding behaviour, and consequently, larval performance. However, under hatchery conditions, the lighting system and illumination protocol are generally based on larval performance (growth and survival) and the lighting energy costs, whereas the particular ecology of a given fish species in most of the cases is rarely taken into account. In the majority of the rearing tanks used for the culture of fish larvae, the incident light is highly directional as the shallow and often clear water body of the fish tank has little absorption and scattering ability in comparison to natural habitats. Therefore, the light environment experienced by larval fish under culture conditions is substantially different from the natural environment that fish have been selectively programmed to develop under (Naas et al. 1996). At present, many hatcheries use either 'true-light' tubes, tungsten filament, fluorescent or metal halide lights, which create bright point light sources with spectral emissions of unnatural wavelengths that may not be detected by fish and/or could potentially compromise their welfare (Boeuf and Le Bail 1999, Migaud et al. 2007). These lights are characterized by lower colour temperatures (2700–3000 K; yellowish-white through red) rather than higher colour temperatures or cool colours (5000 K or more; bluish-white) found in the natural photo-environment (Wolfgang 1992). As an alternative, LED and cathode lights are currently being investigated as they can be tuned to match the target species sensitivities during their developmental stages through narrow bandwidth outputs (Migaud et al. 2007). Although the mechanism by which light spectrum affects larval physiology and behaviour during early development remains still unclear, the possible long-term consequences of early exposure to unsuitable lighting environments are becoming more evident, as proved by the ongoing results of recent research. The development of flexible lighting platforms that allow spectral specific adjustments for European sea bass requirements would certainly improve the culture of this species at early stages of development.

From all environmental factors influencing fish, water **temperature** is considered the "abiotic master factor", as it influences the behaviour, physiology and distribution of aquatic poikilotherms (Fry 1947). Thermal tolerance is also strongly related with the capacity of a developing fish to permanently change its phenotype in order to adapt to the variable temperatures of the water environments (developmental plasticity) (Kinne 1962). Despite fish being able to adapt to a rather wide range of temperatures, they have a preference for a particular temperature which is closely related with the optimal temperature for many biochemical and physiological process such as growth rate, food conversion, immune response and reproduction (Johnson and Kelsch 1998). This preferred temperature may change during the fish's life time since its selection could be influenced by the season of the year (Mortensen et al. 2007), feeding and nutritional status (Van Dijk et al. 2002), health (Golovanov 2006), developmental state and seasonal variations (McCauley and Huggins 1979).

European sea bass tolerates a wide range of temperatures (5–28°C), with 22–24°C being the reported optimal temperature for the species (Claridge and Potter 1983). However, it has been reported that the response to environmental factors varies within different strains; thus, European sea bass juveniles from the Mediterranean Sea are reported to grow quickly under the optimal range, but juveniles from British waters grow better at lower temperatures such as 18°C (Russell et al. 1996). Recently it has been suggested that the best results for rearing European sea bass fry would be at 25°C all year round (Dülger et al. 2012). In the early stages of fish development, temperature is known to have a major impact as suboptimal temperatures can lower fertilization (Brown et al. 1995), egg quality (King et al. 2003), hatching rate (Hokanson et al. 1973), as well as increase embryo deformities (Aegerter and Jalabert 2004). For European sea bass egg incubation and larvae rearing, Moretti et al. (1999) recommended applying the same temperature at which eggs are exposed to during the spawning and fertilization, which in natural conditions takes place in late winter and beginning of spring when temperatures range from 13–15°C. Afterwards, following yolk sac resorption, temperature should be increased

0.5°C/day until reaching 18°C, when the inflation of the swim bladder is completed, and then temperature is increased to 20°C by the fifteenth day. Thus, larval culture follows a very strict protocol where the environmental parameters are rigorously controlled to avoid any possible fluctuation, which in the case of temperature should never exceed 0.5°C within 24 hours. These constant conditions are far from the natural environment of the sea bass, where seasonal and daily oscillations of temperature influence several (if not all) aspects of the developing process. Surprisingly, the experimental research regarding the effect of temperature on European sea bass larvae has also been mostly focused on the effects of constant temperature conditions. Different responses are obtained when exposing European sea bass embryos and larvae to different temperatures within a narrow range (~ 2°C). In their study, Saka et al. (2001) investigated the upper limit of the optimal temperature recommended for incubating European sea bass embryos and found that hatching took place 19 h earlier under 17°C when compared to 15°C (68 vs. 87 h). Higher rearing temperatures (18°C and 21°C) promoted larval growth, although the normal development of the swim bladder was obtained at lower temperatures (12.5°C) (Johnson and Katavic 1984). Whatever the temperature chosen by hatchery managers, European sea bass larvae are reared under constant thermal environments, which are thought to maximize growth, but at the cost of reducing thermal tolerance and phenotypic diversity. Indeed, animals reared under cyclic conditions are more likely to survive environmental changes as temperature oscillations maximized thermal resistance (Woiwode and Adelman 1992, Schaefer and Ryan 2006). The effects of daily thermocycles on the growth, development, behaviour and survival of fish have been poorly studied. Under controlled conditions, in the absence of light cues, when thermocycles are applied, the thermophase usually elicits the same physiological response as the light phase, whereas the cryophase brings out the response corresponding to the dark phase (Rensing and Ruoff 2002). When light/dark and high/low temperature cycles are combined, a stable phase and maximum amplitude can be observed in the rhythms, but when the two synchronizers are phase-shifted, the phase of the rhythm is determined by either stimulus or by both, depending on the relative strength of each one and the sensitivity of the species (Rensing and Ruoff 2002). Thus "optimal temperature cycles", rather than optimal constant temperatures, should be investigated in order to maximise growth performance, survival and larval quality under intensive rearing conditions, as reported for other species (Diana 1984, Baras et al. 2000, López-Olmeda et al. 2006, Blanco-Vives et al. 2010, 2011). Preliminary results on the effects of thermocycles in European sea bass development and rearing practices are promising, as larvae reared at high temperature (21°C) during day hours and at lower temperature (17°C) at night showed a lower proportion of skeletal deformities when compared with larvae reared under constant temperature or under the inverse above-mentioned thermocycle (Villamizar unpublished data). In this context, recent advances reveal the strong impact of daily thermocycles when applied during the early development on fish; therefore, they open an interesting research area for the application of natural thermocycles under commercial intensive hatchery conditions.

Effects of nutrition on larval rearing and performance

Nutrition is considered as one of the most important biotic factors affecting larval performance directly impacting the efficiency of larval production. In this sense, when a particular nutrient is supplied through the diet in an inappropriate or unbalanced level and/or form, larval development and survival could be largely compromised. Information in fish about the effect/requirement of a particular nutrient is fragmentary depending on the nutrient, fish species, fish developmental stage and/or physiologic/ scientific point of view from which the effects or roles of the nutrients are studied. Several reports demonstrated how dietary content of different nutrients has an effect on fish immune competence (Oliva-Teles 2012, Kiron 2012), bone homeostasis (Cahu et al. 2003, Lall and Lewis-McCrea 2007, Boglione et al. 2013), intestinal maturation (Cahu and Zambonino-Infante 2001), and pigmentation pattern (Hamre et al. 2007), among other biological processes. Despite the availability of this additional information about the role of essential nutrients on fish physiology, and the current knowledge on larval nutritional requirements is still limited. From the hatchery managers' perspective, nutrients have been considered as a requirement for fish growth, survival, and an important economic cost in fish production, whereas feed companies have considered them as well as by their availability, digestibility, palatability, feed intake, feed conversion, utilization or their biological function (Glencross et al. 2007).

From a historical point of view, the nutritional requirements for early stages of development were initially assessed in European sea bass similarly to other marine fish species, using enriched live prey (traditionally rotifers and *Artemia*). However, although the nutrient content of live prey generally reflects that of the enrichment emulsion (Monroig et al. 2006), many factors may affect the enrichment efficiency of the emulsions (i.e., live prey density, water temperature, turbulence and/or dissolved oxygen) and finally impact the nutritional value of the diet. Moreover, physiological particularities of rotifers and *Artemia* affect the nutritional value of both live prey, which results in the fact that they are not always consistent with the nutritional profile of the enriching emulsion. For a detailed description of nutritional advantages and disadvantages of live preys see the review of Conceição et al. (2010). Consequently, one of the main objectives of the feed industry during the last two decades has been to develop a compound feed (microdiet) for first-feeding marine fish larvae in order to substitute live prey. However, the formulation of compound diets adequate for fish larval development is not easy to achieve since nutritional requirements during development are not well documented. Diet formulations were first based on data obtained from juveniles that hardly met the specific requirements during larvae development. During the last two years, several studies have been performed in European sea bass in order to better define the optimal contents of nutrients in larval diets based on phenotype determination, which enhanced the knowledge available on their role in larval development and performance. The first microdiet for European sea bass larvae was tested in 1993 by Person Le Ruyet and collaborators. The above mentioned authors reported that their formulated diet sustained good growth and survival in this species from day 40 onwards, whereas at that time the weaning of this species was usually conducted at day 55 in commercial hatcheries (Person Le Ruyet et al. 1993). In 1997, significant growth and good survival (close to that obtained with live prey feeding) were obtained in larvae fed only compound diet from day 20 (Zambonino-Infante et al. 1997). One year later, the same authors reported that 35% of larvae, fed exclusively compound diet from mouth opening, survived at day 28 (Cahu et al. 1998), results that were really good considering the state-of-the-art of larval feeding practices at that time. From this date until the present, microdiets for European sea bass larvae have not ceased to improve, and actually they are currently used for nutritional studies on this species from the first-feeding stage (Table 2), although hatchery managers tend to be more conservative and offer it to larvae at a later stage.

The use of dry feeds during larval development in European sea bass has demonstrated how larvae are able to modulate their physiology in response to a dietary change, as well as to determine the requirements for different nutrients for this species, and by extension, to others where this type of approach cannot be adopted. However, recent information, on the control of the expression of genes involved in developmental processes governing the harmonious development and performance of larvae, is not discussed in this section, since it falls out of the main scope of this chapter. Readers are therefore encouraged to consult the excellent reviews on this topic from Darias et al. (2008), Zambonino-Infante and Cahu (2010) and Mazurais et al. (2012), among others. Thus, the effect of different macro- and micronutrients on European sea bass larval performance is summarized in the following lines, although this section does not intend to provide a detailed revision on nutritional physiology of larvae.

Fish larvae have tremendous growth potential, with relative growth rates much higher than juvenile and adult fish. However, to fully express

Table 2. Ingredients and proximate composition (in %) of various standard microdiets* used in European sea bass larval rearing from the onset of exogenous feeding for experimentally purposes.

Ingredients	Microdiet
Fish meal (LT 94, Norse) [1,2,3]/Squid Powder (Riber & Son)[4]	52.0[1,2,3]/69.0[4]
Protein hydrolysate (CPSP 90, Sopropêche)	11.0[1], 14.0[2]
Soy lecithin	2.0[4], 7.0[1,2,3]
Marine lecithin (LC 60[1,3] or LC40[2], Phosphotech)	14.0[1,3],19.0[2]
Vitamin premix[¥] (DSM[1,2,3], Merck[4])	6.0[4], 7.0[2],8.0[1]
Mineral premix[Ø] (DSM[1,2,3], Merck[4])	2.5[4], 3.0[2], 4.0[1,3]
Betaine/Attractant premix[¶]	1.0[1,2,3]/3.0[4]
Cellulose/Gelatine	4.0[1,3]/3.0[4]
Proximal composition	
Dry matter	87.0[1], 89.7[4], 91.4[2], 92.4[3]
Proteins	63.9[2], 64.7[1],61.5[3], 67.0[4]
Lipids	18.5[2], 15.0[3,4], 21.0[1]
Neutral lipids/phospholipids	5.0/14.0[1]

* Data compiled from the following nutritional studies: Mazurais et al. (2008)[1], Vagner et al. (2009)[2], Darias et al. (2011a)[3] and Betancor et al. (2012)[4]. The commercial name and the main supplier of each ingredient is shown in brackets; however, in the case of soy lecithin where many companies offering this ingredient exist, this information has not been included. Composition per kilogram of the vitamin premix from DMS[¥]: choline chloride 60%, 333 g; vitamin A acetate, (500,000 UI/g) 1 g; vitamin E (500 UI/g) 20 g; vitamin D_3 (500,000 UI/g) 0.96 g; vitamin B_3 2 g, vitamin B_5 4 g; vitamin B_1 200 mg; vitamin B_2 80%, 1 g; vitamin B_6 600 mg; vitamin C 35%, 28.6 g; vitamin B_9 80%, 250 mg; vitamin concentrate B_{12} (10 g/kg), 0.2 g; biotin, 1.5 g; vitamin K_3 51%, 3.92 g; meso-inositol 60 g; cellulose, 542.4 g. Composition of the vitamin premix per 100 g diet from Merck[¥]: cyanocobalamin 0.03 mg; astaxanthin 5.0 mg; folic acid 5.4 mg, pyridoxine-HCI 17.2 mg, thiamin 21.7 mg; riboflavin 72.5 mg, calcium-pantothenate 101.5 mg, p-aminobenzoic acid 145.0 mg, nicotinic acid 290.1 mg; myo-inositol 1,450.9 mg, retinol acetate 0.2 mg, ergocalcipherol 3.6 mg, menadione, 17.3 mg, α-tocopheryl acetate 150.0 mg. Composition per kilogram of the mineral mixture from DSM[Ø]: 90 g KCl, 40 mg KIO_3, 500 g $CaHPO_4$ $2H_2O$, 40 g NaCl, 3 g $CuSO_4$ $5H_2O$,4 g $ZnSO_4$ $7H_2O$, 20 mg $CoSO_4$ $7H_2O$, 20 g $FeSO_4$ $7H_2O$, 3 g $MnSO_4$ H_2O, 215 g $CaCO_3$, 124 g $MgSO_4$ $7H_2O$, and 1 g NaF. Mineral premix supplied per 100 g diet from Merck[Ø]: NaCl, 215.1 mg; $MgSO_4$ $7H_2O$ 677.5 mg; NaH_2PO_4 H_2O 381.5 mg, K_2HPO_4 758.9 mg, $Ca(H_2PO_4)$ $2H_2O$ 671.6 mg, $FeC_6H_5O_7$ 146.9 mg, $C_3H_5O_3$ 0.5Ca 1617.2 mg; $Al_2(SO_4)_3$ $6H2O$ 0.693 mg, $ZnSO_4$ $7H_2O$ 14.8 mg, $CuSO_4$ $5H_2O$ 1.247 mg, $MnSO_4$ H_2O 2.998 mg, KI 0.742 mg, $CoSO_4$ $7H_2O$ 10.706 mg. Attractant premix supplied per 100 g diet[¶]: inosine-5-monophosphate 500.0 mg, betaine 660.0 mg, L-serine 170.0 mg, L-phenylalanine 250.0 mg, DL-alanine 500.0 mg, L-sodium aspartate 330.0 mg, L-valine 250.0 mg, glycine 170.0 mg.

such high growth potential, dietary **protein** must be provided in sufficient quantity and quality. It is generally recommended that artificial diets for fish larvae should have a nitrogen solubility and molecular weight profile similar to that found in wild live food (Carvalho et al. 2003). In addition, the low capacity to digest proteins and the amino acid (AA) requirements for energy production and growth of marine fish larvae, means that AA requirements are likely to be very high and that dietary imbalances will have a burden in terms of nitrogen utilization (Aragão et al. 2004) and, eventually growth and development (Conceição et al. 2003). In this sense, several authors have recommended the inclusion of protein hydrolysates in compound diets for fish larvae, since they enhance the digestibility and nutritional value of the feed. These compounds are more efficiently absorbed and digested by enterocytes compared to high-molecular-weight macromolecules as it has been reported in European sea bass, which is due to the specific digestive features of fish larvae in comparison to juveniles or adults (Zambonino-Infante and Cahu 2010). Furthermore, protein hydrolysates also act as feed attractants as they contain digested protein components such as free amino acids (FAA) and peptides, thus enhancing the palatability and acceptance of the feed (Kasumyan and Døving 2003). The inclusion of protein hydrolysates substituting a fraction of fish meal in microdiets has been proved to enhance growth and survival of fish larvae (Zambonino-Infante et al. 1997), as well as to stimulate their immunological condition (Kotzamanis et al. 2007). However, special attention should be paid with regard to the level of dietary inclusion of protein hydrolysates in microdiets, since an excess of fish meal substitution by these compounds may result in an imbalance in the absorption of certain essential AA, affecting larval performance in terms of growth and survival (Cahu and Zambonino-Infante 1995, Zambonino-Infante et al. 1997, Cahu et al. 1999).

In fish larval nutrition, **lipids** and their constituent fatty acids are probably the most studied nutrients. However, they remain one of the least well-understood and enigmatic nutrients in aquaculture nutrition (Glencross 2009), which might be due in part to their relative complex chemistry and varied functional roles (for review see Tocher 2010). Dietary lipids have been shown to be particularly important for early development of marine finfish larvae (Izquierdo et al. 2003), because they represent the main energy source and a source of polyunsaturated fatty acids (PUFA) and essential fatty acids (EFA), such as eicosapentaenoic acid (EPA, 20:5n-3), docosahexaenoic acid (DHA, 22:6n-3) and arachidonic acid (ARA, 20:4n-6), required for the normal formation of new cell and tissue membranes, organ development, larval growth and morphogenesis (Izquierdo 2005). Lipid requirements of marine fish larvae have been extensively studied during the last two decades and particular attention has been paid to PUFA and phospholipids (PL) (see review in Izquierdo and Koven 2011).

A number of studies have shown that DHA promotes growth more efficiently than EPA and ARA in marine fish larvae. The contribution of DHA to weight gain lies in its structural function in the phospholipids bilayer of the cellular membrane and its influence on membrane fluidity (Izquierdo and Koven 2011). Additionally, particularly abundant in fish, DHA and long chain PUFA play an important role in larval development, especially in neural tissues, such as retina and brain (Benitez-Santana et al. 2012). The optimal level of dietary DHA and EPA components for marine fish is known to be around 3% of dry mass (DM) (Sargent et al. 1999). Atalah et al. (2011a) reported that dietary ARA markedly improves growth in European sea bass larvae, suggesting a requirement for this species of at least 1.2% ARA. In addition, the above mentioned authors found a significant positive correlation between dietary ARA levels and larval survival after handling stress, indicating the importance of this fatty acid in European sea bass larvae response to stressors, as ARA has a regulatory effect on cortisol response. When DHA requirements were satisfied, a significant improvement in larva growth and survival was observed when increased EPA and ARA levels were maintained at a ratio ranging from 3.3 to 4.0 (Atalah et al. 2011). However, a dietary excess of DHA (5%) has been reported to severely increase the incidence of muscular dystrophy (hyaline degeneration and fragmentation of myofibres, and necrosis) and the accumulation of ceroid pigment within hepatocytes in European sea bass larvae, implying that an excessive production of free radicals was present (Betancor et al. 2011).

Recent studies on the effects of dietary phospholipids (PL) on European sea bass larval performance have demonstrated the importance of these compounds in larval diets. In particular, PL have been reported as having important effects on larval development due to the limited capacity of fish larvae to synthesize them *de novo* (Coutteau et al. 1997). First data on this issue indicated that increasing levels of dietary PL (3% to 12% of total lipids) improved larval growth, survival and quality (Cahu et al. 2003). Particularly, fish fed 12% PL showed a mean body weight 16 times higher than those fed 3% PL, whereas their survival rate were 73% and 22%, respectively. In addition, the above mentioned authors concluded that the optimal levels of phosphatidylcholine and phophatidylinositol were 3.5% and 1.6%, respectively. Further results obtained by feeding European sea bass, from first feeding exclusively with a microdiet incorporating 1.1% and 2.5% EPA+DHA related to diet DM, and supplied either by PL or by neutral lipids (NL), indicated that larvae used more efficiently, EFA contained in PL than in NL. In this case, all larvae fed a diet including 2.5% EPA+DHA supplied as NL died before day 37 post hatching; growth was poor in larvae fed a diet including 1.1% EPA+DHA and significantly higher in the groups receiving HUFA in PL. The best growth performance was observed in the group receiving 2.5% EPA+DHA supplied as PL (Villeneuve et al. 2005). Moreover, the same authors showed that a higher EPA+DHA dietary incorporation (5%) did not further improve growth, and significantly reduced larval quality. Further information on the impact of dietary PL levels on fish larval development and physiology can be found in Cahu et al. (2009).

Research on micronutrient dietary requirements has often been initiated in relation to shortcomings and disease problems in commercial larvae production (Moren et al. 2011); however, this tendency has changed during the last decade with the development of microdiets for first-feeding marine fish, and consequently, there exists considerable information on the dietary requirements and effects of vitamins and minerals on fish larvae. In this sense, although micronutrients are incorporated into feeds as premixes, there is scarce information about the effects of different composition and levels of inclusion on European sea larvae. In this sense, most of the available information is centred on vitamins A, E, C and D, and selenium, whereas there is only one study that evaluates the effects of different levels of dietary inclusion of a commercial standard vitamin premix. As a rule of thumb, dietary vitamin and mineral requirements are species specific and could vary as a function of environmental, nutritional, genetic and developmental conditions. Micronutrients should appear in adequate dosages in the diet, and with regard to other nutrients, for fish to achieve an optimal performance.

Vitamin A (VA) is of the most studied nutrients in fish larval nutrition due to its important role as a morphogenetic nutrient that is involved in differentiation, growth and development of cells and tissues and responsible for important skeletal disorders when not correctly dosed in larval diets (Boglione et al. 2013). In European sea bass, excess or deficiency in dietary VA has been correlated with impaired growth and underdevelopment of the digestive system (Villeneuve et al. 2005, 2006, Mazurais et al. 2009), as well as with the development of skeletal deformities and changes in juvenile body shape (Mazurais et al. 2009, Georga et al. 2011). In this sense, Mazurais et al. (2009) evaluated the effects of different levels of VA (0, 5, 10, 15, 25, 35 and 70 mg retinol/kg DM) on European sea bass larval performance and quality and found that larval survival at 45 dph was not affected by the dietary level of VA (*ca.* 60% in all experimental groups), whereas the optimal dietary levels of VA in terms of growth were comprised between 5 mg and 10 mg retinol/kg DM that corresponded to 9,402 and 20,755 IU VA/Kg, respectively. Dietary retinol also had a significant effect on the development of deformities in all the body parts, although the optimum retinol levels with regard to the incidence of skeletal deformities depended on each skeletal structure and the ontogenetic stage at which they developed (Mazurais et al. 2009).

Concerning **vitamins D and C** (ascorbic acid, AA), their suitable dietary level for an optimum larval development of European sea bass is in a restricted range and it has been recently shown that subtle variations could unleash severe physiological disruptions such as delay of intestinal maturation and ossification, and increase the frequency of skeletal deformities (Darias et al. 2011b). In particular, a recent study performed in this issue showed that larvae performed well with 27.6 IU VD_3/g supplemented diet (Darias et al. 2010b); that is 8.4 IU VD_3/g diet more than that recommended for fish larvae and 11.5 times the amount recommended for juveniles by the NRC (2011). The fact that only 14–16 IU VD_3/g above or below the optimal dose (27.6 IU VD_3/g diet) notably increased the incidence of deformities denotes the high sensitivity of European sea bass larvae to this vitamin. Moreover, doses ranging from 42 to 120 IU VD_3/g diet did not proportionally modify the rate of malformations confirming that an excess of VD_3 not only negatively influences the larval development but suppose an economic cost for the aquaculture (Darias et al. 2011b). Regarding AA, European sea bass larvae requires a minimum amount of 15 mg AA/kg diet to survive, 30 mg AA/kg diet to attain maximal growth, and 50 mg/kg diet for an adequate morphogenesis (Darias et al. 2011a), the last being eight times lower than that recommended for fish larvae and corresponding to the amount recommended by the NRC (2011) for most juveniles. Although both an excess and deficiency of VD_3 or AA levels affects larval growth performance and dramatically reduces larval quality, Darias et al. (2010b, 2011a) found that a deficiency of these vitamins is more harmful for developing European bass than excess. Indeed, larvae of this species are highly sensitive to low vitamin D levels than to high vitamin C levels throughout skeletogenesis.

Larval diets are very rich in fatty acids that are very sensitive to peroxidation. In addition, fish larval tissues are known to be very rich in PUFA and require the action of antioxidants to protect them intra- and extracellularly from free radical compounds (Sargent et al. 1997). In this sense, the only available results on the effects of dietary **vitamin E** (VE, α-tocopherol) on European sea bass larvae are testing this fat-soluble vitamin role as antioxidant. Thus, Betancor et al. (2012) fed diets with different ratios of DHA to VE in larvae and found that increasing the level of DHA increased the incidence of muscular degeneration, while adding extra VE at the high DHA levels reduced the incidence of muscular lesions. The authors recommended a ratio of 30 g DHA/kg and 3,000 mg VE/kg as adequate to achieve a good larval performance and to avoid muscular lesions.

Information concerning mineral nutrition of fish is limited compared to most other nutrient groups (NRC 2011) and this becomes even more evident when dealing with fish larvae. The only available information on the effects of minerals on European sea bass larval performance is centred on **selenium** (Se). This is an essential trace element for fish, but also has the smallest window of any element between requirement and toxicity. Betancor et al. (2012) evaluated the larval performance and their oxidative status when Se and VE were included in the diet, at high DHA levels (5%). They found that Se (2,000 mg/kg) proved to enhance the cell antioxidant capacity and tissular protection in muscle, as well as improving larval growth in comparison to those groups of larvae fed high DHA and VE levels (5 g DHA/100 g DM and 300 mg VE/100 g DM). The former authors concluded that when high levels of LC-PUFA are included in European sea bass larvae microdiets, an adequate combination of dietary VE and Se must be included to avoid the appearance of oxidative stress in larval tissues and favour culture performance.

As previously mentioned, vitamins and minerals are added to compound diets as premixes. This means that minerals and trace elements are added in inorganic forms in the dry feeds, while in live prey, they are most often present in organic forms. This will have some consequences for their bioavailability and probably their metabolism in the larval body. Vitamins are also often supplemented in forms other than those naturally present in live prey and little is known on how this affects their bioavailability. Another factor that affects micronutrient nutrition in marine fish larvae is the high leaching rates of water-soluble nutrients from formulated diets, which has lead to higher nutrient incorporation than the real requirements of larvae (Moren et al. 2011). In this context, Mazurais et al. (2008) published the only available study on the effects of different levels of dietary inclusion of a commercial vitamin premix (0.5, 1.5, 2.5, 4 and 8% VM) in European sea bass larvae. They found that growth performance in length and weight was only negatively influenced by diets incorporating less than 2.5% VM, corresponding to a hypovitaminosis, whereas no differences in survival were found among groups. In addition, a lower occurrence of column deformities and a more intense ossification process was found in larvae fed the diets with the higher percentages of vitamin mix. This study indicated that fish larvae require higher dietary VM levels than juvenile fish to achieve their developmental processes correctly. The incorporation of VM (4 and 8%) into larval feeds gave the best results in terms of growth, survival, and also morphogenesis. However, the fact that the percentage of head and column deformities remained significant demonstrates the need to further refine the proportions of certain vitamins (Vitamins A, D, and C) in the standard vitamin mix.

There is a trend towards incorporating dietary components to provide various functional attributes in fish immunocompetence. Those **additives** are intentional inclusions to the feeds, acknowledging their capability to modulate the immune system, alleviate stress or ward off pathogens (Kiron 2012). Additives could be either immunopotentiators or immunosuppressors that attenuate the consequences of pathogenic invasion and contribute to fish welfare. These substances could include intact microbes (e.g., probiotic organisms), microbial cell components, fungal polysaccharide, phytotherapeutic agents and synthetic compounds (extensively reviewed in Kiron 2012). While prebiotic is a selectively fermented ingredient that allows specific changes, both in composition and/or activity in the gastrointestinal microflora that confers benefits upon host well-being and health (Roberfroid

2007); probiotics are microorganisms that are believed to contribute to the well-being of the host organism (Kiron 2012).

A large number of studies have demonstrated the capacity of **probiotics** in avoiding infection by potential pathogens, enhancing the welfare of farmed aquatic species and improving, at the same time, their growth performance (Gatesoupe 2010, Merrifield et al. 2010). There are few studies evaluating the effects of probiotics in European sea bass larvae and early juveniles. Carnevalli et al. (2006) reported that the use of *Lactobacillus delbrueckii delbrueckii* (Gram positive bacteria) administered to larvae in enriched rotifers and *Artemia* at 10^5 bacteria/ml, had positive effects on fish welfare, growth performance, decreased cortisol levels of treated animals and affected the transcription of several genes involved in the regulation of body growth. In addition, early feeding with probiotic-supplemented diet (enriched live prey) stimulated the larval gut immune system and lowered transcription of key pro-inflammatory genes (Picchietti et al. 2009).

Several studies have shown the beneficial effects of live yeast (*Debaryomyces hansenii*) supplementation in terms of gut maturation, enhancement of the antioxidant response, increased growth performance and survival, and prevented vertebral malformations at the first stages of European sea bass development (Tovar et al. 2002, Tovar-Ramírez et al. 2004, 2010). The positive effects of live yeast administration of larval performance might be due to molecules such as growth factors released by the yeast (Tovar-Ramírez et al. 2004). This group of compounds are known to have biological effects on cell proliferation and differentiation, and have received significant attention for their role in the maturational process of the gastrointestinal tract in vertebrates. Even if the intestinal transit is relatively short in fish larvae, it is sufficient for digestion, and likely for the release of active compounds from yeast cells like polyamines, proteases and phosphatases, which could be beneficial for the digestive process (Gatesoupe 2007). In this sense, the inclusion of spermine, a type of polyamine, in microdiets for European sea bass enhanced larval growth and survival, as well as advanced the maturation of their digestive system (Péres et al. 1997).

Conclusions

Although larval rearing is one of the most critical stages for the successful propagation of any species and represents probably one of the major bottlenecks of the whole aquaculture industry, European sea bass larval rearing protocols are well established and most hatcheries have successfully adopted the existing protocols to their facilities and conditions. In this sense, most of the research focused on European sea bass larval production at an industrial scale is focused on producing fast-growing and stress-resistant fry, as well as animals with a low incidence of skeletal deformities with the minimum possible cost. For this purpose, the optimization of biotic and abiotic factors during the rearing process is of extreme importance and a continuous industrial endeavour. However, this process is generally conducted at the hatchery level by hatchery managers and the resultant know-how rarely reaches academic society.

Regarding environmental abiotic factors affecting the overall larval rearing process, recent results have demonstrated the high sensitivity of European sea bass to temperature and light during early development. Natural thermo- and light-cycles of blue spectrum seem to provide the best rearing conditions during the first stages of life. Significantly, the possible long-term consequences of early exposure to unsuitable light and temperature on the performance of adults (e.g., growth, reproduction) must be ascertained, as the improvement of these environmental parameters are key factors that ameliorate the downgrading of product which is associated with economic losses in the aquaculture field. In addition, the role of nutrition and other biotic factors like the microbial community in larval rearing tanks and their impact on gut microbiota deserves further research in gnotobiotic experimental systems, as well as in recirculating aquaculture units. In this sense, nutrition is considered as one of the most important biotic factors affecting larval performance, since the amount and the chemical form of a particular nutrient as supplied to each developmental stage may largely compromise its performance with regards to its immune competence, bone formation, intestinal maturation and pigmentation pattern, among other biological processes. The formulation of compound microdiets for European sea bass larvae during the late nineties has been one of the most important advances in marine fish larval nutrition, since it has allowed the partial or total replacement of live prey from feeding protocols, as well as the evaluation and characterization of the effect of a nutrient on larval performance and morphogenesis. However, improvements are still required for microdiet formulations in order to enhance their administration into rearing tanks, water stability (reduced leaching), as well as their digestibility and attractiveness for fish larvae. In this context, advances have been made in regard to, nitrogen solubility, molecular weight profile and incorporation of protein hydrolysates which have been shown to enhance growth, survival and fish immunological condition. In addition, well balanced dietary lipid content should be provided as the main energy source and the source of PUFA and EFA for cell and tissue membranes homeostasis, organ development, larval growth and morphogenesis. However, since larval diets are very rich in fatty acids, avoiding their peroxidation during feed production, stocking and delivery into fish tanks is imperative. Finally, although different micronutrients are known to be essential for fish development, their dietary requirements are still not known for fish larvae. Moreover, since research works for determination of nutritional requirements are generally performed using unifactorial approaches, where only the effect of a single nutrient is considered, and different nutrients may interact and/or compete during their absorption, metabolism and transport processes inside the organism, further research efforts should be done using multifactorial approaches in which different nutrients are tested at the same time. This approach will warranty a balanced nutritional composition of diets for promoting harmonic larval fish development.

References

Aegerter, S. and B. Jalabert. 2004. Effects of post-ovulatory oocyte ageing and temperature on egg quality and on the occurrence of triploid fry in rainbow trout, *Oncorhynchus mykiss*. Aquaculture 231: 59–71.

Alliot, E., A. Pastoureaud and J. Trellu. 1977. Evolution des activités enzymatiques dans le tube digestif au cours de la vie larvaire du bar (*Dicentrarchus labrax*) variations des protéinogrammes et des zymogrammes. 3rd Meeting of the ICES. Working Group on Mariculture, Brest, France, May 10–13. Actes de Colloques du CNEXO 1: 85–91.

Aragão, C., L.E.C. Conceição, H.J. Fyhn and M.T. Dinis. 2004. Estimated amino acid requirements during early ontogeny in fish with different life styles: gilthead seabream (*Sparus aurata*) and Senegalese sole (*Solea senegalensis*). Aquaculture 242: 589–605.

Atalah, E., C.M. Hernández-Cruz, E. Ganuza, T. Benítez-Santana, R. Ganga, J. Roo, D. Montero and M. Izquierdo. 2011. Importance of dietary arachidonic acid for the growth, survival and stress resistance of larval European sea bass (*Dicentrarchus labrax*) fed high dietary docosahexaenoic and eicosapentaenoic acids. Aquacult. Res. 42: 1261–1268.

Barahona-Fernandes, M.H. 1978. L'elevage intensif des larves et des juveniles du bar (*Dicentrachus labrax* L.). Données biologiques, zootechniques et pathologiques. Theses Université Aix-Marseille II, Marseille.

Barahona-Fernandes, M.H. 1982. Body deformation in Hatchery reared European sea bass *Dicentrarchus labrax* (L.). Types, prevalence and effect on fish survival. J. Fish Biol. 21: 239–249.

Baras, E., C. Prignon, G. Gohoungo and C. Mélard. 2000. Phenotypic sex differentiation of the blue tilapia under constant and fluctuating thermal regimes and its adaptive and evolutionary implications. J. Fish Biol. 57: 210–223.

Barnabé, G. 1974. Mass rearing of the bass *Dicentrarchus labrax* L. pp. 749–754. *In*: J.H.S. Blaxter (ed.). The Early Life History of Fish. Springer Verlag, New York.

Barnabé, G. 1976. Contribution à la connaissance de la biologie du loup, *Dicentrarchus labrax* L. (Poisson Serranidae). Thèse Fac. Sciences Montpellier, Montpellier.

Barnabé, G. 1994. Biological bases of fish culture. pp. 229–333. *In*: G. Barnabé (ed.). Aquaculture: Biology and Ecology of cultivated species. Ellis Horwood Ltd.

Beccaria, C., J.P. Diaz, J. Gabrion and R. Connes. 1990. Maturation of the endocrine pancreas in the sea bass, *Dicentrarchus labrax* L. (Teleostei): an immunocytochemical and ultrastructural study. Gen. Comp. Endocrinol. 78: 80–92.

Beccaria, C., J.P. Diaz, R. Connes and B. Chatain. 1991. Organogenesis of the exocrine pancreas in the sea bass, *Dicentrarchus labrax* L., reared extensively and intensively. Aquaculture 99: 339–354.

Bengtsson, A., B.E. Bengtsson and G Lithner. 1988. Vertebral defects in fourhorn sculpin, *Myoxocephalus quadricornis* L., exposed to heavy metal pollution in the Gulf of Bothnia. J. Fish Biol. 33: 517–529.

Benitez-Santana, T., E. Juarez-Carrillo, M.B. Betancor, S. Torrecillas, M.J. Caballero and M.S. Izquierdo. 2012. Increased Mauthner cell activity and escaping behaviour in seabream fed long-chain PUFA. Br. J. Nutr. 107: 295–301.

Betancor, M.B., E. Atalah, M.J. Caballero, T. Benítez-Santana, J. Roo, D. Montero and M.S. Izquierdo. 2011. α-Tocopherol in weaning diets for European sea bass (*Dicentrarchus labrax*) improves survival and reduces tissue damage caused by excess dietary DHA contents. Aquacult. Nutr. 13: e112–e122.

Betancor, M.B., M.J. Caballero, G. Terova, R. Saleh, E. Atalah, T. Benitez-Santana, J.G. Bell and M. Izquierdo. 2012. Selenium inclusion decreases oxidative stress indicators and muscle injuries in sea bass larvae fed high-DHA microdiets. Br. J. Nutr. 108: 2115–2128.

Blanco-Vives, B., N. Villamizar, J. Ramos, M.J. Bayarri, O. Chereguini and F.J. Sánchez-Vázquez. 2010. Effect of daily thermo- and photo-cycles of different light spectrum on the development of Senegal sole (*Solea senegalensis*) larvae. Aquaculture 306: 137–145.

Blanco-Vives, B., L.M. Vera, J. Ramos, M.J. Bayarri, E. Mananos and F.J. Sánchez-Vázquez. 2011. Exposure of larvae to daily thermocycles affects gonad development, sex ratio, and sexual steroids in *Solea senegalensis*, Kaup. J. Exp. Zool. A 315: 162–169.

Boeuf, G. and P. Le Bail. 1999. Does light have an influence on fish growth? Aquaculture 177: 129–152.

Boglione, C., E. Gisbert, P. Gavaia, P.E. Witten, M. Moren, S. Fontagne and G. Koumoundouros. 2013. A review on skeletal anomalies in reared European fish larvae and juveniles. Part 2: main typologies, occurrences and causative factors. Rev. Aquaculture 5: S121–S167.

Brown, N.P., R.J. Shields and N.R. Bromage. 1995. The effect of spawning temperature on egg viability in the Atlantic halibut (*Hippoglossus hippoglossus*). Aquaculture 261: 993–1002.

Büke, E. 2002. Sea bass (*Dicentrarchus labrax* L., 1781) seed production. Turk. J. Fish. Aqua. Sci. 2: 61–70.

Cahu, C.L. and J.L. Zambonino-Infante. 1994. Early weaning of sea bass (*Dicentrarchus labrax*) larvae with a compound diet: effect on digestive enzymes. Comp. Biochem. Physiol. 109A: 213–222.

Cahu, C.L. and J.L. Zambonino-Infante. 1995. Maturation of the pancreatic and intestinal digestive functions in sea bass (*Dicentrarchus labrax*): Effect of weaning with different protein sources. Fish Physiol. Biochem. 14: 431–437.

Cahu, C.L. and J.L. Zambonino-Infante. 2001. Substitution of live food by formulated diets in marine fish larvae. Aquaculture 200: 161–180.

Cahu, C.L., J.L. Zambonino Infante, A.M. Escaffre, P. Bergot and S.J. Kaushik. 1998. Preliminary results on larval rearing of sea bass (*Dicentrarchus labrax*) without live food. Comparaison with carp (*Cyprinus carpio*) larvae. Aquaculture 169: 1–7.

Cahu, C.L., J.L.Z. Infante, P. Quazuguel and M.M. Le Gall. 1999. Protein hydrolysate vs, fish meal in compound diets for 10-day old sea bass *Dicentrarchus labrax* larvae. Aquaculture 171: 109–119.

Cahu, C., J.L. Zambonino Infante and T. Takeuchi. 2003. Nutritional components affecting skeletal development in fish larvae. Aquaculture 227: 1–4.

Cahu, C.L., E. Gisbert, L.A.N. Villeneuve, S. Morais, N. Hamza, P.A. Wold and J.L. Zambonino-Infante. 2009. Influence of dietary phospholipids on early ontogenesis of fish. Aquacult. Res. 40: 989–999.

Cara, B., F.J. Moyano, J.L. Zambonino and C. Fauvel. 2007. Trypsin and chymotrypsin as indicators of nutritional status of post-weaned sea bass larvae. Journal of Fish Biology 70: 1798–1808.

Carvalho, A.P., A. Oliva-Teles and P. Bergot. 2003. A preliminary study on the molecular weight profile of soluble protein nitrogen in live food organisms for fish larvae. Aquaculture 225: 445–449.

Chatain, B. 1986. La vessie natatoire chez *Dicentrarchus labrax* et *Sparus auratus*. I. Aspects morphologiques du developpement. Aquaculture 53: 303–311.

Chatain, B. 1987. La vessie natatoire chez *Dicentrarchus labrax* et *Sparus auratus*. II Influence des anomalies de developpement sur la croissance de la larve. Aquaculture 65: 175–181.

Chatain, B. 1994. Abnormal swimbladder development and lordosis in sea bass (Dicentrarchus labrax) and sea bream (*Sparus auratus*). Aquaculture 119: 371–379.

Chatain, B. and N. Ounais-Guschemann. 1991. Improved rate of initial swim bladder inflation in intensively reared *Sparus auratus*. Aquaculture 84: 345–353.

Claridge, P.N. and I.C. Potter. 1983. Movements, abundance, age composition and growth of bass, *Dicentrarchus labrax*, in the Severn Estuary and Inner Bristol Channel. J. Mar. Biol. Assoc. 63: 871–879.

Conceição, L.E.C., H. Grasdalen and I. Ronnestad. 2003. Amino acid requirements of fish larvae and post-larvae: new tools and recent findings. Aquaculture 227: 221–232.

Conceição, L.E.C., M. Yúfera, P. Makridis, S. Morais and M.T. Dinis. 2010. Live feeds for early stages of fish rearing. Aquacult. Res. 41: 613–640.

Coutteau, P., I. Geurden, M.R. Camara, P. Bergot and P. Sorgeloos. 1997. Review on the dietary effects of phospholipids in fish and crustacean larviculture. Aquaculture 155: 149–164.

Coves, D. 1985. Etat actuel de l'elevage du loup, *Dicentrarchus labrax*, en ecloserie. Aqua. Revue. 3: 26–30.

Coves, D., G. Dewavrin, G. Breuil and N. Devauchelle. 1991. Culture of sea bass (*Dicentrarchus labrax* L.). pp. 3–20. *In*: J.P. McVey (ed.). Handbook of Mariculture, Vol II: Finfish Culture. CRC Press, Boca Raton, FL.

Cucchi, P., E. Sucré and R. Santos. 2012. Embryonic development of the sea bass *Dicentrarchus labrax*. Helgol. Mar. Res. 66: 199–209.

Daoulas, C., A.N. Economou and I. Bantavas. 1991. Osteological abnormalities in laboratory reared sea-bass (*Dicentrarchus labrax*) fingerlings. Aquaculture 97: 169–180.

Darias, M.J., J.L. Zambonino-Infante, K. Hugot, C.L. Cahu and D. Mazurais. 2008. Gene expression patterns during the larval development of European sea bass (*Dicentrarchus labrax*) by microarray analysis. Mar. Biotechnol. 10: 416–28.

Darias, M.J., O. Lan Chow Wing, C.L. Cahu, J.L. Zambonino-Infante and D. Mazurais. 2010a. Double staining protocol for developing European sea bass (*Dicentrarchus labrax*) larvae. J. Appl. Ichthyol. 26: 280–285.

Darias, M.J., D. Mazurais, G. Koumoundouros, N. Glynatsi, S. Christodoulopoulou, E. Desbruyeres, M.M. Le Gall, P. Quazuguel, C.L. Cahu and J.L. Zambonino-Infante. 2010b. Dietary vitamin D3 affects digestive system ontogenesis and ossification in European sea bass (*Dicentrachus labrax* Linnaeus, 1758). Aquaculture 298: 300–307.

Darias, M.J., D. Mazurais, G. Koumoundouros, C.L. Cahu and J.L. Zambonino-Infante. 2011a. Overview of vitamin D and C requirements in fish and their influence on the skeletal system. Aquaculture 315: 49–60.

Darias, M.J., D. Mazurais, G. Koumoundouros, M.M. Le Gall, C. Huelvan, E. Desbruyeres, Cahu and J.L. Zambonino-Infante. 2011b. Imbalanced dietary ascorbic acid alters molecular pathways involved in skeletogenesis of developing European sea bass (*Dicentrarchus labrax*). Comp. Biochem. Physsiol. A 159: 46–55.

Dekens, M.P., C. Santoriello, D. Vallone, G. Grassi, D. Whitmore and N.S. Foulkes. 2003. Light regulates the cell cycle in zebrafish. Curr. Biol. 13: 2051–2057.

Deplano, M., J.P. Diaz, R. Connes, M. Kentouri-Divanach and E. Cavalier. 1991. Appearance of lipid-absorption capacities in larvae of the sea bass *Dicentrarchus labrax* during transition to the exotrophic phase. Mar. Biol. 108: 361–371.

Devauchelle, N. and D. Coves. 1988. The characteristics of sea bass (*Dicentrarchus labrax*) eggs: description, biochemical composition and hatching performances. Aquat. Liv. Res. 1: 223–230.

Diana, J.S. 1984. The growth of largemouth bass, *Micropterus salmoides* (Lacepede), under constant and fluctuating temperatures. J. Fish Biol. 24: 165–172.

Diaz, J.P., L. Mani-Ponset, C. Blasco and R. Connes. 2002. Cytological detection of the main phases of lipid metabolism during early post-embryonic development in three teleost species: *Dicentrarchus labrax*, *Sparus aurata* and *Stizostedion lucioperca*. Aquat. Living Resour. 15: 169–178.

Diaz, J.P., M. Prié-Granié, M. Kentouri, S. Varsamos and R. Connes. 2003. Development of the lateral line system in the sea bass. J. Fish Biol. 62: 24–40.

Divanach, P., N. Papandroulakis, P. Anastasiadis, G. Koumoundouros and M. Kentouri. 1997. Effect of water currents on the development of skeletal deformities in sea bass (*Dicentrarchus labrax* L.) with functional swimbladder during postlarval and nursery phase. Aquaculture 156: 145–155.

Divanach, P. and M. Kentouri. 2000. Hatchery techniques for specific diversification in Mediterranean finfish culture. Cah. Opt. Mediter. 47: 75–87.

Doroshev, S.I. and T.M. Aronovich. 1974. The effects of salinity on embryonic and larval development of *Eleginus navaga* (Pallas), *Boreogadus saida* (Lepechin) and *Liopsetta glacialis* (Pallas). Aquaculture 4: 353–362.

Downing, G. and M.K. Litvak. 2001. The effect of light intensity and spectrum on the incidence of first feeding by larval haddock. J. Fish Biol. 59: 1566–1578.

Dülger, N., M. Kumlu, S. Türkmen, A. Ölçülü, O. Tufan Eroldoğan, H. Asuman Yilmaz and N. Öçal. 2012. Thermal tolerance of European sea bass (*Dicentrarchus labrax*) juveniles acclimated to three temperature levels. J. Therm. Biol. 37: 79–82.

Estévez, A. and M. Planas. 1987. Producción a gran escala de alevines de lubina (*Dicentrarchus labrax* (L., 1758)). Informes Técnicos de Investigación Pesquera 139: 1–12.

FAO. 2010. Synthesis of Mediterranean Marine Finfish Aquaculture—A marketing and promotion strategy. FAO Studies and Reviews 88, pp. 204.

Finn, R.N. 2007. The physiology and toxicology of salmonid eggs and larvae in relation to water quality criteria. Aquat. Toxicol. 81: 337–354.

Forberg, T., A. Arukwe and O. Vadstein. 2011. A protocol and cultivation system for gnotobiotic Atlantic cod larvae (*Gadus morhua* L.) as a tool to study host microbe interactions. Aquaculture 315: 222–227.

Foster, R.G., M.W. Hankins and S.N. Peirson. 2007. Light, photoreceptors and circadian clocks. *In*: E. Rosato (ed.). Circadian Rhythms: Methods and Protocols. Humana Press, Totowa, NJ, pp. 3–28.

Fry, F.E.J. 1947. Effects of the Environment on Animal Activity. University of Toronto Studies in Biology, Series No. 55, vol. 68, Publ. Ont. Fish. Res. Lab. pp. 1–62.

García-Hernández, M.P. and B. Agulleiro. 1992. Ontogeny of the endocrine pancreas in sea bass (*Dicentrarchus labrax*). An immunocytochemical study. Cell Tissue Res. 270: 339–352.

García-Hernández, M.P., M.T. Lozano and B. Agulleiro. 1994. Ontogeny of the endocrine cells of the sea bass (*Dicentrarchus labrax* L.): an ultrastructural study stomach. Anat. Embryol. 190: 507–514.

García Hernández, M.P., M.T. Lozano, M.T. Elbal and B. Agulleiro. 2001. Development of the digestive tract of sea bass (*Dicentrarchus labrax* L). Light and electron microscopic studies. Anat. Embryol. 204: 39–57.

Gatesoupe, F.J. 2007. Live yeasts in the gut: natural occurrence, dietary introduction, and their effects on fish health and development. Aquaculture 267: 20–30.

Gatesoupe, F.J. 2010. Probiotics and other microbial manipulations in fish feeds: Prospective health benefits. *In*: pp. 541–552. R. Watson and V.R. Preedy (eds.). Bioactive foods in promoting health: Probiotics and prebiotics. Academic, New York.

Georga, I., N. Glynatsi, A. Baltzois, D. Karamanos, D. Mazurais, M.J. Darias, C.L. Cahu, J.L. Zambonino-Infante and G. Koumoundouros. 2011. Effect of vitamin A on the skeletal morphogenesis of European sea bass, *Dicentrarchus labrax* (Linnaeus 1758). Aquacult. Res. 42: 1–19.

Giffard-Mena, I., G. Charmantier, E. Grousset, F. Aujoulat and R. Castille. 2006. Digestive tract ontogeny of *Dicentrarchus labrax*: implication in osmoregulation. Dev. Growth Differ. 48: 139–51.

Girin, M. 1976. La ration alimentaire dans l'elevage du bar (*Dicentrarchus labrax* L). pp. 171–188. *In*: G. Perssone and E. Jaspers (eds.). 10th European Symposium on Marine Biology. IZWO, Wetteren.

Girin, M. 1979. Méthodes de production des juveniles chez trois poissons marins, le bar, la sole et le turbot. Rapp. Sci. Tech. CNEXO 39: 1–202.

Glencross, B.D. 2009. Exploring the nutritional demand for essential fatty acids by aquaculture species. Rev. Aquaculture 1: 71–124.

Glencross, B.D., M. Booth and G.L. Allan. 2007. A feed is only as good as its ingredients—a review of ingredient evaluation for aquaculture feeds. Aquacult. Nutr. 13: 17–34.

Gluckmann, I., F. Huriaux, F. Focant and P. Vandewalle. 1999. Postembryonic development of development of the cephalic skeleton in *Dicentrarchus labrax* (Pisces, Perciformes, Serranidae) Bull. Mar. Sci. 65: 11–36.

Golovanov, V.K. 2006. The ecological and evolutionary aspects of thermoregulation behavior on fish. J. Ichthyol. 46: 180–187.

Gorodilov, Y.N. 2010. The biological clock in vertebrate embryogenesis as a mechanism of general control over the developmental organism. Russ. J. Dev. Biol. 41: 201–216.

Govoni, J.J., G.W. Boehlert and Y. Watanabe. 1986. The physiology of digestion in fish larvae. Env. Biol. Fish 16: 59–77.

Hamre, K., E. Holen and M. Moren. 2007. Pigmentation and eye migration in Atlantic halibut (*Hippoglossus hippoglossus* L.) larvae: new findings and hypotheses. Aquacult. Nutr. 13: 65–80.

Hatziathanasiou, A., M. Paspatis, M. Houbart, P. Kestemont, S. Stefanakis and M. Kentouri. 2002. Survival, growth and feeding in early life stages of European sea bass (*Dicentrarchus labrax*) intensively cultured under different stocking densities. Aquaculture 205: 89–102.

Hokanson, K.E., J.H. McCormick, B.R. Jones and J.H. Tucker. 1973. Thermal requirements for maturation, spawning, and embryo survival of the brook trout, *Salvelinus fontinalis*. J. Fish. Res. Board Can. 30: 975–984.

Holt, J.G., K.A. Webb and B.R. Rust. 2011. Microparticle diets: testing and evaluating success. pp. 353–372. *In*: G.J. Holt (ed.). Larval Fish Nutrition. Wiley-Blackwell, Oxford.

Hurd, M.W. and G.M. Cahill. 2002. Entraining signals initiate behavioural circadian rhythmicity in larval zebrafish. J. Biol. Rhythms 17: 307–314.

Huse, G. and Ø. Fiksen. 2010. Modelling encounter rates and distribution of mobile predators and prey. Prog. Oceanogr. 84: 93–104.

Izquierdo, M.S. 2005. Essential fatty acid requirements in Mediterranean fish species. Cah. Opt. Mediter. 63: 91–102.

Izquierdo, M. and W. Koven. 2011. Lipids. pp. 47–82. *In:* G.J. Holt (ed.). Larval Fish Nutrition. Wiley-Blackwell, Oxford.

Izquierdo, M.S., A. Obach, L. Arantzamendi, D. Montero, L. Robaina and G. Rosenlund. 2003. Dietary lipid sources for seabream and seabass: growth performance, tissue composition and flesh quality. Aquacult. Nutr. 9: 397–407.

Johnson, D.W. and I. Katavic. 1984. Mortality, growth and swim bladder stress syndrome of sea bass (*Dicentrarchus labrax*) larvae under varied environmental conditions. Aquaculture 38: 67–78.

Johnson, D.W. and I. Katavic. 1986. Survival and growth of sea bass (*Dicentrarchus labrax*) larvae as influenced by temperatura, salinity, and delayed initial feeding. Aquaculture 52: 11–19.

Johnson, J.A. and S.W. Kelsch. 1998. Effects of evolutionary thermal environment on temperature preference relationships in fishes. Environ. Biol. Fish. 53: 447–458.

Kasumyan, A.O. and K.B. Døving. 2003. Taste preferences in fishes. Fish and Fisheries 4: 289–347.

King, H.R., N.W. Pankhurst, M. Watts and P.M. Pankhurst. 2003. Effect of elevated summer temperatures on gonadal steroid production, vitellogenesis and egg quality in female Atlantic salmon. J. Fish Biol. 82: 153–167.

Kinne, O. 1962. Irreversible nongenetic adaptation. Comp. Biochem. Physiol. 5: 265–282.

Kiron, V. 2012. Fish immune system and its nutritional modulation for preventive health care. Anim. Feed Sci. Technol. 173: 111–133.

Kitajima, C., T. Watanabe, Y. Tsukashima and S. Fujita. 1994. Lordotic deformation and abnormal development of swimbladders in some hatchery-bred marine physoclistous fish in Japan. J. World Aquacult. Soc. 25: 64–77.

Kjørsvik, E., A. Mangor-Jensen and I. Homefjord. 1990. Egg Quality in Fishes. Adv. Mar. Biol. 26: 71–113.

Kotzamanis, Y.P., E. Gisbert, F.J. Gatesoupe, J.L. Zambonino-Infante and C. Cahu. 2007. Effects of different dietary levels of fish protein hydrolysates on growth, digestive enzymes, gut microbiota, and resistance to *Vibrio anguillarum* in European sea bass (*Dicentrarchus labrax*) larvae. Comp. Biochem. Physiol. A 147: 205–214.

Koumoundouros, G., E. Maingot, P. Divanach and M. Kentouri. 2002. Kyphosis in reared sea bass (*Dicentrarchus labrax* L.): ontogeny and effects on mortality. Aquaculture 209: 49–58.

Kristiansen. 2009. Fastfish, On farm assessment of stress level in fish. Publishable Final Activity Report. FP6 Research Project. http://cordis.europa.eu; last access: 3rd March 2013. pp. 59.

Lall, S.P. and L.M. Lewis-McCrea. 2007. Role of nutrients in skeletal metabolism and pathology in fish, an overview. Aquaculture 267: 3–19.

Lom, J., A.W. Pike and I. Dyková. 1991. *Myxobolus sandrae* Reuss, 1906, the agent of vertebral column deformities of perch *Perca fluviatilis* in northeast Scotland. Dis. Aquat. Org. 12: 49–53.

López-Albors, O., F. Gil, G. Ramírez-Zarzosa, J.M. Vázquez, R. Latorre, A. García-Alcázar, A. Arencibia and F. Moreno. 1998. Muscle development in gilthead sea bream (*Sparus aurata* L.) and sea bass (*Dicentrarchus Labrax* L.): further histochemical and ultrastructural aspects. Anat. Histol. Embryol. 27: 223–229.

López-Olmeda, J.F. and F.J. Sánchez-Vázquez. 2011. Thermal biology of zebrafish (*Danio reiro*). J. Therm. Biol. 36: 91–104.

López-Olmeda, J.F., J.A. Madrid and F.J. Sánchez-Vázquez. 2006. Light and temperature cycles as zeitgebers of zebrafish (*Danio rerio*) circadian activity rhythms. Chronobiol. Int. 23: 537–550.

Madsen L., Arnbjerg and I. Dalsgaard. 2001. Radiological examination of the spinal column in farmed J. rainbow trout *Oncorhynchus mykiss* (Walbaum): experiments with *Flavobacterium psychrophilum* and oxytetracycline. Aquac. Res. 32: 235–241.

Mani-Ponset, L., J.P. Diaz, P. Divanach, M. Kentouri and R. Connes. 1993. Structure de la rétine et potentialities visuelles susceptibles d'influer sur le comportement trophique du loup (*Dicentrarchus labrax*) adulte et en cours de développement. pp. 587. *In:* G. Barnabé and P. Kestemont (eds.). Production, environment and quality: Proceedings of the International Conference Bordeaux Aquaculture '92, EAS Special Publication. Bordeaux, France.

Marchesan, M., M. Spoto, L. Verginella and E.A. Ferrero. 2005. Behavioural effects of artificial light on fish species of commercial interest. Fish. Res. 73: 171–185.

Marino, G., C. Boglione, B. Bertolini, A. Rossi, F. Ferreri and S. Cataudella. 1993. Observations on development and anomalies in the appendicular skeleton of sea bass, *Dicentrarchus labrax* L. 1758, larvae and juveniles. Aquac. Fisher. Manag. 24: 445–456.

Merrifield, D.L., A. Dimitroglou, A. Foey, S.J. Davies, R.T.M. Baker, J. Bøgwald, M. Castex and E. Ringø. 2010. The current status and future focus of probiotic and prebiotic applications for salmonids. Aquaculture 302: 1–18.

Marty, G.D., D.E. Hinton and J.J. Cech. 1995. Oxygen consumption by larval Japanese medaka with inflated and uninflated swimbladders. Trans. Amer. Fish. Soc. 124: 623–627.

Mazurais, D., M.J. Darias, M.F. Gouillou-Coustans, M.M. Le Gall, C. Huelvan, E. Desbruyères, P. Quazuguel, Ch.L. Cahu and J.L. Zambonino-Infante. 2008. Dietary vitamin mix levels influence the ossification process in European sea bass (*Dicentrarchus labrax*) larvae. Am. J. Physiol. Regul. Integr. Comp. Physiol. 294: R520–R527.

Mazurais, D., N. Glynatsi, M.J. Darias, S. Christodoulopoulou, Ch.L. Cahu, J.L. Zambonino-Infante and G. Koumoundouros. 2009. Optimal levels of dietary vitamin A for reduced deformity incidence during development of European sea bass larvae (*Dicentrarchus labrax*) depend on malformation type. Aquaculture 294: 262–270.

Mazurais, D., M. Darias, I. Fernández, C. Cahu, E. Gisbert and J.L. Zambonino-Infante. 2012. Gene expression pattern during European sea bass larvae development: impact of dietary vitamins. pp. 91–111. *In*: M. Saroglia and L. Zhanjiang (eds.). Functional Genomics in Aquaculture. John Wiley & Sons, Inc., Oxford.

McCauley, R.W. and N.W. Huggins. 1979. Ontogenetic and non-thermal seasonal effects on thermal preferenda of fish. Am. Zool. 19: 267–271.

Melotti, P., A. Roncarati, L. Gennari, G. Mosconi and F. Loro. 1993. Performance of wild and reproduced sea bass juveniles, reared in different kinds of tanks and at different stocking densities. pp. 599–603. *In*: Proceedings of the X National Congress of Scientific Association of Animal Production (A.S.P.A.), Bologna, Italy.

Melotti, P., A. Roncarati, F. Loro, L. Gennari and O. Mordenti. 1994. Intensive growing trials of wild and reared sea bass (*Dicentrarchus labrax*). Biol. Mar. Medit. 1: 439–440.

Migaud, H., M. Cowan, J. Taylor and W. Ferguson. 2007. The effect of spectral composition and light intensity on melatonin, stress and retinal damage in post-smolt Atlantic salmon, *Salmo salar*. Aquaculture 270: 390–404.

Monk, J., V. Puvanendran and J.A. Brown. 2006. Do different light regimes affect the foraging behaviour, growth and survival of larval cod (*Gadus morhua* L.)? Aquaculture 257: 287–293.

Monroig, O., J.C. Navarro, F. Amat, P. González and F. Hontoria. 2006. Effects of nauplial density, product concentration and product dosage on the survival of the nauplii and EFA incorporation during *Artemia* enrichment with liposomes. Aquaculture 261: 659–669.

Moren, M., R. Waagbø and K. Hamre. 2011. Micronutrients. pp. 47–82. *In*: G.J. Holt (ed.). Larval Fish Nutrition. Wiley-Blackwell, Oxford.

Moretti, A., M. Pedini, G. Cittolin and R. Guidastri. 1999. Manual on Hatchery Production of Sea bass and Gilthead Seabream. Vol 1. FAO, Rome.

Mortensen, A., O. Ugedal and F. Lund. 2007. Seasonal variation in the temperature preference of Arctic charr (*Salvelinus alpinus*). J. Therm. Biol. 32: 314–320.

Naas, K., I. Huse and J. Iglesias. 1996. Illumination in first feeding tanks for marine larvae. Aquacult. Eng. 15: 291–300.

Navarro, J.C., R.J. Henderson, L.A. McEvoy, M.V. Bell and F. Amat. 1999. Lipid conversions during enrichment of *Artemia*. Aquaculture 174: 155–166.

NRC (National Research Council). 2011. Nutrient requirements of fish. Committee on Animal Nutrition Board on Agriculture National Research Council. National Academy Press (Washington), D.C.

Nebel, C., G. Nègre-Sadargues, C. Blasco and G. Charmantier. 2005. Morphofunctional ontogeny of the urinary system of the European sea bass *Dicentrarchus labrax*. Anat. Embryol. 209: 193–206.

Nehr, M.-O. 1996. Développement d'une méthodologie de production semi-intensive d'alevins en mer dans des enceintes à renouvellement semi-continu: application à l'élevage du loup (*Dicentrarchus labrax*). Thèse de Doctorat, Université Aix-Marseille, Marseille.

Nikaido, S.S. and C.H. Johnson. 2000. Daily and circadian variation in survival from ultraviolet radiation in *Chlamydomonas reinhardtii*. Photochem. Photobiol. 71: 758–765.

Oliva-Teles, A. 2012. Nutrition and health of aquaculture fish. J. Fish Dis. 35: 83–108.

Pando, M.P., A.B. Pinchak, N. Cermakian and P. Sassone-Corsi. 2001. A cell-based system that recapitulates the dynamic light dependent regulation of the vertebrate circadian clock. Proc. Natl. Acad. Sci. USA 98: 10178–10183.

Pavlidis, M., E. Karantzali, E. Fanouraki, C. Barsakis, S. Kollias and N. Papandroulakis. 2011. Onset of the primary stress in European sea bass *Dicentrarchus labrax*, as indicated by whole body cortisol in relation to glucocorticoid receptor during early development. Aquaculture 315: 125–130.

Péres, A., C.L. Cahu and J.L. Zambonino-Infante. 1997. Dietary spermine supplementation induces intestinal maturation in sea bass (*Dicentrarchus labrax*) larvae. Fish Physiol. Biochem. 16: 479–485.

Person Le Ruyet, J., J.C. Alexandre, L. Thebaud and C. Mugnier. 1993. Marine fish larvae feeding: Formulated diets or live prey? J. World Aquacult. Soc. 24: 211–224.

Picchietti, S., A.M. Fausto, E. Randelli, O. Carnevalli, A.R. Taddei, F. Buonocore, G. Scapigliati and Luigi Abelli. 2009. Early treatment with *Lactobacillus delbrueckii* strain induces an increase in intestinal T-cells and granulocytes and modulates immune-related genes of larval *Dicentrarchus labrax* (L.). Fish Shellfish Immunol. 26: 368–376.

Pommeranz, T. 1974. Resistance of plaice eggs to mecanical stress and light. *In*: pp. 397–417. J.H.S. Blaxter (ed.). The Early Life History of Fish. Springer-Verlag, New York.

Prestinicola, L., C. Boglione, P. Makridis, A. Spanò, V. Rimatori, E. Palamara, M. Scardi and S. Cataudella. 2013. Environmental conditioning of skeletal anomalies typology and frequency in gilthead seabream (*Sparus aurata* L., 1758) juveniles. PLoS ONE 8: e55736.

Puvanendran, V. and J.A. Brown. 1998. Effect of light intensity on the foraging and growth of Atlantic cod larvae: interpopulation difference? Mar. Ecol. 167: 207–214.

Ramírez-Zarzosa, G., F. Gil, R. Latorre, A. Ortega, A. García-Alcázar, E. Abellán, J.M. Vázquez, O. López-Albors, A. Arencibia and F. Moreno. 1995. The larval development of lateral musculature in gilthead sea bream *Sparus aurata* and sea bass *Dicentrarchus labrax*. Cell Tissue Res. 280: 217–224.

Rekecki, A., K. Dierckens, S. Laureau, N. Boon, P. Bossier and W. Van den Broeck. 2009. Effect of germ-free rearing environment on gut development of larval sea bass (*Dicentrarchus labrax* L.). Aquaculture 293: 8–15.

Rensing, L. and P. Ruoff. 2002. Temperature effect on entrainment, phase shifting, and amplitude of circadian clocks and its molecular bases. Chronobiol. Int. 19: 807–864.

Roberfroid, M. 2007. Prebiotics: the concept revisited. J. Nutr. 137: 830S–837S.

Roncarati, A., A. Meluzzi, P. Melotti and O. Mordenti. 2001. Influence of the larval rearing technique on morphological and productive traits of European sea bass (*Dicentrarchus labrax* L.). J. Appl. Ichthyol. 17: 244–246.

Ronzani-Cerqueira, V. 1986. L'elevage larvaire intensif du loup (*Dicentrachus labrax* L.). Influence de la lumière, de la densité en proie et de la température sur l'alimentation, sur la digestión et sur les performances zootechniques. These de doctorat, Aix Marseille II, Marseille.

Russell, N.R., J.D. Fish and R.J. Wootton. 1996. Feeding and growth of juvenile sea bass: the effect of ration and temperature on growth rate and efficiency. J. Fish. Biol. 49: 206–220.

Sadler, J., P.M. Pankhurst and H.R. King. 2001. High prevalence of skeletal deformity and reduced gill surface area in triploid Atlantic salmon (*Salmo salar* L.). Aquaculture 198: 369–386.

Saillant, E., A. Fostier, P. Haffray, B. Menu and B. Chatain. 2003. Saline preferendum for the European sea bass, *Dicentrachus labrax*, larvae and juveniles: effect of salinity on early development and sex determination. J. Exp. Mar. Biol. Ecol. 287: 103–117.

Saka, F., K. Firat and H.O. Kamachi. 2001. The development of European sea bass (*Dicentrarchus labrax* L., 1758) eggs in relation to temperature. Turk. J. Vet. Anim. Sci. 25: 139–147.

Sargent, J.R., L.A. McEvoy and J.G. Bell. 1997. Requirements, presentation and sources of polyunsaturated fatty acids in marine fish larval feeds. Aquaculture 155: 119–129.

Sargent, J., G. Bell, L. McEvoy, D.R. Tocher and A. Estevez. 1999. Recent developments in the essential fatty acid nutrition of fish. Aquaculture 177: 191–199.

Scapolo, P.A., A. Veggetti, F. Mascarello and M.G. Romanello. 1988. Developmental transitions of myosin isoforms and organisation of the lateral muscle in the teleost *Dicentrarchus labrax* (L.). Anat. Embryol. 178: 287–295.

Schaefer, J. and A. Ryan. 2006. Developmental plasticity in the thermal tolerance of zebrafish *Danio rerio*. J. Fish. Biol. 69: 722–734.

Shields, R.J. 2001. Larviculture of marine finfish in Europe. Aquaculture 200: 147–160.

Sucré, E., M. Charmantier-Daures, E. Grousset, G. Charmantier and P. Cucchi-Mouillot. 2009. Early development of the digestive tract (pharynx and gut) in the embryos and pre-larvae of the European sea bass *Dicentrarchus labrax*. J. Fish Biol. 75: 1302–22.

Tamai, T.K., V. Vardhanabhuti, N.S. Foulkes and D. Whitmore. 2004. Early embryonic light detection improves survival. Curr. Biol. 14: R104–R105.

Tan-Tue, V. 1976. Etude du développement du tube digestif des larves de bar *Dicentrarchus labrax* (L.). Arch. Zool. Exp. Gén. 117: 493–509.

Trotter, A.J., P.M. Pankhurst and P.R. Hart. 2001. Swimbladder malformation in hatchery-reared striped trumpeter, *Latris lineata* (Latridae). Aquaculture 198: 41–54.

Tocher, D.R. 2010. Fatty acid requirements in ontogeny of marine and freshwater fish. Aquacult. Res. 41: 717–732.

Tovar, D., J. Zambonino, C. Cahu, F.J. Gatesoupe, J.R.Vázquez and R. Lésel. 2002. Effect of live yeast incorporation in compound diet on digestive enzyme activity in sea bass (*Dicentrarchus labrax*) larvae. Aquaculture 204: 113–123.

Tovar-Ramírez, D., J. Zambonino-Infante, C. Cahu, F.J. Gatesoupe and R. Vázquez-Juárez. 2004. Influence of dietary live yeast on European sea bass (*Dicentrarchus labrax*) larval development. Aquaculture 234: 415–427.

Tovar-Ramírez, D., D. Mazurais, J.F. Gatesoupe, P. Quazuguel, C.L. Cahu and J.L. Zambonino-Infante. 2010. Dietary probiotic live yeast modulates antioxidant enzyme activities and gene expression of sea bass (*Dicentrarchus labrax*) larvae. Aquaculture 300: 142–147.

Vagner, M., J.H. Robin, J.L. Zambonino-Infante and J. Person-Le Ruyet. 2007. Combined effects of dietary HUFA level and temperature on sea bass (*Dicentrarchus labrax*) larvae development. Aquaculture 266: 179–190.

Vagner, M., J.H. Robin, J.L. Zambonino-Infante, D.R. Tocher and J. Person-Le Ruyet. 2009. Ontogenic effects of early feeding of sea bass (*Dicentrarchus labrax*) larvae with a range of dietary n-3 highly unsaturated fatty acid levels on the functioning of polyunsaturated fatty acid desaturation pathways. Brit. J. Nutr. 101: 1452–1462.

Van der Meeren, T. and K.E. Jorstad. 2001. Growth and survival of Arcto-Norwegian and Norwegian coastal cod larvae (*Gadus morhua* L.) reared together in mesocosms under different light regimes. Aquac. Res. 32: 549–563.

Van Dijk, P.L.M., G. Staaks and I. Hardewig. 2002. The effect of fasting and refeeding on temperature preference, activity and growth of roach, *Rutilus rutilus*. Oecologia 130: 496–504.

Varsamos, S., R. Connes, J.-P. Diaz, G. Barnabé and G. Charmantier. 2001. Ontogeny of osmoregulation in the European sea bass *Dicentrachus labrax* L. Mar. Biol. 138: 909–915.

Varsamos, S., J.P. Diaz, G. Charmantier, C. Blasco, R. Connes and G. Flik. 2002. Location and morphology of chloride cells during the post-embryonic development of the European sea bass, *Dicentrarchus labrax*. Anat. Embryol. 205: 203–13.

Veggetti, A., F. Mascarello, P.A. Scapolo and A. Rowlerson. 1990. Hyperplastic and hypertrophic growth of lateral muscle in *Dicentrarchus labrax* (L.): An ultrastructural and morfometric study. Anat. Embryol. 182: 1–10.

Villamizar, N., A. García-Alcazar and F.J. Sánchez-Vázquez. 2009. Effect of light spectrum and photoperiod on the growth, development and survival of European sea bass (*Dicentrarchus labrax*) larvae. Aquaculture 292: 80–86.

Villamizar, N., B. Blanco-Vives, H. Migaud, A. Davie, S. Carboni and F.J. Sánchez-Vázquez. 2011. Effects of light during early larval development of some aquacultured teleosts: A review. Aquaculture 315: 86–94.

Villeneuve, L., E. Gisbert, H. Le Delliou, C.L. Cahu and J.L. Zambonino-Infante. 2005. Dietary levels of all-trans retinol affect retinoid nuclear receptor expression and skeletal development in European sea bass larvae. Br. J. Nutr. 93: 791–801.

Villeneuve, L., E. Gisbert, J. Moriceau, C.L. Cahu and J.L. Zambonino. 2006. Intake of high levels of vitamin A and polyunsaturated fatty acids during different developmental periods modifies the expression of morphogenesis genes in European sea bass (*Dicentrarchus labrax*). Br. J. Nutr. 95: 677–687.

Weger, B.D., M. Sahinbas, G.O. Otto, P. Mracek, O. Armant, D. Dolle, K. Lahiri, D. Vallone, L. Ettwiller, R. Geisler, N.S. Foulkes and T. Dickmeis. 2011. The light responsive transcriptome of the zebrafish function and regulation. PloS ONE 6: e17080.

Whitmore, D., N.S. Foulkes and P. Sassone-Corsi. 2000. Light directly sets the circadian clocks of zebrafish organs and cells. Nature 404: 87–91.

Woiwode, J.G. and I.R. Adelman. 1992. Effects of starvation, oscillating temperatures, and photoperiod on the critical thermal maximum of hybrid striped x white bass. J. Therm. Biol. 17: 271–275.

Wolfgang, W. 1992. Determination of correlated color temperature based on a color appearance model. Color Res. Appl. 17: 24–30.

Woolley, L.D. and J.G. Qin. 2010. Swimbladder inflation and its implication to the culture of marine finfish larvae. Rev. Aquaculture 2: 181–190.

Zambonino-Infante, J.L. and C. Cahu. 1994. Development and response to a diet change of some digestive enzymes in sea bass (*Dicentrarchus labrax*) larvae. Fish Physiol. Biochem. 12: 399–408.

Zambonino-Infante, J.L., C.L. Cahu and A. Peres. 1997. Partial substitution of di- and tripeptides for native proteins in sea bass diet improves *Dicentrarchus labrax* larval development. J. Nutr. 127: 608–614.

Zambonino-Infante, J.L. and C.L. Cahu. 1999. High dietary lipid levels enchance digestive tract maturation and improve *Dicentrarchus labrax* larval development. J. Nutr. 129: 1195–1200.

Zambonino-Infante, J.L. and C.L. Cahu. 2010. Effect of nutrition on marine fish development and quality. pp. 103–124. *In*: G. Koumoundouros (ed.). Recent Advances in Aquaculture Research. Transworld Research Network, Kerala.

Zouiten, D., I. Ben Khemis, A.S. Masmoudi, C. Huelvan and C. Cahu. 2011. Comparison of growth, digestive system maturation and skeletal development in sea bass larvae reared in intensive or a mesocosm system. Aquac. Res. 42: 1723–1736.

Foraging Behavior

Sandie Millot,[1,a] *M.-L. Bégout*[1,b] *and David Benhaïm*[2,]*

Introduction

Feeding by fish is dependent upon their sensory capacities to locate food, their ability to capture (linked to swimming), handle and ingest food items, and their physiological and biochemical capacities to digest and transform ingested nutrients. All of these may depend on environmental factors, which are physical, chemical or biological in nature (review in Kestemont and Baras 2001). Here we describe the main traits of foraging behavior both in the wild and in aquaculture for the different life stages of sea bass. Besides feeding, swimming behavior and social interactions closely interact along with other factors such as intrinsic adaptive repertoire and genetic origin or selection processes. Therefore, the chapter first presents foraging behavior in natural environment, then details foraging in aquaculture environment in relation with environmental factors and finally describes other related behavior and illustrates the influence of domestication and selection processes on sea bass behavioral traits.

Foraging Behavior in Natural Environment

Scientific studies of the behavior of sea bass in the wild are scarce. Most of the descriptions available come from Barnabé (1976a, 1978) who worked on western Mediterranean sea bass populations and from Pickett and

[1] Ifremer Place Gaby Coll, BP 7, 17137, L'Houmeau, France.
[a] Email: sandiemillot@yahoo.fr
[b] Email: marie.laure.begout@ifremer.fr
[2] INTECHMER/CNAM, BP 324, 50103 Cherbourg cedex, France.
Email: david.benhaim@cnam.fr
*Corresponding author

Pawson (1994) who worked on Atlantic sea bass populations. Nevertheless, these observations are useful to better understand some behavioral results obtained in captive environment including feeding behavior, social interaction, cognition, etc. (see next sections). Such knowledge is also important for fish rearing and welfare concerns because the understanding of an animal's natural behavior will allow designing captive conditions that cater for its needs.

All the information related to sea bass foraging are synthesized in Table 1. Sea bass are opportunistic and very active predators having a very well developed visual sense (Langridge 2009). At larval stage, sea bass start to feed on zooplankton. At later stages, they adopt a wide range of tactics to find and capture their prey. Generally, juveniles as well as adults hunt as a shoal, taking whatever prey species are seasonally abundant in a particular location (reviewed by Barnabé 1980). Solitary adult hunters are also commonly observed in the wild. If the food items are on the sea-bed, e.g., crabs, bivalve molluscs or polychaete worms, the shoal will spread out and will graze, head down, along the substrate. They are however, more readily observed when they feed on pelagic prey, which they drive upwards towards the surface and attack from below at a steep angle. This feeding may become quite frenzied and sea bass will often be seen to break the water surface in pursuit of their prey among rocks and seaweed. They have also been shown to position themselves in slacker water near piers, overflows and sandbanks, lying in wait for smaller fish to be swept past them by the current. Lastly, an intriguing behavior called "flashing" that could be related to foraging behavior has also been described by several authors (Barnabé 1976a, Pickett and Pawson 1994). Sea bass slowly sink towards the bottom and then suddenly move forward, turn on one side and appear to rub one flank on the substrate, the upper flank producing a sudden silvery flash. This behavior could be intended either to disturb small crustaceans food items buried in a sandy bottom or it could be just an attempt to get rid of ectoparasites.

In close association with feeding, social behavior will change according to the developmental stage of sea bass, i.e., from the larval to the adult stage. It is not clear at an early age whether sea bass form shoals (Pickett and Pawson 1994). Some authors however mentioned post-larvae and fry congregating where physical barriers are present in upper estuaries, creeks and harbors, i.e., a few dozen individuals to many thousands remaining in distinct groups for several years at a time (Aprahamian and Barr 1985, Claridge and Potter 1984, Dando and Demir 1985, Kelley 1988). At the juvenile stage, sometimes as soon as 20 mm of body size, sea bass become a demersal and gregarious species able to orientate in relation to the benthic substrate (Barnabé 1989). The shoaling life is generated by visually stimulating attractions between individuals. This is a crucial event

Table 1. Synthesis of sea bass foraging behaviour both in the wild and in aquaculture reported from the literature.

Behaviour and linked parameters	Stage	Method	Food	References
Hunting behaviour in the wild	Larvae	Stomach contents	Zooplankton	Langridge 2009
Hunting behaviour in the wild • Hunting as a shoal • Flashing behaviour	Juveniles/adults	Visual observations/stomach contents	Pelagic preys : small fish Benthic preys : prawns, crabs, bivalve molluscs, polychaete worms, cuttlefish	Barnabé 1980 Langridge 2009
Hunting behaviour in aquaculture • Location of prey in the field of vision • Movement of orientation towards the prey • Aim and positioning in a S shape • Relaxation and preparation	Larvae	Visual observations	*Artemia salina*	Barnabé 1994
Modal action patterns • Normal/abrupt swimming • Sinking • Resting • Sigmoid posture • Attack	Larvae	Video + software	*Artemia salina*	Georgalas et al. 2007
Feeding behaviour • Bottom creep • Dig • Explore • Ingest • Surface creep	Juveniles	Visual observations	Artificial food	Ajuzie 1998
Self-feeding behaviour • Feed demand • Feeding rhythm • Growth performances • Social structure based on triggering activity	Juveniles/adults	Computerized self-feeder	Artificial food	Covès et al. 1998, 2006 Di-Poï et al. 2007, 2008 Millot et al. 2008, 2011 Millot and Bégout, 2009 Benhaïm et al. 2011, 2012a

Table 1. contd....

Table 1. contd.

Behaviour and linked parameters	Stage	Method	Food	References
Food intake regulation				
• Choose nutrients according to energy requirements	Juveniles/ adults	Manual distribution	Artificial food	Kaushik and Medale 1994
• Feed preference		Self-feeder		Boujard et al. 2004 Paspatis et al. 2002 Rubio et al. 2006, 2008
• Seasonal changes in macronutrient selection pattern		Manual distribution		
• Physical and chemical factors (T°, S‰, O_2, CO_2, pH, water quality)		Manual distribution		Person-Le Ruyet et al. 2004 Pichavant et al. 2000 Conides and Glamuzina 2006 Rubio et al. 2005
Feeding anticipatory activity	Juveniles/ adults	Self-feeder	Artificial food	Boujard et al. 1996 Azzaydi et al. 2007

from the eco-ethological point of view because, at this stage, sea bass can occupy both pelagic and benthic areas. Sea bass can be considered as a ubiquitous species because they adapt to different habitats such as rocky, grassy or even muddy bottoms. The benthic habitat enables sea bass to shelter when threatened by a predator either inside rocky cavities or even in a soft substrate where they are able to bury themselves.

Sea bass reaction to other species of their own size is generally a rapid escape or it can adopt a typical defence posture that involves raising the first dorsal fin and distending the opercula, in order to appear larger and to present as many sharp spines to the aggressor as possible. There is however, little evidence of aggressive behavior between sea bass of similar size in the wild, although they may be territorial when occupying summer feeding areas (Carliste 1961). Sea bass are known as strong swimmers, being able to sustain a high average swimming speed when on migration (maximum speed, U_{crit} can reach 2.7 BLs^{-1} at 20°C, Luna-Acosta et al. 2011), helped by the presence of darker-toned "slow" muscle along the flanks of the fish (Pitcher and Hart 1982).

Foraging Behavior in Aquacultural Environment

Particularities of the larval stage

At the end of the 70s scientists started to develop, in most Mediterranean countries, intensive rearing methods based on a complex hatchery system. The reproduction of sea bass is now fully controlled in European facilities and after hatching, larvae are transferred to specific rearing units (EFSA 2008). Aquaculture and natural habitats are obviously very different. The rearing environment influences a range of behavioral traits (Salvanes and Braithwaite 2006). It is therefore important to understand the behavioral ontogeny of swimming and feeding and the influence of the rearing environment. The first description of behavioral ontogeny of sea bass comes from Barnabé (1976b). This author observed that fish larvae, at hatching, are not yet completely formed e.g., lack of functional eyes and mouth. Swimming activity of the larvae is progressively established (see also Kentouri 1985). In the first three to six days after hatching, and depending on water temperature, the fish larva relies only on its yolk sac reserves as food source. At the end of this period the young fish has developed functional eyes, its mouth has opened and the digestive tract can now assimilate food. Then, its swimming behavior becomes active and the animal is thus able to keep a horizontal position. At this step, the post-larval stage begins and the young fish starts feeding on live preys. Fish are now sensitive to light levels of the order of a few Lux. The larvae move towards the light and tend to swim against the current. They react to prey situated at least 1 cm in front of them and within a field extending laterally through around 40°. Hunting activity is initiated when a suitable sized prey organism comes into view (Fig. 1).

A recent European study (Neophytou 2013) focuses on behavioral ontogeny of sea bass swimming and feeding behavior in captivity in two different rearing technologies: intensive (Papandroulakis et al. 2002) and mesocosm rearing system (Divanach and Kentouri 2000). This work demonstrated that during ontogenesis, individuals go through some clear behavioral changes that are strongly related with their morphological ontogeny and also with the rearing technique (Fig. 2). In particular, from pre-larval to larval stages, swimming activity shifted from an initial spasmodic, intermediated swimming where larvae spent more time resting than acting, to a more continuous, coordinated and energetic swimming under both rearing conditions. Under intensive rearing conditions, "sustained" swimming velocity from just after larvae entered the heterotrophic life showed an increasing trend between the stages studied, with an important increment between the phase from flexion (urostyle torsion) until beginning of metamorphosis (formation of dorsal and anal fin). After that stage, and until the formation of all fins, swimming velocity reduced again, but remained higher than the mean velocity found at the stage from first feeding until flexion. Fish in mesocosm undergo the same morphological changes at the stage from flexion to beginning of metamorphosis but this morphological change does not seem to have a big impact on swimming activity probably because fish have more "space" and are subjected to lesser competition for food.

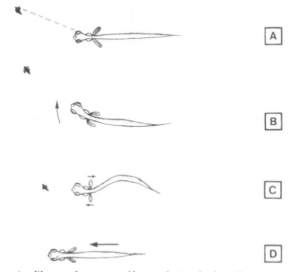

Figure 1. ç-like attack posture of larvae during feeding (Barnabé 1994).

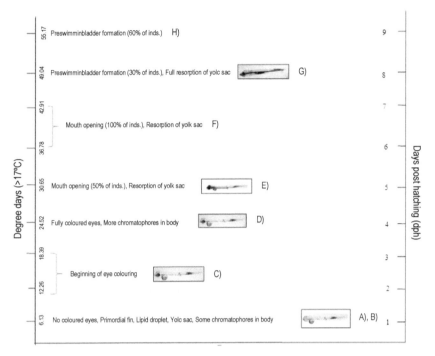

Figure 2. Sequence of morphological changes during autotrophic phase related with behavioural changes (from hatching to 5.3 mm TL). Developmental time is given in dph and corresponding degrees days (d.d. > 17°C). The percentages for the occurrence of the different stages (pre-swim-bladder formation and mouth opening) are given in brackets. **A. 0 dph (hatching):** larvae are drifting from water current in the tank, show negative buoyancy and are distributed in the whole tank. **B. 1 dph:** similar pattern as A but larvae begin counter current swimming far from the aeration and near the tank walls. Twenty percent of individuals start an intermediate swimming away from aeration in all directions, which with duration of 1–2 s, and also appear to have "swirling" downward movements. **C. 2, 3 dph:** similar pattern as B. All individuals have intermediate swimming away from aeration with burst onward movements that have duration of 1–2 s. **D. 4 dph:** similar pattern as C plus a left/right movement of the second half of body (flip/flop). **E. 5 dph:** similar pattern as D but movements are more energic with onwards movements that have duration of 3–7 s, and an onward intermediate swimming away from aeration. All individuals have burst onwards movements that have duration of 0.5–1 s. **F. 6, 7 dph:** similar patterns as E. **G. 8 dph:** similar patterns as F with more energic movements, still drifting and sinking in water column but with more quick returns to swimming endeavours. **H. 9 dph:** similar patterns as G but with a more continuous movement although they still have some negative buoyancy. They appear to have even more energic movements and a counter current swimming near aeration. They begin to demonstrate a prey attack position by adopting a ç-position in their body and they start "searching movements" in the tank in all directions by knocking on walls (from Neofytou 2013).

The efficiency of behavior related to prey searching and hunting is strongly related to the full development of the swim bladder occurring at 10 dph (days post hatching) for larvae from intensive rearing system and at 5 dph for larvae from the mesocosm rearing system, as already shown by Chatain (1986) and Georgalas et al. (2007) respectively. During co-feeding phase (progressive elimination of live prey in the food) larvae continue to show strong preference for live feed, although dry food is inserted a few days before, giving the larvae enough time to recognize the new prey. The co-feeding phase cannot occur earlier than 31 and 32 dph for intensive and mesocosm rearing respectively, thereby demonstrating an adaptation and learning period of larvae for the new food (dry granules) and hence displaying an ontogenetic evolution of feeding behavior. Larvae of both rearing systems tended to swim parallel to the tank walls at the developmental stages of beginning of metamorphosis and formation of all fins (early juveniles). This is a first sign of shoaling behavior which in natural habitats occurs mostly in juveniles (see Section 1) and happens with counter-current positioning.

Lastly, larvae from the intensive rearing system seemed to occupy the whole tank during the autotrophic phase while larvae from the mesocosm rearing system showed a sequence of repositioning in the water column with developmental time. In particular, larvae at the beginning of the autotrophic stage (0 dph) were found in swarms right under the water surface and from 2 dph started to migrate to deeper zones, but still in the first half of the tank. They progressively moved to deeper zones (at 3 dph they occupy up to 2/3 of the water column) and by the end of the autotrophic stage (4 dph) larvae were distributed in the whole tank. While the larvae of the intensive rearing system showed a random orientation towards the whole tank, the larvae of the mesocosm rearing system at the end of the autotrophic phase (4 dph) showed a beginning of positive reaction to natural light (positive phototropism). During heterotrophic phase, larvae of both populations were distributed in the whole tank except for feeding times when individuals moved to upper zones where food was more abundant.

A few studies provided sea bass larvae ethograms either in usual rearing conditions (Ajuzie 1998) or in relation to abiotic factors such as food density (Georgalas et al. 2007) or toxic components (Ajuzie 2008). A recent study, comparing wild caught and domesticated sea bass juveniles, provided swimming behavior characteristics using video analysis methods before and after eliciting a startle response (Benhaïm et al. 2012b, see Section 3). Indeed, fish behavior can be used as an indicator which can give the fish farmer an early warning on changes in environmental or health conditions or could be used for selection or restocking programs. To sum up, Ajuzie (1998) described 29 behavioral units (see Table 2 in Ajuzie 1998) from observations performed on 60-day-old sea bass and with a focus on social interactions between dominant and subordinate fish. This author showed that aggressive behaviors were employed by the dominant fish to gain space, shelter and food. However, aggressiveness was not observed in equal size pair members. The same author also exposed 100-day-old sea bass to both cell-free medium and whole cell cultures of the dinoflagellate *Prorocentrum lima* strain PL2V using six behavioral items: 1. Repeated jumps out of the medium into the air. 2. Pronounced opercular movements. 3. Window creeping. 4. Window pushing near the surface. 5. Surface swims with open mouth. 6. Fast left-right turns. The results showed that fish exposed to the cell-free medium and the whole cell cultures were stressed and behaved abnormally compared to the control fish. Stress-related behaviors included hyperactivity, poor feeding reflexes and abstinence from feeding. Georgalas et al. (2007), analyzed sea bass larvae behavior (10–30 dph) reared with the mesocosm technique in relation to ontogeny and increasing food density, with a particular focus on swimming and feeding activities. These authors used six modal action patterns (MAPs), three classified as activities and three as events. Activities: 1. Normal swimming: according to the definitions given by Rosenthal and Hempel (1970) 2. Sinking: when the larva ceases momentarily "normal swimming" and sinks passively through the water column 3. Resting (Rs): when the larva ceases all its activities and rests immobile on the tank bottom. Events: 1. Abrupt swimming: according to Rosenthal and Hempel (1970), this type of swimming consisted of movements of very short duration often connected with a shift in orientation 2. Sigmoid-posture: the larval body assumed an S-curve which could precede a lunge and prey ingestion 3. Attack: this consisted of the lunge often performed subsequently to the Sigmoid-posture, and it was in many cases followed by movements associated with prey mastication and ingestion. The results revealed that fish swimming activity increased significantly with age whereas resting activities decreased with age and in some cases, also with food density.

Juveniles and adults

Feeding strategies

In order to decide when, how much and by which way fish should be fed, it is essential to know how they behave under different feeding systems. Indeed fish feeding is one of the most important factors in commercial fish farming because the feeding regime may have consequences on growth, feed efficiency and wastage. Moreover, knowledge of the optimum feeding rate is important not only for promoting best growth and feed efficiency, but also for preventing water quality deterioration as a result of excess feeding.

Sea bass are able to regulate their daily food intake based on their nutrient and energy requirements (Kaushik and Médale 1994). Indeed, when sea bass have a choice between feeds formulated for carnivores, herbivores and omnivores, they prefer feed formulated for carnivores which seems the best adapted for them according to their higher growth rate. This also demonstrates the capacity to discriminate between a diet with fish meal and a diet containing exogenous phytase (which increases the acceptance of the plant diet, Paspatis et al. 2002, Fortes-Silva et al. 2011). Boujard et al. (2004) showed also that an increasing dietary lipid level led to a significant decrease in voluntary feed intake without affecting growth rate for sea bass reared in an unrestricted condition and to an increased growth rate for sea bass in a feed-restricted condition. Furthermore, the energy intake changed monthly, the highest values being recorded in May and June and the lowest values in March and April. The results of a study by Rubio et al. (2006, 2008) illustrated seasonal changes in the sea bass macronutrient selection pattern which showed a predominantly proteinic selection during April and lipidic selection in July. This evidence supports the existence of an endogenous rhythm in the seasonal energy regulation and macronutrient selection in fish through post-ingestive mechanisms, probably involving chemosensory detection in the gut and/or post-absorptive mechanisms.

Over all, fish self-regulation does not mean that they will make the most efficient use of the diet (Azzaydi et al. 1998), as the ration size that optimizes feed efficiency is generally lower than that producing the highest growth rate (Talbot 1993). For example, higher feeding efficiencies are seen in sea bass if food access is restricted under certain circumstances (Azzaydi et al. 1998) because reducing their daily ration appears to compel fish to make the best use of the feed they have ingested without affecting their growth. Thus, it appears that the optimum feeding rate for sea bass fingerlings' (3 g) growth reared in sea water and fresh water is 3.0% and 3.5% bw (body weight) day^{-1}, respectively and the optimum feeding rate for feed conversion efficiency is 2.7% bw day^{-1} in sea water and 3.8% bw day^{-1} in fresh water (Eroldogan et al. 2004).

Boujard et al. (1996) and Azzaydi et al. (2007) showed that sea bass are also able to rapidly synchronize their demand-feeding rhythm when they are subjected to a time-restricted feeding schedule. Their food demand appears to be predominantly associated with feed availability, reaching its maximum levels during the hours of reward accompanied by feeding anticipatory activity. Azzaydi et al. (2000) showed that during winter, the highest SGR (Specific Growth rate) and the lowest FCR (Feed Conversion Ratio) were obtained with self-feeder and nocturnal automatic feeding treatments when sea bass showed nocturnal behavior. They further demonstrated that automatic-feeding systems, in which the quantity of feed supplied is modulated in accordance with the natural feeding rhythms of sea bass, may improve growth and feed efficiency (Azzaydi et al. 1999). In conclusion, sea bass should be fed at their preferred time (Boujard et al. 1996, Heilman and Spieler 1999) and in accordance with their nutritional requirements (Sánchez-Vázquez et al. 1998, Yamamoto et al. 2000a, Yamamoto et al. 2000b).

For feeding sea bass in culture conditions, the self-feeding practice has been proved appropriate and is often better than other methods (Andrew et al. 2002, Azzaydi et al. 1998, Boujard et al. 2000, Covès et al. 1998, Divanach et al. 1993, Paspatis et al. 1999). Nevertheless, by comparing a number of automatic feeding systems for sea bass, Azzaydi et al. (1999) concluded that those in which the food is supplied at a defined period coinciding with the time of maximum appetite might produce comparable, and sometimes even better results than self-feeding systems. Other authors have found improved growth when feeding is tailored to the animals feeding rhythm (Azzaydi et al. 1998, Boujard et al. 1995, Hossain et al. 2001, Reddy et al. 1994). Although the beneficial effects of demand-feeding on fish have been largely demonstrated both in cages and tanks, hand feeding is still the most widely used method along with automatic feeders. In the case of semi-intensive systems, fish rely on natural prey and if additional manufactured feed is offered to them, which can lead to differential access to feed, survival and growth of the sea bass population is enhanced (Bégout Anras et al. 2001).

Environmental determinism of sea bass feeding behavior

The diversified character of European coastal aquaculture along with an array of historical and socio-economic factors allow the use of extensive, semi-intensive and intensive systems for most stages of sea bass production (Basurco 2000, 2004). Concerning ongrowing it may take place under intensive, semi-intensive or extensive systems. Intensive systems include (a) net cages (b) land-based flow-through systems and (c) land based recirculation systems. According to Basurco (2000), in Mediterranean countries, 82% of farms use cages, followed by land based intensive raceways or tanks (10%) and semi-intensive production in earth ponds (8%). In cages, feed is distributed by automatic feeders, modulated-automatic feeding systems or hand feeding, and the most variable environmental factors are temperature and oxygen. Land-based intensive systems (either in flow-through or in recirculated water) aim to produce high value fish at high stocking densities (final stocking density of 20–35 kg m^{-3}) in semi-controlled (flow through) or controlled environment (re-circulated) and feeds are usually distributed by automatic feeders. Semi-intensive farming is based on pond culture where ponds are modified environments where it is possible to simulate and accelerate natural processes. This is achieved by controlling water flow, vegetation control, integrating the availability of food, adding some feeds and rearing at low densities (0.2–2 kg m^{-3}). Finally, extensive farming includes coastal lagoon management and *vallicultura*. Around 500,000 hectares of coastal lagoons still survive in the Mediterranean and their exploitation is based on the migratory behavior of sea bass which spend their juvenile and growing phase until reaching sexual maturity in such areas. Such extensive farming makes use of both wild immigrant fish and hatchery-reared juveniles; little environmental control is exerted and only water level and salinity are controlled by hydraulic management in embanked portions of the lagoons.

Depending on the culture method, *physical* and *chemical factors* vary and the effects of some of them on feeding behavior have been investigated and will be described hereafter.

Sea bass is classified as eurythermal fish (Stickney 1994) and is able to tolerate temperature ranging from 2°C to 32°C (Barnabé 1990). In extensive and semi-intensive coastal systems, water temperature can vary significantly, influencing both the duration of production cycles and feeding management. Maximum specific growth rate has been observed at 26°C and maximum feed intake at 27.5°C (under oxygen concentration to saturation, Person-Le Ruyet et al. 2004).

Indeed, dissolved oxygen is of primary significance in fish farms and its concentration is influenced by other abiotic (i.e., temperature, salinity, water quality) and husbandry factors (i.e., stocking density, water renewal). Environmental oxygen levels can directly affect activity, feeding behavior, growth performance, physiology and immune response of sea bass. As an example, long term exposure to oxygen saturation below 80% (at 22°C) impaired feed intake and growth in sea bass (Pichavant et al. 2000), mainly due to a reduced feed intake. CO_2 levels (above 55 mg l^{-1} at 21°C) and pH below 6 also act on swimming, feeding behavior and growth performances which are impaired (Lemarié, personal communication), sometimes in conjunction with low water renewal which also reduces feed intake.

Sea bass is a euryhaline fish capable of tolerating both high saline waters (SW) and freshwater (FW) environments. Extreme tolerance levels for sea bass are reported to be 0‰–5‰ to 60‰ (Jensen et al. 1998) and optimum salinity for feeding and growth of 30‰ has been reported (Conides and Glamuzina 2006). Nevertheless, lowering the salinity level from 25‰ to 7‰ and 0‰ reduced food intake by 27% and 42%, respectively (Rubio et al. 2005). Regarding macronutrient selection, these salinity changes significantly decreased the percentage of carbohydrates intake by 31% and 27%, while increasing that of protein by 30% and 25%, respectively. Fat selection remained unaltered, with an average value of 22% for all tested salinities. Specific growth rate and feed conversion efficiency were affected by macronutrient selection pattern, which in turn was salinity-dependent.

These results indicate the strong influence of salinity on European sea bass food intake and macronutrient selection (Rubio et al. 2005).

Light is a complex ecological factor whose components include colour spectrum (quality), intensity (quantity) and photoperiod (periodicity). In sea bass, photoperiod is recognized as a key factor affecting growth, development, and reproduction. Sea bass exhibits both nocturnal and diurnal feeding behavior being predominantly diurnal in the summer, nocturnal in winter and then returning to diurnal in spring (Sánchez Vázquez and Tabata 1998, Sánchez Vázquez et al. 1995a). Such seasonal changes in feeding behavior, from diurnal to nocturnal, has been described in relation to seasonal variations in photoperiod and water temperature (Bégout Anras 1995, Sánchez Vázquez and Tabata 1998), both under semi-natural or aquaculture conditions. Under constant conditions of water temperature and salinity, but under inversion of light and dark periods and subjected to light: dark pulses, sea bass showed a dualism in their diel feeding pattern. Thus, Sánchez Vázquez et al. (1995b) showed that the diurnal and nocturnal behaviors do not depend exclusively on a circadian phase inversion of the feeding rhythms, as this pattern of behavior is also enhanced under ultradian light: dark pulses.

Finally, sea bass are exposed to disturbances that may cause stress during ordinary farming procedures. Potential stressors can be found in all stages of the production cycle and include feed distribution mode, handling and manipulation, cleaning routines, grading, crowding and confinement, transportation between units, prophylactic measures, and use of chemicals. The presence of predators, boats and divers are also factors of disturbance for ongrowing fish kept in sea cages. The effects on feeding, by some of these factors are documented and are presented in the following sections along with other biological factors.

As regards *biological factors* influencing sea bass feeding behavior, stocking density and group social structure seem to be the most important ones.

A gregarious species such as sea bass shows better tolerance to high densities. For example, Papoutsoglou et al. (1998) and Paspatis et al. (2003) demonstrated that sea bass juvenile's growth rates increase with density whatever the feeding method. Moreover, Paspatis et al. (2003) showed that sea bass feeding rhythm was affected by stocking density, but not by reward levels. Thus, although sea bass presented a typical diurnal activity, fish restricted their feed demands at the first and late hours of daylight (08:00–09:00 and 19:00 h) when held in the lowest stocking density, while in high density they did not. Stirling (1977) showed also that isolated sea bass showed poorer growth than sea bass in pairs or groups (related to stress behavior when the fish is isolated).

Variation in individual growth, a common feature in many cultured fish stocks including sea bass remains a central problem in aquaculture. Growth heterogeneity can be influenced by a wide range of intrinsic and environmental factors (Kestemont et al. 2003). These authors classified all mechanisms as inherent (i.e., having a strong genetic component and expressed to some degree under environmental conditions) or imposed (requiring certain biotic or abiotic conditions to be manifested). Among biotic factors, social interaction has been identified as the major cause of individual variation in growth (Cutts et al. 1998, Jobling et al. 1993) by leading to feeding hierarchies and thus decreasing the growth of the low ranking fish (Koebele 1985). Recently, Benhaïm et al. (2011) showed that grading practice only transiently modifies feed demand behavior and social structure built around the self-feeder, without further improvement in individual growth performances in sea bass.

Studying feeding behavior may contribute to a better understanding of size variation mechanisms. Self-feeders are particularly useful for the study of feeding behavior in fish (Boujard et al. 1992). When coupled with a PIT tag detection antenna, they give the possibility to reveal individual and group feed demand behavior. Such combination of techniques was successfully used in sea bass under a variety of experimental conditions (Benhaïm et al. 2012a, Benhaïm et al. 2011, Covès et al. 2006, Covès et al. 1998, Di-Poï et al. 2007, Di-Poï et al. 2008, Millot et al. 2008, Millot et al. 2011). These systems contributed to a better understanding of the individual behaviors of fish living in groups.

In sea bass an intriguing individual specialization exists when using self-feeders with three coexisting triggering categories: high-triggering (HT), low-triggering (LT) and zero-triggering (ZT) fish (Benhaïm et al. 2012a, Benhaïm et al. 2011, Covès et al. 2006, Di-Poï et al. 2007, Millot et al. 2008). In small populations composed of 50–100 fish, whatever the experimental conditions, the same pattern is always observed: HT fish are very few with only one or two animals being responsible for 80% of the triggering activity under a reward regime of one or two pellets per individual given after each actuation (Covès et al. 2006) or two to three fish responsible for about 45% of the triggering activity under a reward regime equivalent to one pellet per individual (Millot et al. 2008). On the other hand, the ZT status would be attributed to fish that never actuate the device (Covès et al. 2006, Millot et al. 2008) or less than 4% of the time or if they perform a mean triggering activity lower than once a day (Di-Poï et al. 2007). This ZT category represented about 10% of the population (Covès et al. 2006), the rest being composed of individuals that seldom actuated the trigger (LT) (Covès et al. 2006), ≤ 25% of actuations (Millot et al. 2008), 4–15% (Di-Poï et al. 2007), or 0–30% (Covès et al. 2006). This individual specialization in three categories has been shown to be homogenous and stable in time over

200 days (Millot et al. 2008) with HT fish keeping their status during 60 days on average (Millot and Bégout 2009). These authors showed that an HT fish that loses its status rarely recovers it, and becomes LT. They also hypothesized that the frequent change of HT fish in a tank could be regarded as the consequence of an imbalanced group social structure (Millot et al. 2008), pointing out that changes occurred at the time of stressful events or spontaneously but without changing the overall population composition (Millot et al. 2008, Millot and Bégout 2009). The characteristics of HT, LT or ZT fish could be summarized as follows: (1) None of these fish showed differential access to delivered food whichever category they belonged to (Covès et al. 2006) and there was no clear evidence of a link between sex and food demand, i.e., HT fish are either females or males (Covès et al. 2006). (2) In most cases, the three categories exhibited no difference in mean initial or final weights or in mean specific growth rate (Covès et al. 2006, Di-Poï et al. 2007, 2008). (3) The serotonergic turnover of LT and ZT individuals is higher than that of the HTs' one indicating that they could be under social stress due to the high activity of HT individuals (Di-Poï et al. 2007) or that they are subordinate fish (Winberg and Nilsson 1993). (4) Millot et al. (2008) pointed out that the future HT individuals had negative specific growth rate at the beginning of their active period. These authors hypothesized that fish with a negative growth might be searching more for pellets and/or spent more time in the feeding zone, i.e., they had a higher feeding motivation, which in turn, may enhance the self-feeder learning process and hence its actuation. (5) Another study showed however, that sea bass high-triggering status is not only regulated by a negative growth (Benhaïm et al. 2012a). Indeed, two groups of fish of similar mean weight but with either a low or a high coefficient of variation for weight, submitted to a 3-week fasting period in order to induce similar negative specific growth rate, were characterized by the same high-triggering fish before and after the fasting period. High-triggering status could neither be explained by an initial lower SGR nor a sex effect, nor by any of the measured physiological blood parameters. Thus, individuals' triggering activity levels could be related to personality and/or metabolic traits.

In conclusion, when placed under self-feeder condition, the fish personality seems to be a determinant factor in the group feeding behavior. As shown by Di-Poï et al. (2008) and Millot and Bégout (2009) the fish displaying the highest activity in the self-feeder lead to a general food distribution and play a leading role in feeding the entire group. Thus, sea bass group feeding behavior is not the sum of individual feed demand behaviors, but is directed by the rhythm and behavior of a few high-triggering fish. Functional plasticity in this role within individuals indicated that the high-triggering function is essential for the group and not for the individuals themselves.

Another factor which can significantly influence the feeding behavior of sea bass is their level of domestication and selection. Indeed, according to the aquaculture goal to obtain final product in the lowest time possible and with the lowest cost, different procedures of selection (especially based on growth rate) have been applied on sea bass (Dupont-Nivet et al. 2008) but it is only recently that researchers have studied the influence of the domestication and selection level on sea bass behavior. Millot et al. (2010) showed that a first generation of domesticated and selected sea bass exposed to repeated acute stress reacted completely differently from wild fish and showed a significant higher body mass, specific growth rate and body condition factor. In 2011, the same authors showed that the domestication process seemed to improve fish adaptation abilities to acute stress while the process of selection for growth, even if it led to a final better growth, did not seem to improve fish acute stress tolerance (Millot et al. 2011). Furthermore, feeding behavior and thus, its consequences on the fish growth rate is not the only biological trait modified by the processes of domestication and selection (Millot et al. 2011, Millot et al. 2010).

Other Interesting Behaviors

Among behavioral characteristics, several studies have implied that antipredator behavior is highly sensitive to artificial rearing (Berejikian 1995, Dellefors and Johnsson 1995, Fernö and Järvi 1998, Johnsson and Abrahams 1991, Johnsson et al. 1996). This can be approached by comparing wild and domesticated individuals of the same species but little is known about non-salmon fish species, especially at early stages.

A few studies comparing wild-caught and domesticated sea bass juveniles led to similar conclusions. Malavasi et al. (2004) for example showed that in both hatchery reared (113 days old) and wild juveniles, predator exposure elicited a significant increase in the mean level of shoal cohesiveness and mean shoal distance from the predator, and a significant decrease in the mean shoal distance from the bottom. Shoals of wild juveniles, however, aggregated more quickly and reached higher shoal cohesiveness within the first 20 seconds of the stimulus period than shoals of hatchery reared fish. Wild fish also showed a longer mean freezing reaction after predator stimuli than hatchery reared juveniles (Malavasi et al. 2008). The predator inspection behavior was fully developed in both wild and hatchery fish, even if wild fish tended to inspect the predator at a closer distance than hatchery fish.

Benhaïm et al. (2012b) examined swimming behavior characteristics in wild captured and domesticated sea bass juveniles before and after eliciting a startle response at eight different ages (55 to 125 days old in both fish origins) and always on naive individuals. They showed consistent behavioral differences (e.g., higher angular velocity and distance from stimulus point in wild fish) but also similarities between both fish origins (similar response to stimulus actuation: decrease of total distance travelled and mean velocity; increase of angular velocity and immobility). There was also a decrease over time in reactivity and variability in swimming responses among fish of both origins, showing that captivity alone does not fully explain wild fish behavior. This shows that ontogenic modifications are likely interplaying. Finally, Millot et al. (2009a) showed that when exposed to a new risky situation, wild fish were generally bolder than selected fish but showed a decrease in risk taking over time, contrary to selected fish which showed a constant increase in their risk-taking behavior. Over all, domestication reduces flight response behavior and Millot et al. (2009b) further showed that only one generation of captivity could be sufficient to obtain fish presenting the same behavioral characteristics than fish reared since at least two generations. Moreover, this study also highlighted that selection for growth seemed to select fish characterized by a bolder personality and potentially better adapted to the rearing environment.

Recent studies (Benhaïm 2011, Benhaïm et al. 2013b) provided the first insight into sea bass cognition in experimental conditions using several types of mazes, e.g., the ability of individuals to learn to discriminate between two two-dimensional objects associated with a reward. Cognition includes perception, attention, memory formation and executive functions related to information processing, such as learning and problem solving (Brown et al. 2007). Some fish species are known to display a rich array of sophisticated behaviors; several studies have shown that they have long-term memories and that learning plays a crucial role in their behavioral development (Brown et al. 2007). However, there are very few studies on sea bass (only Millot et al. (2009a) showed that sea bass could memorize a challenge applied one month apart). Benhaïm et al. (2013a) provided the first insight into the impact of total plant-based diet (PB)—introduced at an early stage of sea bass development—on behavioral traits, using a maze learning challenge, and confirmed the effect of this diet on cortisol release in response to stress. These authors found a lot of behavioral similarities between fish categories, i.e., self-feeding behavior in the first 30 days, swimming activity and learning performances in a maze. The maze test procedure induced the production of high concentrations of cortisol, indicating acute stress in both groups of fish during testing. Plasma cortisol concentration was higher in fish fed a classic marine diet than in fish fed a total plant-based diet, suggesting that the PB diet may affect the short-term release of cortisol. These results are promising regarding the development potential of such sustainable aquaculture strategy.

Among other trails of research to better understand general sea bass behavior and how the rearing process can affect it, numerous studies are currently running on fish coping style and appraisal ability. Fish coping style could be defined as the suite of behavioral and physiological responses of an individual that characterize its reactions to a range of stressful situations (Koolhaas et al. 1999). A distinction is often made between proactive (active coping, or 'fight-flight') and reactive (passive coping, or 'conservation-withdrawal') responses. A proactive stress coping style is behaviorally characterized by a high level of active avoidance, aggression and other behavioral patterns, indicating an active attempt to counteract the stressful stimulus. Reactive coping style, on the other hand, involves immobility and low levels of aggression. Sea bass coping style characterization is being explored in the frame of EU project FP7 COPEWELL. Further, the large individual variation in terms of physiology and behavior under averse conditions suggest also that stress responses do not depend exclusively on the situation to which the individual is exposed, but also on the cognitive evaluation that the individual makes of the situation, i.e., on the way the stressor is appraised (Lazarus 1991). There is growing evidence that the cognitive abilities of fish are more complex than previously assumed and that fish, like higher vertebrates, are capable of a cognitive, memory/experienced based, evaluation of the environment (appraisal) with subsequent expression of affective states or emotions such as pain and fear. This facet of behavior is also under exploration within FP7 COPEWELL.

Conclusion

Foraging behavior is a complex trait having both inherited components and others modified by the environment. Obviously, there are still gaps in present knowledge on sea bass on this topic, especially when compared with salmonids, but knowledge is gained step by step over the last few years. It is, however, necessary to develop further research on this topic in order to develop effective systems for protecting the welfare and optimizing the culture performances of the sea bass that has now become a leading species in Mediterranean aquaculture.

Actually, the need to better understand sea bass behavior started very recently (in the seventies) along with its domestication. This domestication process is, however, so recent that sea bass could be considered, according to some authors, as exploited captives, or being on the threshold of becoming domesticated (Balon 2004). Behavioral traits are good indicators of this process because they are likely the first to be affected, sometimes as soon as the first generation (Vandeputte and Prunet 2002, Bégout Anras and Lagardère 2004, Huntingford 2004). Studies first focused on natural environment to provide the initial necessary knowledge to try keeping sea bass in a captive environment. In their natural environment, juveniles and adult sea bass hunt (usually as a shoal) sea-bead and pelagic preys. In aquacultural environment, their foraging behavior is dependent on their rearing conditions. Indeed, sea bass can be reared under extensive, semi-intensive or intensive systems, held in net cages or in land-based tanks under flow-through or recirculation systems and fed by hand, automatically or by self feeders, showing the high plasticity of this species (involving learning and memory). All these factors will determine the general environmental conditions of the fish (i.e., temperature, oxygen concentration, salinity, density, noise, predation, husbandry stressors, etc.) and consequently alter their feeding behavior. All the studies conducted on the subject concluded that whatever the rearing conditions, it is essential, in order to reach a good specific growth rate, to ensure a good food conversion ratio, and to preserve welfare (limiting, as much as possible, all sources of stress) to feed the fish according to their endogenous rhythm (i.e., nocturnal feeding during the winter) and to their nutritional requirements (i.e., higher lipidic needs in summer).

Recently, a lot a work has been done on this species, e.g., selection programs on growth and other traits of commercial interest, and first attempts to investigate the influence of the domestication and selection level on fish behavior at different life stages. Further, ongoing research aims to link foraging behavior with other traits such as cognition and personality, thereby bringing new insights to research programs targeting fish selection based upon personality traits in order to improve adaptation for aquaculture or restocking purposes.

References

Ajuzie, C.C. 1998. Aspects of behavior in European Sea bass Juveniles. Aqua. Mag. 37–44.

Ajuzie, C.C. 2008. Toxic *Prorocentrum lima* induces abnormal behaviour in juvenile sea bass. J. Appl. Phycol. 20: 19–27.

Andrew, J.E., C. Noble, S. Kadri, H. Jewell and F.A. Huntingford. 2002. The effect of demand feeding on swimming speed and feeding responses in Atlantic salmon *Salmo salar* L., gilthead sea bream *Sparus aurata* L. and European sea bass *Dicentrarchus labrax* L. in sea cages. Aquac. Res. 33: 501–507.

Aprahamian, M.W. and C.D. Barr. 1985. The growth, abundance and diet of 0-group sea bass, *Dicentrarchus labrax*, from the Severn Estuary. J. Mar. Biol. Ass. UK 65: 169–180.

Azzaydi, M., J.A. Madrid, F.J. Sánchez-Vázquez and F.J. Martínez. 1998. Effect of feeding strategies (automatic, ad libitum demand feeding and time-restricted demand-feeding) on feeding rhythms and growth in European sea bass (*Dicentrarchus labrax* L.). Aquaculture 163: 285–296.

Azzaydi, M., F.J. Martínez, S. Zamora, F.J. Sánchez-Vázquez and J.A. Madrid. 1999. Effect of meal size modulation on growth performance and feeding rhythms in European sea bass (*Dicentrarchus labrax* L.). Aquaculture 170: 253–266.

Azzaydi, M., F.J. Martínez, S. Zamora, F.J. Sánchez-Vázquez and J.A. Madrid. 2000. The influence of nocturnal vs. diurnal feeding under winter conditions on growth and feed conversion of European sea bass (*Dicentrarchus labrax* L.). Aquaculture 182: 329–338.

Azzaydi, M., V.C. Rubio, F.J. Martinez, F.J. Sanchez Vazquez, S. Zamora and J.A. Madrid. 2007. Effect of time-restricted feeding on annual rhythms of demand-feeding and food anticipatory activity in european sea bass. Chronobiol. Inter. 24: 859–874.

Balon, E.K. 2004. About the oldest domesticates among fishes. J. Fish Biol. 65: 1–27.

Barnabé, G. 1976a. Contribution à la connaissance de la biologie du loup (*Dicentrarchus labrax* L.) de la région de Sète. Thèse 3ème Cycle, Univ. Sc. Techn. Languedoc, Montpellier. pp. 160.

Barnabé, G. 1976b. Elevage larvaire du loup (*Dicentrarchus labrax* (L.); Pisces, Serranidae) à l'aide d'aliment sec composé. Aquaculture 9: 237–252.

Barnabé, G. 1978. Etude dans le milieu naturel et en captivité de l'écoéthologie du Loup *Dicentrarchus labrax* (L.) (Poisson Serranidae) à l'aide de nouvelles techniques. Annales Sciences Naturelles, Zoologie, Paris, 12e série, 20: 423–502.

Barnabé, G. 1980. Exposé synoptique des données biologiques sur le loup, *Dicentrarchus labrax* (Linné, 1758). Synopsis FAO Pêches 126: 70.

Barnabé, G. 1989. L'élevage du loup et de la daurade. In Aquaculture. Lavoisier—Tec & Doc, Paris.

Barnabé, G. 1990. Rearing bass and gilthead bream. pp. 647–686. *In*: G. Barnabé (ed.). Aquaculture, Vol. 2. Ellis Horwood, Sussex, England.

Barnabé, G. 1994. Biological basis of fish culture. pp. 314–315. *In*: G. Barnabé (ed.). Aquaculture, Biology and ecology of cultured species. New York: Ellis Horwood, Sussex, England.

Barnabé, G. 1994. Biological basis of fish culture, pp. 214–315. *In*: G. Barnabé (ed.). Aquaculture: Biology and Ecology of Cultured Species. Ellis Horwood, New York, USA.

Basurco, B. 2000. Offshore mariculture in Mediterranean countries. Options Méditerranéennes, Series B, 30: 9–18.

Basurco, B. 2004. Introduction to Mediterranean aquaculture. Options Méditerranéennes, Series B, 49: 9–13.

Bégout Anras, M.L. 1995. Demand-feeding behaviour of sea bass kept in ponds: diel and seasonal patterns, and influences of environmental factors. Aquac. Int. 3: 186–195.

Bégout Anras, M.-L. and J.P. Lagardère. 2004. Measuring cultured fish swimming behaviour: first results on rainbow trout using acoustic telemetry in tanks. Aquaculture 240: 175–186.

Bégout Anras, M.-L., M. Beauchaud, J.-E. Juell, D. Covès and J.-P. Lagardère. 2001. Environmental factors and feed intake: Rearing systems. pp. 157–188. *In*: D. Houlihan, T. Boujard and M. Jobling (eds.). Food Intake in Fish. Blackwell Science, Oxford.

Benhaïm, D. 2011. Caractérisation de l'adaptation comportementale des téléostéens en élevage: plasticité et effets de la domestication. PhD Thesis, La Rochelle.

Benhaïm, D., S. Péan, M.-L. Bégout, B. Brisset, D. Leguay and B. Chatain. 2011. Effect of size grading on sea bass (*Dicentrarchus labrax*) juvenile self-feeding behaviour, social structure and culture performance. Aquat. Living Resour. 24: 391–402.

Benhaïm, D., M.-L. Bégout, S. Péan, B. Brisset, D. Leguay and B. Chatain. 2012a. Effect of fasting on self-feeding activity in juvenile sea bass (*Dicentrarchus labrax*). Appl. Anim. Behav. 136: 63–73.

Benhaïm, D., S. Péan, G. Lucas, N. Blanc, B. Chatain and M.-L. Bégout. 2012b. Early life behavioural differences in wild caught and domesticated sea bass (*Dicentrarchus labrax*). Appl. Anim. Behav. 141: 79–90.

Benhaïm, D., M.-L. Bégout, S. Péan, M. Manca, P. Prunet and B. Chatain. 2013a. Impact of a plant-based diet on behavioural and physiological traits in sea bass (*Dicentrarchus labrax*). Aquat. Living Resour. 26: 121–131.

Benhaïm, D., M.-L. Bégout, G. Lucas and B. Chatain. 2013b. First Insight into Exploration and Cognition in Wild Caught and Domesticated Sea Bass (*Dicentrarchus labrax*) in a Maze. PLoS ONE 8(6): e65872.

Berejikian, B.A. 1995. The effects of hatchery and wild ancestry and experience on the relative ability of steelhead trout fry (*Oncorhynchus mykiss*) to avoid a benthic predator. Can. J. Fish. Aquat. Sci. 52: 2476–2482.

Boujard, T., X. Dugy, D. Genner, C. Gosset and G. Grig. 1992. Description of a modular, low cost, eater meter for the study of feeding behavior and food preferences in fish. Physiol. Behav. 52: 1101–1106.

Boujard, T., A. Gelineau and G. Corraze. 1995. Time of single daily meal influences growth performance in rainbow trout, *Oncorhynchus mykiss* (Walbaum). Aquac. Res. 26: 341–349.

Boujard, T., M. Jourdan, M. Kentouri and P. Divanach. 1996. Diel feeding activity and the effect of time-restricted self-feeding on growth and feed conversion in European sea bass. Aquaculture 139: 117–127.

Boujard, T., A. Gélineau, G. Corraze, S. Kaushik, E. Gasset, D. Coves and G. Dutto. 2000. Effect of dietary lipid content on circadian rhythm of feeding activity in European sea bass. Physiol. Behav. 68: 683–689.

Boujard, T., A. Gélineau, D. Covès, G. Corraze, G. Dutto, E. Gasset and S. Kaushik. 2004. Regulation of feed intake, growth, nutrient and energy utilisation in European sea bass (*Dicentrarchus labrax*) fed high fat diets. Aquaculture 231: 529–545.

Brown, C., K. Laland and J. Krause. 2007. Fish Cognition and Behaviour. Blackwell Publishing Ltd.

Carliste, D.B. 1961. Intertidal territory in fish. Journal of Animal Behaviour 9: 106–107.

Chatain, B. 1986. La vessie natatoire chez *Dicentrarchus labrax* et *Sparus auratus*. Aquaculture 53: 303–311.

Claridge, P.N. and I.C. Potter. 1984. Abundance, movements and size of gadoids (Teleostei) in the Severn Estuary. J. Mar. Biol. Ass. UK 64: 771–790.

Conides, A.J. and B. Glamuzina. 2006. Laboratory simulation of the effects of environmental salinity on acclimation, feeding and growth of wild-caught juveniles of European seabass *Dicentrarchus labrax* and gilthead seabream, *Sparus aurata*. Aquaculture 256: 235–245.

Covès, D., E. Gasset, G. Lemarié and G. Dutto. 1998. A simple way of avoiding feed wastage in European seabass, *Dicentrarchus labrax*, under self-feeding conditions. Aquat. Living Resour. 11: 395–401.

Covès, D., M. Beauchaud, J. Attia, G. Dutto, C. Bouchut and M.L. Bégout. 2006. Long-term monitoring of individual fish triggering activity on a self-feeding system: An example using European sea bass (*Dicentrarchus labrax*). Aquaculture 253: 385–392.

Cutts, C.J., N.B. Metcalfe and A.C. Taylor. 1998. Aggression and growth depression in juvenile Atlantic salmon: the consequences of individual variation in standard metabolic rate. J. Fish Biol. 52: 1026–1037.

Dando, P.R. and N. Demir. 1985. On the spawning and nursery grounds of bass, *Dicentrarchus labrax*, in the Plymouth area. J. Mar. Biol. Ass. UK 65: 159–168.

Dellefors, C. and J.I. Johnsson. 1995. Foraging under risk of predation in wild and hatchery-reared juvenile sea trout (*Salmo trutta* L.). Nord. J. Freshw. Res. 70: 31–37.

Di-Poï, C., J. Attia, C. Bouchut, G. Dutto, D. Covès and M. Beauchaud. 2007. Behavioral and neurophysiological responses of European sea bass groups reared under food constraint. Physiol. Behav. 90: 559–566.

Di-Poï, C., M. Beauchaud, C. Bouchut, G. Dutto, D. Covès and J. Attia. 2008. Effects of high food demand fish removal in groups of juvenile sea bass (*Dicentrarchus labrax*). Can. J. Zool. 86: 1015–1023.

Divanach, P. and M. Kentouri. 2000. Hatchery techniques for specific diversification in Mediterranean finfish larviculture. Recent advances in Mediterranean Aquaculture finfish diversification 75–87.

Divanach, P., M. Kentouri, G. Charalambakis, F. Pouget and A. Sterioti. 1993. Comparison of growth performance of six Mediterranean fish species reared under intensive farming conditions in Crete (Greece), in raceways with the use of self feeders. pp. 285–297. *In*: G. Barnabé and P. Kestemont (eds.). Production, Environment and Quality, Bordeaux Aquaculture '92. EAS Spec. Publ., Vol. 18. Ghent, Belgium.

Dupont-Nivet, M., M. Vandeputte, A. Vergnet, O. Merdy, P. Haffray, H. Chavanne and B. Chatain. 2008. Heritabilities and GxE interactions for growth in the European sea bass (*Dicentrarchus labrax* L.) using a marker-based pedigree. Aquaculture 275: 81–87.

EFSA. 2008. Scientific report of EFSA prepared by Working Group on seabass/seabream welfare on Animal Welfare Aspects of Husbandry Systems for Farmed European seabass and gilthead seabream. Annex I to The EFSA Journal 844: 1–89.

Eroldogan, O.T., M. Kumlu and M. Aktas. 2004. Optimum feeding rates for European sea bass *Dicentrarchus labrax* L. reared in seawater and freshwater. Aquaculture 231: 501–515.

Fernö, A. and T. Järvi. 1998. Domestication genetically alters the antipredator behaviour of anadromous brown trout (*Salmo trutta*)—a dummy predator experiment. Nord. J. Freshw. Res. 74: 95–100.

Fortes-Silva, R., F.J. Sanchez Vazquez and F.J. Martinez. 2011. Effects of pretreating a plant-based diet with phytase on diet selection and nutrient utilization in European sea bass. Aquaculture 319: 417–422.

Georgalas, V., S. Malavasi, P. Franzoi and P. Torricelli. 2007. Swimming activity and feeding behaviour of larval European sea bass (*Dicentrarchus labrax* L.): Effects of ontogeny and increasing food density. Aquaculture 264: 418–427.

Heilman, M.J. and R.E. Spieler. 1999. The daily feeding rhythm to demand feeders and the effects of timed meal-feeding on the growth of juvenile Florida pompano, *Trachinotus carolinus*. Aquaculture 180: 53–64.

Hossain, M.A.R., G.S. Haylor and C.M. Beveridge. 2001. Effect of feeding time and frequency on the growth and feed utilization of African catfish *Clarias gariepinus* (Burchell 1822) fingerlings. Aquac. Res. 32: 99–104.

Huntingford, F.A. 2004. Implications of domestication and rearing conditions for the behaviour of cultivated fish. J. Fish Biol. 65: 122–142.

Jensen, M.K., S.S. Madsen and K. Kritiansen. 1998. Osmoregulation and salinity effects on the expression and activity of Na.K-ATPase in the gills of European sea bass, *Dicentrarchus labrax* (L.). J. Exp. Zool. 282: 290–300.

Jobling, M., E.H. Jorgensen, A.M. Arnesen and E. Ringo. 1993. Feeding, growth and environmental requirements of arctic charr : a review of aquaculture potential. Aquac. Int. 1: 20–46.

Johnsson, J.I. and M.V. Abrahams. 1991. Interbreeding with domestic strain increases foraging under threat of predation in juvenile steelhead trout (*Oncorhynchus mykiss*): an experimental study. Can. J. Fish. Aquat. Sci. 48: 243–247.

Johnsson, J.I., E. Petersson, E. Jönsson, B. Björnsson and T. Järvi. 1996. Domestication and growth hormone alter antipredator behaviour and growth patterns in juvenile brown trout, *Salmo trutta* Can. J. Fish. Aquat. Sci. 53: 1546–1554.

Kaushik, S. and F. Médale. 1994. Energy requirements, utilization and supply to salmonids. Aquaculture 124: 81–97.

Kelley, D.F. 1988. The importance of estuaries for sea-bass *Dicentrarchus labrax* (L.). J. Fish Biol. 33(Suppl A): 25–33.

Kentouri, M. 1985. Comportement larvaire de 4 Sparidés méditerranéens en élevage: *Sparus aurata, Diplodus sargus, Lithognathus mormyrus, Puntazzo puntazzo* (Poissons téléostéens). Thèse de Doctorat ès Sciences, Université de Sciences et Techniques du Languedoc, Montpellier.

Kestemont, P. and E. Baras. 2001. Environmental factors and feed intake: mechanisms and interactions. pp. 131–156. *In*: D. Houlihan, T. Boujard and M. Jobling (eds.). Food Intake in Fish. Blackwell Science—COST Action 827, Oxford.

Kestemont, P., S. Jourdan, M. Houbart, C. Mélard, M. Paspatis, P. Fontaine, A. Cuvier, M. Kentouri and E. Baras. 2003. Size heterogeneity, cannibalism and competition in cultured predatory fish larvae: biotic and abiotic influences. Aquaculture 227: 333–356.

Koebele, B.P. 1985. Growth and the size hierarchy effect: an experimental assessment of three proposed mechanisms; activity differences, disproportional food acquisition, physiological stress. Environ. Biol. Fishes 12: 181–188.

Koolhaas, J.M., S.M. Korte, S.F. De Boer, B.J. Van Der Vegt, C.G. Van Reenen, H. Hopster, I.C. De Jong, M.A.W. Ruis and H.J. Blokhuis. 1999. Coping styles in animals: current status in behavior and stress-physiology. Neurosci. Biobehav. Rev. 23: 925–935.

Langridge, K.V. 2009. Cuttlefish use startle displays, but not against large predators. Anim. Behav. 77: 847–856.

Lazarus, R.S. 1991. Progress on a cognitive-motivational-relational theory of emotion. Am. Psychol. 46: 819–834.

Luna-Acosta, A., C. Lefrançois, S. Millot, B. Chatain and M.L. Bégout. 2011. Physiological response in different strains of sea bass (*Dicentrarchus labrax*): Swimming and aerobic metabolic capacities. Aquaculture 317: 162–167.

Malavasi, S., V. Georgalas, M. Lugli, P. Torricelli and D. Mainardi. 2004. Differences in the pattern of antipredator behaviour between hatchery-reared and wild European sea bass juveniles. J. Fish Biol. 65: 143–155.

Malavasi, S., V. Georgalas, D. Mainardi and P. Torricelli. 2008. Antipredator responses to overhead fright stimuli in hatchery-reared and wild European sea bass (*Dicentrarchus labrax* L.) juveniles. Aquac. Res. 39: 276–282.

Millot, S. and M.-L. Bégout. 2009. Individual fish rhythm directs group feeding: a case study with sea bass juveniles (*Dicentrarchus labrax*) under self-demand feeding conditions. Aquat. Living Resour. 22: 363–370.

Millot, S., M.-L. Bégout, J. Person-Le Ruyet, G. Breuil, C. Di-Poï, J. Fievet, P. Pineau, M. Roué and A. Sévère. 2008. Feed demand behavior in sea bass juveniles: effects on individual specific growth rate variation and health (inter-individual and intergroup variation). Aquaculture 274: 87–95.

Millot, S., M.-L. Bégout and B. Chatain. 2009a. Risk-taking behaviour variation over time in sea bass *Dicentrarchus labrax*: effects of day–night alternation, fish phenotypic characteristics and selection for growth. J. Fish Biol. 75: 1733–1749.

Millot, S., M.-L. Bégout and B. Chatain. 2009b. Exploration behaviour and flight response toward a stimulus in three sea bass strains (*Dicentrarchus labrax* L.). Appl. Anim. Behav. 119: 108–114.

Millot, S., S. Péan, D. Leguay, A. Vergnet, B. Chatain and M.-L. Bégout. 2010. Evaluation of behavioral changes induced by a first step of domestication or selection for growth in the European sea bass (*Dicentrarchus labrax*): A self-feeding approach under repeated acute stress. Aquaculture 306: 211–217.

Millot, S., S. Péan, B. Chatain and M.-L. Bégout. 2011. Self-feeding behavior changes induced by a first and a second generation of domestication or selection for growth in the European sea bass, *Dicentrarchus labrax*. Aquat. Living Resour. 24: 53–61.

Neophytou, M. 2013. Study of the behavioral ontogeny of larvae and early juveniles of the European sea bass (*Dicentrarchus labrax*) (Pisces: Moronidae) in two different rearing systems (intensive and mesocosm techniques). PhD Thesis, University of Crete.

Papandroulakis, N., D. Papaioannou and P. Divanach. 2002. An automated feeding system for intensive hatcheries. Aquacult. Eng. 26: 13–26.

Papoutsoglou, S.E., G. Tziha and A. Athanasiou. 1998. Effects of stocking density on behavior and growth rate of European sea bass (*Dicentrarchus labrax*) juveniles reared in a closed circulated system. Aquacult. Eng. 18: 135–144.

Paspatis, M., C. Batarias, P. Tiangos and M. Kentouri. 1999. Feeding and growth responses of sea bass (*Dicentrarchus labrax*) reared by four feeding methods. Aquaculture 175: 293–305.

Paspatis, M., D. Maragoudaki and M. Kentouri. 2002. Feed discrimination and selection in self-fed European sea bass *Dicentrarchus labrax*. Aquac. Res. 33: 509–514.

Paspatis, M., T. Boujard, D. Maragoudaki, G. Blanchard and M. Kentouri. 2003. Do stocking density and feed reward level affect growth and feeding of self-fed juvenile European sea bass? Aquaculture 216: 103–113.

Person-Le Ruyet, J., K. Mahé, N. Le Bayon and H. Le Delliou. 2004. Effects of temperature on growth and metabolism in a Mediterranean population of European sea bass, *Dicentrarchus labrax*. Aquaculture 237: 269–280.

Pichavant, K., J. Person-Le-Ruyet, N. Le Bayon, A. Sévère, A. Le Roux, L. Quéméner, V. Maxime, G. Nonnott and G. Boeuf. 2000. Effects of hypoxia on growth and metabolism of juvenile turbot. Aquaculture 188: 103–114.

Pickett, G.D. and M.G. Pawson. 1994. The sea bass—Biology, Exploitation and conservation. Fish and Fisheries Series. Chapman & Hall, London.

Pitcher, T.J. and P.J.B. Hart. 1982. Fisheries Ecology. Chapman and Hall, London.

Reddy, P.K., M.N. Khan and T. Boujard. 1994. Effect of the daily meal time on the growth of rainbow trout fed different ration levels. Aquac. Int. 2: 165–179.

Rosenthal, H. and H. Hempel. 1970. Experimental studies in feeding and food requirements of herring larvae (*Clupea harengus* L.). pp. 344–364. *In*: J.H. Steel (ed.). Marine Food Chains. University of California Press, Berkeley.

Rubio, V.C., F.J. Sánchez-Vázquez and J.A. Madrid. 2005. Effects of salinity on food intake and macronutrients selection in European sea bass (*Dicentrarchus labrax* L.). Physiol. Behav. 85: 333–339.

Rubio, V.C., F.J. Sánchez-Vázquez and J.A. Madrid. 2006. Influence of nutrient preload on encapsulated macronutrient selection in European sea bass. Physiol. Behav. 89: 662–669.

Rubio, V.C., F.J. Sánchez-Vázquez, S. Zamora and J.A. Madrid. 2008. Endogenous modification of macronutrient selection pattern in sea bass (*Dicentrarchus labrax* L.). Physiol. Behav. 95: 32–35.

Salvanes, A.G.V. and V. Braithwaite. 2006. The need to understand the behaviour of fish reared for mariculture or restocking. ICES J. Mar. Sci. 63: 346–354.

Sánchez-Vázquez, F.J. and M. Tabata. 1998. Circadian rhythms of demand-feeding and locomotor activity in rainbow trout. J. Fish Biol. 52: 255–267.

Sánchez-Vázquez, F.J., J.A. Madrid and S. Zamora. 1995a. Circadian rhythms of feeding activity in sea bass, *Dicentrarchus labrax* L.: dual phasing capacity of diel demand-feeding pattern. J. Biol. Rhythms 10: 256–266.

Sánchez-Vázquez, F.J., S. Zamora and J.A. Madrid. 1995b. Light-Dark and food restriction cycles in sea bass; Effect of conflicting zeitgebers on demand-feeding rhythms. Physiol. Behav. 58: 705–714.

Sánchez-Vázquez, F.J., T. Yamamoto, T. Akiyama, J.A. Madrid and M. Tabata. 1998. Selection of macronutrients by goldfish operating self-feeders. Physiol. Behav. 65: 211–218.

Stickney, R.R. 1994. Principles of Aquaculture. John Wiley and Sons, New York.

Stirling, H.P. 1977. Growth, food utilization and effect of social interaction in the European bass *Dicentrarchus labrax*. Mar. Biol. 40: 173–184.

Talbot, C. 1993. Some aspects of the biology of feeding and growth in fish. Proc. Nut. Soc. 52: 403–416.

Vandeputte, M. and P. Prunet. 2002. Génétique et adaptation chez les poissons : domestication, résistance au stress et adaptation aux conditions de milieu. INRA Prod. Anim. 15: 365–371.

Winberg, S. and G.r.E. Nilsson. 1993. Roles of brain monoamine neurotransmitters in agonistic behaviour and stress reactions, with particular reference to fish. Comparative Biochemistry and Physiology Part C: Pharmacology, Toxicology and Endocrinology 106: 597–614.

Yamamoto, T., T. Shima, H. Furuita, M. Shiraishi, F.J. Sánchez-Vázquez and M. Tabata. 2000a. Self-selection of diets with different amino acid profiles by rainbow trout (*Oncorhynchus mykiss*). Aquaculture 187: 375–386.

Yamamoto, T., T. Shima, T. Unuma, M. Shiraishi, T. Akiyama and M. Tabata. 2000b. Voluntary intake of diets with varying digestible energy contents and energy sources, by juvenile rainbow trout *Oncorhynchus mykiss*, using self-feeders. Fish. Sci. (Tokyo) 66: 528–534.

Food Intake Regulation

José Miguel Cerdá-Reverter

Introduction

Feeding is the way in which heterotroph animals incorporate organic nitrogen and carbon for growth. Most heterotroph organisms also obtain energy in this way. Together with reproduction, feeding is a key function in evolution since metabolic energy obtainment and storage make an important contribution to survival rates. This evolutionary relevance suggests that feeding regulation should be highly redundant and conservative. Redundancy confers backup systems on functional regulation, which, while preserving the function, leads to greater complexity of the same, increasing the difficulty of its study. On the other hand, evolutionary conservation of regulatory mechanisms makes fish an excellent model for studying mammalian species. However, the opposite is also true, that mammalian species are an excellent model for fish and fish research can take full advantage of previous experimental effort focusing on mammalian obesity.

Sea bass is an opportunistic hunter that takes advantage of almost any available food item in the column water. It is not a specialized feeder but mostly feeds on cephalopods, crustaceans and fish in the wild. Studies on feeding behavior in the wild are scarce and usually provide biased information. Such information on food intake quantity, frequency and efficiency in the wild is mostly inferred from analyses of vacuity, which are severely conditioned by the sampling method used (Pickett and Pawson

Instituto de Acuicultura de Torre de la Sal, Consejo Superior de Investigaciones Científicas (IATS-CSIC), Torre de la Sal s/n, Ribera de Cabanes, 12595, Castellon Spain.
Email: cerdarev@iats.csic.es; jm.cerda.reverter@csic.es

1994). The main bulk of our knowledge of feeding behavior and food intake regulation in sea bass comes from captive animals but this information may not necessarily reflect the situation in the wild. For example, in the wild, sea bass experience long feeding restriction periods since food availability fluctuates and they are accustomed to long inter-meal periods, which can be measured in days or weeks (Attia et al. 2012). Subsequent voracity could suggest that fish, including sea bass have no systems to control food intake and that they eat when food it is available, but nothing could be further from the truth, as any angler will tell you (Pickett and Pawson 1994).

In culture conditions, where food may be permanently available, sea bass finely regulate food intake and keep their body weight within a narrow interval. An easy experiment using demand feeders demonstrated that when the recompense level (quantity of delivered food) of the demand-feeding system changes, the sea bass will vary the number of feeding system activations accordingly to meet previous feeding levels (Cerdá-Reverter, unpublished data). Compensatory feeding is another sound demonstration on the regulation of food intake in sea bass (Rubio et al. 2010). Most fish exhibit compensatory growth episodes after long fasting periods. This catch-up growth is attained by increasing daily food intake levels after fasting periods, by increasing food conversion efficiency, or by both, until fish have recovered the lost growth (Ali et al. 2003).

This chapter does not aim to review all the systems involved in the control of food intake, but only those which have been studied in sea bass. In addition, we will also review the relationship of food intake with some physiological processes and environmental factors which are known to modulate food intake in sea bass. Excellent reviews on the central and peripheral systems controlling food intake in fish have been already published (Lin et al. 2000, Volkoff et al. 2005, Gorissen et al. 2006, Volkoff et al. 2009, Canosa et al. 2012).

Central and Peripheral Regulators of Food Intake

Food intake is one of the end points of feeding behavior and, like every behavioral regulation, takes place at central level. The central nervous system (CNS) retrieves and integrates information from environment and peripheral systems (body) to elaborate a coordinated response that affects feeding behavior and, in many circumstances, food intake levels. After food intake, the gastrointestinal tract releases satiety signals that travel via nervous (*vagus* nerve) or endocrine loops to inform the CNS about the arrival and assimilation of organic material and energy. This short-term regulation is highly variable as it depends on many environmental, physiological and behavioral factors and, in many cases, is not well-correlated with the energy needs of the animal. However, the energy balance is regulated on a long-term basis and in a hierarchical way. This long-term regulation depends on energy status/storage in the animal and is superimposed on short-term regulation. In physiological terms, long-term regulatory systems are able to modify the sensitivity of the mechanisms regulating energy intake on a short-term basis. Therefore, food intake inhibitory systems will be less sensitive to gastrointestinal signals when energy storage is depleted.

The CNS uses both anabolic and catabolic pathways to regulate food intake and to create a peripheral environment in agreement with the energy balance. Therefore, activation of an anabolic pathway stimulates food intake but also yields a peripheral environment that promotes energy storage. The opposite is also true for the activation of catabolic pathways, i.e., the inhibition of food intake but promotion of energy catabolism. These central pathways use different neuropeptides to promote their effects and many times the pathways can be traced based on the localization of neurotransmitters and the neuropeptides involved in the control of food intake (reviewed in Volkoff et al. 2005, 2009). Only some central or peripheral factors have been studied in sea bass, and these will be reviewed in this section.

Melatonin

Melatonin is mainly produced in the pineal organ and retinal photoreceptors. The precursor, serotonin, is acetylated by the aralkylamine N-acetyltransferase (AANAT) and then methylated by the hydroxyindole O-methyltransferase (HIOMT) to yield melatonin in the pineal gland. The synthesis and secretion of melatonin follows a rhythmic pattern, reaching a zenith during the night-time and nadir during the light-time. Therefore, melatonin is a key molecule for the vertebrate circadian system, including fish, and is related to many functions which have a common rhythmic expression (see Chapters 2.1 and 2.2). Therefore, melatonin has been associated with all major cyclic functions in vertebrates, including feeding behavior. Animals do not eat throughout the whole 24-hour period but are usually active either during the day or night. During the active period animals select specific time windows to search for and capture their food, and such behavior has been genetically fixed under the influence of stable causal agents, e.g., predators, feed availability, feeding optimization, etc. In addition, some animals, including the sea bass, have been seen to display a dual pattern of feeding behavior, alternating diurnal and nocturnal feeding activity. The mechanisms involved in control of such feeding phase inversion remain unknown (see Chapter 1.2). However, melatonin has also been detected in the gastrointestinal tract of a number of vertebrates, suggesting that this hormone could play a role in the regulation of food ingestion. In fact, gastrointestinal levels of melatonin increase after feeding and intestinal motility is reduced, probably delaying gastric emptying in some fish species (Vera et al. 2007). Obviously, pineal organ and/or the synthesis/secretion of melatonin have been thought to be responsible for fish feeding rhythms.

Studies in several fish species, including sea bass, have demonstrated that melatonin administration inhibits food intake and body weight gain (Pinillos et al. 2001, Rubio et al. 2004, de Pedro et al. 2008, Piccinetti et al. 2010), but only when it is administrated peripherally, at least in the goldfish (Pinillos et al. 2001). Intraperitoneal administration reduced muscle glycogen stores and increased liver lipid mobilization in goldfish (de Pedro et al. 2008). At the central level, melatonin increased dopamine metabolism without modifying serotoninergic activity, suggesting that melatonin may mediate its effect at the central level by activating the catecholaminergic system. In fact, unpublished data have demonstrated that activation of the dopaminergic system by the chronic administration of L-DOPA severely inhibits food intake in the sea bass (Leal et al. 2013). Circulating leptin and ghrelin, as well as hypothalamic neuropeptide Y (NPY) levels, remain unaltered in the goldfish. However, experiments in zebrafish demonstrated that chronic melatonin treatment stimulated NPY and melanocortin 4 receptor subtype (MC4R) but inhibited ghrelin, NPY and cannabinoid receptor expression (Piccinetti et al. 2010). Only one paper has focused on the effect of melatonin on sea bass food intake and macronutrient selection. Doses of 2.5 mg/g of melatonin inhibited food intake by 34% but also modified the pattern of macronutrient selection by decreasing the carbohydrate intake (see Chapter 1.2). Since the gastrointestinal tract shows rhythmic changes in melatonin levels, it may be thought that food availability could drive circadian synchronization. Therefore, the arrival or processing of food in the gastrointestinal tract could induce melatonin secretion, which would act as an appetite -regulating factor (Rubio et al. 2004).

Serotonin (5-HT)

Serotonin or 5-hydroxytryptamine (5-HT) is synthesized form tryptophan by the action of tryptophan hydroxylase (TPH) and subsequent descarboxylation by the 5-hydroxitryptophan decarboxylase (DDC). In the CNS, after secretion to synapses, the excess of serotonin is taken up again by specific serotonin transporters (SERT) and metabolized into 5-hydroxy-indoleacetic acid (5-HIAA) by monoamine oxidase (MAO) and aldehyde dehydrogenase (ALDH).

The bulk of body serotonin ($\approx$95%) comes from the enterochromaffin tissue of the gastrointestinal tract. When the food reaches the gastrointestinal tract, the enterochromaffin tissue releases serotonin, which initiates the luminal secretion of sodium chloride and fluids through the activation of intrinsic submucosal primary afferent neurons. These neurons also project to the myenteric plexus, activating peristaltic reflexes. Serotonin can also activate extrinsic sensory neurons with cell bodies in the vagus nerve and dorsal root ganglia that convey visceral sensory information to the CNS, e.g., pain, nausea, satiety. In the CNS, the serotoninergic system seems to be remarkably well conserved among vertebrate species. Serotonin neurons are clustered in two main groups—rostral (mesencephalic) and caudal (metencephalic)—which project to the main forebrain structures, including the hypothalamic that control food intake (see Tecott 2007, for a review). In the fish CNS, there are also two main localizations for the 5-HT system in the teleost brain, one anterior, which covers the nuclei associated with the ventricle and its recess in the caudal hypothalamus and one posterior in the brainstem. The forebrain group includes neurons in the pretectal and preoptic area, posterior tuberculum, remarkably in the paraventricular organ (PVO), and caudal hypothalamus. The posterior group integrates neurons of the raphe, medulla oblongata and spinal cord (reviewed by Cerdá-Reverter and Canosa 2009 and Lillesaar 2011). In the periphery, as in mammalian species, 5-HT is localized abundantly in the intestine but, at least in trout, serotonin is mainly found in the intestine wall and hardly at all in the mucosa where the endocrine cells are located. This suggests the presence of a nervous plexus with a very high concentration of 5-HT (Caamaño-Tubío et al. 2007).

Both central and peripheral serotonin are involved in the control of food intake. In vertebrates, the global activation of central serotonin circuits seems to suppress food intake and fish are not an exception (de Pedro et al. 1998, Ruibal et al. 2002). In the goldfish, de Pedro et al. (1998) demonstrated that the intracerebroventricular administration of 5-HT induced a short-term inhibition of food intake, although this effect was absent when the neurotransmitter was administrated intraperitoneally. These results argue against the possible role of 5-HT in the regulation of food intake in goldfish since serotonin does not easily cross the blood-brain barrier. In contrast, the peripheral administration of fenfluramine induced a severe short-term inhibition de que? in rainbow trout (Ruibal et al. 2002). This drug readily crosses the blood-brain barrier to stimulate serotonin synaptic release and block its re-uptake into the presynaptic terminals (Tecott 2007). Studies in the sea bass have demonstrated that orally administrated 5-HT affects food intake levels and macronutrient selection (Rubio et al. 2006). In their experiment, when 5-HT was administrated together with encapsulated diets to avoid the organosensory properties of food, a decrease in the food intake levels was accompanied by a decrease or increase in fat or protein intake, respectively. Orally administrated 5-HT crosses the intestinal wall into the bloodstream since 5-HT plasma levels increased after feeding. However, the authors did not check the potential increase in central serotonin levels since the possibility of 5-HT crossing the blood-brain barrier in fish cannot be fully discarded. Independently, this result suggests that 5-HT could also have a peripherally mediated effect on, and indeed play a central role in, sea bass food intake. This peripheral effect could be explained by a modulating effect on gastrointestinal motility or by the interaction of 5-HT with other regulatory hormones. In fact, 5-HT has been shown to increase gastrointestinal motility in several fish species (Holmgren and Olsson 2009, Velarde et al. 2010). In summary, it seems that the activation of peripheral or central serotonin systems suppresses food intake.

Cholecystokinin (CCK)

CCK is a brain-gut peptide which, together with gastrin, constitutes a family of peptides characterized by the common C-terminus of Trp–Met–Asp–Phe–NH2. The structure of the C-terminus octapeptide of CCK (CCK-8) has been well conserved during evolution and is identical in tetrapods, with only one amino acid substitution in fish. The CCK-8 is the main CCK peptide produced in the CNS, while longer peptides such as CCK-58, CCK-33 and CCK-22 are also found in peripheral tissues and circulation (Cerdá-Reverter and Canosa 2009). Peripherally, CCK is synthesized and released from the enteroendocrine cells in response to the ingestion of food and acts as a postprandial satiety signal or a short-term meal reducing factor. CCK elicits multiple effects on the gastrointestinal system including the regulation of gut motility, gallbladder contraction, pancreatic enzyme, secretion gastric emptying and gastric acid secretion. Most of these actions are thought to be due to endocrine interactions of circulating CCK that has entered into the bloodstream. Peripherally administered CCK reduces meal size, eliciting the same sequence of behavior that animals display when they terminate meals. However, chronic administration of CCK has no effect on body weight since animals compensate by increasing meal frequency. CCK is thought to acts via vagal pathways due to the presence of CCK receptors in the vagal sensory fibers. Consistently, vagotomy or selective damage of vagal afferences eliminates the CCK satiety signal (reviewed by Strader and Woods 2005, Wynne et al. 2005).

In fish, CCK is also involved in gastrointestinal regulation, i.e., gut motility and gastric emptying, gallbladder contraction and gastric acid and pancreatic secretion (reviewed by Holmgrem and Olssen 2009). Both peripheral and central CCK administration inhibit food intake in several fish species (reviewed by Volkoff et al. 2005) but also, peripheral antagonist administration stimulates food intake in trout and sea bass (Gélineau and Boujard 2001, Rubio et al. 2008). Fasting increases central and peripheral expression levels of CCK in several fish species (Volkoff et al. 2005) and the protein:lipid relation is able to modify CCK plasma levels in trout (Jönsson et al. 2006). Several studies have also been suggested that central CCK and CCK binding sites could be involved in the chemosensory detection by the gustatory and olfactory systems (Himick and Peter 1994, 1996). In the sea bass, the inclusion of CCK-8 in gelatin capsules dose-dependently reduced food intake levels and the ingestion of all three macronutrients, i.e., protein, lipid and carbohydrate, without affecting their relative proportions in the diet. In contrast, high doses of proglumide, a non-specific CCK antagonist, stimulated food intake by increasing protein and carbohydrate intake but not fat ingestion levels. Finally, blocking the receptors with proglumide eliminated the effects CCK on sea bass food intake.

Environment and Food Intake

Sea bass feeding is seasonal and closely depends on environmental conditions. As in ectotherm animals, feeding mainly depends on the thermal range in which they develop their life cycle. Some works have focused on seasonal fluctuations in food intake levels in laboratory or intensive farming conditions (see Kavadias et al. 2003), in which the animals were hand-fed to apparent satiation or using self-feeding systems. Invariably, the studies demonstrated that sea bass eat more during the summertime and less during the winter period, following the thermal cycle. Such fluctuation delimits both growing and resting or non-growing periods. Gonadal recrudescence starts during the growing period (September) but maturation and spawning takes place during the resting phase, coinciding with the coldest temperatures (Zanuy and Carrillo 1985).

Frequently, peaks of food intake do not match maximal annual temperatures and there is some controversy about the time of the year when sea bass eat most. Carrillo and Zanuy (1985) reported that maximal food-intake levels occur during August, coinciding with the highest temperatures of the year. Their experiments were run indoor ($40°7'$ N, $0°9'$E) and animals were hand-fed to apparent satiation once a day between 17.00 h and 19.00 h. Using self-feeding systems and rearing fish in earth ponds ($46°9'$ N, $1°9'$ W), Begout Anras (1995) reported maximal feeding demands from June to July. These then gradually declined from August to October with particularly low feeding activity during this last month. During the whole experimental period animals basically self-fed during the light time and not during the night period. Therefore, feeding activity is positively correlated with temperature, day length, atmospheric pressure, and negatively correlated with dissolved oxygen, east-west wind direction and rain. In a similar experiment using self-feeding systems and outdoor tanks ($37°50'$ N, $0°46'$ W), Sánchez-Váquez et al. 1998 reported maximal self-demands during October and November (water temperature of

19–22°C), during the vitellogenic period, far from the acrophase found in the photoperiod (June) and warm water temperature (27°C in August). An abrupt change in the daily feeding rhythms was also reported, sea bass self-demanding predominantly during the day light in summer and autumn but during the night time in winter, and returning to diurnalism in spring (Sanchez-Vazquez et al. 1998).

Temperature

Fish are ectothermic animals that do not regulate body temperature, while internal temperature fluctuates with that of the water. Fish exhibit limits in the tolerance of temperature and can be classified into stenotherms or eurytherms. Stenotherms only tolerate a narrow range of temperature but eurytherms have a wider range. The optimal temperature range can be defined as the range over which feeding occurs and where there are no signs of abnormal behavior linked to thermal stress. Temperature governs metabolic rate by influencing molecular activation of the metabolic chain. Therefore, it controls most physiological processes that directly or indirectly module feeding behavior and food intake (see Kestemont and Baras 2001).

Temperature is the major factor controlling food intake level in ectotherms but thermal effects interact with both external (oxygen, salinity, food availability) and internal biological factors (age, genetic). For example, when feeding or oxygen availability is restricted, the optimal temperature at which fish grow best is commonly lower than when no restriction in the above parameters exist (Jobling 1996). In addition, there are intraspecific differences for feeding and the growth/temperature relationship according to the geographic region (Imsland and Jonassen 2001).

Sea bass is a eurytherm species. Studies in sea bass using three different acclimatization temperatures in an eastern Mediterranean strain have demonstrated that the thermal tolerance ranged from 4.10°C–6.77°C to 33.23°C–35.95°C (Dülger et al. 2012). The threshold below which sea bass feeding ceases seems to be close to 7°C (Pickett and Pawson 1994). Because sea bass is of great importance for Mediterranean aquaculture, several works have studied the optimal temperature for feeding and food efficiency in juvenile sea bass (Hidalgo et al. 1987, Russell et al. 1996, Person-Le Ruyet et al. 2004).

Person-Le Ruyet and collaborators (2004) used a western Mediterranean strain and six different experimental temperatures covering the range typical of Mediterranean coastal waters throughput the year (13°C–29°C). Maximal food intake levels were recorded at 26°C–29°C (1.4%–1.5% BW x day^{-1}) as estimated by feeding fish twice a day to apparent satiation. In cooler water (13°C), food intake was markedly depressed by about 50%, inducing a drop in the specific growth rate of 65% compared with the maximal growth temperature (26°C). Therefore, optimal temperature for maximal food intake and growth rate were similar and strictly dependent on temperature. However, the thermal dependency of food efficiency was not so rigorous and maximal levels were reached between 19°C and 29°C. Experiments in our laboratory focusing on seasonal effects (winter *vs.* late spring) on the feeding response to sampling stress provided similar results using self-feeding systems. During the winter (11°C–12°C) sea bass self-demanded about 0.8% BW x day^{-1} but during late spring (18°C–23°C) fish demanded about 1.2% BW x day^{-1} (Rubio et al. 2010).

Salinity

Sea bass are euryhaline, meaning that animals of all ages tolerate a wide range of salinity. The direct transfer from sea water to fresh water will kill 100% of the animals but all of them will survive when transferred directly to dilute sea water with a salinity of 4 ‰. Similarly, they will survive when transferred gradually from sea water to fresh water (Pickett and Pawson 1994). The high tolerance of the species to low salinity levels suggests the sea bass could be cultured in freshwater or diluted sea water and some studies have focused on the effects of salinity on food intake and growth of this valuable species. Dendrinos and Thorpe (1985) studied the effects of a range of salinity values (0.5, 5, 10, 20, 25, 30, 33 ‰) on food intake, growth and food conversion efficiency at 19°C in juvenile sea bass of an Atlantic population. Fish at lowest salinity (0.5 ‰) lost the appetite and died in the first 20 days. Average growth increased with increasing salinity form 5 ‰ to 30 ‰, but only a slight increase in salinity of 3 ‰ (33 ‰) resulted in fish which are on average 21% thinner after one year. Fitted curves concluded that salinity causes a proportional increase of the food intake equal to 0.0206 x salinity (‰) but food conversion efficiency and protein conversion efficiency reach maximum values at 25 ‰ and 30 ‰, respectively. The authors concluded that optimum salinity for the growth of juvenile sea bass reared at 19°C was 30 ‰. Fish reared at 25 ‰ still showed better performance than those cultured at 33 ‰.

Similar results were found when studying the effects of a salinity range (8 ‰, 18 ‰, 28 ‰, 32.87 ‰) in juvenile sea bass from a Mediterranean population (Conides and Glamuzina 2006). Daily growth rates were highest at 28 ‰, while maximal food intake was recorded at 18 ‰. In all the studies, maintenance diets are higher at lower salinity levels since fish need to invest energy in osmoregulatory processes. In addition, Rubio and co-workers reared Mediterranean juvenile sea bass in three different salinities (0 ‰, 7 ‰ and 25 ‰). By using encapsulated food, the authors demonstrated that total food intake gradually decreased as salinity was lowered from 25 ‰ to

7‰ and then to 0‰. Macronutrient intake also was affected by the salinity level since carbohydrate intake decreased and protein intake increased in response to falling salinity. Fat intake remained unaltered. Similarly, growth rate and food conversion efficiency tended to decrease but the authors found no significant differences (Rubio et al. 2005).

Oxygen

There is no doubt that severe hypoxia negatively affects fish growth, although a constant level below a critical range is also considered to reduce food intake, food conversion efficiency and growth. Limited food intake and growth seem to be an adaptation to reduced oxygen capture to constrain the metabolic rate, i.e., a decreased food intake reduces the energy demanded for metabolism and therefore oxygen requirements (Jobling 1994). Basically, reduced food intake during hypoxia is responsible for growth retardation, but increased locomotor activity or cardio-respiratory adaptations may increase energy expenditure (see Thetmeyer et al. 1999). Oxygen tension is the driving force governing oxygen uptake, so that oxygen saturation (roughly proportional to tension) may be more critical than concentration for limiting fish feeding and growth. In some species, critical oxygen levels for growth (levels below which growth ceases) have been determined and a range of 50%–70% air saturation seems to be typical (Jobling 1994).

An acute decrease in water oxygen levels may occur in intensive fish farming when high densities are used. Also, coastal waters are exposed to daily fluctuations of oxygen saturation resulting from biotic and abiotic changes, e.g., respiration and photosynthesis, temperature, ocean currents, etc. Fish kept in net cages or tanks are not able to escape from these adverse conditions and the availability of dissolved oxygen may be a limiting factor for fish farming. Experiments in sea bass have demonstrated that this species tolerates both hypoxia (40% saturation) and fluctuating oxygen conditions (40%–86% saturation). Hypoxia (40% saturation) significantly decreased food intake level to 76% of the daily ration. Similarly, fish exposed to oscillating oxygen conditions (40%–86% saturation) ate 85% of the daily ration. This reduction limited sea bass growth but no differences were observed in the food conversion efficiency (Thetmeyer et al. 1999). Similar results were found by Pichavant and collaborators (2001) working with three oxygen levels, i.e., 3.2 mg/l, 4.5 mg/l and 7.4 mg/l. The animals reared at the highest oxygen level (considered as normoxic conditions) were fed to satiation but also with the same ration as fish kept at 3.2 mg/l. Mass gain under hypoxia was reduced but growth of fish kept in normoxia and fed with the same ration as fish reared at 3.2 mg/l was similar. Once again, feed efficiency was unaffected by oxygen levels.

Since the levels of dissolved oxygen are one of the major limiting factors in land-based fish farming, especially in warm locations, a condition of mild hyperoxia is often maintained through the addition of oxygen. Depending on the method used to dissolve the oxygen in the water, the gas pressure may be severely affected. Therefore, it is important to distinguish between normbaric and hyperbaric hyperoxia. Normbaric hyperoxia stimulates food intake and growth without any effect on food conversion efficiency in the sea bass. In contrast, hyperbaric hyperoxia stimulates food intake but decreases food conversion efficiency, resulting in a decreased specific growth rate (Lemarié et al. 2011).

Ammonia

Water quality is another key limiting factor in land-based farming, especially when using recirculation systems. Ammonia and urea are the two main nitrogenous products excreted by teleost fish, with ammonia representing around 75%–90% of nitrogen excretion. Ammonia is mainly excreted as un-ionized from NH_3 that ionizes to NH_4^+ in sea water depending on pH, temperature and salinity. NH_3 is soluble in lipids and 300–400 times more toxic than NH_4^+. In seawater, ammonia is measured as total ammonia nitrogen, which integrates un-ionized ammonia nitrogen and NH_4^+ nitrogen. There are few studies on the chronic effects of ammonia in seawater fish but safety levels, normally extrapolated from LC50 (lethal concentration for 50% of the population), range between 0.05 mg/l–0.2 mg/l un-ionized ammonia nitrogen. In the sea bass, the LC50–96 hours has been reported to be 1.7 mg/l un-ionized ammonia nitrogen (40 mg/l total ammonia nitrogen). Levels for seabream and turbot are around 2.5 mg/l–2.6 mg/l un-ionized ammonia nitrogen (57 mg/l–59 mg/l total ammonia nitrogen) showing that the sea bass is more sensitive to environmental ammonia levels (Person-Le Ruyet et al. 1995).

In the sea bass, weight gains are negatively correlated with ammonia levels. Lemarié et al. (2004) demonstrated that mean weights in control fish increased 3.4-fold over a 55-day feeding period, while fish exposed to higher ammonia levels only increased their mean weights 1.8-fold. In this experiment, food was delivered by contact self-feeders, in which fish had to push the triggering system or rod to obtain food reward. However, random activations increase with the locomotor activity of the fish. Paradoxically, authors reported a sharp increase in triggering activity with increasing ammonia levels. The authors argued that this increase correlate with changes in the locomotion behavior (disorientation and erratic swimming), and was related with erratic or random activations of the self-feeder.

Sex Steroids and Food Intake

Several studies have demonstrated that peripheral endocrine information from the reproductive tissue is involved in the physiological control of appetite in mammals (reviewed by Eckel 2004). In fish, data regarding the involvement of the sex steroids in the control of food intake and growth are controversial. Both androgen and estrogen have been shown to promote feeding and growth in several species but they seem to act as feeding/growth deterrent agents in other species (see Leal et al. 2009, for references).

Our preliminary experiments screening different systems for *in vivo* steroid release in sea bass suggested that estrogens had a potent anorexigenic effect. Animals treated with silastic implants containing 17-b estradiol did not want to eat after hand-feeding (B. Muriach and J.M. Cerdá Reverter unpublished data). This effect was quantified by treating animals with silastic implants of testosterone and 17-b estradiol (Leal et al. 2009). Both testosterone and 17-b estradiol implanted fish ate much less (around 50%) than control animals (Fig. 1). This severe reduction in food intake induced a drastic reduction in the sea bass growth, although food intake was not the only cause, since food conversion efficiency was severely reduced (72% in control *vs.* 40%–45% in treated animals). Interestingly, animals implanted with testosterone ate and converted similarly to those treated with 17-b estradiol and grew more in length, but not in weight. This suggests the existence of anabolic effects of testosterone on sea bass bone and muscle growth. Why does testosterone inhibit food

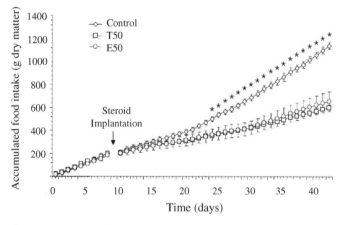

Figure 1. Effects of testosterone (T) and 17b-estradiol (E) on accumulative food intake in the sea bass. Animals were treated with silastic implants containing T or E at 50 mg/g body weight (BW). Asterisks indicate significant differences after ANOVA followed by Tukey's multiple range test (P<0.05). Adapted from Leal et al. 2009.

intake but stimulate linear growth? Since testosterone can be aromatized into estradiol, is it possible that the effects of testoterone are mediated by estradiol after aromatization? To answer this question, we designed a new experiment using non-aromatizable androgen, 11-keto and rostenedione, the precursor of the main fish androgen (11k-testosterone). Fish treated with non-aromatizable androgen self-demanded similar quantities of food as control fish. Likewise, growth and feed conversion efficiency did not exhibit significant differences compared with control animals. Therefore, the effects of testosterone on food intake, conversion efficiency and growth seem to be mediated by testosterone aromatization to estradiol. If aromatizable androgens and estrogens can reduce food intake and growth, it is plausible that reproductive anorexia of the sea bass can be mediated, in part, by the increase in plasma steroids. Obviously, the minimal temperatures of the year are the main vector driving reproductive anorexia but sex steroids could also contribute. To solve this question, an experiment was designed in which male sea bass were gonadectomized and pit-tagged during the reproductive season. Both controls and orchiectomized fish were then reared together in a 2000 L tank and the following summer were split up into twelve 500 L tanks provided with self-feeders. Feeding behavior and food intake levels were then recorded during the whole reproductive season (September–March). No significant differences in food intake levels (Fig. 2) or daily feeding rhythms were detected between orchiectomized and control fish (E. Leal, B. Fernández-Durán and J.M. Cerdá-Reverter unpublished data). It can be concluded that, at least in males, gonadal steroids do not play an important role in reproductive anorexia and it is likely that

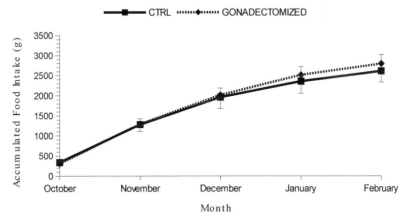

Figure 2. Effects of orchiectomy on cumulative food intake in the sea bass during its reproductive cycle. No differences between treatments were detected during the experimental period [ANOVA followed by Tukey's multiple range test ($P<0.05$), E. Leal, B. Fernández-Durán and J.M. Cerdá-Reverter unpublished data].

temperature effects on food intake are superimposed. It was not possible to repeat the experiment with females because of scarcity in culture. However, reproductive anorexia is probably more potent in females, especially during the maturation and spawning period (January–March). This may only be a personal interpretation, but, to our knowledge, there are no studies focusing on sex-associated food intake levels and feeding behavior.

It is well-known that the replacement of fish meal and fish oils by plant counterparts in aquafeeds is a priority for the long-term sustainability of aquaculture industries (Drew et al. 2007, Gatlin III et al. 2007, Glencross et al. 2007). Soybean meal is the main source of vegetable protein at present in animal diets. However, a large number of studies in fish have shown that a high dietary percentage of soybean meal result in decreased growth and feed efficiency (reviewed by Drew et al. 2007). Although non-steroidal estrogenic substances are widely distributed among potential plant-derived feeds (Francis et al. 2001) and fish diets (Pelissero and Sumpter 1992, Matsumoto et al. 2004), these estrogenic substances induce behavioral changes (Clofelter and Rodríguez 2006) and alterations in the reproductive physiology of fish when fed commercial diets (Pelissero and Sumpter 1992). It is therefore evident that these estrogenic substances present in the plant derived feeds have adverse effects on fish food intake and growth. Because of this, we have developed a sensitive system to detect estrogenic components in aquafeeds, raw materials and waste water effluents. Eleven out of thirty two analyzed aquafeeds were able to induce the activation of sea bass estrogen receptor 1 *in vitro* (Esr1, Quesada et al. 2012). Following the above results showing that estrogen severely affects sea bass growth, aquaculturists should be aware of the raw materials used to make fish feed, especially when these diets are destined for broodstock. In general, the results suggest that the food-intake levels and growth rates of sea bass may be sensitive to the substitution of fish meal by plant-derived prime materials containing estrogenic compounds.

Stress and Food Intake

It is evident that the sea bass is very susceptible to stressors, which severely curtail food intake levels, feed conversion efficiency and, by extension, growth (Rubio et al. 2010, Leal et al. 2011). Aquaculturists working with sea bass are aware of this and try to avoid getting too close to sea cages or tanks if it is not necessary. Biometric samplings or fish grading severely affect sea bass physiology and producers often overlook grading protocols in sea cages. In fact, a routine in our lab for feeding experiments is to pre-anesthetize animals in the home tanks before netting them for biometric and or blood sampling. In our hands, sea bass are fasted for two to three days when biometric sampling are made in the absence of pre-anesthesia

(V.C. Rubio and J.M Cerdá-Reverter unpublished results). But even when using pre-anesthesia protocols, sea bass will not eat during the sampling day and will reduce their food intake levels during the post-sampling day (Rubio et al. 2010).

We evaluated the sensitivity of the mechanisms controlling feeding behavior to the routine physical stressors involved in sea bass culture such as human proximity, tank cleaning and sampling protocols. Although fish were visually isolated to reduce the influence of human interference, feeding activity was reduced when operators were working in the culture facilities. All three physical stressors, including mere human proximity (64% inhibition *vs.* control fish only during the day of the stressor), reduced the food intake of the sea bass. The inhibitory effect on food intake seems to be graded in accordance with the intensity of the stressor. For example, the cleaning protocol involving partial emptying and brushing of the tanks resulted in a reduction in self-feeding of about 84% for two consecutive days (Rubio et al. 2010). This cleaning protocol induces a significant increase of cortisol plasma levels two hours after the stressor ceased but this increase was resumed after eight hours (Leal et al. 2011). We used the same cleaning protocol to study whether repetitive physical stressors were able to induce sustained effects on sea bass physiology. To this end, four tanks were cleaned once a week (Monday at 10 am) or three times a week (Monday, Wednesday and Friday at 10 am), whereas the four control tanks were never cleaned. Animals were subjected to this experimental protocol for 32 days. Acute physical stress imposed by cleaning once or three times a week significantly reduced daily and cumulative food intake levels as well as food conversion efficiency. Control fish were heavier but not longer and, accordingly, exhibited a higher condition factor (CF) [BW/Length (L)^3] than that shown by fish stressed once a week. All three growth rates, i.e., g(BW), g(L) and g(CF) were higher in control fish but only g(BW) and g(L) showed significant differences (Table 1). Similarly, the hepatosomatic index (HSI) and mesenteric fatty index (MFI) were higher in control fish than in the treated fish although differences did not reach statistical relevance. The temporal feeding pattern of fish was similar in all treatments and demands were mainly recorded during the light phase. Control fish exhibited a significant increase in food intake between 8.00 h and 10.00 h, when they consumed approximately 28% of their daily feed intake. During this time interval, stressed fish ate approximately 10% of the total daily feed (Fig. 3).

Repeated acute stressors are thought to alter behavior mainly by reducing feeding activity during the stress period, which is associated with decreased growth rates (Millot et al. 2010 and references therein). In our experiment, fish postponed their maximal demands until after the stressor had terminated. In some way, fish learnt the time when the stressor would take place and adapted their feeding behavior to avoid the stressful

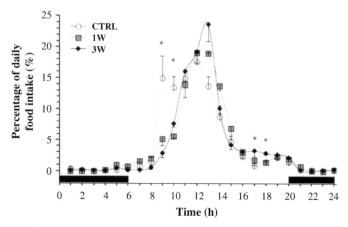

Figure 3. Daily food intake rhythms in sea bass subjected to repetitive physical disturbance. The cleaning protocol was always performed at 10.00 a.m. and involved draining and brushing the tanks with the animals inside. In 1W and 3W treatments the cleaning protocol was performed once or three times a week, respectively. Values are the average of food demanded every 60 min expressed as percentage of the total food intake for at least three groups of nine fish each. Black and white areas at the bottom of the graphs represent the dark and light phases, respectively, of the photoperiod. Asterisk indicates significant differences within the same hour after ANOVA followed by Tukey's multiple range test (P<0.05). Adapted from Leal et al. 2011.

conditions even when the stressor was not present. The delay in feeding behavior when fish are acutely and repeatedly stressed could be of adaptive importance but also could be of substantial importance for coupling the feeding protocols to typical routines of sea bass farming. For example, one practical consequence of these results would be to avoid feeding feed sea bass at the time when cleaning is regularly carried out, even on the days when the protocol is not applied.

The effect of stressors is mainly mediated by the hypothalamic-sympathetic chromaffin cell (HSC) and the hypothalamic-pituitary-interrenal axes (HPI). The activation of these axes contributes to restore homeostasis after the stressor by mobilizing food stores, thus making energy available to overcome stressful condition. Sustained reallocation of metabolic energy away from growth processes compromises the performance capacity of the fish during chronic stress. Cortisol is the main glucocorticoid in fish and the end product of HPI axis activation. Increased corticosteroid plasma levels is one of the most evolutionary conserved stress responses and it is commonly used as an indicator of the degree of stress experienced.

Cortisol is thought to mediate many effects of stressors on physiological, metabolic and behavioral processes (see Chapter 2.10). Therefore, in the following experiment, we tested whether the administration of exogenous cortisol could simulate the results obtained in the repetitive physical stress experiment. Three tanks were fed the control diet (CTRL), three tanks with cortisol-containing food at 200 mg/g food (C200) and the remaining three tanks with cortisol-enriched diet at 500 mg/g food (C500). The administration of the dietary cortisol induced a significant decrease in the daily food intake levels, which led to a dose-response reduction in the cumulative food intake. Feeding conversion efficiency was also significantly reduced by cortisol. Differences in food intake and food conversion efficiency resulted in heavier and longer fish in the control group than in those treated with cortisol. Animals treated with the highest cortisol dose (C500) lost weight, displaying a negative specific growth rate. Similarly, g_BW in animals treated with the lowest cortisol dose (C200) was 77.6% lower than in control fish. The condition factor of the control animals increased throughout the experiment but the cortisol-treated animals lost body condition, probably, at least partially, due to a severe decrease in the mesenteric fat content. Therefore, the decreased (C200) or suspended (C500) growth in cortisol-treated fish was probably a combined result of decreased food intake and efficiency levels and even increased energy expenditure as suggested by the severe reduction in body fat levels.

As in the repetitive physical stress experiment, sea bass exhibited a diurnal feeding rhythm with acrophases around midday. Cortisol administration slightly modified the food demand rhythms in the sea bass, since treated fish exhibited a significant food-demand peak during the last period of the photophase, which was absent in the control fish. This increased feeding activity lengthened the demand period in treated fish towards the scotophase. Differences in the feeding rhythms of sea bass caused by chronic stress induced by stocking density have been reported. With crowding, sea bass tend to lengthen the feeding activity period, surpassing the nocturnal restriction (Paspatis et al. 2003).

The data indicate that chronic stress induced by high stock densities has negative effects on sea bass performance. However, there is some controversy in the literature. Stock densities up to 45 kg/m^3 did not affect the energy status of sea bass and their sensitivity to subsequent acute crowding stressor (Di Marco et al. 2008). Reduced daily food intake and specific growth rate were observed at densities of 100 kg/m^3 but FCE was unaltered (d'Orbcastel et al. 2010, Sammouth et al. 2009). However, experiments with similar sized animals demonstrated negative effects on FCE at 50 kg/m^3, whereas growth performance was reduced at 75 kg/m^3 (Santos et al. 2010).

Conclusion

Probably, the main conclusion that can be reached after exploring the bibliography concerning the mechanisms controlling food intake in sea bass is that our knowledge of the same is scarce. Both practical and basic approaches are required. Many basic questions are still unresolved; for example, we do not know whether females eat more than males or whether the conversion efficiency is better in males than in females. Photoperiod effects on food intake have not been essentially separated from the effects of temperature. Sex-specific dietary preferences might exist that can improve growth/reproductive performance of both males and females.

Stress response is a major concern in sea bass production. Food intake levels and, by extension, growth performance are highly sensitive to stressful conditions and more basic information about the central and peripheral systems regulating food intake are required. An important advance would be achieved if we could segregate low stress phenotypes displaying reduced feeding responses to stressors. Finally, the sea bass is sensitive to the inclusion of antinutritional components in the diet and special attention should be paid to any new raw materials included in sea bass diets.

Acknowledgements

The authors were financially supported by the Spanish Ministry of Education and Science (AGL2010-22247-C03-01, CSD2007-00002) and the Generalitat Valenciana (Prometeo/2010/006).

References

Ali, M., A. Nicieza and R.J. Wootton. 2003. Compensatory growth in fishes: a response to growth depression. Fish. 4: 147–190.

Attia, J., S. Millot, C. Di-Poï, M.-L. Bégout, C. Noble, F.J. Sánchez-Vázquez, G. Terova, M. Saroglia and B. Damsgard. 2012. Demand feeding and welfare in farmed fish. Fish Physiol. Biochem. 38: 107–118.

Begout Ánras, M.-L. 1995. Demand-feeding behavior of sea bass kept in ponds: diel and seasonal patterns, and influences of environmental factors. Aquacult. Internat. 3: 186–195.

Caamaño-Tubío, R.I., J. Pérez, S. Ferreiro and M. Aldegunde. 2007. Peripheral serotonin dynamics in the rainbow trout (*Oncorhynchus mykiss*). Comp. Biochem. Physiol. 145: 245–55.

Canosa, L.F., P. Gómez-Requeni and J.M. Cerdá-Reverter. 2012. Integrative crosstalk between food intake and growth function. *In*: S. Polakoff and T.W. Moon (eds.). Trout: From Physiology to Conservation. Nova Science Publishers. in press.

Cerdá-Reverter, J.M. and L.F. Canosa. 2009. Neuroendocrine systems of the fish brain. pp. 3–74. *In*: N.J. Bernier, G. Van Der Kraak, A.P. Farrell and C.J. Brauner (eds.). Fish Neuroendocrinology, Fish Physiology. Academic Press, London.

Clofelter, E.D. and A.C. Rodríguez. 2006. Behavioral changes in fish exposed to phytoestrogens. Environ. Pollut. 144: 833–839.

Conides, A.J. and B. Glamuzina. 2006. Laboratory simulation of the effects of environmental salinity on acclimatization, feeding and growth of wild-catch juveniles of European sea bass *Dicentrarchus labrax* and gilthead sea bream, *Sparus aurata*. Aquaculture 256: 235–245.

d'Orbcastel, E.R., G. Lemarie, G. Breuil, T. Petochi, G. Marino, S. Triplet, G. Dutto, S. Fivelstad and J.L. Coeurdacier. 2010. Effects of rearing density on sea bass (*Dicentrarchus labrax*) biological performance, blood parameters and disease resistance in a flow through system. Aquat. Liv. Res. 23: 109–117.

De Pedro, N., M.L. Pinillos, A.I. Valenciano, M. Alonso-Bedate and M.J. Delgado. 1998. Inhibitory effect of serotonin on feeding behaviour in goldfish: Involvement of CRF. Peptides 19: 505–511.

De Pedro, N., R. Martínez-Álvarez and M.J. Delgado. 2008. Melatonin reduces body weight in goldfish (*Carassius auratus*): effects on metabolic resources and some feeding regulators. J. Pineal Res. 45: 32–39.

Dendrinos, P. and J.P. Thorpe. 1985. Effects of reduced salinity on growth and body composition in the European bass *Dicentrarchus labrax* (L.). Aquaculture 49: 333–358.

Di Marco, P., A. Priori, M.G. Finoia, A. Massari, A. Mandich and G. Marino. 2008. Physiological responses of European sea bass *Dicentrarchus labrax* to different stocking densities and acute stress challenge. Aquaculture 275: 319–328.

Drew, M.D., T.L. Borgeson and D.L. Thiessen. 2007. A review of processing of feed ingredients to enhance diet digestibility in finfish. Anim. Feed Sci. Tech. 138: 118–136.

Dülger, N., M. Kumlu, S. Türkmen, A. Ölcülü, O.T. Eroldogan, H.A. Yilmaz and N. Öcal. 2012. Thermal tolerance of European sea bass (*Dicentrarchus labrax*) juveniles acclimated to three temperature levels. J. Thermal Biol. 37: 79–82.

Eckel, L.A. 2004. Estradiol: a rhythmic, inhibitory, indirect control of meal size. Physiol. Behav. 82: 35–41.

Francis, G., H.P.S. Makkar and K. Becker. 2001. Antinutritional factors present in plant-derived alternate fish feed ingredients and their effects in fish. Aquaculture 199: 197–227.

Gatlin III, D.M., F.T. Barrows, P. Brown, K. Dabrowski, T.G. Gaylord, R.W. Hardy, E. Herman, G. Hu, A. Krogdahl, R. Nelson, K. Overturf, M. Rust, W. Sealey, D. Skonberg, E.J. Souza, D. Stone, R. Wilson and E. Wurtele. 2007. Expanding the utilization of sustainable plant products in aquafeeds: a review. Aquacult. Res. 38: 551–579.

Gélineau, A. and T. Boujard. 2001. Oral administration of cholecystokinin receptor antagonist increase feed intake in rainbow trout. J. Fish Biol. 58: 716–724.

Glencross, B.D., M. Booth and G.L. Allan. 2007. A feed is only as good as its ingredients—a review of ingredient evaluation strategies for aquaculture feeds. Aquac. Nutr. 13: 17–34.

Gorissen, M., G. Flik and M. Huising. 2006. Peptides and proteins regulating food intake: a comparative view. Animal Biol. 56: 447–473.

Hidalgo, F., E. Alliot and H. Thebault. 1987. Influence of water temperature on food intake, food efficiency and gross composition of juvenile sea bass *Dicentrarchus labrax*. Aquaculture 64: 199–207.

Himick, B.A. and R.E. Peter. 1994. CCK/gastrin-like immunoreactivity in brain and gut, and CCK suppression of feeding in goldfish. Am. J. Physiol. 267: R841–851.

Himick, B.A., S.R. Vigna and R.E. Peter. 1996. Characterization of cholecystokinin binding sites in goldfish brain and pituitary. Am. J. Physiol. 271: R137–143.

Holmgren, S. and C. Olsson. 2009. The neuronal and endocrine regulation of gut function. pp. 467–512. *In:* N.J. Bernier, G. Van Der Kraak, A.P. Farrell and C.J. Brauner. [eds.]. Fish Neuroendocrinology, Fish Physiology. Academic Press, London.

Imsland, A.K. and Jonassen. 2001. Regulation of growth in turbot (*Scophthalmus maximus* Rafinesque) and Atlantic halibut (*Hippoglossus hippoglossus* L.): Aspects of environment x genotype interactions Rev. Fish Biol. Fish 11: 71–90.

Jobling, M. 1994. Fish Bioenergetics. Chapman and Hall, London.

Jobling, M. 1996. Temperature and growth: Modulation of growth rate via temperature. pp. 225–253. *In:* C.M. Wood and D.G. McDonald. [eds.]. Global warning: Implications for fresh water and marine fish. Society for experimental biology seminar series. Cambridge University Press, Cambridge.

Jönsson, E., A. Forsman, I.E. Einarsdottir, B. Egnér, K. Ruohonen and B.T. Björnsson. 2006. Circulating levels of cholecystokinin and gastrin-releasing peptide in rainbow trout fed different diets. Gen. Comp. Endocrinol. 148: 187–194.

Kavadias, S., J. Castritsi-Catharios and A. Dessypris. 2003. Annual cycles of growth rate, feeding rate, food conversion, plasma glucose and plasma lipids in a population of European sea bass (*Dicentrarchus labrax* L.) farmed in floating marine cages. J. Appl. Ichthyol. 19: 29–34.

Kestemont, P. and E. Baras. 2001. Enviromental factors and feed intake: Mechanism and interactions. pp. 131–156. *In*: D. Houlihan, T. Boujard and M. Joblin. [eds.]. Food intake in fish. Blackwell Science, Oxford.

Leal, E., E. Sánchez, B. Muriach and J.M. Cerdá-Reverter. 2009. Sex Steroid-Induced Food Intake Inhibition in the Sea Bass (*Dicentrarchus labrax*). J. Comp. Physiol. B. 179: 77–86.

Leal, E., B. Fernández-Durán, R. Guillot, D. Ríos and J.M. Cerdá-Reverter. 2011. Stress-induced effects on feeding behaviour and growth performance of the sea bass (*Dicentrarchus labrax*): a self-feeding approach. J. Comp. Physiol. B. 181: 1035–1044.

Leal, E., B. Fernández-Durán, M.J. Agulleiro, M. Conde-Sieira, J.M. Míguez and J.M. Cerdá-Reverter. 2013. Effects of dopaminergic system activation on feeding behavior and growth performance of the sea bass (*Dicentrarchus labrax*): a self-feeding approach. Horm. Beh. 64: 113–121.

Lemarié, G., A. Dosdat, D. Covés, G. Ditto, E. Gasset and J. Person-Le Ruyet. 2004. Aquaculture 229: 479–491.

Lemarié, G., C. Diesen Hosfeld, G. Breuil and S. Fivestald. 2011. Effects of hyperoxic water conditions under different total gas pressure in European sea bass (*Dicentrarchus labrax*). Aquaculture 318: 191–198.

Lillesaar, C. 2011. The serotonergic system in fish. J. Chem. Neuroanat. 4: 294–308.

Lin, X., H. Volkoff, Y. Narnaware, N.J. Bernier, P. Peyon and R.E. Peter. 2000. Brain regulation of feeding behavior and food intake in fish. Comp. Biochem. Physiol. A. 126: 415–434.

Matsumoto, T., M. Kobayashi, T. Moriwaki, S. Kawai and S. Watabe. 2004. Survey of estrogenic activity in fish feed by yeast estrogen-screen assay. Comp. Biochem. Physiol. C. 139: 147–152.

Millot, S., S. Pean, D. Leguay, A. Vergnet, B. Chatain and M.L. Begout. 2010. Evaluation of behavioral changes induced by a first step of domestication or selection for growth in the European sea bass (*Dicentrarchus labrax*): A self-feeding approach under repeated acute stress. Aquaculture 306: 211–217.

Paspatis, M., T. Boujard, D. Maragoudaki, G. Blanchard and M. Kentouri. 2003. Do stocking density and feed reward level affect growth and feeding of self-fed juvenile European sea bass? Aquaculture 216: 103–113.

Pelissero, C. and J.P. Sumpter. 1992. Steroid and steroid-like substances in fish diets. Aquaculture 107: 283–301.

Person-Le Ruyet, J. and H.C.L. Quemener. 1995. Comparative acute ammonia toxicity in marine fish and plasma ammonia response. Aquaculture 136: 181–194.

Person-Le Ruyet, J., K. Mahé, N. Le Bayon and H. Le Delliou. 2004. Effects of temperature on growth and metabolism in the Mediterranean population of European sea bass, *Dicentrarchus labrax*. Aquaculture 237: 269–280.

Piccinetti, C., B. Migliarini, I. Olivotto, G. Coletti, A. Amici and O. Carnevali. 2010. Appetite regulation: The central role of melatonin in *Danio rerio*. Hormones Behav. 58: 780–785.

Pichavant, K., J. Person-Le Ruyet, N. Le Bayon, A. Severe, A. Le Roux and G. Boeuf. 2001. Comparative effects of long-term hypoxia an growth, feeding and oxygen consumption in juvenile turbot and Europen sea bass. J. Fish Biol. 59: 875–883.

Pickett, G.D. and M.G. Pawson. 1994. Sea bass: biology, exploitation and conservation Chapman and Hall, London.

Pinillos, M.L., N. De Pedro, A.L. Alonso-Gómez, M. Alonso-Bedate and M.J. Delgado. 2001. Food intake inhibition by melatonin in goldfish (*Carassius auratus*). Physiol. Behav. 72: 629–634.

Quesada, A., A. Valdehita, Fernández-Cruz, E. Leal, E. Sánchez, M. Martín-Belinchón, J.M. Cerdá-Reverter and J.M. Navas. 2012. Assessment of estrogenic and thyrogenic activities in fish feeds. Aquaculture 338–341: 172–180.

Rubio, V.C., J.A. Sánchez-Vázquez and J.A. Madrid. 2004. Oral administration of melatonin reduces food intake and modifies macronutrient selection in European sea bass (*Dicentrarchus labrax*). J. Pineal. Res. 37: 42–47.

Rubio, V.C., F.J. Sánchez-Vázquez and J.A. Madrid. 2005. Effects of salinity on food intake and macroniutrient selection. Physiol. Behav. 85: 333–339.

Rubio, V.C., F.J. Sánchez-Vazquez and J.A. Madrid. 2006. Oral serotonin administration affects the quantity and the quality of macronutrient selection in European sea bass (*Dicentrarchius labrax*). Physiol. Behav. 87: 7–15.

Rubio, V.C., F.J. Sánchez-Vázquez and J.A. Madrid. 2008. Role of cholecystokinin and its antagonist proglumide on macronutrient selection in European sea bass *Dicentrarchus labrax* L. Physiol. Behav. 93: 862–869.

Rubio, V.C., E. Sánchez and J.M. Cerdá-Reverter. 2010. Compensatory feeding in the sea bass after fasting and physical stress. Aquaculture 298: 332–337.

Ruibal, C., J.L. Soengas and M. Aldegunde. 2002. Brain serotonin and the control of food intake in trout (*Oncorhynchus mykiss*): effects of changes in plasma glucose levels. J. Comp. Physiol. 188: 479–484.

Russell, N.R., J.D. Fish and R.J. Wootton. 1996. Feeding and growth of juvenile sea bass: the effect of ration and temperature on growth rate and efficiency. J. Fish Biol. 49: 206–220.

Sammouth, S., E.R. d'Orbcastel, E. Gasset, G. Lemarie, G. Breuil, G. Marino, J.L. Coeurdacier, S. Fivelstad and J.P. Blancheton. 2009. The effect of density on sea bass (*Dicentrarchus labrax*) performance in a tank-based recirculating system. Aquacult. Engineering 40: 72–78.

Sánchez-Vázquez, F.J., M. Azzaydi, F.J. Martínez, S. Zamora and J.A. Madrid. 1998. Annual rhythms of demand-feeding activity in sea bass: evidence of a seasonal phase inversion of the diel feeding pattern. Chronobiol. Int. 15: 607–622.

Santos, G.A., J.W. Schrama, R.E.P. Mamauag, J.H.W.M. Rombout and J.A.J. Verreth. 2010. Chronic stress impairs performance, energy metabolism and welfare indicators in European sea bass (*Dicentrarchus labrax*): The combined effects of fish crowding and water quality deterioration. Aquaculture 299: 73–80.

Strader, A.D. and S.C. Woods. 2005. Gastrointestinal hormones and food intake. Gastroenterology 128: 175–191.

Teccot, L.H. 2007. Serotonin and the orchestration of energy balance. Cell Metab. 6: 352–361.

Thetmeyer, H., U. Waller, K.D. Black, S. Inselmann and H. Rosenthal. 1999. Growth of European sea bass (*Dicentrarchus labrax* L.) under hypoxic and oscillating oxygen conditions. Aquaculture 174: 355–367.

Velarde, E., M.J. Delgado and A.L. Alonso-Gómez. 2010. Serotonin-induced contraction in isolated intestine from a teleost fish (*Carassius auratus*): Characterization and interactions with melatonin. Neurogastroenterol. Motil. 22: e364–373.

Vera, L.M., N. De Pedro, E. Gómez-Milan, M.J. Delgado, M.J. Sánchez-Muros, J.A. Madrid and F.J. Sánchez-Vázquez. 2007. Feeding entrainment of locomotor activity rhythms, digestive enzymes and neuroendocrine factors in goldfish. Physiol. Behav. 90: 518–524.

Volkoff, H., L.F. Canosa, S. Unniappan, J.M. Cerdá-Reverter, N.J. Bernier, S.P. Kelly and R.E. Peter. 2005. Neuropeptides and the control of food intake in fish. Gen. Comp. Endocrinol. 142: 3–19.

Volkoff, H., S. Unniappan and S.P. Kelly. 2009. The endocrine regulation of food intake. pp. 421–465. *In:* N.J. Bernier, G. Van Der Kraak, A.P. Farrell and C.J. Brauner [eds.]. Fish Neuroendocrinology, Fish Physiology. Academic Press, London.

Wynne, K., S. Stanley, B. McGowan and S. Bloom. 2005. Appetite control J. Endocrinol. 184: 291–318.

Zanuy, S. and M. Carrillo. 1985. Annual cycles of growth, feeding rate, gross conversion efficiency and hematocrit levels of sea bass (Dicentrarchus-labrax l) adapted to 2 different osmotic media. Aquaculture 44: 11–25.

Nutrition and Dietary Selection

Rodrigo Fortes da Silva,[1] Francisco Javier Sánchez Vázquez[2] and Francisco Javier Martínez López[2,]*

Introduction

Aquatic animal nutrition is a relatively new science. Although it has many similarities to terrestrial animal nutrition, fish nutrition research is confronted with a wide variety of challenges not encountered in research on terrestrial animals. These challenges are primarily due to the aquatic media that require special considerations related to the delivery of nutrients, monitoring of food intake, collection and quantification of waste products, and the unique physiology of various species of fish (NRC 2011).

The classic way to assess the nutritional requirements of cultured fish is to feed diets composed of a mixture of practical ingredients that provide the desired amino-acid balance, essential fat acids, energy, vitamins and minerals. However, in some cases, for example in low-cost formulations that involve ingredient substitutions, it may be necessary to supplement certain nutrients, e.g., amino acids in purified form to produce a suitable amino acid profile (Ambardekar et al. 2009).

Aquaculture continues to grow at a relatively high and constant rate over the past decades (FAO 2010). The rapid expansion of aquaculture and advances in fish farming techniques has increased the demand for balanced diets. However, increasing demand, uncertain availability, and high price has made it necessary to search for alternative food sources (Ai et al. 2007). In farming European sea bass (*Dicentrarchus labrax*) as for other aquaculture

[1]Agricultural Science, Biological and Environmental Center, University of Bahia Recôncavo, Brazil.
[2]Dep. Physiology. Fac. Biology. University of Murcia. Spain.
*Corresponding author

species, the cost of feed is the largest factor of production, and, therefore, it is important to minimize the cost of the feed while maintaining high growth rate of the fish. Our challenge is to supply adequate amounts of nutrients and energy in a balanced manner through diverse ingredients available today or those yet to be tapped or developed. Fish nutrition research has played a significant role in meeting these challenges over the past decades, and new behavioral concepts are being considered when investigating fish appetite and food intake regulation in aquaculture (Jobling et al. 2012).

This chapter is divided into two sections. The first one (A) deals with basic and practical aspects, comprising protein, energy, lipid and carbohydrate nutrition. The requirements of each macronutrient and its replacement with sustainable food sources will be reviewed. In addition, micronutrient requirements and the use of probiotics is discussed. The second section (B) is devoted to dietary selection and the ability of sea bass to detect and self-select different food types. Both macro- and micro-nutrient selection will be considered, as well the basic mechanisms responsible for the control of food intake and dietary selection.

Protein

Protein and amino acids requirements

Protein is the main limiting nutrient to formulate cost-effective diets for fish because of its high cost in the market. In past years, the optimum protein level in the diets for sea bass juveniles was estimated to be around 50% (Peres and Oliva-Teles 1999a). However, some discrepancy on the optimum dietary protein level for sea bass was observed for several authors. Pérez et al. (1997) observed an optimum growth with a 45% protein diet. Today, according to NRC (2011), the optimal level of crude protein for sea bass on dry-matter basis is 40%. This discrepancy may also be due to advances in fish nutrition research.

A wealth of protein sources for use in practical diets for European sea bass, as fish meal, soy protein concentrates, corn gluten meal (Dias et al. 2005) is readily available for the aquafeed industry. However, there are few studies regarding amino acid availability from these products for European sea bass, because many studies were conducted using pelleted purified reference diets (Guimarães et al. 2008), when the industries use extrusion process to manufacture the diets. Thus, more studies are needed on protein digestibility and amino acid availability of ingredients submitted to feed processing used by industries, because fish have amino acids requirement instead of protein (Guimarães et al. 2008). Basically the denaturation exposes sites for enzyme to attack and may thus make the protein more digestible. Anti-nutritive factors such as trypsin inhibitors (high in soya), lectins and others may be destroyed with extrusion process. Another issue is that the amino acid requirement of fish is frequently determined with diets of different palatability; therefore, a variation in feed intake is neglected if not superimposed on changes in efficiencies in diet utilization (Dabrowski et al. 2007).

Protein deposition in organisms is dictated by specific templates genetically determined, but modulated by exogenous factors, e.g., environment and diet (NRC 2011). The essentiality of various amino acids for fish can be determined either by feeding trials involving the successive deletion of each amino acid in the diet or by isotopic-labeling studies. Along with 10 amino acids, fish have a nonspecific requirement for a source of amino groups for the synthesis of nonessential amino acids (Table 1). Amino acid requirement of European sea bass is well established by NRC (2011) (Table 2).

Table 1. Essential and nonessential amino acids for fish.

Essential	Nonessential
Arginine	Alanine
Histidine	Asparagine
Isoleucine	Aspartate
Leucine	Cysteine[a]
Lysine	Glycine
Methionine	Glutamine
Phenylalanine	Proline
Threonine	Serine
Tryptophan	Tyrosine[a]

[a] Conditionally essential
Source: Adapted from NRC (2011)

Source of protein: replacement of fish meal by plant proteins

The total or partial replacement of fish meal in diets for cultured fish species has been the subject of numerous studies (e.g., Refstie and Storebakken 2001, Dias et al. 2005). A variety of plant ingredients have been evaluated as alternative sources of protein: soy, seeds of lupin, rapeseed, cotton seed, carob seeds, sunflower seeds, seeds of peas. Despite some variability between and within species of fish, the majority of "carnivorous" fish studies tend to show that the partial replacement (30% to 40%) of protein from fish on a diet with a source of vegetable protein is acceptable (Kaushik et al. 2004), but at higher levels, the growth performance and nutritive utilization of diet are depressed (Robaina et al. 1995, Refstie et al. 2000, Dias et al. 2005).

Table 2. Amino acid body composition (g/16g N) and amino acid requirement of European sea bass.

Amino acids	Amino acid composition	Amino acid requirements
Alanine	6.8	NT
Arginine	7.5	7.5
Asparate	9.5	-
Cysteine	1.0	-
Glutamate	15.5	-
Glycine	8.1	-
Histidine	2.6	NT
Isoleucine	4.3	NT
Leucine	7.1	NT
Lysine	7.9	2.2
Methionine	2.7	NT
Phenylalanine	4.3	NT
Proline	5.3	-
Serine	4.5	-
Treonine	4.4	1.2
Tryptophan	NT	NT
Tyrosine	3.9	-
Valine	4.7	4.7

Source: Adapted from NRC (2011)

It has also been shown that the almost total replacement of fish meal by a mixture of vegetable protein sources is possible with very good growth rates (Kaushik et al. 1995, 2004). Alliot et al. (1979b) and Ballestrazzi et al. (1994) reported that with sea bass, the substitution of fishmeal by vegetable protein reduced growth and protein efficiency ratios when the incorporation of proteins of soy or corn gluten exceeded the level of 20%–30%. The poor growth results of sea bass fed these diets have been linked to a reduction in voluntary consumption of diet (Dias 1999, Kaushik 2002), in addition to the reduction of growth caused by a limiting supply of sulphur amino acids and the scarcity of certain minerals (Tibaldi and Tulli 1998, Tibaldi et al. 2006).

Studies with sea bass and other species of fish fed to apparent satiety have also shown that the palatability is not affected, or even increases in response to graded levels of soybean diets with adequate levels of methionine (Alliot et al. 1979a, Robaina et al. 1995, Oliva-Teles et al. 1998, Krogdahl et al. 2003, Tibaldi et al. 2006). Sea bass fed with diets containing very high levels of protein concentrate of soy or corn gluten meal as the sole source of protein showed a decrease in voluntary feed intake, which was improved by supplementation with a mixture of attractants (Dias et al. 1997). But other data indicate that when the same protein sources are used replacing approximately 60% of fish meal, appropriately supplemented with limitant amino acids such as lysine or methionine, the cited mix is not necessary (Tibaldi et al. 1999, Kaushik et al. 2004). The addition of arginine in diets of sea bass with high levels of gluten proved to be beneficial in enhancing growth and significantly improved the efficiency of N retention and gain, but was ineffective in juveniles fed with fish meal as the main source of N (Tulli et al. 2007); this is in agreement with Kaushik et al. (2004).

The nutritional value of complete diets for sea bass may be affected differently depending on the processing of soybean used to replace different amounts of protein from fish meal, resulting in a higher rate of substitution with soybean meal treated with enzyme, with a reduced content of oligosaccharides and anti-nutritional factors (Tibaldi et al. 2006).

Other studies have shown the beneficial effects of the use of hydrolysates of proteins rather than protein in diets for larvae (Zambonino Infante et al. 1997, Cahu et al. 1999). Zambonino Infante et al. (1997) initially indicated that the best rate of incorporation of fish protein hydrolysate (FPH) is 20% of the total nitrogen supply, since this diet improves growth and survival of the larvae of sea bass, in comparison with groups fed diets without FPH, or at a rate of 40% replacement. A similar trend was observed by Cahu et al. (1999), who found that 25% replacement of the diet with the commercial FPH facilitated the development of sea bass larvae, while the 50% and 75% rates of substitution led to a reduction in the growth of larvae (Kotzamanis et al. 2007). Diets with rates of low substitution of fishmeal by the FPH (10% of the total ingredients) produced improvement in the growth of sea bass larvae with respect to diets with higher rates of replacement, of 19% (Kotzamanis et al. 2007). Larvae nutrition is further discussed in another chapter in this book.

The source of dietary protein affects the regulation of the metabolism of lipid in fish (Dias et al. 2005). These authors show that in sea bass, replacement of fish meal by soy protein or corn gluten meal depresses the activities of enzymes implicated in the metabolism of lipid compared with fish fed with fish-based diet. Fish fed with corn gluten diet showed greater activity of fatty acid synthetase (FAS, EC 2.3.1.38), a situation that could be related to the imbalance of amino acids in the diet, particularly a deficiency of lysine, which leads to greater catabolism of other amino acids, confirmed by the increase in the excretion of ammonia N, previously observed by Ballestrazzi et al. (1994).

Kaushik et al. (2004) observed a significant increase in the fat content with increasing levels of fishmeal replacement. A similar increase in whole body energy content was observed. The high retention of fat and energy values clearly indicate that there was increased lipogenesis with increasing levels of substitution, without any effect on the use of nitrogen, but Tibaldi et al. (2006) noted otherwise.

One final point to consider when using vegetable ingredients as important dietary sources of protein, is the bioavailability of phosphorus (P), since these ingredients usually have lower levels of phosphorus that animal by-products and also, above all, in the form of the phytate, which is unaffordable to the majority of the fish species (Lall 1991). Fortes-Silva et al. (2011b) confirmed that the European sea bass fed with a diet with vegetable source or diet with vegetable source plus phytase (1500 FTU/kg) did not show improved growth compared with fish that were fed a diet with fish meal. However, the addition of phytase significantly improved the nutritional properties of the plant-based diet, increasing the retention of P and Ca in bone.

Energy

Optimal protein utilization can be achieved by increasing dietary energy concentration through the inclusion of non-protein energy sources, such as lipids and carbohydrates. In the proper proportions, these nutrients can improve protein utilization, thus reducing nitrogen excretion and enhancing the quality of fish farm effluent discharge. Thus, the energy requirement should be expressed in terms of the amount of protein in the diet.

The gross energy (GE) is the energy released by complete oxidation of a particular food. GE is determined by measuring the total heat produced when food is burnt in the calorimetric bomb. This value is expressed in kcal: energy required to increase from 14.5°C to 15.5°C, a sample of 1 kg of water. The digestible energy (DE) is considered as the energy that has been ingested and not eliminated in the feces of animals, given by the formula: DE = (ingested GE) - (excreted GB). The metabolizable energy (ME) is calculated by the difference in digestible energy and energy excreted in urine and gills, given by the formula: ME = DE–(E urine + E gills). The net energy (NE) is the energy used for animal production (development, weight gain and reproduction), given by: NE =ME–Eic, where "ic" = caloric increment (produced on processes of digestion and absorption).

The protein sparing effect of lipid and carbohydrates treated in species of farmed fish, such as European sea bass, appears to be influenced by the level of of diet rather than the nature of the sources of energy without proteins (Watanabe 1982, Cho and Kaushik 1990). The best growth combined with low nitrogen losses was achieved with a level of crude protein of 43% (on dry matter) when the level is at least 21 kJ/g of dry matter (Dias et al. 1998). The optimal DP/DE ratio in the literature ranges from 19 mg (Dias et al. 1998) to 25–30 mg/kJ (Métailler et al. 1981) of digestible protein by kJ of digestible energy. Lupatsch et al. (2001) estimated the needs of protein and energy for the maximum growth at 28.3 mg DP/kJ DE, for fish of 50 g, and at 25.2 mg DP/kJ DE, for fish of 150 g.

For high levels of protein (55%), higher content of digestible energy seems to improve the performance of growth. Peres and Oliva-Teles (1999b) confirmed a hyperphagia in animals fed diets with high content digestible energy. According to NRC (2011), the energy requirement of European sea bass is 4,000 Kcal/Kg diet (dry matter) of digestible energy, with a relationship of energy/protein around 10.

Lipid

Classic observations (Alliot et al. 1979b) pointed to higher growth rates in sea bass fed with diets containing a level of dietary lipid of 12% compared with those fed a diet containing 8%. However, other studies (Métailler et al. 1981, Pérez et al. 1997, Tibaldi et al. 1991, Morales and Oliva-Teles 1995, Peres and Oliva-Teles 1999b) did not show any beneficial effect of increasing the level of lipids. Dias et al. (1998) and Lanari et al. (1999) suggest that dietary lipids increased up to 18%–19% improves utilization of protein in sea bass. Peres and Oliva-Teles (1999b) suggest that very high levels (30% lipid) lead to a depression of growth in sea bass. Lipid levels for optimum growth of juveniles of sea bass that range between 12% and 24% of the diet have been proposed (Oliva-Teles 2000, Dias et al. 2003).

According to Dias et al. (1998) the beneficial effects of an increase in the level of lipid from 10% to 18% in diets of sea bass are significant only with a low protein diet (40% DM), but not with a high protein diet (50% DM) (Peres and Oliva Teles 1999b). The increase in the level of dietary lipid from 12% up to 24% improved utilization of energy, according to other authors (Morales and Oliva-Teles 1995, Dias et al. 1998). The retention of lipids decreased significantly as dietary lipid levels increased, suggesting that an increasing proportion of lipids were used to produce energy. The inclusion of 30% lipids did not affect the rate of growth, but significantly reduced the effectiveness of the protein and energy retention.

The fatty acid composition of neutral lipids in the sea bass, as in many other teleosts, reflects the profile of the diet of fatty acids. However, the sea bass is prone to having a high deposition of fat in the liver (Kaushik 2002).

Geurden et al. (1997a,b) have identified the importance of the phospholipids of the diet for growth and composition of lipid tissues of larvae of sea bass when going to eat compound dry diets. Supplementation with 1–2% of phospholipids improves significantly the growth of juvenile sea bass, as well as the retention of fatty acids. Larval growth and survival were very low with a diet with phospholipids of 2.7%, however with a diet containing 11.6% of phospholipids, the final weight was 32 mg, and survival was 73% (Cahu et al. 2003). On the other hand, in the diet with the lowest level of phospholipids, 35 % of the larvae showed malformations, while only 2% of the larvae fed the diet with 11.7% of phospholipids exhibit malformations (Cahu et al. 2003). These authors suggest that a dietary level of 1.6% of phosphatidylinositol, used since the first feeding, is suitable for the prevention of deformities during the development of sea bass.

There is also evidence of an effect of nutritional conditioning in the metabolism of lipids in juveniles fed with a diet deficient in HUFA (0.5% DM n-3 HUFA), by a positive modulation of transcription of $\Delta 6D$, and a greater ability to regulate the fatty acids of phospholipids when they are fed a diet deficient in HUFA during the larval stage (Vagner et al. 2007). For further information see the chapter on larvae development.

Replacement of fish oil with vegetable oil in the diet

Sea bass presents a "marine" pattern in the metabolism of the 18:3n-3, EPA and DHA (Mourente and Dick 2002). In addition, and unlike other species studied, the nutritional regulation of hepatocyte fatty acid desaturation is minimal. In contrast, in hepatocytes of sea bass a different pattern was observed from the one seen in other species, so that DHA appears to be the most abundant product of desaturation and elongation of the 18:3n-3 (Mourente et al. 2005a). Thus, the partial replacement (60%) of fish oil with a mixture of colza, linseed and palm oils in diets for European sea bass did not significantly compromise the growth and/or the survival of the fish for 64 weeks (Mourente et al. 2005b). These authors suggested that vegetable oils such as canola oil, linseed oil and olive oil can potentially be used as a partial substitute of dietary fish oil on the culture of the European sea bass during the growth phase, without compromising the rates of growth or health of fish. However the inclusion of these oils significantly reduced levels of EPA and DHA, and increased those of 18:1n-9 (OA), 18-2 n-6 (LA) and 18: 3n-3 (LNA) in tissues lipids. This fact is overcome with later feeding on a diet of finishing that contains 100% fish oil for 14 additional weeks. Although the levels of OA, the LNA in the flesh remains high, the levels of arachidonic acid (20:4n - 6) (ARA) and DHA n-3 HUFA can be fully restored after the diet of finish (Mourente and Bell 2006).

A higher level of substitution (80%), however, may lead to lower growth (Montero et al. 2003, Montero et al. 2005, Menoyo et al. 2004, Izquierdo et al. 2005, Richard et al. 2006) and total replacement compromises the performance of growth, revealing a nutritional deficiency. This is consistent with Skalli and Robin (2004), determining the requirement of n-3 HUFA around 0.7% of the diet in dry matter (or 18% of total lipids in the diet) of sea bass. This slowdown in growth can be correlated with the significant decrease in n-3 HUFA levels observed in the flesh of fish fed on vegetable diet (Oliva-Teles 2000, Parpoura and Alexis 2001, Skalli and Robin 2004). Another reason could be linked to the content low of arachidonic acid from vegetable diet compared with the diet of fish (Castell et al. 1994).

Polyunsaturated fatty acids

The nutritional aspects of essential fatty acids (EFAs) in fish have been widely investigated (Sargent et al. 2002, Tocher 2003). The series of highly unsaturated n-3 fatty acids (HUFA n-3) are nutrients needed for marine fish (Kanazawa et al. 1979), especially eicosapentaenoic acid (20:5n–3) (EPA) and docosahexaenoic acid (22:6n–3) (DHA). These fatty acids are the main components of polar lipids in fish, ensuring the functionality of the membrane (Sargent et al. 2002). The HUFA are involved in numerous physiological processes, including the adaptation of the fish to environmental factors such as temperature (Hazel et al. 1992, Farkas et al. 1994). However, the regulation of the fluidity of the membrane may involve other components, such as the ratios of monounsaturated fatty acid/saturated (Wodtke and Cossins 1991) or cholesterol/phospholipids (Robertson and Hazel 1996).

In juveniles of European sea bass, the dietary minimum requirement of n-3 HUFA to sustain maximum growth is 0.7% dry (MS) (Skalli and Robin 2004), which indicates that only some marine foods (fish meal or oil) may comply with this requirement. With dry formulated diets, Coutteau et al. (1996) suggest a requirement of around 1% of n-3 PUFA to sea bass. The interaction between temperature and n-3 HUFA of diet has been investigated in juveniles of European sea bass (Person-Le Ruyet et al. 2004) and it has been shown that a deficiency during three months of n-3 HUFA does not drastically affect the adaptive capacity of the fish at a high temperature (29°C).

However, egg and larval stages of fish probably have greater demands of n-3 HUFA, due to the preponderance in its visual and neural tissue that predominates in the early stages of development. Therefore, any deficiency in these particular fatty acids can cause congenital abnormalities in the nervous system and may affect its success as a visual predator at the beginning of the first feeding (Bell et al. 1995a,b).

Thrush et al. (1993) and Bell et al. (1997) showed that sea bass fed exclusively 'fresh fish' were consistently better that those fed on artificial diets containing oil of maize and oil of fish as a lipid component. As well as a lower ratio DHA:EPA, artificial diets were characterized by a comparatively low proportion AA:EPA of < 0,1. These results confirm that in addition to providing sufficient DHA, it is imperative to maintain sufficient levels of

AA in relation to the EPA in the formulation of artificial diets for marine broodstock (Bruce et al. 1999).

European sea bass fatty acid nutrition has been investigated for broodstock and has stressed the importance of highly unsaturated fatty acids n-3 and n-6 (Bell et al. 1997, Bruce et al. 1999) and larval stages (Navarro et al. 1997). The data indicate that marine fish oil, rich in long-chain HUFA is very essential for improving the quality of the eggs, as in the majority of other marine fish (Kaushik 2002).

The requirements also depend on the level of dietary lipids. Glencross et al. (2002) suggest that EFA should be defined as a percentage of the total fatty acids in conjunction with the amount of lipids in the diet. Skalli and Robin (2004) estimated that the requirement of 0.70% n-3 HUFA DM should be considered in experimental conditions of a DHA:EPA ratio of 1.5 and a diet that contains 18% of lipids.

Fish are not a rich natural source of conjugated linoleic acid, but in several species, the dietary supplementation resulted in high levels of deposition in muscle (Twibell et al. 2000, Twibell et al. 2001, Berge et al. 2004, Kennedy et al. 2005, Bandarra et al. 2006, Valente et al. 2007a,b). Valente et al. (2007a,b) results indicate that, as freshwater species, the European sea bass can incorporate CLA liver and muscle up to 2%, contributing to the production of a functional food.

Carbohydrates

The complexity of carbohydrate molecules affects their use by fish. For example, the starch proved to be used more efficiently than glucose in species of freshwater and marine (Enes et al. 2006a,b).

The European sea bass does not appear capable of dealing efficiently with high dietary levels of carbohydrate (30%). According to the data of Enes et al. (2010), regulation of the metabolism of carbohydrates was only effective at levels of the diet low in carbohydrates (10%). Alliot et al. (1979b) observed a protein sparing effect for glucose, although depression of growth occurred at high levels of inclusion in the diet. Despite the differences in digestibility of carbohydrates, Alliot et al. (1984) showed a similar growth of sea bass fed diets including maltose or starch. Although the effect of gelatinization of the starch on growth performance is somewhat discordant, it is well established that the gelatinization of starch improves its digestibility (Dias et al. 1998, Peres and Oliva-Teles 2002). Also, in sea bass, the digestibility of cornstarch (waxy) cooked was greater than normal cornstarch (Enes et al. 2006a).

The partial substitution of dietary protein for carbohydrate, in the form and right levels, spares protein without affecting the performance of growth in fish farming, which has also been proven in sea bass (Pérez et al.

1997, Dias et al. 1998, Dias et al. 2003, Peres and Oliva-Teles 2002, Enes et al. 2006b, Pérez-Jiménez et al. 2007). This may be related to glucose as the oxidative substrate choice for blood and nerve tissue cells and also because the carbohydrate present in the diet can depress the gluconeogenic activity, through oxidative pathways of amino acids (Cowey et al. 1977, Sanchez-Muros et al. 1996).

A protein sparing effect of the starch pregelatinizado was observed in sea bass, (Spyridakis et al. 1986) at a rate of 11% inclusion and by Hidalgo and Alliot (1988) at a higher level (15%). The use of cooked starch (up 18.1% inclusion rate) allows to use lower lipid levels in diets of sea bass (Lanari et al. 1999). For the European sea bass, there is evidence that the incorporation of 20% to 25% digestible carbohydrates in the diet does not affect growth or food efficiency (Lanari et al. 1999, Peres and Oliva-Teles 2002).

Vitamins and Minerals

The quantitative data of European sea bass vitamin requirements remains low. The requirements of the water-soluble vitamins do not differ from those established for salmonids (Kaushik et al. 1998).

Fournier et al. (2000) have indicated in juvenile sea bass, a much lower requirement of vitamin C than the study cited by Kaushik (2002) for growth, but more to maximize the concentration of ascorbate in the liver.

Ascorbic acid (AA) is involved in the occurrence of skeletal deformities in fish, most of them associated with scurvy and represented chiefly by atrophy of opercle and abnormal development of the cartilage in the gills (Soliman et al. 1986, Halver 1989, Hilomen-Garcia 1997). Many of these shortcomings of AA signs can be attributed to the deterioration of collagen and formation of cartilage of support in most tissues (Halver et al. 1975, Terova et al. 1998). The larvae need a minimum amount of dietary AA to survive during the larval period (Darias et al. 2011). AA levels higher than those required for growth are needed to meet the demands of other nutrients

Table 3. Vitamin requirement values for diverse physiological conditions of *D. labrax*.

Vitamin	Requirement	Physiological conditions	Authors
C	20 mg kg^{-1} dry matter	in weaning (35 days)	Merchie et al. 1996
C	<50 mg kg^{-1} diet	for juvenile growth	Kaushik 2002
C	30 mg kg^{-1} diet	for maximun larval growth	Darias et al. 2011
A	35 mg kg^{-1} dry matter	for skeletal development	Villeneuve et al. 2005
E	500 mg kg^{-1} diet	against peroxidative damages	Messager et al. 1992

such as vitamin E. AA regenerates vitamin E in the liver and amounts of AA above the necessary in diet produce an increase of the protection of fish from the deficiency of vitamin E (Hamre et al. 1997).

It was also noted that different levels of AA induce different percentage and type of malformations. There were specific deformities for low and high levels of AA (Darias et al. 2011). The number of vertebrae was also modulated by AA levels in European sea bass larvae. Some specimens fed with low levels of AA lacked a vertebra, while some larvae that fed the highest level tested AA presented one extra vertebrae.

The AA in the diet influenced the degree of bone mineralization and the amount of cartilage formation. Skeletal elements that developed early also were similarly affected by both low and high levels of AA, while the skeletal elements formed later were less sensitive to the amount of this vitamin.

The levels of dietary retinol for the optimal development of European sea bass larvae depend largely on the evaluated skeletal elements (Mazurais et al. 2009). The frequency of deformities of the jaw and the hyoid is minimized with levels of retinol in diet below 9402 IU kg^{-1} DM, but the vertebral elements and fins develop better when levels of retinol are equal or higuer to 9402 IU kg^{-1} DM. This study suggests that vitamin A of maternal origin is sufficient to maintain normal development of European sea bass until the total consumption of lipid reserves (5–8 days after the end of the larval stage of yolk sac) (Koumoundouros et al. 2001) and suggests an effect of age post-hatching on the response of the European sea bass larvae to dietary retinol. The composition of diets (especially the levels of vitamin A and fatty acids) for the larvae of European sea bass has a particularly decisive effect on skeletal malformations within 13 days post hatching (Villeneuve et al. 2006).

Tocopherols are widely distributed in animal tissues and act as potent antioxidants. Probably, its presence in biological membranes represents an important defense against reactive oxygen species (ROS) damage system and membrane lipid peroxidation (Glascott and Farber 1999). The importance of vitamin E (mainly α-tocopherol) plays an important role in biological processes (Guerriero et al. 2004). There is experimental evidence on the different effects of deficiency of vitamin E and its degree of severity in different animal species (Montero et al. 2001, Sahin et al. 2002, Guerriero et al. 2004). An important relationship has been found between vitamin E in plasma and body mass in both sexes of farmed *D. labrax*. Given that the consumption of vitamin E depends on the entity of oxidative phenomena, it could be that the fish with body of lesser mass have higher rates of production of ROS (Enes et al. 2010). These authors have suggested that vitamin E might be involved in different aspects of the reproduction of sea bass.

The interaction between vitamin D and C is evident in the sea bass (Darias et al. 2011). Vitamin D works through the receiver of vitamin D (VDR), inducing the expression of various unions of calcium and protein transport in the intestine to stimulate active calcium absorption, thus preserving the normocalcemia and indirectly, keeping up bone mineralization. In addition, vitamin D also acts directly on the osteoblasts, the resident bone forming cells of the skeleton, to inhibit proliferation, modulate the differentiation and regulate the extracellular matrix mineralization (Sutton et al. 2005).

Sea bass larvae are extremely sensitive to the dietary level of vitamin D_3 and show the impact of vitamin D_3 of the diet in the ontogenesis of the digestive system of sea bass. The results clearly indicated that levels of UI 27.6/g induce a faster maturation of the digestive function but vitamin D_3 requirements during the development of sea bass larvae vary depending on the state of differentiation of bone cells. Thus, on the one hand, a diet deficient in vitamin D_3 produces a delay in growth, development of the digestive system and esqueletogenesis, causing the appearance of disorders of the skeleton at the end of the larval period. On the other hand, higher doses of 27.6/g IU do not induce better larval performance. The appropriate level of vitamin D_3 for optimal larval development is within a restricted range and variations could trigger serious physiological disorders such as delayed maturation of intestinal functions with negative effects on the absorption of Ca^{+2} and bone mineralization, leading to the appearance of skeletal deformities (Darias et al. 2010).

There is a clear lack of data about the mineral requirements of European sea bass. The availability of phosphorus has received some attention (Santinha et al. 1996). Bioavailability is greater in animal feed than in vegetables. Kaushik (2002) cited unpublished data suggesting that the requirements would be about 0.6%. The possible improvement of the availability of phosphorus in plants feeds by exogenous phytase has also been demonstrated (Oliva-Teles et al. 1998, Fortes-Silva et al. 2011b).

Probiotics

Probiotics are microbial cells that manage to enter and stay alive in the gastrointestinal tract, with the aim of improving the health and the quality of farmed fish (Gatesoupe 1999).

Tovar et al. (2002) incorporated living yeast in pellets supplied to larvae of sea bass, from the mouth opening phase. *Listeria debaryomyces* produced a greater maturity of the digestive tract in marine fish larvae. The survival of the larvae and their skeletal development are enhanced by diets that contain yeast. Tovar-Ramírez et al. (2004) also noted a marked effect on the growth of fish larvae for the first time. The incorporation of large amounts of inactive yeast in diet of fish was tested as an alternative source of protein, but no significant improvement in growth was reported (Métailler and Huelvan 1993, Oliva-Teles and Gonçalves 2001). The beneficial effects of yeast on sea bass larvae development could be attributed to the role of polyamines (spermine and spermidine) in promoting intestinal maturation and increasing the capacity of enterocytes to absorb nutrients. Bardócz et al. (1993) argued that the requirements of polyamines are higher in young animals during periods of intensive growth. However, the results of Tovar-Ramírez et al. (2004) indicate that yeast *D. Listeria* population should not exceed approximately 1.1×10^4 colony units forming (CFU) per larva to optimum effect in the development of sea bass.

Lactic acid bacteria, *Lactobacillus delbrueckii delbrueckii*, has also been used as probiotic in juvenile sea bass to test stress responses and growth factors (IGF-I and myostatin) (Carnevalli et al. 2006). Fish treated with this probiotic showed significantly lower cortisol, as well as increased IGF-I and decreased miostatin expression levels, which explained the higher growth performanced observed with respect to the control.

DIETARY SELECTION

Nutritional Framework: A New Tool to Investigate Fish Nutrition

Animals eat food items containing dozens of different types of molecules that provide nutrients for their survival, growth and reproduction. Thus, most animals developed an extraordinary diversity of means and challenges to get adequate nutrients for their growth. The application of new behavioural concepts will be required to understand why fish behave the way they do when displaying a particular feeding behaviour and dietary selection, and their translation into aquaculture solutions (Raubenheimer et al. 2012). As feeding strategies have evolved, the animals have acquired mechanisms of food intake regulation (e.g., stomach distension and perception of the energetic content of the food) and even complex mechanisms such as "specific hunger" to regulate the intake of particular micronutrients. Hypothetically, on one hand an animal adapted to feeding on a single type of food (i.e., "invariant nutrient composition") just needs to regulate the amount of food it has to eat to ensure a sound nutritional regulation, for example using stretch receptors of the intestine (Simpson and Raubenheimer 2001). On the other hand, an animal that has evolved in an environment spatially and temporally heterogeneous, composed of numerous types of foods that vary greatly in their composition, is forced to feed on a combination of (suboptimal) diets, regulating the total amount of each food or nutrient consumed over time. These two feeding scenarios have been depicted in Fig. 1. Some authors prefer the term "nutritional wisdom",

Nutritional wisdom

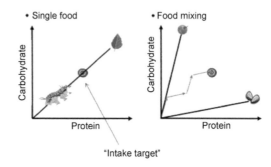

Figure 1. Nutritional framework. In nature, there are different nutritional scenarios: a single food that provides an adequate proportion of nutrients (left), where the animals simply regulate the intake of food, and food mixing (right), where in order to meet the nutritional requirements the animal must choose between two or more unbalanced diets. In the latter case, some rules of compromise will be established, so that excess of carbohydrate intake might increase with the extent to which protein is under eaten and *vice-versa*.

for the ability of the fish to regulate nutrient intake (Sánchez-Vázquez et al. 1994, Sánchez-Vázquez et al. 1995a, Simpson and Raubenheimer 2001, Fortes-Silva et al. 2010, Fortes-Silva et al. 2011a). This term was first used with reference to animals that actively search for specific food, and thus maintain homeostasis. This concept was also used to explain the appetite of cattle for sodium (Katz 1937). Thus, an animal is viewed as moving through a multidimensional nutrient space, which is bounded by axes representing each required nutrient and within which lie optimal points of intake and nutrient allocation ("targets") (Simpson and Raubenheimer 1997). In insects, such compensatory responses indicate that even if the insects were 'hardwired' to leave time for other activities and to minimize time spent feeding so as to avoid predation, and accepting that time is required for the efficient digestion of food, the insects were clearly regulating intake rather than maximizing consumption (Simpson et al. 2004).

Innovative technologies for fish farming often include multiple feeding alternatives, especially in farms that employ multistage fish growth processes. Using the animal as guide to develop new, nutritionally balanced diets adds a new concept to the description of complex systems such as aquaculture farms. Moreover, this innovative thinking seeks to answer the following questions from the perspective of the animal: what woud the animal like to eat?

The emergence of concerns for optimal conditions for fish has also stimulated interest in theoretical and philosophical problems related to fish welfare and whether fish are, in fact, sentient creatures. According to Volpato et al. (2007), the identification of the internal state of welfare is still a challenge. Moreover, according to these authors the productivity measures (e.g., biomass produced) may not match welfare requirements. Practical diets for fish generally tend to contain low or high percentages of nutrients to reduce feed costs while sustaining growth performance (Borghesi et al. 2008, Zhao et al. 2009). Since the internal welfare state of an organism cannot be defined in physiological terms, preference (dietary choice) tests may be the only means to determine the wellbeing of fish (Dawkins 2006).

How Can we Study Dietary Selection in Fish?

Studies on nutrient requirements are dificult and vary in quality due to the scientific approach and nutrient analysis (Waagbo 2010). Classic research on nutrition has faced challenges with feed processing and storage stability, live feed enrichment, leaching, bioavailability, nutrient interactions and a situation-dependent increase in requirements. Such conditions may lead to suboptimal nutrition with potential consequences for fish (Waagbo 2010).

The use of self-feeders in nutrition research provides an easier and less expensive way to investigate the food supply strategies (Alanara 1996) and the nutritional target of fish (Sánchez-Vázquez et al. 1999, Aranda et al. 2000). How fish regulate the intake and the use of macronutrients, depending on developmental stages and their environmental conditions, is of special importance to aquaculturists, since this knowledge enables them to avoid inefficient feeding regimes and poor growth of fish (Atienza et al. 2004). In a self-feeding system, it is assumed that the fish themselves accurately control the feeding level by actuating a trigger in the water. Thus, in such a system there is no need for any feedback control since any decrease in feeding motivation (appetite) is immediately manifested as a reduction in the actuation of the trigger (Alanara 1996). Basically, a pendulum (trigger) is attached to the gate, the tip of which extends down into the water. Once the fish actuates (presses or bites) a trigger positioned in or slightly above the water, food is delivered (Fig. 2).

Research on self-feeding activity was pioneered using sea bass (Sánchez-Vázquez et al. 1994). The signal from the control unit to the feeder can be fed to a counter or computer, making it possible to register the feeding activity in each rearing unit. One prerequisite for a well-functioning self-feeding system is that a single trigger actuation provides more than one fish with food; i.e., one fish releases food for several others. The resulting level of reward must correspond with the number of food particles that the fish are capable of catching and digesting before the food moves out of reach

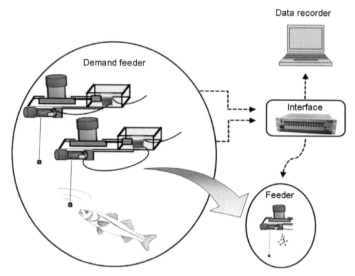

Figure 2. Schematic diagram showing the arrangement of the string sensor used to activate self-feeders and the connections between food demand sensor, interface-microcomputer and dispenser.

(e.g., passes out of the rearing system) (Alanara 1996). There are, however, several disadvantages with this type of system: the gate mechanism often remains open or gets jammed; feeding can easily be induced by accident, for example, by fish swimming through the gate by mistake or by wave action (Shepherd and Bromage 1989); it is difficult to finely adjust the size of each food portion. However, the unintentional activation may be reduced or avoided by suspending the sensor above the water surface (Boujard et al. 1991) or by protecting it with a cylindrical screen (Covès et al. 1998).

Another question that should be raised is the percentages of macronutrient use in self-selection experiments. Studies with self-feeders can only provide fish with pellet feed in pairs of macronutrients with different chemosensory properties. If fish are simultaneously provided with three feeds made of paired macronutrients (e.g., 50% protein + 50% fat; 50% protein + 50% carbohydrate; 50% fat + 50% of carbohydrate), all potential diets that fish can compose selecting from experimental diets is 50% of each macronutrient (Aranda et al. 2000) (Fig. 3).

Several mechanisms, involving both learning and physiological processes, have been proposed to explain the existence of food preferences. With regard to learning processes, it can be argued that once a food item is consumed, animals learn to associate its orosensory properties (taste, smell and texture) with its postingestive consequences. However, when the orosensorial characteristics of food are masked, food selection persists (Rubio et al. 2003). A feasible methodology to provide different macronutrients

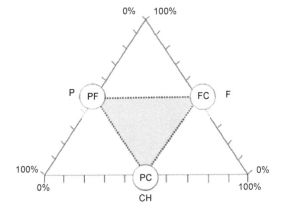

Figure 3. Schematic illustration of fish fed with paired macronutrients. Big circles represent relative proportion of macronutrient of the experimental diets. The gray areas indicate the composition (given as the percentage of each macronutrient) of all possible diets that fish can select combining experimental diets.

(Carbohydrate, fat and protein) with the same flavour and texture is that of packaging them into gelatine capsules. Again, this type of research was pioneered using sea bass. In order to test their ability to discriminate and differentially ingest macronutrients individually packaged into capsules, sea bass were fed with colour-coded capsules containing either protein, fat or carbohydrate in a floating container (Rubio et al. 2003) (Fig. 4).

It is also possible that the post-absorptive effects of ingesting an unbalanced diet somehow prompt brain centres, through still unknown mechanisms, to select different colour capsules (Rubio et al. 2001). Yet, the use of such a feeding system is restricted to a few species that can swallow whole capsules without masticating and tasting its content before ingestion (Ruohonen et al. 1997, Ruohonen and Grove 2001). However, existing data clearly indicate that European sea bass fulfils this requirement and, therefore, can be used as a suitable model for macronutrient self-selection studies using gelatine capsules (Rubio et al. 2003).

Both methodologies (self-feeders and capsules) lead to the conclusion that sea bass use their nutritional wisdom for dietary selection. An important reference point in any attempt to understand physiological, behavioural or evolutionary aspects of nutrition is the animal's nutritional requirements. These are depicted models as points in multi-dimensional nutrient space, where each dimension (axis) represents a required nutrient. Each food is represented by a linear nutritional *rail* that projects into nutrient space from the origin at an angle which is defined by the balance of the nutrients it contains. Thus, a nutritionally balanced food (i.e., one that contains the required proportion of the various nutrients) passes through the intake target (Simpson and Raubenheimer 1997). This is depicted in Fig. 5.

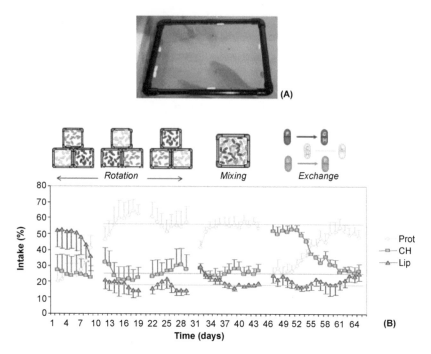

Figure 4. Illustration of dietary selection in fish. Different macronutrients encapsulated into three different colour capsules (A). Fish are able to discriminate and evaluate the quality of encapsulated diets through postingestive processes and establish a percentage of selection (B). This information can be used by the fish to develop an associative learning between the colour and nutrient content of gelatine capsules.

Color image of this figure appears in the color plate section at the end of the book.

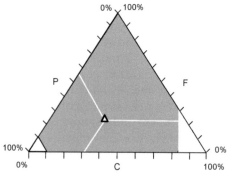

Figure 5. Schematic illustration of nutritional target by fish fed with encapsulated macronutrients. Provides examples of three-dimensional regulation for protein (P), fat (F) and carbohydrate intake (CH). The gray areas indicate the composition (given as the percentage of each macronutrient) of all possible diets that fish can select combining experimental diets. The white areas indicate the percents of other substances (e.g., minerals and vitamins) or wall of the capsules that fish cannot select.

Macro-/Micro-Nutrient Self-selection & Response to Nutritional Challenges

Using demand feeders and encapsulated macronutrients, the ability of European sea bass to choose the appropriate proportions of each macronutrient (protein, fat and carbohydrate) to self compose a nutritionally adequate diet has been observed in the following studies which applied different challenges to sea bass (Table 4).

Aranda et al. (2000) reviewed the use of four different experimental techniques in which sea bass could choose between two or three different diets through self-feeders, which gradually increased the complexity and potential range of selection to design their own diet in accordance with their requirements. First, sea bass were allowed to select between two complete diets differing in the proportion of protein (52%–58%): this showed their capacity to distinguish between two diets made of the same ingredients. Next, two incomplete diets, containing a fixed amount of protein (56%) and lacking either fat or carbohydrate, were made available. Three mixed diets made up of pairs of macronutrients (protein-carbohydrate, protein-fat or fat-carbohydrate) were tested in the next experiment and, finally, three diets containing only one macronutrient (protein fat or carbohydrate) were made available to fish. Taking into account the selection made by the fish in the first three experiments, in which macronutrient selection was statistically different, protein was the main macronutrient chosen by fish (278.15 kJ/kg BW/day, on average), followed by fat and carbohydrate (162.85 and 64.56 kJ/kg BW/day, respectively). According to these authors, the results reveal the ability of sea bass to select an appropriate diet from experimental

Table 4. Macronutrient self-selection by the sea bass.

Method		Nutrient delivery		Macronutrient (Protein, P; Fat, F; Carbohydrate, C)	Authors
Self-feeders	Capsules	Pure	Pairs		
√			√	58.8% P; 19.4% F; 21.8% C	Aranda et al. 2000
√			√	51% P, 32.5% F, 16.5% C	Aranda et al. 2001
	√	√		55% P, 22% F, 23% C	Rubio et al. 2003
√			√	65.3%P, 26.2% F, 8.4% C	Vivas et al. 2003
	√	√		37% P, 44% F, 19% C	Rubio et al. 2005a
	√	√		3.5 P, 1.8 F, 2.8C [1]	Rubio et al. 2005b
	√	√		38%P, 44%F, 18% C 46% P, 34% F, 20% C 33%P, 51%F, 16% C	Rubio et al. 2006a
	√	√		43% P, 40% F, 17%C	Rubio et al. 2006b
√			√	63%P, 16%F, 21% C	Vivas et al. 2006

[1] capsules 100 g BW/day

diets containing two or three macronutrients and suggest that the proposed methodology is a powerful tool for studying the differing nutritional needs of different species of fish.

The influence of two different fasting periods on macronutrient self-selection during refeeding was studied in sea bass (Aranda et al. 2001). Each aquarium was provided with three self-feeders containing three different diets made up of two macronutrients: protein-fat, protein-carbohydrate, and carbohydrate-fat. The fish selected a baseline diet of 51% protein, 16.5% carbohydrate, and 32.5% fat, in terms of digestible energy. The fish were then deprived of food for six days, after which they were permitted to refeed. When food demands stabilized, they were subjected to total food deprivation for another two weeks. The total energy demanded from all the macronutrients increased significantly during the first day after both fasting periods. The proportion of macronutrient self-selection after fasting periods remained unchanged in respect to the baseline, except for protein energy which remained high in the two first days of refeeding after the second fasting period (the 51% of total demanded energy from protein during the baseline period rising to 60%). In short, after two short fasting periods of different length, sea bass showed a compensatory ingestion of energy, with a slightly increased demand from protein after the longest fasting period.

Vivas et al. (2003) studied the effect of restricted availability of protein and fat on the feeding behaviour and macronutrient selection. According to those authors, sea bass select a diet consisting of 65.3% P, 26.2% F and 8.4% C. The composition of the self-selected diet did not differ when sea bass selected between the three self feeders (pairs of macronutrient) or protein provided separately. Moreover, when fish were deprived of protein for two weeks they were unable to sustain their previous energy intake. During fat restriction, the intake of the P+CH diet did not increase above the previous levels. Thus, the total energy intake was reduced, indicating that sea bass failed to regulate energy intake if fat was not available. This result explains the feeding habit of sea bass, because protein and fat are the most important dietary components for this specie.

For the first time, the role of the food orosensory properties on protein (P), fat (F) and carbohydrate (CH) self-selection was investigated in sea bass with a new methodology using gelatine capsules containing pure macronutrients (Rubio et al. 2003). In a sequence of experimental phases, sea bass were fed a pelleted complete diet, an encapsulated complete diet or a combination of separately encapsulated pure macronutrients. The composition of selected diet was 55% P, 23% CH and 22% F in terms of macronutrient percentage.

The impact of salinity on diet self selection has also been investigated in sea bass (Rubio et al. 2005b). Indeed, salinity is one of the most relevant environmental parameters with regard to fish physiology, modifying food intake and growth performance in many fish species. The aim of this study was to determine the effects of three salinity levels (25‰, 7‰ and 0‰) on total food intake and encapsulated macronutrient selection in sea bass. Regarding macronutrient selection, these salinity changes significantly decreased the percentage of CH intake by 31% and 27%, while increasing that of P by 30% and 25%, respectively. Fat selection remained unaltered, with an average value of 22% for all tested salinities (Rubio et al. 2005b).

Previous studies have also focused on the question of whether fish are able to detect and select micro nutritnts and food additives (Table 5). According to Hidalgo et al. (1988), European sea bass was not only able to select a balanced diet in methionine, but also detected levels of the nutrient considered adequate for the nutrition of the species. These authors conducted a study about self-selection of amino acids. The experimental design was to provide to animals, five different diets offered simultaneously to each of three different groups of fish under self feeding conditions. Diets differed only in methionine content (0.30, 0.65, 1.00, 1.35 and 1.70% of diet). The results revealed that small fish (0.2 g weight) prefer 1.35% methionine (considered as optimum). The authors concluded that sea bass can detect and discriminate small doses of methionine, showing preferences easily demonstrable with the self-feeding method.

According to Martínez et al. (2004), nutrient requirements of fish have been classically determined through the analytical investigation of the dose-response relationship, where graded levels of nutrients are included in diets and tested in parallel on fish at similar (but human prescribed) daily feeding rations and frequencies. Although not disputed, the dose-response methodology may also present some gaps and its effectiveness can be increased if associated to data of diet selectivity with demand feeders. When submitted to restricted diets, fish have no other choice than to adapt to the imposed feeding regimen, often based on the least bad response (Martínez et al. 2004). Those authors showed the effect of dietary taurine supplementation on survival, growth performance and feed preference of European sea bass fry when fed a fish/soybean meal-based diet with demand feeders. The taurine derives from methionine via cysteine and is not considered to be among the 10 indispensable amino acids. However, the ability of fish to synthesize taurine is species dependent and possibly affected by the stage of development. In the first experiment, fish were imposed one of four diets supplemented with taurine (0, 0.1, 0.2 or 0.3% taurine on a dry weight basis), while in the second they had simultaneous free access to the four diets supplemented with taurine. The results indicate that sea bass fry require a 0.2% taurine in the diet for better growth and have a clear preference for this nutrient level when fish meal and soybean meal are the primary sources of protein. According to Martínez et al. (2004), the positive correlation between growth performance and nutrient preference is discussed as a method for improving fry food formulation.

Some of the main concerns about food used in aquaculture are the antinutritional effects, such as phytic acid or phytate, present in vegetable flours. The presence of this component in the diet can cause a decrease in consumption and consequently weight loss among fish. Nevertheless, the enzyme phytase is capable of hydrolyzing phytic acid and promoting the release of chelated minerals in the diet. Fortes-Silva et al. (2011b) performed two experiments with the aim of examining the capacity of groups of European sea bass to differentiate between isoproteic diets containing soybean meal (SM0), soybean meal+1500 FTU/kg of phytase (SM1) and fish meal (FM) to detect any preference and also the effect of each diet on nutrient utilization and growth. Fish were divided into three self-selection treatments based on different diets (FM vs. SM0, T1); (FM vs. SM1, T2) and (SM0 vs. SM1, T3) using self-feeders (Experiments 1). Other experimental designs were performed to analyze body composition (P and Ca retention) and growth parameters (Experiment 2). In this case, fish were fed a single diet (FM or SM0 or SM1) in the same laboratory condition as Experiment 1. Averaged over all groups of Experiment 1, the preference tests demonstrated the capacity of European sea bass to discriminate between a diet with FM and a diet containing exogenous phytase SM1, their final choices being 15.8% SM0 vs. 84.2% FM; 26.7% SM1 vs. 73.3% FM; 9.8% SM0 vs. 90.2% SM1. In Experiment 2, the partial replacement of fish meal by soybean meal in diets (SM0 and SM1) resulted in reduced weight gain and lower specific growth rate (SGR), while, P and Ca retention increased significantly in fish fed diets SM1 and FM. The authors believe that the metabolic consequences of the plant meal intake (SM0) negatively affected the feed acceptance in the studied species, and that exogenous phytase increased the acceptance of the plant diet. These results of mineral utilization by sea bass had a behavioral relationship with the beneficial effect of fish meal diet and also with phytase added in the commercial diet, which is consistent with similar finding in phytase preference trials performed in other fish species (Fortes-Silva et al. 2010).

Table 5. Micronutrients and additives self-selection by the sea bass.

Method		Micronutrients and substances	Authors
Self-feeder	Capsules		
X		Methionine	Hidalgo et al. 1988
X		Taurine	Martínez et al. 2004
X		Phytase	Fortes-Silva et al. 2011

Regulation of Food Intake: What Food Signals use Fish to Make Dietary Selection?

The basic mechanism proposed to control the intake of protein in fish has been described in mammals in which the protein can be detected by gastrointestinal receptors during digestion (Rubio et al. 2003, Almaida-Pagán et al. 2006, Fortes-Silva et al. 2011a) and amino acids that can be detected in the liver (Bellinger et al. 1996) after digestion. These receptors would cause signals (neural and hormonal activity), informing brain centers about the nutritional properties of food, thus modifying eating behavior. With this information, called "nutritional reward", the animals learn to associate eating too much or too little of a particular nutrient with its metabolic consequences at post-ingestive and post-absorptive levels (Forbes 2001). The metabolic consequences due to consumption of an unbalanced diet drive the individual to have a characteristic behavior of diet selection. According to Peters and Harper (1984), protein intake in young rats is not regulated at a constant proportion of total calories, but is controlled between a minimum level that will support rapid growth through the availability of essential amino acids and a maximum that, if exceeded, the animal would suffer some substantial metabolic consequences. Likewise, according to Simpson and Raubenheimer (2001), a fish that meets its goal of each specific nutrient intake will provide its tissues with optimal concentrations of nutrients for proper growth and reproduction. Thus, there are strong evidences in the literature that most vertebrates have some ability to select a balanced diet of nutrients by post-ingestive mechanisms in an attempt to control the adverse effects of an unbalanced diet. The existence of these mechanisms, as well as the robustness of the behavioural outcomes, implies that selection has operated on these animals to avoid both nutrient surpluses and déficits (Simpson et al. 2004).

Nutrient preloads are used to study food intake regulation because they are known to reduce food intake and modify macronutrient selection patterns in mammals. Rubio et al. (2006a) conducted a study with the objective to determine the effect of orally administered macronutrient preloads of protein (P), fat (F) or carbohydrate (CH) on the subsequent macronutrient selection, using for the purpose, feed consisting of CH, P or F packaged separately in gelatin capsules. According to those authors, the macronutrient preloads left the total food intake unaltered, but caused differential changes in the pattern of macronutrient selection. The CH preload increased the selection of CH (39%) and decreased that of P (20%), independently of the fish's previous nutritional preferences. The F preload induced an F increase (32%) and a P decrease (18%) in P-preferring fish, but not in F-preferring fish in which the macronutrient selection pattern remained unaffected. The P preload stimulated F selection by 42% in P-preferring fish, but left the macronutrient selection pattern unchanged in F-preferring fish. Thus, oral macronutrient preloads affected the pattern of macronutrient selection in sea bass, acting by post-ingestive mechanisms.

In all vertebrates, the regulation of consumption, appetite, and body weight is a complex phenomenon that involves elaborate interactions between the brain and peripheral signals. The brain, especially the hypothalamus, produces key factors that either stimulate (orexigenic) or inhibit (anorexigenic) the intake of food. These factors can be directly related to search or rejection of a particular food and turn in feeding behavior. The hypothalamus is continually informed about nutritional, energetic and environmental status of the body by anorexigenic and orexigenic messages of central and peripheral systems. The peripheral feedback signals include nerve impulses, GIT peptides, leptin, cortisol, glucose and insulin. These signals are integrated into the hypothalamus, as monoamines and neuropeptides transmit signals from the central system (Kulczykowska and Sanchez-Vazquez 2010).

The effect of regulatory peptides on dietary selection has also been investigated in sea bass. For instance, serotonin (5-HT) reduces food intake in mammals and fish and modifies the macronutrients selection pattern in mammals. Rubio et al. (2006b) conducted a study with the objective to determine the effect of orally administered serotonin (0.1, 0.5 and 2.5 mg kg BW^{-1}) into gelatine capsules on the subsequent macronutrient selection of sea bass, using for this purpose gelatine capsules including carbohydrates, protein, or lipids separately. The results showed that the voluntarily ingested 5-HT was released into the plasma of fish, reaching a level two times greater than the controls, 45 minutes after the ingestion of a capsule containing 2.5 mg kg BW^{-1} of 5-HT. The indoleamine, at doses of 0.1, 0.5 and 2.5 mg kg BW^{-1}, produced a reduction in total food intake of 31%, 49% and 37%, respectively, compared to the baseline, modifying the macronutrient selection pattern. The percentage of fat selected was significantly reduced whereas the percentage of protein significantly increased after administration of highest dose, but no changes were observed in the proportion of carbohydrate for any 5-HT doses. The authors confirm that the oral administration of 5-HT affected both amount of food intake and pattern of macronutrients selected.

In fish, a series of peptides homologous to mammals have been isolated or their sequence deduced from cloned cDNA sequences. These peptides include cholecystokinin, CCK (Peyon et al. 1998), bombesin (Volkoff et al. 1999), neuropeptide Y (Blomqvist et al. 1992, Cerdá-Reverter et al. 2000), melanin concentrating hormone (Baker et al. 1995), galanin (Anglade et al. 1994, Unniappan et al. 2002, Wang and Conlon 1994), corticotropin releasing factor (Bombardelli et al. 2006) and orexins (Kaslin et al. 2004). Information on the role of these neuropeptides in the control of food intake and its mechanism of action as well as the regulation of eating behavior will be discussed elsewhere in this book.

Conclusion

The review presented here shows a new approach to the study of nutrition and the importance of sea bass behavior in relation to endogenous effects provided by food. Using the animal as our "guide" to formulate diets, we can use the vast complexity of nutritional space to reach optimized solutions of diet formulation, consumption regulation and welfare in sea bass.

References

Ai, Q., K. Mai, W. Zhan, W. Xu, B. Tan, C.H. Zhang and H. Li. 2007. Effects of exogenous enzymes (phytase, non-starch polysaccharide enzyme) in diets on growth, feed utilization, nitrogen and phosphorus excretion of Japanese seabass, *Lateolabrax japonicus*. Comp. Biochem. Physiol. Part A 147: 502–508.

Alanara, A. 1996. The use of self-feeders in rainbow trout (*Oncorhynchus mykiss*) production. Aquaculture 145: 1–20.

Alliot, E., A. Pastoureaud, J. Pelaez Hudlet and R. Métailler. 1979a. Utilisation des farines végétales et des levures cultivées sur alcanes pour l'alimentation du bar *Dicentrarchus labrax*. Influence sur la croissance et la composition corporelle. pp. 229–238. *In*: E. Halver and K. Tiews (eds.). Proc. World Symp. on Finfish Nutrition and Fishfeed Technology, Hamburg, Germany, vol. I. Heenemann, Berlin.

Alliot, E., A. Pastoureaud and J. Nedelec. 1979b. Etude de l'apport calorique et du rapport calorico-azoté dans l'alimentation du bar *Dicentrarchus labrax*. Influence sur la croissance et la composition corporelle. pp. 241–255. *In*: J.E. Halver and K. Tiews (eds.). Proceedings of the World Symposium on Finfish Nutrition and Fishfeed Technology. Heenemann Verlagsgesellschaft, Berlin, Germany.

Alliot, E., A. Pastoureaud and H. Thebault. 1984. Amelioration des formules d'aliments artificiels chez le loup (*Dicentrarchus labrax*): II. Utilisation dês glucides. Rech. Biol. Aquac. 1: 87–94.

Almaida-Pagán, P.F., C.V. Rubio, P. Mendiola, J. De Costa and J.A. Madrid. 2006. Macronutrient selection through post-ingestive signals in sharpsnout seabream fedgelatine capsules and challenged with protein dilution. Physiol. Behav. 88: 550–558.

Ambardekar, A.A., R.C. Reigh and M.B. Williams. 2009. Absorption of amino acids from intact dietary proteins and purified amino acid supplements follows different time-courses in channel catfish (*Ictalurus punctatus*). Aquaculture 291: 179–187.

Anglade, I., Y. Wang, J. Jensen, G. Tramu, O. Kah and J.M. Conlon. 1994. Characterization of trout galanin and its distribution in trout brain and pituitary. J. Comp. Neurol. 350: 63–74.

Aranda, A., F.J. Sánchez-Vázquez, S. Zamora and J.A. Madrid. 2000. Self-design of diets by means of self-feeders: validation of procedures. J. Physiol. Biochem. 56: 155–166.

Aranda, A., F.J. Sánchez-Vázquez and J.A. Madrid. 2001. Effect of short-term fasting on macronutrient self-selection in sea bass. Physiol. Behav. 73: 105–109.

Atienza, M.T, S. Chatzifotis and P. Divanach. 2004. Macronutrient selection by sharp snout seabream (*Diplodus puntazzo*). Aquaculture 232: 481–491.

Baker, B., A. Levy, L. Hall and S. Lightman. 1995. Cloning and expression of melanin-concentrating hormone genes in the rainbow trout brain. Neuroendocrinology 61: 67–76.

Ballestrazzi, R., D. Lanari., E. D'Agaro and A. Mion. 1994. The effect of dietary protein level and source on growth, body composition, total ammonia and reactive phosphate excretion of growing sea bass (*Dicentrarchus labrax*). Aquaculture 127: 197–206.

Bandarra, M., M.L. Nunes, A.M. Andrade, J.A.M. Prates, S. Pereira, M. Monteiro, P. Rema and L.M.P. Valente. 2006. Effect of dietary conjugated linoleic acid on muscle, liver and visceral lipid deposition in rainbow trout juveniles (*Oncorhynchus mykiss*). Aquaculture 254: 496–505.

Bardócz, S., G. Grant, D.S. Brown, A. Ralph and A. Pusztai. 1993. Polyamines in food implications for growth and health. J. Nutr. Biochem. 4: 66–71.

Bell, J.G., J.D. Castell, D.R. Tocher, F.M. MacDonald and J.R. Sargent. 1995a. Effects of different dietary arachidonic acid: docosahexaenoic acid ratios on phospholipid fatty acid compositions and prostaglandin production in juvenile turbot (*Scophthalmus maximus*). Fish Physiol. Biochem. 14: 139–151.

Bell, J.G., B.M. Farndale, M. Bruce, J.M. Navas and M. Carrillo. 1997. Effects of broodstock dietary lipid on fatty acid composition of eggs from sea bass (*Dicentrarchus labrax*). Aquaculture 149: 107–119.

Bell, M.V., R.S. Batty, J.R. Dick, K. Fretwell, J.C. Navarro and J.R. Sargent. 1995b. Dietary deficiency of docosahexaenoic acid impairs vision at low-light intensities in juvenile herring (*Clupea harengus* L.). Lipids 30: 443–449.

Bellinger, L.L., F.E. Williams, Q.R. Rogers and D.W. Gietzen. 1996. Liver denervation attenuates the hypophagia produced by an imbalanced amino acid diet. Physiol. Behav. 59: 25–929.

Berge, G.M., B. Ruyter and T. Asgard. 2004. Conjugated linoleic acid in diets for juvenile Atlantic salmon (*Salmo salar*); effects on fish performance, proximate composition, fatty acid and mineral content. Aquaculture 237: 365–380.

Blomqvist, A.G., C. Soderberg, I. Lundell, R.J. Milner and D. Larhammar. 1992. Strong evolutionary conservation of neuropeptide Y: sequences of chicken, goldfish, and Torpedo marmorata DNA clones. Proc. Natl. Acad. Sci. 89: 2350–2354.

Bombardelli, R.A., M.A. Syperreck and E.A. Sanches. 2006. Hormônio Liberador De Gonadotrofinas em Peixes: Aspectos Básicos E Suas Aplicações. Arq. Ciên. Vet. Zool. 9: 59–65.

Borghesi, R., L. Potz, M. Oetterer and J.E.P. Cyrino. 2008. Apparent digestibility coefficient of protein and amino acids of acid, biological and enzymatic silage for Nile tilapia (*Oreochromis niloticus*). Aquacult. Nutr. 14: 242–248.

Boujard, T., Y. Moreau and P. Luquet. 1991. Entrainment of the circadian rhythm of feed demand by infradial cycles of light–dark alternation in Hoplosternum littorale (Teleostei). Aquat. Living Resour. 4: 221–225.

Bruce, M., F. Oyen, G. Bell, J.F. Asturiano, B. Farndale, M. Carrillo, S. Zanuy, J. Ramos and N. Bromage. 1999. Development of broodstock diets for the European Sea Bass (*Dicentrarchus labrax*) with special emphasis on the importance of n-3 and n-6 highly unsaturated fatty acid to reproductive performance. Aquaculture 177: 85–97.

Cahu, C., J.L. Zambonino-Infante, P. Quazuguel and M.M. Le Gall. 1999. Protein hydrolysate vs. fish meal in compounds diets for 10-day-old seabass *Dicentrarchus labrax* larvae. Aquaculture 171: 109–119.

Cahu, C., J. Zambonino-Infante and T. Takeuchi. 2003. Nutritional components affecting skeletal development in fish larvae. Aquaculture 227: 245–258.

Carnevalli, O., L. de Vivoa, R. Sulpizioa, G. Gioacchinia, I. Olivottoa, S. Silvib and A. Crescib. 2006. Growth improvement by probiotic in European sea bass juveniles (*Dicentrarchus labrax* L.), with particular attention to IGF-1, myostatin and cortisol gene expression. Aquaculture 258: 430–438.

Castell, J.D., J.G. Bell and D.R. Tocher. 1994. Effects of purified diets containing different combinations of arachidonic and docosahexaenoic acid on survival, growth and fatty acid compositionof juvenile turbot (*Scophthalmus-maximus*). Aquaculture 128: 315–333.

Cerdá-Reverter, J.M., G. Martinez-Rodriguez, S. Zanuy, M. Carrillo and D. Larhammar. 2000. Molecular evolution of the neuropeptide Y (NPY) family of peptides: cloning of three NPY-related peptides from the sea bass (*Dicentrarchus labrax*). Regul. Pept. 95: 25–34.

Cho, C.Y. and S.J. Kaushik. 1990. Nutritional energetics in fish: energy and protein utilization in rainbow trout *Salmo gairdneri*. World Rev. Nutr. Diet. 61: 132–172.

Coutteau, P., G.V. Stappen and P. Sorgeloos. 1996. A standar experimental diet for the study of fatty acid requirements of weaning and first ongrowing stages of the European seabass *Dicentrarchus labrax* L.: comparison of extruded and extruded/coated diets. Arch. Anim. Nutr. 49: 49–59.

Covès, D., E. Gasset, G. Lemarié and G. Dutto. 1998. A simple way of avoiding feed wastage in European seabass, *Dicentrarchus labrax*, under self-feeding conditions. Aquat. Living Resour. 11: 395–401.

Cowey, C.B., M. De la Higuera and J.W. Adron. 1977. The effect of dietary composition and of insulin on gluconeogenesis in rainbow trout (*Salmo gairdneri*). Br. J. Nutr. 38: 385–395.

Dabrowski, K., M. Arslan, B.F. Terjesen and Y. Zhang. 2007. The effect of dietary indispensable amino acid imbalances on feed intake: Is there a sensing of deficiency and neural signaling present in fish? Aquaculture 268: 136–142.

Darias, M.J., D. Mazurais, G. Koumoundouros, N. Glynatsi, S. Christodoulopoulou, C. Huelvan, E. Desbruyeres, M.M. Le Gall, P. Quazuguel, C.L. Cahu and J.L. Zambonino-Infante. 2010. Dietary vitamin D_3 affects digestive system ontogenesis and ossification in European sea bass (*Dicentrachus labrax* Linnaeus 1758). Aquaculture 298: 300–307.

Darias, M.J., D. Mazurais, G. Koumoundouros, M.M. Le Gall, C. Huelvan, E. Desbruyeres, P. Quazuguel, C.L. Cahu and J.L. Zambonino-Infante. 2011. Imbalanced dietary ascorbic acid alters molecular pathways involved in skeletogenesis of developing European sea bass (*Dicentrarchus labrax*). Comp. Biochem. Physiol. Part A 159: 46–55.

Dawkins, M.S. 2006. Through animal eyes: What behaviour tells us. Appl. Anim. Behav. Sci. 100: 4–10.

Dias, J. 1999. Lipid deposition in rainbow trout (*Oncorhynchus mykiss*) and European seabass (*Dicentrarchus labrax* L.): nutritional regulation of hepatic lipogenesis. Dr thesis, Univ. Porto (Portugal) and Univ. Bordeaux I (France). pp. 190.

Dias, J., E.F. Gomes and S.J. Kaushik. 1997. Improvement of feed intake through supplementation with an attractant mix in European seabass fed plant-protein rich diets. Aquat. Living Resour. 10: 385–389.

Dias, J., M.J. Alvarez, A. Diez, J. Arzel, G. Corraze, J.M. Bautista and S.J. Kaushik. 1998. Regulation of hepatic lipogenesis by dietary protein/renergy in juvenile European sea bass (*Dicentrarchus labrax*). Aquaculture 161: 169–186.

Dias, J., J. Arzel, P. Aguirre, G. Corraze and S. Kaushik. 2003. Growth and hepatic acetyl coenzyme-A carboxylase activity are affected by dietary protein level in European seabass (*Dicentrarchus labrax*). Comp. Biochem. Physiol. Part B 135: 183–196.

Dias, J., M.J. Alvarez, J. Arzel, G. Corraze, A. Diez, J.M. Bautista and S.I. Kaushik. 2005. Dietary protein source affects lipid metabolism in the European seabass (*Dicentrarchus labrax*). Comp. Biochem. Physiol. A. Mol. Integr. Physiol. 142: 19–31.

Enes, P., S. Panserat, S. Kaushik and A. Oliva-Teles. 2006a. Effect of normal and waxy maize starch on growth, food utilization and hepatic glucose metabolism in European sea bass (*Dicentrarchus labrax*) juveniles. Comp. Biochem. Physiol. Part A 143: 89–96.

Enes, P., S. Panserat, S. Kaushik and A. Oliva-Teles. 2006b. Rapid metabolic adaptation in European sea bass (*Dicentrarchus labrax*) juveniles fed different carbohydrate sources after heat shock stress. Comp. Biochem. Physiol. Part A 145: 73–81.

Enes, P., J. Sanchez-Gurmaches, I. Navarro, J. Gutiérrez and A. Oliva-Teles. 2010. Role of insulin and IGF-I on the regulation of glucose metabolism in European sea bass (*Dicentrarchus labrax*) fed with different dietary carbohydrate levels. Comparative Biochemistry and Physiology Part A 157: 346–353.

FAO. 2010. The State of World Fisheries and Aquaculture. Food and Agriculture Organization of the United Nations, Rome, Italy.

Farkas, T., I. Dey, C.S. Buda and J.E. Halver. 1994. Role of phospholipid molecular species in maintaining lipid membrane structure in response to temperature. Biophys. Chemist. 50: 147–155.

Forbes, J.M. 2001. Consequences of feeding for future feeding. Comp. Biochem. Physiol. A 128: 463–70.

Fortes-Silva, R., F.J. Martínez, M. Villaroel and F.J. Sánchez-Vázquez. 2010. Daily feeding patterns and self-selection of dietary oil in Nile tilapia. Aquacul. Res. 42: 157–160.

Fortes-Silva, R., F.J. Martínez and F.J. Sánchez-Vázquez. 2011a. Macronutrient selection in Nile tilapia fed gelatin capsules and challenged with protein dilution/restriction. Physiol. Behav. 102: 356–360.

Fortes-Silva, R., F.J. Sánchez-Vázquez and F.J. Martínez. 2011b. Effects of pretreating a plant-based diet with phytase on diet selection and nutrient utilization in European sea bass. Aquaculture 319: 417–422.

Fournier, V., M.F. Gouillou-Coustans and S. Kaushik. 2000. Hepatic ascorbic acid saturation is the most stringent response criterion for determining the vitamin C requirement of juvenile European seabasss (*Dicentrarchus labrax*). J. Nutr. 130: 617–620.

Gatesoupe, F.J. 1999. The use of probiotics in aquaculture. Aquaculture 180: 147–165.

Geurden, I., P. Coutteau and P. Sorgeloos. 1997a. Effect of dietary phospholipid supplementation on growth and fatty acid composition of European sea bass (*Dicentrarchus labrax*) and turbot (*Scophtalmus maximus* L.) juveniles from weaning onward. Fish Physiol. Biochem. 16: 259–272.

Geurden, I., P. Coutteau and P. Sorgeloos. 1997b. Increased docosahexaenoic acid levels in total and polar lipid of European sea bass (*Dicentrarchus labrax*) post larvae fed vegetable or animal phospholipids. Mar. Biol. 129: 686–272.

Glascott Jr., P.A. and J.L. Farber. 1999. Assessment of physiological interaction between vitamin E and vitamin C. Methods Enzymol. 300: 78–88.

Glencross, B.D., D.M. Smith, M.R. Thomas and K.C.Williams. 2002. Optimising the essential fatty acids in the diet for weight gain of the prawn, *Penaeus monodon*. Aquaculture 204: 89–99.

Guerriero, G., R. Ferro, G.L. Russo and G. Ciarcia. 2004. Vitamin E in early stages of sea bass (*Dicentrarchus labrax*) development. Comp. Biochem. Physiol. Part A 138: 435–439.

Guimaraes, I.G., L.E. Pezzato and M.M. Barros. 2008. Amino acids availability and protein digestibility of several proteion source for Nile tilapia, *Oreochromis niloticus*. Aquacult. Nutr. 14: 396–404.

Halver, J.E. 1989. The vitamins. pp. 31–109. *In*: J.E. Halver (ed.). Fish Nutrition, 2nd edn. Academic Press, San Diego, CA.

Halver, J.E., R.R. Smith, B.M. Tolbert and E.M. Baker. 1975. Utilization of ascorbic acid in fish. Ann. N.Y. Acad. Sci. 258: 81–102.

Hamre, K., R. Waagbo, R.K. Berge and O. Lie. 1997. Vitamin C and E interact in juvenile Atlantic salmon. Free Radic. Biol. Med. 22: 137–149.

Hazel, J.R., S.J. McKinley and E.E. Williams. 1992. Thermal adaptation in biological membranes: interacting effects of temperature and pH. J. Comp. Physiol. B 162: 593–601.

Hidalgo, F. and E. Alliot. 1988. Influence of water temperature on protein requirement and protein utilization injuvenile sea bass, *Dicentrarchus labrax*. Aquaculture 72: 115–129.

Hidalgo, F., M. Kentouri and P. Divanach. 1988. The utilisation of a self-feeder as a tool for the nutritional study of sea bass, *Dicentrarchus labrax*—preliminary results with methionine. Aquaculture 68: 177–190.

Hilomen-Garcia, G.H. 1997. Morphological abnormalities in hatchery-bred milkfish. *Chanos chanos* Forsskal fry and juveniles. Aquaculture 152: 155–166.

Izquierdo, M.S., D. Montero, L. Robaina, M.J. Caballero, G. Rosenlund and R. Ginès. 2005. Alterations in fillet fatty acid profile and flesh quality in gilthead seabream (*Sparus aurata*) fed vegetable oils for a long term period. Recovery of fatty acid profiles by fish oil feeding. Aquaculture 250: 431–444.

Jobling, M., A. Alanara, C. Noble, F.J. Sanchez-Vazquez, S. Kadri and F. Huntingfrd. 2012. Appetite and feed intake. pp. 183–219. *In*: Aquaculture and Behaviour. F. Huntingfors, M. Jobling and S. Kadri (eds.). Wiley-Blackwell.

Kanazawa, A., S.I. Teshima and K. Ono. 1979. Relationship between essential fatty acid requirements of aquatic animals and the capacity for bioconversion of linolenic acid to highly unsaturated fatty acids. Comp. Biochem. Physiol. Part B 63: 295–298.

Kaslin, J., J.M. Nystedt, M. Ostergard, N. Peitsaro and P. Panula. 2004. The orexin/hypocretin system in zebrafish is connected to the aminergic and cholinergic systems. J. Neurosci. 24: 2678–2689.

Katz, D. 1937. Animals and Man. Studies in Comparative Psychology. Longmans, Green & Co., London.

Kaushik, S.J. 2002. European sea bass, *Dicentrarchus labrax*. pp 28–39. *In*: C.D. Webster and C.E. Lim (eds.). Nutrient Requirements and Feeding of Finfish for Aquaculture. CABI Publishing, UK.

Kaushik, S.J., J.P. Cravedi, J.P. Lalles, J. Sumpter, B. Fauconneau and M. Laroche. 1995. Partial or total replacement of fishmeal by soybean protein on growth, protein utilisation, potential estrogenic or antigenic effects, cholesterolemia and flesh quality in rainbow trout. Aquaculture 133: 257–274.

Kaushik, S.J., M.F. Gouillou-Coustans and C.Y. Cho. 1998. Application of the recommendations on vitamin requeriments of finfhis by NCR (1993) to salmoids and seabass using practical and purified diets. Aquauculture 161: 463–474.

Kaushik, S.J., D. Covès, G. Dutto and D. Blanc. 2004. Almost total replacement of fish meal by plant protein sources in the diet of a marine teleost, the European sea bass, *Dicentrarchus labrax*. Aquaculture 230: 391–404.

Kennedy, S.R., P.J. Campbell, A. Porter and D.R. Tocher. 2005. Influence of dietary conjugated linoleic acid (CLA) on lipid and fatty acid composition in liver and flesh of Atlantic salmon (*Salmon salar*). Comp. Biochem. Physiol. Part B 141: 168–178.

Kotzamanis, Y.P., E. Gisbert, F.J. Gatesoupe, J.L. Zambonino-Infante and C. Cahu. 2007. Effects of different dietary levels of fish protein hydrolysates on growth, digestive enzymes, gut microbiota, and resistance to Vibrio anguillarum in European sea bass (*Dicentrarchus labrax*) larvae. Comp. Biochem. Physiol. Part A 147: 205–214.

Koumoundouros, G., P. Divanach, L. Anezaki and M. Kentouri. 2001. Temperature-induced ontogenetic plasticity in sea bass (*Dicentrarchus labrax*). Mar. Biol. 139: 817–830.

Krogdahl, A., A.M. Bakke-McKellep and G. Baeverfjord. 2003. Effects of graded levels of standard soybean meal on intestinal structure, mucosal enzyme activities, and pancreatic response in Atlantic salmon (*Salmo salar* L.). Aquacult. Nutr. 9: 361–371.

Kulczykowska, E. and F.J. Sánchez-Vázquez. 2010. Neurohormonal regulation of feed intake and response to nutrients in fish: aspects of feeding rhythm and stress. Aquac. Res. 41: 654–667.

Lall, S.P. 1991. Digestibility, metabolism and excretion of dietary phosphorus in fish. pp. 21–30. *In*: C.B. Cowey and C.Y. Cho (eds.). Nutritional Strategies and Aquaculture Waste. University of Guelph, Canada.

Lanari, D., B.M. Poli, R. Ballestrazzi, P. Lupi, E. D'Agaro and M. Mecatti. 1999. The effects of dietary fat and NFE levels on growing European sea bass (*Dicentrarchus labrax* L.): Growth rate, body and fillet composition, carcass traits and nutrient retention efficiency. Aquaculture 179: 351–364.

Lupatsch, I., G.W. Kissil and D. Sklan. 2001. Optimization of feeding regimes for European sea bass *Dicentrarchus labrax*: A factorial approach. Aquaculture 202: 289–302.

Martínez, J.B., S. Chatzifotis, P. Divanach and T. Takeuchi. 2004. Effect of dietary taurine supplementation on growth performance and feed selection of sea bass *Dicentrarchus labrax* fry fed with demand-feeders. Fish. Sci. 70: 74–79.

Mazurais, D., N. Glynatsi, M.J. Darias, S. Christodoulopoulou, C.L. Cahu, J.L. Zambonino-Infante and G. Koumoundouros. 2009. Optimal levels of dietary vitamin A for reduced deformity incidence during development of European sea bass larvae (*Dicentrarchus labrax*) depend on malformation type. Aquaculture 294: 262–270.

Menoyo, D., M.S. Izquierdo, L. Robaina, R. Gines, C.J. Lopez-Bote and J.M. Bautista. 2004. Adaptation of lipid metabolism, tissue composition and flesh quality in gilthead sea bream (*Sparus aurata*) to the replacement of dietary fish oil by linseed and soybean oils. Br. J. Nutr. 92: 41–52.

Merchie, G., P. Lavens, Ph. Dhert, M.G.U. Gomez, H. Nelis, A. De Leenheer and P. Sorgeloos. 1996. Dietary ascorbic acid requirements during the hatchery production of turbot. J. Fish Biol. 49: 573–583.

Messager, J.L., G. Stéphan, C. Quentel and F.B. Laurencin. 1992. Effects of detary oxidized fish oil and antioxidant deficiency on histopathology, haematology, tissue and plasma biochemistry of seabass *Dicentrarchus labrax*. Aquat. Living Resour. 5: 205–214.

Métailler, R. and C. Huelvan. 1993. Uitilisation des levures dans l'alimentation du juvénile de bar (*Dicentrarchus labrax*). pp. 945–948. *In*: S.J. Kaushik and P. Luquet (eds.). Fish Nutrition in Practice Les Colloques, vol. 61. Institut National de la Recherche Agronomique, Paris.

Métailler, R., J.F. Aldrin, J.L. Messager, G. Mevel and G. Stephan. 1981. Feeding of European seabass (*Dicentrarchus labrax*): role of protein level and energy source. J. World Maricult. Soc. 12: 117–118.

Montero, D., L. Tort, L. Robaina, J.M. Vergara and M.S. Izquierdo. 2001. Low vitamin E in diet reduces stress resistance of gilthead seabream (*Sparus aurata*) juveniles. Fish Shellfish Immunol. 11: 473–490.

Montero, D., T. Kalinowski, A. Obach, L. Robaina, L. Tort, M.J. Caballero and M.S. Izquierdo. 2003. Vegetable lipid sources for gilthead seabream (*Sparus aurata*): effects on fish health. Aquaculture 225: 353–370.

Montero, D., L. Robaina, M.J. Caballero, R. Gines and M.S. Izquierdo. 2005. Growth, feed utilization and flesh quality of European sea bass (*Dicentrarchus labrax*) fed diets containing vegetable oils: a time-course study on the effect of re-feeding period with a 100% fish oil diet. Aquaculture 248: 121–134.

Morales, A.E. and A. Oliva-Teles. 1995. Resultados preliminares sobre la utilización metabólica de dietas con dos niveles lipidicos en lubina (*Dicantrarchus labrax*). pp. 570–575. *In*: F.C Orvay and A.C. Reig (eds.). Proc. of the V National Congress on Aquaculture, Sant Carles de la Ràpita, Spain.

Mourente, G. and J.R. Dick. 2002. Influence of partial substitution of dietary fish oil by vegetable oils on the metabolism (desaturationand h-oxidation) of [1-14C]18:3n-3 in isolated hepatocytes of European sea bass (*Dicentrarchus labrax* L.). Fish Physiol. Biochem. 26: 297–308.

Mourente, G. and J.G. Bell. 2006. Partial replacement of dietary fish oil with blends of vegetable oils (rapeseed, linseed and palm oils) in diets for European sea bass (*Dicentrarchus labrax* L.) over a long term growth study: Effects on muscle and liver fatty acid composition and effectiveness of a fish oil finishing diet. Comp. Biochem. Physiol. Part B 145: 389–399.

Mourente, G., J.R. Dick, J.G. Bell and D.R. Tocher. 2005a. Effect of partial substitution of dietary fish oil by vegetable oils on desaturation and h-oxidation of [1-14C]18:3n-3 (LNA) and [1-14C]20:5n-3 (EPA) in hepatocytes and enterocytes of European sea bass (*Dicentrarchus labrax* L.). Aquaculture 248: 173–186.

Mourente, G., J.E. Good and J.G. Bell. 2005b. Partial substitution of fish oil with rapeseed, linseed and olive oils in diets for European sea bass (*Dicentrarchus labrax* L.): effects on flesh fatty acid composition, plasma prostaglandins E_2 and F_{2v}, immune function and effectiveness of a fish oil finishing diet. Aquacult. Nutr. 11: 25–40.

Navarro, J.C., L.A. McEvoy, M.V. Bell, F. Amat, F. Hontoria and J.R. Sargent. 1997. Effect of different dietary levels of docosahexaenoic acid (DHA, 22:6n-3) on the DHA composition of lipid classes in sea bass larvae eyes. Aquacult. Int. 5: 509–516.

National Research Council (NRC). 2011. Nutrient Requirements of fish and shrimp. pp. 6.

Oliva-Teles, A. 2000. Recent advances in European sea bass and gilthead sea bream nutrition. Aquacult. Int. 8: 477–492.

Oliva-Teles, A. and P. Gonçalves. 2001. Partial replacement of fishmeal by brewers yeast (*Saccharomyces cerevisiae*) in diets for sea bass (*Dicentrarchus labrax*) juveniles. Aquaculture 202: 269–278.

Oliva-Teles, A., J.P. Pereira, A. Gouveia and E. Gomes. 1998. Utilisation of diets supplemented with microbial phytase by seabass (*Dicentrarchus labrax*) juveniles. Aquat. Living Resour. 11: 255–259.

Parpoura, A.C.R. and M.N. Alexis. 2001. Effects of different dietary oils in sea bass (*Dicentrarchus labrax*) nutrition. Aquacult. Int. 9: 463–476.

Peres, H. and A. Oliva-Teles. 1999a. Influence of temperature on protein utilization in juvenile European seabass (*Dicentrarchus labrax*). Aquaculture 170: 337–348.

Peres, H. and A. Oliva-Teles. 1999b. Effect of dietary lipid level on growth performance and feed utilization by European sea bass juveniles (*Dicentrarchus labrax*). Aquaculture 179: 325–334.

Peres, M.H. and A. Oliva-Teles. 2002. Utilization of raw and gelatinized starch by European sea bass (*Dicentrarchus labrax*) juveniles. Aquaculture 205: 287–299.

Pérez, L., H. Gonzalez, M. Jover and J. Fernández-Carmona. 1997. Growth of European sea bass fingerlings (*Dicentrarchus labrax*) fed extruded diets containing varying levels of protein, lipid and carbohydrate. Aquaculture 156: 183–193.

Pérez-Jiménez, A., M.J. Guedes, A.E. Morales and A. Oliva-Teles. 2007. Metabolic responses to short starvation and refeeding in *Dicentrarchus labrax*. Effect of dietary composition. Aquaculture 265: 325–335.

Person-Le Ruyet, J., A. Skalli, B. Dulau, N. Le Bayon, H. Le Delliou and J.H. Robin. 2004. Does dietary n-3 highly unsaturated fatty acids level influence the European sea bass (*Dicentrarchus labrax*) capacity to adapt to a high temperature? Aquaculture 242: 571–588.

Peters, J.C. and A.E. Harper. 1984. Influence of dietary protein level on protein self selection and plasma and brain amino acid concentrations. Physiol. Behav. 33: 783–90.

Peyon, P., X.W. Lin, B.A. Himick and R.E. Peter. 1998. Molecular cloning and expression of cDNA encoding brain preprocholecystokinin in goldfish. Peptides 19: 199–210.

Raubenheimer, D., S. Simpson, F.J. Sanchez-Vazquez, F. Huntingfrd, S. Kadri and M. Jobling. 2012. Nutrition and diet choice. Aquacul. Behav. 150–182.

Refstie, S. and T. Storebakken. 2001. Vegetable protein sources for carnivorous fish: potential and challenges. Recent Adv. Anim. Nutr. Australia 13: 195–203.

Refstie, S., Ø.J. Korsøen, T. Storebakken, G. Baeverfjord, I. Lein and A.J. Roem. 2000. Differing nutritional responses to dietary soybean meal in rainbow trout (*Onchorynchus mykiss*) and Atlantic salmon (*Salmo salar*). Aquaculture 190: 49–63.

Richard, N., G. Mourente, S. Kaushik and G. Corraze. 2006. Replacement of a large portion of fish oil by vegetable oils does not affect lipogenesis, lipid transport and tissue lipid uptake in European seabass (*Dicentrarchus labrax* L.). Aquaculture 261: 1077–1087.

Robaina, L., M.S. Izquierdo, F.J. Moyano, J. Socorro, J.M. Vergara, D. Montero and H. Fernandez-Palacios. 1995. Soybean and lupin seed meals as protein sources in diets for gilthead seabream (*Sparus aurata*): nutritional and histological implications. Aquaculture 130: 219–233.

Robertson, J.C. and J.R. Hazel. 1996. Membranes constraints to physiological function at different temperatures; does cholesterol stabilize membranes at elevated temperatures? pp. 25–49. *In*: C.M. Woods and D.G. McDonald (eds.). Society for Experimental Biology, Seminar Series 61: Global Warming: Implications for Freshwater and Marine Fish. Cambridge University Press.

Rubio, V.C., F.J. Sánchez-Vázquez and J.A. Madrid. 2001. Macronutrient self-selection in European sea bass using macroencapsulated nutrients. Fourth and Final Workshop of COST 827 action on Voluntary Food Intake in Fish, Reykjavik, pp. 44.

Rubio, V.C., F.J. Sánchez-Vázquez and J.A. Madrid. 2003. Macronutrient selection through postingestive signals in sea bass fed on gelatine capsules. Physiol. Behav. 78: 795–803.

Rubio, V.C., F.J. Sánchez-Vázquez and J.A. Madrid. 2005a. Fish macronutrient selection through post-ingestive signals: Effect of selective macronutrient deprivation. Physiol. Behav. 84: 651–657.

Rubio, V.C., F.J. Sánchez-Vázquez and J.A. Madrid. 2005b. Effects of salinity on food intake and macronutrient selection in European sea bass. Physiol. Behav. 85: 333–339.

Rubio, V.C., F.J. Sánchez-Vázquez and J.A. Madrid. 2006a. Influence of nutrient preload on encapsulated macronutrient selection in European sea bass. Physiol. Behav. 89: 662–669.

Rubio, V.C., F.J. Sánchez-Vázquez and J.A. Madrid. 2006b. Oral serotonin administration affects the quantity and the quality of macronutrients selection in European sea bass *Dicentrarchus labrax* L. Physiol. Behav. 87: 7–15.

Ruohonen, K. and D.J. Grove. 2001. *In situ* study of gastric emptying using X-radiography and macrocapsules. Aquac. Res. 32: 491–493.

Ruohonen, K., D.J. Grove and J.T. McIlroy. 1997. The amount of food ingested in a single meal by rainbow trout offered chopped herring, dry and wet diets. J. Fish Biol. 51: 93–105.

Sahin, K., N. Sahin and M. Onderci. 2002. Vitamin E supplementation can alleviate negative effects of heat stress on production, egg quality, digestibility of nutrients and egg yolk mineral concentrations of Japanese quails. Res. Vet. Sci. 73: 307–312.

Sánchez-Muros, M.J., L. García-Rejón, J.A. Lupiáñez and M. De la Higuera. 1996. Long-term nutritional effects on the primary liver and kidney metabolism in rainbow trout (*Oncorhynchus mykiss*). 2. Adaptive response of glucose-6-phosphate dehydrogenase activity to high carbohydrate/low-protein and high-fat/noncarbohydrate diets. Aquacult. Nutr. 2: 193–200.

Sánchez-Vázquez, F.J., M. Martínez, S. Zamora and J.A. Madrid. 1994. Design and performance of an accurate demand feeder for the study of feeding behaviour in sea bass, *Dicentrarchus labrax* L. Physiol. Behav. 56: 789–794.

Sánchez-Vázquez, F.J., S. Zamora and J.A. Madrid. 1995a. Circadian rhythms of feeding activity in sea bass, *Dicentrarchus labrax* L.: dual phasing capacity of diel demand-feeding pattern. J. Biol. Rhythms 10: 256–66.

Sánchez-Vázquez, F.J., S. Zamora and J.A. Madrid. 1995b. Light-dark and food restriction cycles in sea bass: effect of conflicting zeitgebers on demand feeding rhythms. Physiol. Behav. 58: 705–714.

Sánchez-Vázquez, F.J., T. Yamamoto, T. Akiyama, J.A. Madrid and M. Tabata. 1999. Macronutrient self-selection through demand-feeders in rainbow trout. Physiol. Behav. 66: 45–51.

Santinha, P.J.M., E.F.S. Gomes and J.O. Coimbra. 1996. Effects of protein level of the diet on digestibility and growth of gilthead sea bream, *Sparus aurata* L. Aquacult. Nutr. 2: 81–87.

Sargent, J.R., D.R. Tocher and J.G. Bell. 2002. The lipids. pp. 181–257. *In*: J.E. Halver and R.W. Hardy (eds.). Fish Nutrition. Academic Press, New York.

Shepherd, J. and N. Bromage. 1989. Intensive Fish Farming. Blackwell, Oxford. pp. 78–83.

Simpson, S.J. and D. Raubenheimer. 1997. Geometric analysis of macronutrient selection in the rat. Appetite 3: 201–213.

Simpson, S.J and D. Raubenheimer. 2001. A framework for the study of macronutrient intake in fish. Aquac. Res. 32: 421–432.

Simpson, S.J., R.M. Sibly, K.P. Lee, S.T. Behmer and D. Raubenheimer. 2004. Optimal foraging when regulating intake of multiple nutrients. Anim. Behav. 68: 1299–1311.

Skalli, A. and J.H. Robin. 2004. Requirement of n-3 long chain polyunsaturated fatty acids for European sea bass (*Dicentrarchus labrax*) juveniles: growth and fatty acid composition. Aquaculture 240: 399–415.

Soliman, A.K., K. Jauncey and R.J. Roberts. 1986. The effect of varying forms of ascorbic acid on the nutrition of juvenile tilapia (*Oreochromis niloticus*). Aquaculture 52: 1–10.

Spyridakis, P., R. Metailler and J. Gabaudan. 1986. Proteine et amidon dans l'alimentation du juvenile de bar ou loup (*Dicentrarchus labrax*). ICES, C.M. 1986/F: 30, pp. 11.

Sutton, A.L., X. Zhang, T.I. Ellison and P.N. Macdonald. 2005. The 1, 25(OH)$_2$D$_3$-regulated transcription factor MN1 stimulates vitamin D receptor-mediated transcription and inhibits osteoblastic cell proliferation. Mol. Endocrinol. 19: 2234–2244.

Terova, G., M. Saroglia, Z.G. Papp and S. Cecchini. 1998. Dynamics of collagen indicating amino acids, in embryos and larvae of sea bass (*Dicentrarchus labrax*) and gilthead sea bream (*Sparus aurata*), originated from broodstocks fed with different vitamin C content in the diet. Comp. Biochem. Physiol. Part A 121: 111–118.

Thrush, M., J.M. Navas, J. Ramos, N. Bromage, M. Carrillo and S. Zanuy. 1993. The effect of artificial diets on lipid class and total fatty acid composition of cultured sea bass eggs (*Dicentrarchus labrax*). Actas IV Congreso Nacional de Acuicultura. 37–42.

Tibaldi, E. and F. Tulli. 1998. Partial replacement of fish meal with soybean products in diets for juveniles sea bass (*Dicentrarchus labrax*). Proceeding of Recent Advances in Finfish and Crustacean Nutrition: VII. International Symposium on Nutrition and Feeding of Fish, Las Palmas de Gran Canaria, Spain. pp. 149.

Tibaldi, E., F. Tulli, R. Ballestrazzi and D. Lanari. 1991. Effect of dietary protein metabolizable energy ratio and body size on the performance of juvenile seabass. Zootec. Nutr. Anim. 17: 313–320.

Tibaldi, E., F. Tulli and M. Amerio. 1999. Feed intake and growth responses of sea bass (*D. labrax*) fed differentplant-protein sources are not affected by supplementation with a feeding stimulant. pp. 752–754. *In*: G. Piva, G. Bertoni, S. Satoh, P. Bani and L. Calamari (eds.). Recent Progress in Animal Production Science: I. Proc. A.S.P.A. XIII Congress, Piacenza, Italy, 21–24 June 1999. Assn. Sci. Anim. Production, Italy.

Tibaldi, E., Y. Hakim, Z. Uni, F. Tulli, M. de Francesco, U. Luzzana and S. Harpaz. 2006. Effects of the partial substitution of dietary fish meal by differently processed soybean meals on growth performance, nutrient digestibility and activity of intestinal brush border enzymes in the European sea bass (*Dicentrarchus labrax*). Aquaculture 261: 182–193.

Tocher, D.R. 2003. Metabolism and functions of lipids and fatty acids in teleost fish. Rev. Fish. Sci. 11: 107–184.

Tovar, D., J.L. Zambonino-Infante, C. Cahu, F.J. Gatesoupe, R. Vázquez-Juárez and R. Lésel. 2002. Effect of live yeast incorporation in compound diet on digestive enzyme activity in sea bass larvae. Aquaculture 204: 113–123.

Tovar-Ramírez, D., J.L. Zambonino Infante, C. Cahu, F.J. Gatesoupe and R. Vázquez-Juárez. 2004. Influence of dietary live yeast on European sea bass (*Dicentrarchus labrax*) larval development. Aquaculture 234: 415–427.

Tulli, F., C. Vachot, E. Tibaldi, V. Fournier and S.J. Kaushik. 2007. Contribution of dietary arginine to nitrogen utilisation and excretion in juvenile sea bass (*Dicentrarchus labrax*) fed diets differing in protein source. Comp. Biochem. Physiol. Part A 147: 179–188.

Twibell, R.G., B.A. Watkins, L. Rogers and P.B. Brown. 2000. Effects of dietary conjugated linoleic acids on hepatic and muscle lipids in hybrid striped bass. Lipids 35: 155–161.

Twibell, R.G., B.A. Watkins and P.B. Brown. 2001. Dietary conjugated linoleic acids and lipid source alter fatty acid composition of juvenile yellow perch, *Perca flavescens*. J. Nutr. 131: 2322–2328.

Unniappan, S., X. Lin, L. Cervini, J. Rivier, H. Kaiya, K. Kangawa and R.E. Peter. 2002. Goldfish ghrelin: molecular characterization of the complementary deoxyribonucleic acid, partial gene structure and evidence for its stimulatory role in food intake. Endocrinology 143: 4143–4146.

Vagner, M., J.L. Zambonino Infante, J.H. Robin and J. Person-Le Ruyet. 2007. Is it possible to influence European sea bass (*Dicentrarchus labrax*) juvenile metabolism by a nutritional conditioning during larval stage? Aquaculture 267: 165–174.

Valente, L.M.P., N.M. Bandarra, A. Figueiredo-Silva, P. Rema, S. Vaz-Pires, J.A.M. Martins and M.L. Prates e Nunes. 2007a. Conjugated linoleic acid in diets for large size rainbow trout (*Oncorhynchus mykiss*): effects on growth, chemical composition and sensory attributes. Brit. J. Nutr. 97: 289–297.

Valente, L.M.P., N.M. Bandarra, A.C. Figueiredo-Silva, A.R. Cordeiro, R.M. Simões and M.L. Nunes. 2007b. Influence of conjugated linoleic acid on growth, lipid composition and hepatic lipogenesis in juvenile European sea bass (*Dicentrarchus labrax*). Aquaculture 267: 225–235.

Villeneuve, L., E. Gisbert, H. Le Delliou, C.L. Cahu and J.L. Zambonino-Infante. 2005. Dietary levels of all-trans retinol affect retinoid nuclear receptor expression and skeletal development in European sea bass larvae. Br. J. Nutr. 93: 791–801.

Villeneuve, L.A.N., E. Gisbert, J. Moriceau, C.L. Cahu and J.L. Zambonino. 2006. Intake of high levels of vitamin A and polyunsaturated fatty acids during different developmental periods modifies the expression of morphogenesis genes in European sea bass (*Dicentrarchus labrax*). Br. J. Nutr. 95: 677–687.

Vivas, M., F.J. Sánchez-Vázquez, B. García García and J.A. Madrid. 2003. Macronutrient self-selection in European sea bass in response to dietary protein or fat restriction. Aquac. Res. 34: 271–280.

Vivas, M., V.C. Rubio, F.J. Sánchez-Vázquez, C. Mena, B. García García and J.A. Madrid. 2006. Dietary self-selection in sharpsnout seabream (*Diplodus puntazzo*) fed paired macronutrient feeds and challenged with protein dilution. Aquaculture 251: 430–437.

Volkoff, V.H., J.M. Bjorklund and R.E. Peter. 1999. Stimulation of feeding behavior and food consumption in the goldfish, *Carassius auratus*, by orexin-A and orexin-B. Brain Res. 846: 204–209.

Volpato, G.L., E. Gonçalves-de-Freitas and M. Fernandes-de-Castilho. 2007. Insights into the concept of fish welfare. Dis. Aquat. Org. 75: 165–171.

Waagbo, R. 2010. Water-soluble vitamins in fish ontogeny. Aquac. Res. 41: 733–744.

Wang, Y. and J.M. Conlon. 1994. Purification and characterization of galanin from the phylogenetically ancient fish, the bowfin (*Amia calva*) and dogfish (*Scyliorhinus canicula*). Peptides 15: 981–986.

Watanabe, T. 1982. Lipid nutrition in fish. Comp. Biochem. Physiol. Part B 73: 3–15.

Wodtke, E. and A.R. Cossins. 1991. Rapid cold-induced changes of membrane order and delta-9-desaturase activity in endoplasmic reticulum of carp liver: a time-course study of thermal adaptation. Biochim. Biophys. Acta 1064: 343–350.

Zambonino Infante, J.L., C. Cahu and A. Péres. 1997. Partial susbsitution of di-and tripeptides for native proteins in seabass diet improves *Dicentrarchus labrax* larval development. J. Nutr. 127: 608–614.

Zhao, H., R. Jiang, M. Xue, S. Xie, X. Wu and L. Guo. 2009. Fishmeal can be complementely replaced by soy protein concentration by increase feeding frequency in Nile tilapia (*Oreochromis niloticus* GIFT strain) 284 less than 2g. Aquacult. Nutr. 16: 648–653.

Pathology

Ariadna Sitjà-Bobadilla,[1,]* *Carlos Zarza*[2] *and Belén Fouz*[3]

Introduction

Fish may carry different pathogens (virus, bacteria, parasites, fungi) without any disease sign or any negative effect on their physiology or survival. This is a frequent scenario in natural fish populations, in which a balance between the host and the pathogen has been reached as a result of millions of years of co-evolution. Farming practices alter this equilibrium, favouring the emergence of diseases and posing a major problem for the aquaculture industry. These practices are likely to select for fast-growing, early-transmitted, and hence probably more virulent pathogens (Mennerat et al. 2010). Farmed fish are also constantly exposed to different types of stressors such as bad or poor culture conditions, inadequate diets and other anthropogenic factors that might compromise their performance and survival (Fig. 1).

The culture of European sea bass (ESB) is a strong farming industry in Europe, mainly in the Mediterranean region, with 126,240 tonnes produced in 2011 (APROMAR 2012). Intensive rearing and international trading have helped in extending prophylactic measures (vaccination programs) against some bacteria; still, many pathogens pose a threat for the production of this fish species. Although some pathogens are devastating as unique disease

[1]Instituto de Acuicultura Torre de la Sal (IATS-CSIC), Torre de la Sal, s/n, 12595 Ribera de Cabanes, Castellón, Spain.
Email: ariadna.sitja@csic.es
[2]Tassal Group Limited, G.P.O. Box 1645, Hobart Tasmania Australia 7001.
Email: carlos.zarza@tassal.com.au
[3]Departamento de Microbiología y Ecología. Universidad de Valencia, Spain, Doctor Moliner, 50. 46100, Burjasot, Valencia, Spain.
Email: belen.fouz@uv.es
*Corresponding author

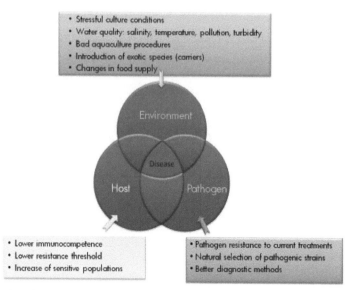

- Stressful culture conditions
- Water quality: salinity, temperature, pollution, turbidity
- Bad aquaculture procedures
- Introduction of exotic species (carriers)
- Changes in food supply

Environment

Disease

Host

Pathogen

- Lower immunocompetence
- Lower resistance threshold
- Increase of sensitive populations

- Pathogen resistance to current treatments
- Natural selection of pathogenic strains
- Better diagnostic methods

Figure 1. Venn diagram showing the three types of factors involved in a disease. For each of them, a set of possible modulators are in action.

Color image of this figure appears in the color plate section at the end of the book.

agents, frequently, pathological surveys find different etiological agents contributing to the disease outbreak, and therefore multiple diagnostic methods should be applied to obtain a realistic view of the situation. Massive establishment of net cages has contributed to the tonnage increase, but it is also responsible in part, for the extension and spread of some diseases. Net cages offer favourable conditions for pathogen dispersal for several reasons: 1) non-controlled water, 2) introduction of fish from other geographical areas, 3) access to intermediate hosts, 4) contact with wild fish (sometimes natural or reservoir hosts), 5) a good surface for biofouling, and, therefore, for the settlement and attachment of parasitic stages, and 6) non-controlled fish effluents (mucus casts, faeces, dead animals, escapees) which allow pathogen life cycles to keep going. In any case, each type of facility (extensive ponds, recirculating systems, cages, open flow tanks) determines the presence and abundance of the different pathogens and their impact.

Fish diseases are controlled under the European directive 2006/88/EC and by the directive 93/140/EC which allows for the removal of fishery products seriously deteriorated by the action of a pathogen. However, none of the known pathogens producing mortalities in ESB cultures are included in the EU legislation or in the Office International des Epizooties (OIE) list of notifiable fish diseases.

Pathologies Due to Bacteria

Bacteria represent the most common group of pathogens in cultured fish worldwide, some of them acting as primary pathogens, but others colonize and cause disease in damaged hosts (Austin and Austin 2007). In recent years, fish-borne zoonotic bacteria are also acquiring relevance in aquaculture (Lehane and Rawlin 2000, Ghittino et al. 2003, Jacobs et al. 2009). In Table 1, bacteria reported from both wild and cultured ESB are listed. Although the bacterial population associated to ESB is highly diverse, this section will focus only on those bacteria relevant either for cultured stocks or for their zoonotic implications. In Figs. 2 and 3, a selection of images of these and other bacterial pathogens listed in Table 1 can be found.

Vibriosis

Infections caused by several members of the genus *Vibrio*, as well as the related genera *Moritella* and *Photobacterium*, are known as vibriosis and can affect numerous cultured and wild marine and fresh/brackish water fish. They are the classical example of stress-borne diseases (Noga 2011) and therefore, minimizing those stress factors responsible for the development of the disease is essential to achieve good results in therapy and for long-term management. Vibrios causing disease in ESB (Table 1) are gram-negative, rod-shaped, halophilic and facultatively anaerobic bacteria (Buller 2004).

Vibrio anguillarum

This is the most common fish-pathogenic vibrio and causes a fatal disease in more than 50 fish species. It is responsible for relevant economic losses in marine cultures worldwide (Toranzo et al. 2005, Actis et al. 2011, Noga 2011), ESB being the species most affected in Mediterranean waters (Le Breton 1999). In fact, this bacterium has caused problems in Europe since 1977 in cultured ESB (Baudin-Laurencin 1981), with the highest incidence in Turkey and Greece. *V. anguillarum*, as well other pathogenic vibrios, grows rapidly at temperatures between 25°C and 30°C and appears in both mortality and routine fish surveys. Infection occurs mostly through penetration of the fish skin, injuries or damaged mucous layer being typical portals of entry for the pathogen (Austin and Austin 2007).

Up to a total of 23 O serotypes (European serotype designation) are recognized, being O1 and O2 (subgroups O2α and O2β) mainly associated with mortalities in farmed and wild ESB (Pedersen et al. 1999). Genetic studies have confirmed the existence of two separated clonal lineages within the major pathogenic serotypes, corresponding to the North European and South European isolates, respectively (Toranzo et al. 2005).

Table 1. Bacteria infecting wild (W) and cultured (C) European sea bass.

Group/ Genus	Species/ Subspecies	Host Habitat	Geographic Location	Reference
Vibrio	*V. anguillarum*	C	Spain, Greece, Italy, Croatia, France, Portugal	Baptista et al. 1999, Le Breton 1999, Ghittino et al. 2003, Toranzo et al. 2005
	V. alginolyticus	C	France	Le Breton 1999, Ghittino et al. 2003
	*V. vulnificus biotype 2**	C	Greece, Tunisia, Spain (E)	Ghittino et al. 2003, Fouz et al. 2010
	V. harveyi	C	Spain, France	Le Breton 1999, Pujalte et al. 2003
	V. ordalii	C	Mediterranean area	Le Breton 1999
Photobacterium	*Ph. damselae* subsp. *piscicida* (formerly *Pasteurella piscicida*)	C, W	Croatia, Cyprus, Italy, France, Greece, Israel, Malta, Morocco, Spain, Turkey, Portugal	Magariños et al. 1992, Candan et al. 1996, Le Breton 1995, Le Breton 1999, Nitzan et al. 2001, Toranzo et al. 2005,
	Ph. damselae subsp. *damselae* (formerly *Vibrio damsela*)*	C, W	Mediterranean area, Spain (E)	Fouz et al. 1992, Le Breton 1999
Tenacibaculum	*T. maritimum* (formerly *Flexibacter maritimus*)	C, W	Cyprus, France, Greece, Spain, Italy, Malta	Bernardet et al. 1994, Bernardet 1998, Santos et al. 1999, Salati et al. 2005, Toranzo et al. 2005, Kolygas et al. 2012
	T. dicentrarchi sp. nov.	C	Spain	Piñeiro-Vidal et al. 2012
Mycobacterium (Gram +)	*M. marinum**	C	Greece, Italy, Israel, Spain, France, Turkey	Colorni 1992, Colorni et al. 1993, 1996, Ghittino et al. 2003, Toranzo et al. 2005
	M. spp. *(chelonae, fortuitum...)*	C	Israel	Colorni 1992, Colorni et al. 1996
Piscirickettsiosis/ Rickettsia-like organisms (RLO)	*P. salmonis*	C	Mediterranean area, Greece, Italy	Le Breton 1999, Steiropoulos et al. 2002, Athanassopoulou et al. 2004, McCarthy et al. 2005
Streptococcus (Gram +)	*S. iniae**	C, W	Israel, Spain, Turkey, Eastern Med. area	Zlotkin et al., 1998, Colorni et al. 2002, Ghittino et al. 2003, Toranzo et al. 2005

Chlamydia-like organism, CLO	*Chlamydia*-like organism, CLO	C	France, Spain	Paperna et al. 1981, Crespo et al. 2001
Aeromonas	*A. salmonicida* subsp. *salmonicida*	C	France, Greece, Spain, Portugal, Turkey	Le Breton 1999
	A. hydrophila complex	C	Spain, Portugal	Baptista et al. 1998, Toranzo et al. 2005
Pseudomonas	*P. anguilliseptica*	C	France, Turkey	Berthe et al. 1995
Edwarsiella	*E. tarda**	C	Spain	Blanch et al. 1990
Nocardia (Gram +)	*Nocardia* sp.	C		Vigneulle et al. 1992, Le Breton 1999

*Potencial zoonotic agent. E, Experimental infection.

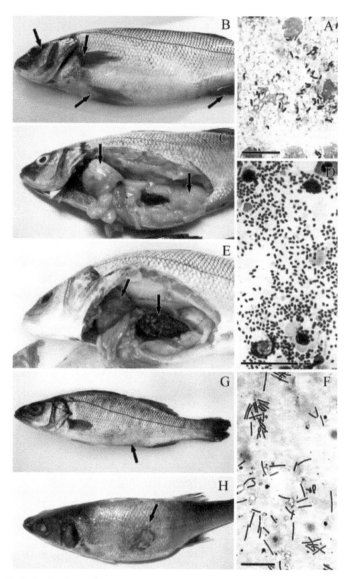

Figure 2. Pathologies due to bacteria in European sea bass. (A) Rod shaped bacillus of *Vibrio anguillarum* in a fuchsine-stained smear. (B) External and (C) internal typical signs of vibriosis. Note the haemorrhagic fins, head (in B), liver and digestive tract (in C) (arrows). (D) Bacillus of *Photobacterium damselae* subsp. *piscicida* in a methylene blue-stained smear. (E) Internal clinical signs of photobacteriosis, with the typical yellowish nodules in the spleen and haemorrhages in the liver (arrows). (F) *Tenacibaculum maritimum* filamentous bacteria in a methylene blue-stained smear. (G-H) External lesions and ulcers typical of tenacibaculosis (arrows). Bar scales = 10 µm. All illustrations are original from the authors (CZ).

Color image of this figure appears in the color plate section at the end of the book.

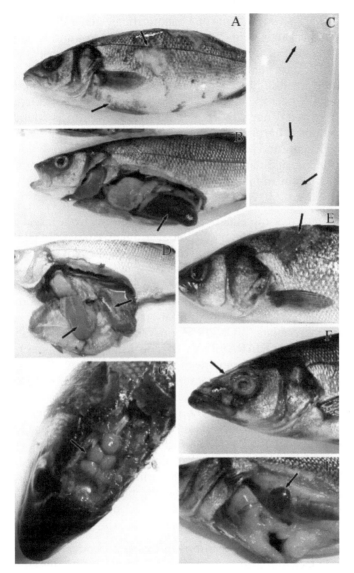

Figure 3. Pathologies due to bacteria (A-E) and virus (F-H) in European sea bass. External (A) and internal (B) clinical signs of mycobacteriosis, Note the external lesions and ulcers (arrows in A), and the granulomata in the spleen (arrow in B). (C) Colonies of *Mycobacterium marinum* grown in Lowenstein-Jensen agar. (D) Clinical internal signs of streptococosis, note the haemorrhages in the digestive tract (arrows). (E) Nocardiosis external lesions, note the large ulcer (arrow). (F-H) Viral Nervous Necrosis (VNN) external (F) and internal (G-H) clinical signs, note the head erosions (F, arrow), the haemorrhages in the encephalus (G, arrow) and the inflamed swim bladder (H, arrow). All illustrations are original from author CZ.

Color image of this figure appears in the color plate section at the end of the book.

Clinical signs. Fish show signs of a haemorrhagic generalized septicaemia. External clinical signs (Fig. 2B) include: a) weight loss and lethargy; b) red spots on the ventral and lateral areas and at the base of fins; c) abdominal distension; d) dark skin lesions that can ulcerate and bleed and; e) corneal opacity and exophthalmia. Internal signs include visceral petechiation, splenomegaly, necrotic enteritis producing yellow, mucoid exudate and liquefactive renal necrosis (Fig. 2C). The bacterium is found in high concentrations in the blood and haematopoietic organs (spleen and kidney) (Fig. 2A). In acute and severe epizootics, the infection spreads very fast and most of the infected fish die without showing any clinical signs (Toranzo et al. 2005, Austin and Austin 2007).

Diagnosis. Since *Vibrio* species can secondarily invade fish, definitive diagnosis requires the isolation of the bacterium from target tissues to ascertain that it is the primary ethiological agent. Conventional diagnostic procedures include interpretation of clinical and histological signs, culture of the pathogen on a general or selective medium (such as thiosulphate citrate bile salts sucrose, TCBS, agar) and analysis of standard morphological and biochemical tests. However, other assays are needed for confirmation: serological techniques (slide agglutination or ELISA) or nucleic acid-based ones. Commercial kits (BIONOR Mono-Va-kit, a latex agglutination-based assay or Bionor AQUARAPID-Va test, an ELISA-based assay) for a fast diagnosis are available (Romalde et al. 1995, González et al. 2004), but they do not allow the identification of serotypes. Nucleic acid-based techniques such as polymerase chain reaction (PCR) amplification allow specific and sensitive identification of the pathogen but they cannot distinguish serotypes. Conventional PCR-based approaches (for example, using gene *rpo*N, which codes for the sigma factor 54, as target) (Frans et al. 2011) have been developed for the accurate detection of the pathogen in infected tissues. PCR strategies based on monitoring the amplification reaction in real time or loop-mediated isothermal amplification method (LAMP) will probably replace conventional PCR ones for detection of *V. anguillarum* in the laboratory and in the field (Frans et al. 2011).

Risk factors. Losses caused by vibriosis are highly dependent on environmental stressors. In fact, the bacterium can survive and multiply in the environment, but some major predisposing risk factors can precipitate outbreaks or vary them from acute to chronic. The high temperature (more than 15°C), the poor quality of water (high organic matter content, low oxygen concentration, pollution) and biological stressors (e.g., population density, presence of other micro- or macro-organisms, bad handling practices) are important inducers of vibriosis. Moreover, some strains can cause disease without any predisposing stress. Juvenile ESB and adults are especially susceptible to *Vibrio* infections (Austin and Austin 2007, Frans et al. 2011).

Control and treatments. Therapy with antibiotics (oxytetracicline, potentiated sulfonamides or quinolones) has been extensively used in the treatment of classical vibriosis (Noga 2011) in cultured ESB and it is still required for the containment of outbreaks, in spite of its drawbacks (environmental contamination and fish/strain resistances). Currently, vaccination with bacterins by immersion, injection and oral routes, is the most efficient strategy to protect fish against vibriosis. Multiple commercial vibriosis vaccines provide good protection for population at risk (Angelidis et al. 2006). Most consist of inactivated strains of serotypes O1 or O1+O2α in their formulations (e.g., MICROViB, ALPHA MARINE®, ALPHA Dip® Vibrio and Aqua-Vac Vibrio) (Angelidis et al. 2006), although a bacterin (GAVA-3) covers also, the antigenic subgroup O2β (Toranzo et al. 2005). Recently, a DNA vaccine based on the major outer membrane protein has shown promising results in protecting Asian sea bass (Kumar et al. 2007). In ESB and other marine fish, *V. anguillarum* vaccines are being administered by bath in 1–2 g fish, two immersions being necessary in the optimized protocol with a month's interval. Since mortalities occur mainly in unvaccinated fish stocks, it can be concluded that the *V. anguillarum* vaccines/vaccination programmes work properly. Alternative preventive measures, such as the use of antimicrobial peptides, application of probiotics, nutritional immunostimulation and quorum sensing inhibitors are also being assayed. Finally, prevention is also based on the application of hygienic measures, such as good management practices.

Vibrio spp.

Other *Vibrio* species have been sporadically isolated in novel vibriosis outbreaks in ESB: *V. ordalii, V. alginolyticus, V. harveyi* and *V. vulnificus* biotype 2 (VVBT2) (Table 1). They are considered opportunistic invaders, responsible for ulcerative and haemorrhagic lesions in skin and fins. Cause for concern is VVBT2, since one of its serovars (ser E) acts as a zoonosis agent causing infection in humans, hence representing a potential health hazard for fish farmers. Biochemical, serological or genetic methods allow to identify these pathogens, even at serotype level, for example the multiplex PCR developed for VV (Sanjuán and Amaro 2007). Control of novel infections is only based on the delivery of medicated feed (with the appropriate antibiotic), with the exception of that caused by VVBT2 which can be prevented by using a specific bacterin, Vulnivaccine, developed for eels (Fouz et al. 2001).

Photobacterium damselae subsp. *damselae*

This subspecies (formerly *Vibrio damsela*) has been described as an etiological agent of sporadic pathologies (also known as "digestive vibriosis") in cultured ESB (Table 1). Although infection can be controlled with the appropriate antibiotic, it represents a potential health hazard for fish farmers because this bacterium can also cause infection in humans.

Photobacteriosis

The non-motile, Gram-negative, halophilic rod bacterium *Ph. damselae* subsp. *piscicida*. (*Pdp*) (formerly *Pasteurella piscicida*) is the causative agent of photobacteriosis, one of the most serious bacterial diseases in warm marine aquaculture (Toranzo et al. 2005, Daly and Aoki 2011). Although gilthead sea bream is the species most affected in Mediterranean waters, this disease has caused economic losses in ESB culture in the Mediterranean European countries since 1990 (Table 1) (Baudin Laurencin et al. 1991, Magariños et al. 1992, Romalde and Magariños 1997), with the highest incidence in Turkey and Greece.

The subspecies is biochemical and antigenically homogeneous, with two genetic groups, one comprising the Mediterranean/European isolates and the other one the Japanese and USA isolates (Toranzo et al. 2005). Both groups also differ in the plasmid content (Magariños et al. 1992). This pathogen, which appears in both mortality and routine surveys of cultured fish, is one of the few causing massive mortalities in wild fish (Daly and Aoki 2011).

Clinical signs. *Pdp* causes a bacteremia which takes acute or chronic form (Noga 2011). In the acute one, diseased fish can die in the bottom of tanks or net cages with few visible clinical signs. Blood channels in the secondary gill filaments are sometimes observed, as well as slight darkening, skin ulcers and abdominal distension. Splenomegaly, visceral petechiation and necrotic enteritis are also typical internal signs of the disease (Fig. 2E). In the chronic form, there are white-spotted lesions within internal viscera, particularly the spleen and kidney (Fig. 2E), with focal necrosis and extracellular bacterial colonies that incite a chronic inflammatory response. The disease, traditionally named as pasteurellosis, is also known as pseudotuberculosis because of the characteristic presence of white nodules in viscera.

Diagnosis. The gross pathological internal signs together with the microscopic examination of abundant small rods in Gram preparations from internal viscera (Fig. 2D) can be used for the preliminary diagnosis. This first step must be confirmed using culture methods or with a fluorescent antibody technique. In the first case, after isolating of typical colonies in enriched media such as blood agar containing sodium chloride (0.5%–1%), the bacterium can be presumptively identified on the basis of standard biochemical tests, displaying a characteristic profile (2005004) in the API-20 E system. However, a serological confirmation employing specific polyclonal antisera is needed (Magariños et al. 1992, Bakopoulos et al. 1997). Commercial diagnostic kits from BIONOR AS, based on slide agglutination, ELISA or EIA based methods, allow a rapid detection of the bacterium in fish tissues, even in healthy carrier fish (Toranzo et al. 2005). In the last decade, several DNA-based detection protocols have been developed but only a multiplex PCR that target the 16S rRNA (16S rDNA) and ureC genes allows to discriminate the subspecies *piscicida* and *damselae* (Osorio and Toranzo 2002).

Risk factors. Juvenile and adult ESB are especially susceptible to photobacteriosis. Severe mortalities occur usually when water temperatures are above 18°C–20°C, but also at 25°C in summer time. Below this temperature, fish can harbour the pathogen as subclinical infection for long periods (Magariños et al. 2001). The bacterium can survive as viable but not culturable cells for long periods, maintaining its virulence properties (Magariños et al. 1994). These aspects make its detection difficult in the farm environment, spreading the infection easily in new populations of stocked fish. The mode of transmission of the disease is unknown but fish-to-fish contact and oral transmission have been suggested. The pathogen usually invades the gills or enters fish through the gastrointestinal tract. Since the bacterium can also be transmitted vertically through the ovarian and seminal fluids from apparently healthy broodstock, the application of specific and sensitive serological and molecular tools is very important (Osorio and Toranzo 2002). Poor water quality, deficient sanitary practices and bad handling practices increases the impact of this pathogen.

Control and treatments. Many chemotherapeutic agents (such as ampicillim, florfenicol, quinolones, oxytetracycline and trimethoprim-sulfamethoxazole) show activity against *Pdp* and have been widely used for successful bath or oral treatments of photobacteriosis (Daly and Aoki 2011). As a direct consequence, multiple drug-resistant strains have appeared (Magariños et al. 1992, Laganà et al. 2011). Moreover, immunostimulants such as β-glucans and lysozyme seem to increase the resistance of fish to the disease (Noya et al. 1995). Several types of commercial vaccines against photobacteriosis are available in the market, but their efficacy is dependent on the fish species, fish size, vaccine formulation and combined use of immunostimulants (Romalde and Magariños 1997). In ESB and other marine fishes, specific bacterins are being employed by bath in 1 g–2 g fish, two immersions with a month's interval being necessary to avoid the high economic losses caused by this disease. Moreover, divalent bacterins against pasteurellosis and vibriosis (e.g., ALPHA Dip® 2000 and ALPHA Ject® 2000), administered by immersion and booster, provide significant protection to ESB against the disease (Gravningen et al. 1998, Afonso et al. 2005). The early diagnosis of the pathogen is crucial in order to implement adequate measures to prevent the vertical and horizontal transmission of the disease in the farms.

Tenacibaculosis

Tenacibaculum maritimum

This species, formerly *Flexibacter maritimus*, a Gram-negative and filamentous bacterium showing gliding motility, is one of the main causative agents of tenacibaculosis (marine flexibacteriosis) in cultured and wild fish in Europe, Japan and USA. It is an opportunistic pathogen that causes extensive skin and gill damage with subsequent systemic infection. The disease, considered a significant obstacle for ESB culture in Mediterranean countries (Table 1) (Toranzo et al. 2005), is also named as "gliding bacterial diseases of sea fish", "eroded mouth syndrome" and "black patch necrosis" (Bernardet 1998). The pathogen requires media having optimally around 30% seawater (Wakabayashi et al. 1986, Noga 2011). The *F. maritimus* medium (FMM) proved to be the most appropriate for the successful recovery of the pathogen from fish tissue after incubation at 20°C to 25°C for 48 hours to 72 hours, showing typical colonies, pale yellow and flat with uneven edges (Pazos et al. 1996). The species is phenotypically homogeneous, but at least three major O-serogroups and two main clonal lineages have been described which seem to be related to the host species (Avendaño-Herrera et al. 2006).

Clinical signs. Affected fish show an ulcerative condition, with eroded and hemorrhagic mouth, skin ulcers, frayed fins, and tail rot (Fig. 2G-H) (Santos et al. 1999, Avendaño-Herrera et al. 2006). The systemic condition can also be established, involving different internal organs (spleen, liver and kidney). Moreover, fish become anorexic immediately post-infection.

Diagnosis. The clinical signs together with the microscopic examination of abundant long rods in wet mounts or Gram preparations from gills or skin lesions (Fig. 2F) can be used for a rapid presumptive diagnosis (Noga 2011). This first step must be confirmed with the isolation of typical colonies on appropriate medium and subsequent identification by means of specific molecular DNA-based methods, such as PCR protocols using the 16S rRNA gene as target (Toyama et al. 1996, Avendaño-Herrera et al. 2004). Miniatured phenotypic systems, such as API ZYM, can also be used since most strains display a characteristic profile (Santos et al. 1999). The PCR protocols are very sensitive, detecting low bacterial levels in mucus samples, and specific, distinguishing the species from other phylogenetic and phenotypically similar ones.

Risk factors. The prevalence and severity of the disease increase at temperatures above 15°C and salinities of 30‰ to 35‰. Its onset is also influenced by a variety of environmental (physical and chemical stress), biological (bad handling, poor feeding and high density) and host-related factors (skin surface condition). Both adults and juveniles of ESB may be affected, but the disease is more severe in younger fish (2 g–80 g), progressing rapidly from early stages to advanced ulcerative lesions. The pathogen can be isolated from sediment, the surface of tanks and from water exposed to infected stocks (Santos et al. 1999). However, seawater does not seem to be an important route of transmission of *T. maritimum* since its growth and survival is inhibited by natural aquatic microbiota. The bacterium can be part of the autochthonous populations of the fish skin and its long-term survival in the aquatic environment requires fish mucus as a reservoir (Avendaño-Herrera et al. 2006). Although the primary sites of infection are body surfaces, especially the gills (Bernardet 1998), the bacterium also enters the host via food (Chen et al. 1995). The loss of the epithelial surface (characteristic of the disease) benefits the entry for other secondary pathogens.

Control and treatments. The pathogen is susceptible to various chemotherapeutic agents, bath treatments being more effective than the oral ones. Amoxicillin and trimethoprim are effective, but enrofloxacin is the most useful compound, in spite of the appearance of resistant strains (Avendaño-Herrera et al. 2006). The use of disinfectants, such as hydrogen peroxide (bath treatment at 240 ppm) or manipulation of temperature and/or salinity can be used to reduce its morbidity. Avoiding overcrowding and overfeeding help reduce the incidence of the disease, but immunoprophylaxis would be the best way to prevent it. The specific bacterin "FM-95" is the only monovalent vaccine on the market to prevent mortalities caused by *T. maritimum* in turbot (Toranzo et al. 2005). The vaccine is administered by bath to fish of 1 g–2 g followed by a booster injection at 20 g–30 g size. Divalent formulations to simultaneously prevent tenacibaculosis/vibriosis or tenacibaculosis/streptococcosis are also available (Toranzo et al. 2005). No information about ESB vaccination with these vaccines is currently available.

Tenacibaculum dicentrarchi

This bacterium is a novel species recently isolated from diseased ESB (Table 1) showing the typical skin lesions of tenacibaculosis (Piñeiro-Vidal et al. 2012). Studies on the epizootiology and pathogenicity of this novel species must be performed to reveal its importance as a causative agent of tenacibaculosis.

Mycobacteriosis

Mycobacterium marinum, is a Gram-positive, aerobic and acid-fast, non-motile rod, which intracellularly parasitizes phagocytes. It is considered the primary causative agent of fish mycobacteriosis (or fish tuberculosis), an important disease affecting around 200 wild and cultured fin fish (Lewis and Chinabut 2011). Other *Mycobacterium* spp. such as *M. fortuitum*, *M. chelonae* and *M. poriferae* have been associated with this disease in fish populations (Colorni 1992, Colorni et al. 1996, Le Breton 1999, Jacobs et al. 2009, Lewis and Chinabut 2011). Mycobacteriosis represents a significant threat for ESB (especially adults) cultured on the Mediterranean coast and the Red Sea coast of Israel (Table 1) (Colorni 1992, Colorni et al. 1993, Colorni et al. 1996, Toranzo et al. 2005), but it is not associated with large-scale mortality in wild fish. Most fish pathogenic mycobacteria can also cause disease in humans, especially in immunocompromised persons (Ghittino et al. 2003), *M. marinum* being a major zoonotic organism.

Clinical signs. The course of the disease is typically chronic and may not produce clinical signs. Diseased ESB are cachectic and show abdominal swelling. All fish tissues may be involved, including eyes, gills, visceral organs, musculature, and fins. External signs are nonspecific and include lethargy, pale gills, skin necrosis, scale loss accompanied by haemorrhagic lesions penetrating the musculature in advanced cases and pigmentary changes (Fig. 3A). Gross internal signs include enlargement of the spleen, kidney and liver and greyish-white nodules (granulomas or tubercles) in internal organs (Fig. 3B) (Lewis and Chinabut 2011). In humans, *M. marinum* produces granulomatous lesions in skin and peripheral deep tissues (Lewis et al. 2003, Petrini 2006) and the disease is commonly known as "fish tank- or swimming pool-granuloma", or "fish-fancier's finger".

Diagnosis. Diagnosis is primarily based on the clinical and histological signs together with the identification of the bacterium by the visualization of acid-fast rods in spleen and kidney smears. Secondly, the bacterium should be isolated from internal organs using specific media (Loewenstein-Jensen medium) (Fig. 3C) and presumptively distinguished on the basis of phenotypic testing. However, the slow and poor growth exhibited by the majority of the mycobacteria requires a DNA-based method for fast and reliable identification (Knibb et al. 1993, Herbst et al. 2001). A PCR technique using the 16S rRNA as target gene, coupled with direct sequencing of the amplified fragment, is highly specific and sensitive for the detection of *M. marinum* and other species in fish tissues and also in the blood (Colorni et al. 1993, Knibb et al. 1993, Talaat et al. 1997), being a useful non-destructive method to screen broodstock. It is very important to use a combined phenotypic and genotypic approach for identifying *Mycobacterium* species.

Risk factors. Although transmission of *Mycobacterium* spp. in fishes is poorly understood, water and associated biofilms are their natural habitats. Feeding fish with infected trash fish is probably related to disease acquisition, the gut, rather than the gills, being one of the primary sites of infection of *M. marinum* (Harriff et al. 2007). Since aquatic mycobacteria pose significant zoonotic concern, handling of infected ponds or cages or manipulating infected fish can represent a potential hazard for fish farmers (Ghittino et al. 2003). Thus, the significance of the current epizootic to human health, as well as the potential adverse effects on fish stocks, must be considered.

Control and treatments. Currently, there is no widely accepted curative therapy or medical prophylaxis to control fish mycobacteriosis. Moreover, antibiotic resistance appears to be highly dependent on the fish species and bacterial strain. The affected fish, asymptomatic for long periods, are rendered unmarketable. Prevention is based on the application of hygienic measures as well as good management practices, avoiding overcrowding and overfeeding. Moreover, feeding fish with trash fish is not recommended. Unfortunately, control of fish mycobacteriosis requires destruction of affected stock and disinfection of holding tanks and plumbing (Noga 2011). A recombinant vaccine has been developed, showing short-term protection against *M. marinum* in striped bass (Pasnik and Smith 2005). Antibiotic therapy is effective for *M. marinum* infections in humans, although surgical excision is often recommended.

Pathologies Due to Virus

Viral Nervous Necrosis Viruses (VNNV) cause the highest ecological and economic impacts in wild and farmed ESB. This viral disease is a major emergent problem, expanding in host and geographic range, without yet having effective treatments. Thus, in order to minimize its dispersal and pathogenic action, all fish in infected farms have to be destroyed and fish movements to and from infected areas have to be avoided. This section will concentrate only on this virus group, but other viral diseases can be found in Table 2.

Viral Nervous Necrosis (VNN)

VNN viruses cause lesions in the central nervous system (CNS) known as viral nervous necrosis (VNN). The OIE also terms it "Viral Encephalopathy and Retinopathy" (VER) and considers it a "significant disease". It was first

Table 2. Virus infecting wild (W) and cultured (C) European sea bass.

Family/ Genus	Viral Species	Host Habitat	Geographic Location	Reference
Nodaviridae/ betanodavirus	Viral Nervous Necrosis Viruses (*VNNV*)	C, W	Croatia, Cyprus, France, Greece, Israel, Italy, Malta, Spain, Turkey, Portugal	Le Breton et al. 1997, Le Breton 1999, Bovo et al. 1999, Breuil et al. 2001, Skliris et al. 2001, Ghittino et al. 2003, Athanosopoulou et al. 2003, Hodneland et al. 2011
Birnaviridae/ aquabirnavirus	Infectious Pancreatic Necrosis Viruses (*IPNV*)	C*	France	Le Breton 1999
Retroviridae	Viral Erythrocytic Infection Viruses (*VEIV*)	C	Spain	Pintó et al. 1991

Viral species and family/genus have been designed according to the vs. 7 of the International Committee on taxonomy of Viruses (ICTV 2009), except for VNNV (a general term).
* Isolated from healthy larvae.

described in 1990, but the etiological agent was not isolated until 1996, by means of a cell line named SSN-1 from *Chana striatus* (striped snake-head) (Frerichs et al. 1996). The disease has become a major problem in the culture of larval and juvenile marine fish worldwide. In fact, it has been reported from a wide range of wild and farmed fin fish species (more than 35), mostly from the marine environment. It often leads to rapid mortality generally during early fry and fingerling stages (Munday et al. 2002), but it also causes significant morbidity and mortality in market-size fish (Le Breton et al. 1997), being a significant obstacle for the ESB culture (Table 2).

VNNV belong to the *Betanodavirus* genus within the Nodaviridae family (ICTV 2009), and are the smallest single-stranded RNA virus affecting fish (25–34 nm in diameter) (Munday et al. 1992). They are non-enveloped and icosahedral, with a single coat protein. Among the five genetic groups described within betanodavirus, the red-spotted grouper nervous necrosis virus (RGNNV) is the most frequently isolated from ESB (Nishizawa et al. 1997). Temperature might be more important than host specificity for the distribution of these various genotypes (Korsnes et al. 2005).

Clinical signs. The first clinical signs are erosions and ulcers in the head and eyes (Fig. 4F), with occasional enucleation. These lesions are probably caused by the fish rubbing against the nets as a consequence of the un-coordinated and erratic locomotion, which include spiral swimming in circles and around their longitudinal axis, followed by periods of sudden, fast swimming. The characteristic lesions in the CNS include vacuolization and necrosis of the brain tissue and nuclear layers in the eye retina, causing varying degrees of neurological disorders in fish (Munday et al. 1992), from a loss of equilibrium and failure of muscular control to visual dysfunction. Hyperinflammation of swimbladder and severe congestion of brain are usually observed (Fig. 4G, H) in affected fish.

Diagnosis. VNN can be preliminary diagnosed on the basis of clinical signs and the light microscopic aspect of the brain and/or retina. It must be confirmed with the isolation of the virus in the SSN-1 cell-line together with the direct detection of the virus in fish tissues (brain, spinal cord and eyes or ovarian fluid at spawning) by immunological or genetic methods. The best procedures, recommended by the OIE (2011), include reverse-transcriptase PCR (RT-PCR) protocols to amplify a fragment from the coat protein gene (Gómez et al. 2004, Cutrín et al. 2007). Real-time RT-PCR assays show higher sensitivity (10–100 times) than standard ones (Hodneland et al. 2011). Moreover, serological or molecular assays allow to detect asymptomatic carrier fish that otherwise could have been classified as free of infection (Breuil et al. 2000, Hodneland et al. 2011). The detection of the virus at an early phase in the production cycle helps prevent the disease and slows down its spread to new locations.

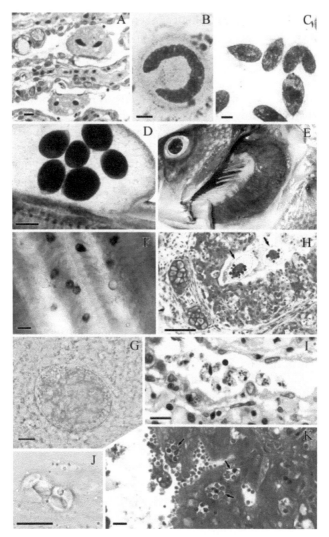

Figure 4. Protozoan parasites of European sea bass. (A-C) *Trichodina* sp. in toluidine blue stained section of gills (A, arrows), in Giemsa-stained smear (B) note the horseshoe-shaped macronucleus. (C) *Philasterides dicentrarchi* Giemsa-stained smear. (D-E) *Cryptocaryon irritans* in the gills. (F-H) *Amyloodinium ocellatum* in fresh smears of gills (F-G), and in a toluidine blue-stained section of gills (H), note the dark-stained amyloopectin granules, (arrows). (I) *Cryptobia branchialis* in a toluidine-blue stained section of gills. (J) Mature oocysts of *Eimeria dicentrarchi* in a fresh smear of intestine. (K) *Cryptosporidium molnari* in a toluidine-blue stained section of stomach invading all the epithelium. Note the groups of oocysts, arrows. Scale bars: 200 μm (D); 50 μm (F, H); 10 μm (A-C, G-K). Illustrations courtesy of: Dr. Ariel Diamant (National Center for Mariculture, Israel) (D), Dr. Panos Varvarigos (VetCare, Greece) (F), Dr. Pilar Alvarez-Pellitero (IATS-CSIC) (G, J). Illustrations original from the authors: ASB (A-C, H, I, K), CZ (E).

Color image of this figure appears in the color plate section at the end of the book.

Risk factors. The disease causes severe mortalities in acute infections (from 50 up to 100%) in larval and juvenile ESB (Breuil et al. 2001), these values decreasing with fish age. Moreover, the virus might also produce a persistent infection, giving raise to asymptomatic carriers where no clinical signs are detectable. The fish susceptibility to the virus is also dependent on temperature and strain. The prevalence and severity of the disease increase at temperatures above 25°C and distinct genogroups possess different replication temperatures (around 25°C or below 15°C) (Korsnes et al. 2005).

Betanodavirus can be transmitted both by horizontal and vertical routes. Horizontal transmission by contact between diseased and naïve fish or through seawater is the most important dispersal mechanism (Péducasse et al. 1999, Hodneland et al. 2011). The presence of betanodavirus in crabs, shrimps and Mediterranean mussels has been demonstrated, and may represent a reservoir for this infection in cultured fish (Gómez et al. 2008). The vertical transmission has been demonstrated experimentally from infected brood fish of ESB to its larval offspring (Breuil et al. 2002).

Control and treatments. There are no efficacious treatments or prophylactic measures for controlling VNN. Most epizootics could be prevented by applying the appropriate biosecurity means: disinfection and quarantine, elimination of carriers by screening the broodstock and antisepsis of eggs. Moreover, to reduce the possibilities of transmission via utensils, vehicles or human activities, disinfectants such as acid peroxygen, ozone or sodium hypochloride should be used (Frerichs et al. 2000, Noga 2011). No vaccine is commercially available for VNN, although several vaccination strategies are currently being assayed (Gómez-Casaddo et al. 2011). In fact, promising results have been obtained with an oral recombinant protein vaccine (capsid) encapsulated and administered using live *Artemia* (Lin et al. 2007), simultaneous bath immunization with an aquabirnavirus and inactivated betanodavirus (Yamashita et al. 2005) and also on a nano-encapsulated inactivated vaccine in grouper larvae (*Epinephelus coioides*) (Kai and Chi 2008).

Pathologies Due to Parasites

Parasites constitute an emerging group of pathogens in cultured fish and outbreaks due to them have been reported worldwide increasingly. Parasites also constitute emerging markers of fish populations (Mattiuci et al. 2007), are biomarkers of pollutants in water (Ferreira et al. 2010) and indicators of the health of an ecosystem. Fish borne zoonotic parasites are also acquiring worldwide relevance in aquaculture (Lima dos Santos and Howgate 2011). Tables 3 and 4 gather the main protozoan and metazoan parasites reported from wild and cultured ESB. The diversity of the parasitofauna is so high that this section will concentrate only on those parasites of relevance, either for cultured stocks or for their zoonotic implications.

Protozoan parasites

In this section, only the two most relevant protozoans causing diseases in ESB will be broached, though in Fig. 4 a selection of images of these and other protozoan parasites present in Table 3 can be found.

Ciliates

The main ciliate disease in ESB, cryptocaryonosis, is due to the holotrich, obligate ectoparasite *Cryptocaryon irritans*. It is also named "marine white spot disease". This parasite is eurixenic and cosmopolitan, but there are differences in host susceptibility. There are also strain differences among different *Cryptocaryon* isolates from various parts of the world, but all of them are considered to be *C. irritans* (Yambot et al. 2003). The life cycle is holoxenic, with stages on and off the fish. The trophont lives embedded in the host's skin, and feeds on tissue debris and body fluids; the mature trophont leaves the host, adheres to the substrate (bottom of the tank, pond, etc.), encysts and begins to reproduce (tomont); after several divisions, the cyst ruptures, releasing the infective stage (tomites). It invades the epithelium of the skin and gills of fish and affects the physiological function of these organs, provoking high mortalities. More information can be found in the review by Dickerson (2006).

Clinical signs. Infected fish have small white spots, nodules, or patches on their fins, skin, or gills (Fig. 4E). They may also have eroded fins, cloudy eyes, pale gills, increased mucus production, changes in skin color, and they may scratch, swim abnormally, act lethargic, or breathe more rapidly as if in distress.

Diagnosis. Cryptocaryon infection cannot be diagnosed unequivocally by the clinical signs, since spots are not always visible and other diseases can cause similar signs. Microscopic evaluation of skin, fin, and gill fresh smears and identification of the trophont-spherical to pear-shaped, ciliated, opaque, rolling (Fig. 4D) or of any other life stages is required to verify infection. Stained smears may help to discern the morphology of the macronucleus (segmented and not horseshoe-shaped). Recently, the transcriptomic analysis of this parasite has identified potential genes for the development of diagnostic and control strategies for cryptocaryonosis (Lokanathan et al. 2010).

Table 3. Protozoan parasites infecting wild (W) and cultured (C) European sea bass.

Group	Species	Host Site	Host Habitat	Geographic Location	Reference
Amoeba	*Neoparamoeba* sp.	Gills	C	Portugal, Spain	Dyková et al. 2000, Santos et al. 2010
Ciliophora	*Cryptocaryon irritans*	Gills	C	Israel, Spain, Croatia, Turkey, Greece	Diamant et al. 1991, Álvarez-Pellitero et al. 1993, Rigos et al. 2001, Mladineo 2006, Çağırgan 2009
	Ephelota sp.	Gills	W	Portugal	Santos 1996
	Philasterides dicentrarchi	Gills, brain	C	France Portugal	Dragesco et al. 1995, Santos et al. 2010
	Trichodina sp.	Gills	C,W	Spain, France, Israel, Norway, Portugal, Italy	Paperna and Baudin-Laurencin 1979, Paperna 1984, Álvarez-Pellitero et al. 1993, Santos 1996, 2000, Duarte et al. 2000, Sterud 2002, Fioravanti et al. 2006
Flagellata	*Amyloodinium ocellatum*	Gills	C, W	Spain, Turkey, Croatia, Israel, Portugal, Italy	Álvarez-Pellitero et al. 1993, Paperna, 1980a, Çağırgan and Tokşen 1996, Duarte et al. 2000, Fioravanti et al. 2006, Mladineo 2006
	Colponema sp.	Gills	C	France, Israel	Paperna and Baudin-Laurencin 1979, Paperna 1984
	Cryptobia branchialis	Gills	C	Spain, Italy	Álvarez-Pellitero et al. 1993, Fioravanti et al. 2006
	Cryptobia sp.	Gills	W	Norway	Sterud 2002
	Ichthyobodo sp.	Gills, skin	C, W	Spain, Norway, Portugal	Santos 1996, Sterud 2002, Duarte et al. 2000, Álvarez-Pellitero 2004
Apicomplexa	*Cryptosporidium molnari*	Estomach	C	Spain	Álvarez-Pellitero and Sitjà-Bobadilla 2002
	Eimeria bouixi	Piloric ceca	C, W	France, Italy	Daoudi and Marques 1987, Fioravanti et al. 2006
	Eimeria dicentrarchi	Intestine	C, W	France, Spain	Daoudi and Marques 1987, Álvarez-Pellitero et al. 1993
	Eimeria sp.	Intestine, liver	C, W	Croatia, Portugal	Santos 1996, Čož -Rakovac et al. 2002
	Hemogregarina sp.	Blood	C	Spain	Álvarez-Pellitero et al. 1993
Microsporea	*Loma* sp.	Intestine	C	Italy	Caffara et al. 2010
	Not determined	Intestine	C	Spain	Álvarez-Pellitero et al. 1993
	Not determined	Muscle	C	Italy	Fioravanti et al. 2006

Risk factors. Cryptocaryonosis is especially virulent under recirculation conditions, or with bad water flows or in earth ponds, where the cystic stages can survive. Water temperature is critical for its development, and ranges between 15°C and 30°C. If the immune status of the fish is compromised or if environmental factors are not optimal, *Cryptocaryon* infection will be even more explosive and harmful. In ESB, outbreaks due to ESB are more frequent in earth ponds.

Treatment and control. Because the trophont is embedded within the skin, it is relatively protected from any potential treatments. In any case, copper sulfate penthahydrate appears to be the most effective treatment to date, though important environmental and consumer safety issues forbid its use in aquaculture stocks. Chloroquine, formaline and hyposalinity have also some effectiveness, and drying and desiccation of ponds is also required (Colorni 1987). Targeted studies to characterize the best immunogenic antigens of *C. irritans* that confer protection in other fish species than ESB have been ongoing for a number of years. Several articles have reported that immunization with theronts, trophonts and tomonts of *C. irritans*, and the latest immunization tests with a DNA vaccine are very promising (Priya et al. 2012). However, no commercial vaccines are currently available.

Flagellates

Amyloodinium ocellatum is a dinoflagellate with a direct, but triphasic life cycle. The stationary or feeding stage, the trophont, remains attached to the skin and gill surface by root-like structures. When the trophont matures, it detaches from the host, forming a reproductive "cyst" or tomont in the substrate. This tomont divides, forming many free-swimming dinospores that can infect a new host. This parasite causes serious morbidity and mortality in aquacultured brackish and marine warm-water fish worldwide and it is considered very pathogenic for larval ESB where mortalities can reach up to 90% in Adriatic ESB cultures (Sarusic 1990). In Turkey, *Amyloodinium* infection in pond-cultured ESB of 15–20 g produced 100% mortality (Çağırgan and Tokşen 1996).

Clinical signs. The clinical signs of amyloodiniosis are proportional to the epithelial damage invoked by the parasite in the epithelial cells of the skin and gills, and include anorexia, depression, dyspnea and pruritis. The gills are the primary site of infection, which appear hyperplasic, haemorrhagic and necrotic. In heavy infections, skin, fins and eyes can also be involved, with the fish presenting a dusty appearance, which is the reason for calling it "velvet disease", but in most cases, fish die without obvious gross skin lesions. Death is usually attributed to anoxia, but osmoregulatory impairment and secondary bacterial infections may also contribute to the weakness and death of the fish.

Diagnosis. Presumptive diagnosis may be made from the gross clinical signs of the fish, but it must be confirmed by microscopic examination of fresh skin and gills scrapings (Fig. 4 F, G) taken from fish still living or immediately after death, as parasites detach shortly after death. Staining of smears with dilute Lugol's iodine helps to visualize the parasite, since the iodine reacts with the starch of the parasite. Histopathology can also be used for diagnosis (Fig. 4H), but possibly many trophonts dislodge during fixation, making it difficult to estimate the severity of the infection. A PCR assay based on consensus *Amyloodinium*-specific oligonucleotide primers has been developed, which detects as few as 10 dinospores/ml of water (Levy et al. 2007).

Risk factors. Temperature and salinity are the primary environmental modulators of the pathogenicity, with greater virulence at higher temperatures. The temperature tolerance of *A. ocellatum* isolates varies among isolates, but typically ranges from about 16°C to 30°C; salinity tolerance is also variable from 50 ppt to 12 ppt. The parasite cannot reproduce below 15°C, but the life cycle often may resume when the temperature becomes higher. Other risk factors such as low dissolved oxygen and young fish ages have been associated with some outbreaks.

Treatment and control. Noga (2012) has reviewed a number of drugs that have been used to treat *A. ocellatum*-infected fish (copper, chloroquine, hydrogen peroxide, formalin, lasalocid), but none unequivocally cure fish; rather, they only control the disease, because most act only against the infective dinospore and are generally ineffective against the trophont or tomont. For routine disinfectant in aquaculture facilities (tanks, pipes, nets, etc.), benzalkonium chloride is effective, but its use is not approved in all countries. Lowering salinity or temperature delays, but does not prevent infection. Thus, given the lack of a safe, effective therapeutant for the control of *Amyloodinium*, avoidance is extremely important for preventing outbreaks and recurring problems. The parasite is easily introduced with infected fish or water, and once in a facility, it may be spread by aerosolization, fomites, or subclinically infected fish. Thus, quarantine and disinfecting of incoming water with ultraviolet irradiation or ozone (for killing dinospores) are essential, especially in re-circulating systems. The strong evidence for a protective immune response against this parasite holds promise for eventual development of protective vaccines.

Metazoan parasites

In this section, only the most relevant metazoans causing diseases in ESB will be broached, though in Figs. 5 and 6 a selection of these and other metazoan parasites present in Table 4 can be found.

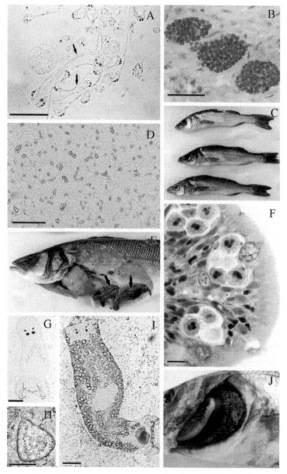

Figure 5. Metazoan parasites of European sea bass. (A) Fresh smear of bile with trophozoites and spores of *Ceratomyxa labracis*. Note the long appendages of the spores (arrows). (B) *Sphaerospora dicentrarchi* in a PAS-stained section of intestine. (C) Juveniles infected by *S. dicentrarchi*. Note the depressed abdomen and the poor dorsal musculature. (D) *Sphaerospora testicularis* in a fresh smear of seminal fluid. (E) Necrotic testes (arrows) infected by *S. testicularis* at the end of the spawning season. (F) *Enteromyxum leei* in a toluidine blue-stained section of intestine. (G-J) *Diplectanum aequans* juvenile (G), egg (H), adult (I), affected gills (J). Scale bars: 100 μm (I); 50 μm (A, D, H); 20 μm (B); 10 μm (F, G). Illustrations original from the authors: ASB (A, B, D, F-I), CZ (C, F, J).

Color image of this figure appears in the color plate section at the end of the book.

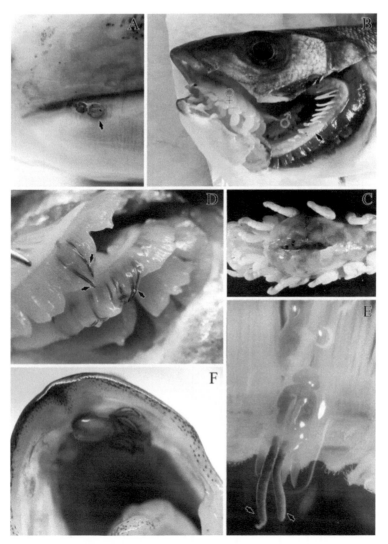

Figure 6. Crustacean parasites of European sea bass. (A) *Caligus minimus* at the base of the mouth. (B) *Cerathotoa oestroides*. Note the female (♀) and the male (♂) in the oral cavity and the larvae (arrow) in the gill arch. (C) Gravid female of *C. oestroides* full of larvae. (D-E) *Lernanthropus kroyeri*. The ovigerous sacs of females are visible out of the gill filaments (arrows). Illustrations courtesy of: Dr. Ivona Mladineo (Institute of Oceanography & Fisheries, Croatia) (C) and Dr. Panos Varvarigos (VetCare, Greece) (D-F). Illustrations original from the author CZ (A, B).

Color image of this figure appears in the color plate section at the end of the book.

Table 4. Metazoan parasites infecting wild (W) and cultured (C) European sea bass (I).

Group	Species	Host Site	Host Habitat	Geographic Location	Reference
Myxozoa	*Ceratomyxa diplodae*	Gall bladder	C, W	Spain, Italy, Portugal, Greece	Álvarez-Pellitero and Sitjà-Bobadilla 1993, Santos 1996, Rigos et al. 1999, Fioravanti et al. 2006
	Ceratomyxa labracis	Gall bladder	C, W	Spain, Italy, Portugal, Croatia	Álvarez-Pellitero and Sitjà-Bobadilla 1993, Santos 1996, Fioravanti et al. 2006, Mladineo 2006
	Enteromyxum leei	Intestine	C	Israel, Italy, Greece	Álvarez-Pellitero 2004, Fioravanti et al. 2006, Sitjà-Bobadilla et al. 2007
	Henneguya sp.	Heart	C	France	Siau and Sakiti 1981
	Kudoa sp.	Int. muscle	C	Greece	Rigos et al. 1999
	Myxobilatus sp.	Spleen, kidney, u. bladder	W	Spain, Portugal	Álvarez-Pellitero et al. 1993, Santos 1996
	*Sphaerospora dicentrarchi**	Systemic	C, W	Spain, Italy, Portugal, Croatia, Greece	Sitjà-Bobadilla and Álvarez-Pellitero 1993a, Santos 1996, Rigos et al. 1999 Fioravanti et al. 2004, 2006, Mladineo 2006
	Sphaerospora testicularis	Testes	C, W	Spain, Italy	Sitjà-Bobadilla and Álvarez-Pellitero 1993b, Fioravanti et al. 2004, 2006, Fernandes et al. 2007
Monogenea	*Diplectanum aequans*	Gills, skin	C, W	Spain, France, Turkey, Greece, Croatia, Norway, Portugal, Italy	Silan and Maillard 1986, González-Lanza et al. 1991, Santos 1996, Čož-Rakovac et al. 2002, Sterud 2002, Mladineo 2004, Fioravanti et al. 2006, Vagianou et al. 2006a, Öktener et al. 2009
	Diplectanum laubieri	Gills, skin	C, W	Spain, France	Lambert and Maillard 1975, González-Lanza et al. 1991
	Serranicotyle labracis	Gills, skin	C, W	France, Portugal, Greece	Maillard et al. 1988, Silan and Maillard 1989, Santos 1996, Ragias et al. 2004

*This parasite was wrongly described as *Myxobolus* sp. in early reports.

Table 4. Metazoan parasites infecting wild (W) and cultured (C) European sea bass (II).

Group	Species	Host Site	Host Habitat	Geographic Location	Reference
Digenea	*Acanthostomum imbutiforme*	Rectum, intestine	W, C*	Italy, France, Spain	Maillard 1976a, Giavenni 1988, Lozano et al. 2001
	Brachipallus crenatus	GI tract	W	Norway	Sterud 2002
	Bucephalus baeri	Rectum	W	France, Norway, Portugal	Maillard 1976a, Santos 1996, Sterud 2002
	Bucephalus labracis	Piloric caeca	W	France, Tunisia, Norway	Maillard 1976a, Sterud 2002, Ben Abdallah and Maamouri 2005
	Cainocreadium labracis	Intestine	C*, W	France, Norway, Portugal, Spain, Italy	Maillard 1976a, Giavenni 1988, Fernández et al. 1989, Santos 1996, Sterud 2002
	Derogens varicus	GI tract	W	Norway, Portugal	Santos 1996, Sterud 2002
	Hemiurus communis	Stomach	W	Portugal	Santos 1996
	Hemiurus sp.	GI tract	W	Spain	Ortega et al. 1991
	Helicometra fasciata	Digestive tract	W	Spain	Ortega et al. 1991
	Labatrema minimus	GI tract	C*, W	France, Norway, Portugal, Italy, Spain	Maillard 1976b, Giavenni 1988, Santos 1996, Lozano et al. 2001, Sterud 2002
	Lecithochirium gravidum	GI tract	W	Portugal	Ortega et al. 1991, Santos 1996
	Phrosorhynchus crucibulum	Fins, kidney gills, heart	W	Portugal	Santos 1996
	Riphiidocotyle sp.	GI tract	W	Spain	Ortega et al. 1991
	Tomoniella imbutiforme	Intestine	W	Norway, Portugal	Santos 1996, Sterud 2002
	Timoniella praeteritum	Piloric caeca	W	France, Norway, Portugal	Maillard 1976a, Santos 1996, Sterud 2002

Table 4. contd....

Table 4. contd.

Group	Species	Host Site	Host Habitat	Geographic Location	Reference
Nematoda	*Anisakis simplex*	GI tract, muscle	W	Northwest Atlantic (FAO area 27)	Ortega et al. 1991, Bernardi et al. 2011
	Anisakis pegreffii	Mesenteries, stomach	W	Sardinia	Culurgioni et al. 2011
	Anisakis sp.	GI, abdominal cavity	W	Norway	Sterud 2002
	Hysterothylacium aduncum	GI	W	Norway	Sterud 2002
	Hysterothylacium sp.	GI tract	W	Spain	Ortega et al. 1991
	Philometra sp.	gills	W	Croatia	Čož-Rakovac et al. 2002
Cestoda	*Botriocephalus scorpii*	GI tract	W	Spain	Ortega et al. 1991
	Botriocephalus sp.	GI tract	W	Norway	Sterud 2002
	Triaenophorus sp.	Liver	C	Croatia	Čož-Rakovac et al. 2002
Acanthocephala	*Telosentis molini*	Intestine	W	Portugal	Santos 1996
Mesomycetozoa	*Ichthyophonus* sp.**	Heart, spleen, kidney, liver	C	Spain	Franco-Sierra et al. 1997

* broodstock
** This species is included in this table for structural purposes, since its taxonomic ascription is controversial. It has changed from protozoans to fungi, and now is in the class Mesomycetozoa which is included within Opisthokonta and not in fungi.

Table 4. Metazoan parasites infecting wild (W) and cultured (C) European sea bass (III).

Group	Species	Host Habitat	Geographic Location	Host Site	Reference
Crustacea	*Anilocra physodes*	C	Spain, France, Greece	Gills, skin	Álvarez-Pellitero 2004
	Caligus apodus	C	Greece	Skin, oral cavity	Ragias et al. 2004
	Caligus dicentrarchi	W	Portugal	Skin	Santos 1996
	Caligus minimus	C, W	Israel, Greece Norway, Ireland, Portugal, Italy, Turkey	Gills, skin, oral cavity	Paperna 1980b ,1984, Santos 1996, Caillot et al. 1999, Holmes and O'Connor 1999, Sterud 2002, Vagianou et al. 2006a, Fioravanti et al. 2006, Çağırgan 2009
	Caligus mugilis	C	Greece	Skin, oral cavity	Ragias et al. 2004
	Caligus pageti	W, C	Portugal, Greece	Skin, oral cavity	Santos 1996, Ragias et al. 2004
	Cerathotoa collaris	W	Lebanon	Oral and branchial cavities	Bariche and Trilles 2008
	Ceratothoa oestroides	C, W	Croatia, Greece, Turkey	Gills, skin	Sarusic 1999, Horton and Okamura 2001, Mladineo 2002, Vagianou et al. 2006b
	Colobomatus labracis	C, W	France, Tunisia, Portugal	Gills	Raibaut et al. 1979, Santos 1996
	Emetha audouini	C	Greece	Skin, oral and branchial cavities	Papapanagiotou et al. 1999
	*Lepeophtheirus salmonis**	W	Scotland	Skin	Pert et al. 2006
	Lernaea sp.	W	Croatia	Skin, muscle	Čož-Rakovac et al. 2002
	Lernaeolophus sultanus	C	Greece	Skin, mouth	Varvarigos 2007
	Lernanthropus kroyeri	C, W	Corsica Turkey, Greece, Egypt, Norway, Tunisia, Portugal	Gills	Cabral et al. 1984, Caillot et al. 1999, Sterud 2002, Bahri et al. 2002, Santos 1996, Vagianou et al. 2006a, Tokşen 2007b, Antonelli et al. 2012
	Nerocila orbignyi	C, W	Corsica, Turkey	Skin	Bragoni et al. 1984, Horton and Okamura 2001

* peripatetic host

Myxozoans

Myxozoans constitute a numerous group of obligate, microscopic, metazoan endoparasites affecting mainly fish species. Myxozoan life cycles include alternate hosts (invertebrates and vertebrates), but for ESB the life cycle of the thus far reported myxozoan species is unknown. Mass mortalities and important pathologies related to myxozoan infections have been recorded in freshwater and marine fish. In ESB, several myxozoan species have been described, but only the two sphaerosporids and *Enteromyxum leei* will be described in detail.

Sphaerospora dicentrarchi (Fig. 5B) and *S. testicularis* (Fig. 5D) represent ubiquitous pathogens of cultured ESB in the western Mediterranean basin. *S. dicentrarchi*, described for the first time in Spain (Sitjà-Bobadilla and Álvarez-Pellitero 1992a) is a systemic species now considered one of the most common parasites of ESB with records both in Mediterranean and Atlantic waters reaching 100% prevalence in wild populations. On the contrary, *S. testicularis* (Sitjà-Bobadilla and Alvarez-Pellitero 1990), is specific of testis, is less prevalent that *S. dicentrarchi* and has very rarely been found in wild populations (2.5%). The southern-most record of this parasite is at the Canary Islands, in which a 20% prevalence of infection has been found in ESB escaped from farms (Toledo-Guedes et al. 2012). The main pathological effect of this testicular myxozoan is parasitic castration, which can have important effects in broodstocks (Sitjà-Bobadilla 2009).

  E. leei (Fig. 5F) is becoming a universal problem infecting more than 46 marine fishes and the geographical distribution comprises the Canary Islands, the Mediterranean and Red Sea and Western Japan. Although it has a low degree of host specificity, its pathogenic capacity is host-specific. This parasite can be transmitted from fish to fish, making its dispersal in cultured stocks very easy. The prevalence and mortality due to this myxozoan in ESB is lower than in sparids under farming conditions (Varvarigos 2003, Fioravanti et al. 2006). More information on *Enteromyxum* can be found in the review by Sitjà-Bobadilla and Palenzuela (2012).

Clinical signs. S. dicentrarchi seems to be harmless in light infections, becoming a serious problem at high intensity, mainly under stressful conditions and particularly in fry (Sitjà-Bobadilla and Alvarez-Pellitero 1993a,b, Rigos et al. 1999). Severe enteritis in heavily infected ESB was especially observed at the end of summer (Fioravanti et al. 2004). A low-grade mortality was sometimes observed, often associated with bacterial infections (vibriosis) or bad environmental conditions. Fish with heavy infections appear with depressed abdomen and poorly developed dorsal musculature (Fig. 5C).

  S. testicularis infected males are easily detected during the spawning season because of their abdominal distension produced by the accumulation of ascites or gonad hypertrophy, and the watery quality of the seminal fluid. The genital pore can also appear enlarged and reddish. However, during the resting period they remain completely asymptomatic, and only at necropsy do infected testes appear necrotic, hardened and much larger than non-infected ones, due to the presence of large granulomata that reduce the amount of germinal tissue for the next spawning season (Fig. 5E).

The effects of *E. leei* on ESB are milder that in sparids. Experimentally infected fish did not show any of the typical disease signs of enteromyxosis (loss of appetite, poor food conversion rates and difficulties to reach commercial size in the final, severe emaciation with epiaxial muscle atrophy), no mortalities were recorded, and animals eliminated the parasite 15 weeks later (Sitjà-Bobadilla et al. 2007). However, there is evidence of stunted growth in some strains of experimentally infected young ESB (A. Diamant unpublished data), in which the intestinal epithelium is invaded by the parasite, sometimes with detachment from the lamina propria and eosinophilic and lymphocytic infiltration in the submucosa.

Diagnosis. The main routine diagnostic method for myxozoans is the examination of fresh smears or histological sections of the target tissues: gall bladder, swim bladder and intestine for *S. dicentrarchi*, testes and seminal fluid for *S. testicularis* and the posterior intestine or the rectum for *E. leei*. For the latter, when the parasite is in a latent location, or low numbers of parasites with a patchy distribution are present, the infection may be missed. In such cases, molecular methods such as PCR (Palenzuela et al. 2004, Yanagida et al. 2005) and *in situ* hybridization are conclusive (Cuadrado et al. 2007).

Risk factors. For *S. dicentrarchi*, extensive and intensive systems showed similar values of prevalence, indicating the presence of a common risk factor (an alternate host, such as benthic annelid) in these environments. The infection also increases during the warm season and in older animals. For *S. testicularis* the main risk factor is the season, as the parasite follows the reproductive cycle of the host. For *E. leei* the main risk factors are year-round elevated water temperatures, poor water exchange and/or re-intake of contaminated effluent water, recirculation systems, and a prolonged culture period necessary for production of large fish.

Control and treatment. There are no approved antiparasitic preparations for myxozoans in general, and those tested experimentally, mainly coccidiostats, have had relative success. The combination of salinomycin and amprolium significantly reduced prevalence, intensity and mortality in *E. leei*-infected sharpsnout sea bream, without apparent toxic effects (Golomazou et al. 2006). Other drugs such as fumagillin tested against *S. testicularis* and *E. leei* have toxic effects on the host or increase the mortality rates (Sitjà-Bobadilla and Alvarez-Pellitero 1992b, Golomazou et al. 2006). Management measures can also be applied for *E. leei* infections, such as culling of dead fish, avoiding placement of new stocks in the vicinity of infected ones, and constant survey of the epidemiologic situation of the stocks.

Monogeneans

They are skin and gill ectoparasites causing important pathologies worldwide. In cultured ESB, *Diplectanum aequans* is the most relevant (Figs. 5G-I). Although its distribution is almost coincident with that of the host, its impact is higher in Mediterranean waters, particularly in Greece and Turkey. This parasite may produce not only direct morbidity and mortality, but also predispose fish to secondary fungal, bacterial and/or viral infections resulting in high mortalities. The histopathological damage to gills is notorious (Olivier 1977, Dezfuli et al. 2007), especially when the infection levels reach thousands of individuals per fish.

Clinical signs. D. aequans feeds on skin and gill epithelial cells. They attach to the fish surface with the haptor, provoking irritation, mucus hypersecretion, hyperplasia of the gill epithelium, disruption and fusion of secondary gill lamellae, haemorrhages, marked erosion and inflammation of the epithelium (Fig. 5J). All this leads to a respiratory dysfunction and eventually death when the intensity of infection is very high. Heavily infected fish are lethargic and swim close to the surface with open mouths.

Diagnosis. Diagnosis is very easy in fresh scrapings from gills at low magnifications at the microscope, and does not require sacrificing the fish. The precise species identification may need higher magnification and a certain degree of expertise to differentiate this species from *D. laubieri* on the basis of the shape, size, and number of hamuli and hooks on the haptor, and male copulatory organ of adult parasites.

Risk factors. The life cycle is direct and therefore fish to fish transmission is easy and especially favoured by high density culture conditions in sea cages. Juvenile ESB and broodstocks are especially susceptible to this monogenesis. Poor water quality (high turbidity, high organic matter content, low oxygen concentration) (Papoutsoglou et al. 1996), poor sanitary practices (tanks, pipes net cleaning) and bad handling practices such as pooling of non-surveyed stocks notably increases the proliferation and impact of this monogenean. High water temperature is another risk factor, which is a determinant for the development of the parasite. Adults produce high numbers of eggs which float in water currents and easily entangle in net cages. The parasite can survive winter conditions in the cages and spreads easily to new stocked animals. Net biofouling not only provides a suitable place for the entanglement of eggs, but also prevents a high water flow, which is essential for animals with diminished gill function.

Control and treatments. Formalin is the main chemical used in sea cages, but adult parasites can survive even concentrations of 1,500 ppm (Cruz e Silva et al. 2000). Baths of 6 ppm of Rafoxanid for 48 hours (Cognetti-Varriale et al. 1992), 0.15 ppm of the organophosphate trichlorfon for 2 days (Cognetti-Varriale et al. 1991) or 1 ppm azametifos for two hours (Tokşen 2007a) are effective. However, immersing treatments of large stocks are costly, impractical and stressful to fish. Recently, trichlorfon has been reported to reduce the prevalence of infection up to 52% when cultured ESB were fed for seven days with a dose of 50 mg/Kg of fish. However, this drug should be used only during acute outbreaks, due to the risk of antihelmintic resistance (Tokşen et al. 2012). In any case, chemical treatments have to be synchronized with net and tank cleaning (either with disinfectants or by dehydration), especially with the introduction of new animals in the facilities.

Crustaceans

Parasitic crustaceans produce serious economic losses in farmed fish worldwide. For ESB, among the different ectoparasitic crustaceans listed in Table 4, the isopods *Ceratothoa oestroides* and *Nerocila orbignyi* are the most relevant and comprise a frequent problem in Mediterranean waters (Trilles et al. 1989, Öktener and Trilles 2004). Both are euryxenic species, reported in many wild and cultured fish species and have a holoxenic (direct) life cycle. *C. oestroides* is a fastidious and persistent parasite that seasonally inflicts serious losses in ESB fingerlings and juveniles in Turkish and Greek sea cages (mortalities up to 15%). The adults are found in pairs and occur mainly in the buccal cavity, while the infective larval stages and the juveniles are present also in the opercular cavity, on the fish head, behind the eye or behind the operculum, above the lateral line, on the caudal fin and the caudal peduncle (Figs. 6B, C). On the other hand, *N. orbingyi* attaches to the skin, fins and gills. Adults of both isopod species are haematophagous (feed on blood).

Clinical signs. Larval stages of *C. oestroides* cause considerable damage to the skin around the head, the eyes and the gill epithelium. The infected fish are usually apathetic and anorexic and may show respiratory distress. Parasitized fish are more susceptible to secondary bacterial pathogens, such as *Aeromonas* spp., *Tenacibaculum* spp., *Vibrio* spp., to lymphocystis and *Rickettsia*-like infections and this may lead to severe escalation of mortality (Vagianou et al. 2006b). The adults cause anaemia, with significantly lower erythrocyte counts, haematocrit and haemoglobin values (Horton and Okamura 2003). *N. orbignyi* also produces a decrease in body condition, weight, blood protein, lipids and triglycerides levels, but an increase in blood levels of urea. It also provokes hypochromic macrocytic anemia with increase in eosinophils, neutrophils and a decrease in lymphocytes (Bragoni et al. 1983). In addition, adult *C. oestroides* isopods can cause considerable damage to the mouth tissues with their biting and sucking mouth parts, or their copulation activity. Their large size (up to 6 cm in length) may cause atrophy of the tongue, dysplasia of teeth and slackening of the cartilaginous tissues leading to a "bag-shaped" lower jaw. The presence of large adult parasites in the buccal cavity interferes with feeding, causes chronic stress and results in growth retardation and a predisposition to bacterial and/ or endo-parasitic invasions. A negative correlation was found between the abundance of *C. oestroides* and host weight (Mladineo et al. 2010). At harvest, fish that have survived parasitism also have developed the "bag-shaped" jaw. Although, parasitism by isopods poses no risk for the consumer, the repelling characteristics of infected fish reduce the marketability of the fish product. The cost of rejects may run high in cases of heavy infestations (usually 1%, but also up to 25% of prevalence among harvest-size fish).

Diagnosis. Isopod infections are confirmed by gross observation of the parasites on the skin, mouth, or in the gill chamber of the fish. *C. oestroides* adults in the mouth are easier to detect as they are usually 4 cm–5 cm long. In addition, they often produce characteristic lesions. Thus, fish infected by larval stages may be recognized by their hemorrhagic and necrotic head tissues.

Risk factors. Fecundity and hatching rate of the isopoda increases with water temperature (Mladineo 2003) and therefore these parasites proliferate mostly during the summer, peaking during July and August, when the prevalence of infection in the cages may exceed 50% (Vagianou et al. 2009). Other risks factors are excessive fish densities, weak sea currents, proximity of vulnerable fry with on-growers carrying adult isopods. It has also been hypothesized that large numbers of feral fish around the cages may transfer the infection to cultured fish. However, a recent molecular study has shown that the parasites isolated from wild and farmed fish on the two most productive Adriatic fish farms showed genetic heterogeneity, contradicting the widely accepted hypothesis of cross-contamination between wild and cultured fish (Mladineo et al. 2009).

Control and treatment. Preventive measures are very important. They include adequate management of the stocks, such as avoiding overcrowding in fry holding nets, avoiding placing young fish in close proximity with adult ones probably harbouring adult parasites in reproductive phase, locating cages in deep sites with sufficient water currents and fallowing of sites. In some farms, manual delousing with a small blunt forceps, while fish are anaesthetised for vaccination results in a sharp drop of fish retaining adult isopods, with the subsequent sharp reduction in the number of larval isopods.

Because of the sheltered location in the buccal cavity of fish, chemotherapeutics are not as effective for some adult parasites as for isopod larvae infections. The range of available drugs in Mediterranean fish is very narrow, with only a few drugs efficient against isopods. Preliminary work with emamectin benzoate has had disappointing results in ESB (Athanassopolou unpublished results). For the pyrethroid deltamethrin, the minimal *in vitro* dose that kills *C. oestroides* adults in two hours is 0.05 mg/litre (Athanassopoulou et al. 2001) and 3 ppb of Alphamax™ for 30 minutes is efficacious and safe for treating small ESB (<10 g) in sea cages against juvenile isopods. Larger ESB parasitized with adult *Ceratothoa* were effectively and safely treated with baths of 7.5 ppb deltamethrin for 30 minutes and with diflubenzuron administered in feed (Bouboulis et al. 2004). Extensive empirical/commercial data exists on the efficacy of cypermethrin against all stages of the isopods: 5–10 ppb for 60 min (Excis™) on all age classes of ESB at many different locations. In all bath treatments, a tarpaulin is required as well as a continuous supply of oxygen, especially during the warm season. Field data on environmental implications or about the potential acquisition of resistance by the isopods against these compounds is lacking.

Nematodes

Anisakiasis is the most important human fish-borne zoonosis in the Mediterranean basin caused by larval stages of the family Anisakidae, such as *Pseudoterranova*, *Contracaecum* and *Anisakis*. Anisakid larvae have been detected worldwide in a large variety of fish belonging approximately to 200 fish species (Abollo et al. 2001). The life cycle of these cosmopolitan parasites involves marine mammals (cetaceans, pinnipeds) or fish eating birds as definitive hosts and various aquatic invertebrate species, and teleosteans and cephalopods are intermediate and/or paratenic hosts. Infection with anisakids can affect the commercial value of fish, particularly when larvae are located in the musculature and thus represent important economical loss for the fishing industry. The incidence of anisakiasis in the human population is increasing with the growing trend for consumption of raw, under-cooked seafood (Rosales et al. 1999).

In wild ESB from Northeast Atlantic Ocean (F.A.O area 27), anisakid larvae were found in 95% and 42.5% of viscera and edible muscle, respectively (Bernardi 2009). A more detailed study on the prevalence of *Anisakis simplex sensu stricto* in the same area showed a lower incidence: the visceral prevalence was 0.95% and the main localization of the larvae was under the gastric serosa, whereas the main localization in edible parts was in belly muscles, with a prevalence of 0.425% (Bernardi et al. 2011). The prevalence in Mediterranean ESB was believed to be lower (50%) (Culurgioni et al. 2011), but a recent report from the Mediterranean coast of Egypt has shown a high prevalence of Anisakidae larvae in viscera (76.7%) (Morsy et al. 2012).

Clinical signs. Nematode infections do not usually threaten fish survival. Larval stages are normally encysted in fish tissues with no detectable effects and without any apparent tissue reaction. Some reports show that the parasite can cause a localized, the parasite can cause a localized host inflammatory response following the death of individual nematodes (Hauck and May 1977, Beck et al. 2008). In experimental infections with *A. pegreffii*, some ESB showed hemorrhages and/or irregular neoformations within the celomic cavity (Macrì et al. 2012). Significance of infection to host survival is unknown.

Diagnosis. The detection of anisakid larva in fish can be done by visual inspection of the visceral cavity and fillets. Whitish larvae between 10 mm to 40 mm long appear coiled like a watch spring in capsules of irregular shape on the surface of the viscera. Muscle fillets can be screened using candling techniques. Dead nematodes in the flesh fluoresce in the presence of ultraviolet light. More sensitive methods include digestion of viscera and muscle with pepsin solutions. The identification to the genus level can be done morphologically from specimens fixed in 70% ethanol and cleared with glycerine, but the unequivocal identification to the species level is currently done with molecular/genetic methods such as DNA sequence analysis and allozyme markers (Mattiucci and Nascetti 2008, Chen et al. 2008a).

Risk factors. The rate of development of nematodes in seawater is temperature dependent. Furthermore, lower environmental temperatures reduce the numbers of larvae that penetrate into fish flesh. Larval worms in fish tend to be more prevalent in areas where the various hosts occur in large numbers such as in inshore waters. It has been hypothesized that massive evisceration and discard of fisheries in large freezer vessels has efficiently feed-back the cycle of anisakids in the wild food chain. As fish get infected when ingesting infected invertebrates or other small fish, the probability of infection in aquacultured fish should be lower. In fact, a recent report on cultured ESB in Southeast Spain has shown a null prevalence of anisakids (Peñalver et al. 2010). Thus, the consumption of farmed ESB carries a minimal human risk of exposure to these nematodes in this region. However, some aquaculture practices do not avoid the access to these intermediates hosts, and prevalence of infection by anisakid larvae in viscera can be as high as 50% in some maricultured fish in China (Chen et al. 2008b).

Control and treatment. None feasible for fish.

Non-infectious, Multifactorial, Anthropogenic Diseases

Diseases due to non-infectious factors can have a high economical impact in the production cycle of aquacultured species; however, little attention is paid to them. Some of them are produced by routine aquaculture practices and others are due to environmental disturbances. Here we describe the most relevant and emerging pathological conditions observed during the last years in the clinical practice of ESB farming in the Mediterranean area, which are depicted in Fig. 7. Aquaculture practices that induced skeletal and muscle deformities during larval development, and pathologies associated with nutritional problems are explained in other sections of this book.

Pathologies induced by aquaculture practices

Stress, traumatic lesions and secondary infections

ESB has a higher physiological stress response compared to other Mediterranean aquacultured fish species (Fanouraki et al. 2011), which makes it very sensitive to any stressful situation associated with intensive aquaculture practices. A whole chapter is devoted to stress in this book. Daily farming operations in nurseries and on-growing facilities have always associated stress and mechanical lesions. High stocking densities (Person-Le Ruyet and Le Bayon 2009), not careful handling, grading and vaccination processes can cause traumatic injuries in the skin of the fish, particularly in the fins (Figs. 7A-C). These external lesions can cause direct mortality if severe or, most frequently, can be colonised by secondary bacteria such as *Tenacibaculum* sp. (Figs. 2F–H), this being the most commonly diagnosed bacterial diseases in farmed ESB. Transportation of ESB fry and fingerlings from the nursery to the on-growing farm by truck or well boat can be especially stressful and traumatic if fish have to travel long distances, particularly if water quality is not controlled. Adverse weather conditions, strong currents, storms and lightings, not infrequent in off-shore fish farming, can also be responsible for external lesions and even mortalities due to nervous movements of fish against the nets.

In addition, the immunodepression associated with post-stress situation makes that true or opportunistic pathogens present in the environment can easily infect the fish. Thus, outbreaks of photobacteriosis in juvenile ESB are commonly reported a few weeks after introduction in the on-growing facility in Mediterranean farms.

Some mitigation practices can be applied to minimize the impact of stressful farming practices, such as starvation before the operation, avoidance of manipulation at high water temperatures and increasing the recovery time according to the stress intensity. During transport, sedation

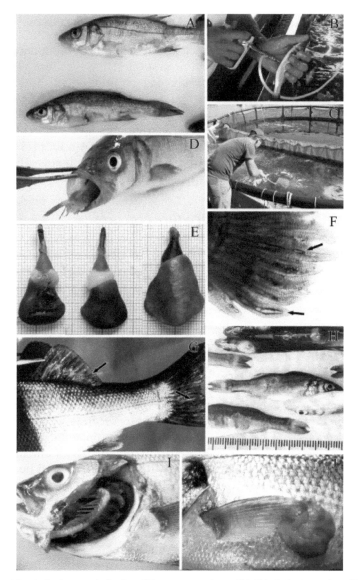

Figure 7. Non-infectious pathologies of European sea bass. (A) Skin lesions produced by bad handling during routine aquaculture procedures, like vaccination (B) or sedation, treatments and samplings in tarpaulins (C). (D) Cannibalism. (E) Hearts of fish affected by SMS (left, centre) compared with a normal heart from a wild fish (right). (F-G) Gas bubble disease, air bubbles are visible in the gill capillaries (F, arrows) and the fins (G, arrows). (H) Typical skin lesions due to nefrocalcinosis. (I) Gill affected by jellyfish. (J) Fibropapilloma. Illustrations courtesy of Dr. Panos Varvarigos (VetCare, Greece) (B, C, F-H). Illustrations original from the author CZ (A, D, E, I, J).

Color image of this figure appears in the color plate section at the end of the book.

of the fish is desirable, since oxygen consumption and CO_2 and NH_3 production are all decreased. More information on adequate handling of fish during transportation can be found in the review by Southgate (2008). Treatment with oral antibiotics is often necessary to control the secondary bacterial infections.

Sudden Mortality Syndrome (SMS)

SMS if often used to describe a condition characterised by hyperacute mortalities observed in ESB farmed in floating cages in the Mediterranean Sea, generally after severe stressful situations, like deficient, wavy cage movements, net changes, harvests or storms. Sometimes the stressor that initiates the problem is difficult to elucidate. After the stress period, fish exhibit acute hyperactivity by fast swimming that leads to apparent exhaustion and finally death, generally with open mouth and opercula. Different units of the same farm can be affected in the same event and SMS susceptible fish are generally larger than 150 g. Mortality can reach 0.5% to 5% in each affected unit in two to three days, and then mortality goes back to normal if no secondary infections occur.

Clinical observations suggest that ESB affected by SMS have cardiac morphological changes similar to those reported by Poppe et al. (2003) in Atlantic salmon and rainbow trout. ESB that suffer SMS have smaller hearts with more rounded ventricles and fat deposits in the epicardium than wild ESB (Fig. 7E). Completely abnormal heart morphology has also been reported in a 1.4 kg individual that died suddenly in a net pen in France (Gamperl and Farrell 2004). During the stressful event, oxygen demand is much higher, and probably abnormal hearts cannot work efficiently under this pressure. The cause of the altered heart morphology is still unknown, but it could be multifactorial. Some factors already pointed out are genetic selection, rapid growth and a sedentary life style (Poppe et al. 2003).

Cannibalism

In ESB hatcheries and nurseries, cannibalism can be a major problem that causes high mortalities (Fig. 7D). High densities in tanks of ungraded populations associated with incorrect feeding practices seem to be the origin of the disorder (Katavic et al. 1989). Typically, large fish attack their small siblings, swallowing them and also depriving them from food access. This behaviour quickly increases the difference in size within the affected population and mortalities of the smaller fish can reach 25% if corrective measures are not applied. During severe episodes of cannibalism, secondary infections by *Tenacibaculum* sp. can also happen due to colonisation of the provoked micro-traumatisms.

This problem should be prevented by a combination of proper grading of the population and careful and correct feeding management, with frequent feedings per day distributed uniformly in the tanks.

Gas bubble disease (GBD)

Inadequate degassing of pumped water into inland facilities induces a super-saturation of dissolved gases in water which provokes typical external signs in affected fish. In earth ponds with excessive organic matter, plants and algae can increase oxygen input through photosynthesis. Gases in supersaturated water will try to escape out of the water into any medium where the gas saturation level is lower. GBD damages the fish's tissue, causing tiny gas bubbles (like tiny blisters) in the capillaries in the gills, skin (fins) and eyes of the animal, restricting the blood flow and forming haemorrhages and clots (Figs. 7F, G). Fish often show signs of swimming upside down or vertically, sometimes looking as if they are gasping for air at the surface. The condition affects small fish more, as the membranes that the gas has to permeate through are thinner. Large fish can be affected, but only at higher supersaturation levels. Gas supersaturation levels need only to be >103% to put very small fish at risk. GBD can be prevented by daily controls of dissolved gases and by the implementation of bubble disrupters at the inflow of tanks. Once the clinical signs appear, fish recover when placed with water with normal gas levels (Lemarié et al. 2011).

Nefrocalcinosis

Nefrocalcinosis is induced in nursery units when fry are overcrowded while liquid oxygen is injected in their water to sustain respiration and metabolism. However, excessive amounts of carbon dioxide are also produced, the water pH drops, fish blood acidifies and renal calculi (often $CaCO_3$) are formed. These may be diagnosed by X-ray mammograms. When the kidney osmoregulatory function is impaired, skin ulcerations can appear (Varvarigos 2008) (Fig. 7H).

Harmful algae blooms (HABs) and jellyfish

Gill disorders and mortalities associated to HABs and jellyfish are a significant challenge in fish farming worldwide. Recently, episodes of acute mortality in ESB associated with jellyfish-inflicted gill damage have been reported in Spain (Baxter et al. 2011). Probably, these events happened in the past and were misdiagnosed, and only with specialized investigations conducted during the last years has the origin of the problem probably been elucidated.

Juvenile ESB populations farmed in floating cages are the main target. Mortality is generally associated with clinical signs of asphyxia and gill damage, with excess of mucus production (Fig. 7I). Structures compatible with jellyfish were found by histological examination of affected gills. In addition, different species of harmful jellyfish were found in the water around affected farms (Baxter et al. 2011). Generally, secondary bacterial gill infections by *Tenacibaculum* and *Vibrio* sp. developed quickly, which can aggravate the mortality and make the diagnosis of the primary cause difficult.

HABs and jellyfish surveillance programs have to be applied in the fish farms, including proper training of farm technicians in algae and jellyfish identification, in order to increase the diagnostic capacity of these gill disorders.

Tumours

Laboratory experiments have demonstrated that certain viruses, chemicals, inherited characteristics and radiation can cause neoplasms in fish. The cause of tumours in wild fish are more difficult to ascertain. In certain fish populations there is strong evidence that chemical pollutants are important, whereas in others neoplasms occur sporadically. Extensive information on neoplasias can be found in the review by Grizzle and Goodwin (2010). Neoplasia is a rare finding in farmed ESB. A multicentric infiltrative lipoma was decribed recently in Italy in a single farmed 400 g fish (Marino et al. 2011). In this case, the tumoral masses invaded the underlying musculature and deformed the skin surface in the lateral side of the fish. Ramos and da Conceição Peleteiro (2003) described a case of benign fibropapilloma in the skin of one ESB broodstock held under laboratory conditions. This type of tumour has been frequently found in market size (400 g) ESB and gilthead sea bream, farmed in floating cages in Spain (Fig. 7J). Fibropapillomas are neoplastic proliferations in the epidermal and dermal layer of the skin.

In general, these tumours are benign for the fish, but they make fish unmarketable. Thus, the potential economical impact is high, and their causes should be studied to avoid their presence in ESB farmed populations.

Future Research Perspectives

In spite of the latest advancements in our knowledge of the biology of fish pathogens and the pathology of the infections, the fight against them is far from being won. Some of the infectious diseases described here are seriously threatening ESB farming in the Mediterranean. Current farming practices in off-shore cages lack the basic biosecurity regulations present in other salmon farming countries, such as area management agreements between companies, compulsory class separation or fallowing. This deficient health management means continuous presence of pathogens that are transferred easily from market size to juveniles that cohabitate in the same farm lease. Unfortunately, monogeneans, photobacteriosis or VNN will continue causing significant economic impact if these deficiencies are not addressed in the future.

Another refrain is the scarcity of antiparasitic and antibiotic preparations registered for its use in Mediterranean fish, and more specifically for ESB. The absence of licensed antiparasitic compounds or official minimal residue levels (MRLs) leads to the extrapolation of information and procedures from coldwater species, mainly salmonids. This can cause problems since treatment conditions are very different in these species in terms of environmental (temperature, pH, stability, toxicity to other aquatic animals) and individual fish factors (safety, metabolism, stress, residues, etc.). Therefore, research on this area is still much needed.

Another holding point is the scarcity of commercial vaccines; in fact, there is only one bacterial vaccine specifically developed for ESB (in many cases vaccines developed for other host species are used), and no vaccines are available for virus and parasites. Therefore, this is an essential issue in future research and alternative approaches are urgently needed. Strategies based on the exploitation of the fish immune system are just beginning to be considered. One approach is the use of immunostimulators, which now are being included in many aquafeeds, but their efficacy has to be determined for each pathogen. Furthermore, genetic selection, which is based on the innate resistance of certain fish strains, is the youngest and still least exploited strategy in aquaculture. Finally, when no adequate therapies are available, the development, standardization and harmonisation of diagnostic procedures, as well as the improvement of the knowledge of the biology of the pathogen and associated risk factors, are urgently needed to avoid pathogen dispersal and to define prophylactic measures in the cultures.

Acknowledgements

The authors are indebted to Dr. Panos Varvarigos (VetCare, Greece), Dr. Ariel Diamant (National Center for Mariculture, Israel), Dr. Ivona Mladineo (Institute of Oceanography & Fisheries, Croatia), Fernando Sanz (Skretting, Spain) and Professor Pilar Álvarez-Pellitero (founder of the Fish Pathology group of IATS-CSIC) for kindly providing illustrations and information on ESB pathologies. Part of the information gathered in this chapter has been generated along research projects funded by different Spanish Ministries: MAR89-0557-C02-01, AGF92-0199-C02-01, AGL-2002-04075-C02-01, AGL2009-13282-C02-01, AGL2011-29639, Consolider 2009-00006 and by the European Union through FEDER funds, the project MYXFISHCONTROL (QLRT-2001-00722) and TOK action PATMED (MTKD-CT-2004-0145019). Additional funding was obtained from the Generalitat Valenciana through the projects REVIDPAQUA (ISIC/2012/003) and NISAAM (PROMETEO 2010/06).

References

Abollo, E., C. Gestal and S. Pascual. 2001. Anisakis infestation in marine fish and cephalopods from Galician waters: an updated perspective. Parasitol. Res. 87: 492–499.

Actis, L.A., M.E. Tolmasky and J.H. Crosa. 2011. Vibriosis. pp. 570–605. *In:* P.T.K. Woo and D.W. Bruno (eds.). Fish Diseases and Disorders, Vol. 3. CABI Publishing.

Afonso, A., S. Gomes, J. da Silva, F. Marques and M. Henrique. 2005. Side effects in seabass (*Dicentrarchus labrax* L.) due to intraperitoneal vaccination against vibriosis and pasteurellosis. Fish Shellfish Immunol. 19: 1–16.

Álvarez-Pellitero, P. 2004. Report about fish parasitic diseases. pp. 103–130. *In:* P. Álvarez-Pellitero, J.L. Barja, B. Basurco, F. Berthe and A.E. Toranzo (eds.). Mediterranean aquaculture diagnostic laboratories: results of the survey on Mediterranean aquaculture diagnostic laboratories conducted within the framework of the CIHEAM/FAO network on "Technology of Aquaculture in the Mediterranean". Options Méditerranéennes, B/49 Etudes et Recherches, Options Méditerranéennes. CIHEAM/ FAO Publications, France.

Álvarez-Pellitero, P. and A. Sitjà-Bobadilla. 1993. Pathology of Myxosporea in marine fish culture. Dis. Aquat. Org. 17: 229–238.

Alvarez-Pellitero, P. and A. Sitjà-Bobadilla. 2002. *Cryptosporidium molnari* n. sp. (Apicomplexa: Cryptosporidiidae) infecting two marine fish species, *Sparus aurata* L. and *Dicentrarchus labrax* L. Int. J. Parasitol. 32: 1007–1021.

Álvarez-Pellitero, P., A. Sitjà-Bobadilla and A. Franco-Sierra. 1993. Protozoan parasites of wild and cultured European sea bass, *Dicentrarchus labrax*, from the Mediterranean area. Aquac. Res. 24: 101–108.

Angelidis P., D. Karagiannis and E.M. Crump. 2006. Efficacy of a *Listonella anguillarum* (syn. *Vibrio anguillarum*) vaccine for juvenile sea bass *Dicentrarchus labrax*. Dis. Aquat. Org. 71: 19–29.

Antonelli, L., Y. Quilichini and B. Marchand. 2012. *Lernanthropus kroyeri* (Van Beneden and Hesse 1851) parasitic Copepoda (Siphonostomatoidae, Lernanthropidae) of European cultured sea bass *Dicentrarchus labrax* (Linnaeus 1758) from Corsica: ecological and morphological study. Parasitol. Res. 110: 1959–1968.

APROMAR. 2012. La acuicultura marina en España. pp. 1–88. Available at http://www.apromar.es/informes.asp.

Athanassopoulou, F., D. Bouboulis and B. Martinsen. 2001. *In vitro* treatments of deltamethrin against the isopod parasite *Ceratothoa oestroides*, a pathogen of seabass *Dicentrarchus labrax* L. Bull. Eur. Assoc. Fish Pathol. 21: 26–29.

Athanassopoulou, F., C. Billinis, V. Psychas and K. Kariplogou. 2003. Viral encephalopathy and retinopathy of *Dicentrarchus labrax* (L.) farmed in fresh water in Greece. J. Fish Dis. 26: 361–365.

Athanassopoulou, F., D. Groman, T. Prapas and O. Sabatakou. 2004. Pathological and epidemiological observations on rickettsiosis in cultured sea bass (*Dicentrarchus labrax* L.) from Greece. J. Appl. Ichthyol. 20: 525–529.

Austin, B. and D.A. Austin. 2007. Bacterial Fish Pathogens: Diseases of Farmed and Wild Fish, 4th ed. Springer-Praxis Publishing, New York-Chichester.

Avendaño-Herrera, R., B. Magariños, A.E. Toranzo, R. Beaz and J.L. Romalde. 2004. Comparative evaluation of species-specific polymerase chain reaction primer sets for the routine identification of *Tenacibaculum maritimum*. Dis. Aquat. Org. 62: 75–83.

Avendaño-Herrera, R., A.E. Toranzo and B. Magariños. 2006. Tenacibaculosis infection in marine fish caused by *Tenacibaculum maritimum*: a review. Dis. Aquat. Org. 71: 255–266.

Bahri, L., J. Ben Hamida and O.K. Ben Hassine. 2002. Use of the parasitic copepod, *Lernanthropus kroyeri* (Van Beneden 1851) as a bio-indicator of two fish populations, *Dicentrarchus labrax* (Linnaeus, 1758) and *Dicentrarchus punctatus* (Bloch 1792) (Moronidae) in Tunisian inshore areas. Int. J. Crust. Res. 75: 253–267.

Bakopoulos, V., D. Volpatti, E. Papapanagiotou, R. Richards, M. Galleotti and A. Adams. 1997. Development of an ELISA to detect *Pasteurella piscicida* in cultured and "spiked" fish tissue. Aquaculture 156: 359–366.

Baptista, T., J. Costa and F. Soares. 1999. Patologías más comunes en Dorada (*Sparus aurata*) y Lubina (*Dicentrarchus labrax*) registradas en las piscifactorías al sur del Río Tajo durante 1998. Revista AquaTIC, n° 7, June 1999.

Bariche, M. and J.P. Trilles. 2008. *Ceratothoa collaris* (Isopoda: Cymothoidae) new to the eastern Mediterranean, with a redescription and comments on its distribution and host specificity. J. Mar. Biol. Assoc. UK 88: 85–93.

Baudin-Laurencin, F. 1981. Fish *vibrio* strains antisera in France. *In*: Int. Symp. on fish Biologics: Serodiagnostics and Vaccines, Leetown, W. Va., U.S.A. Develop. Biol. Standard. 49: 257–259.

Baudin-Laurencin, F., J.F. Pepin and J.C. Raymond. 1991. First observation of an epizootic of pasteurellosis in farmed and wild fish on the French Mediterranean coast. *In*: Proceedings of the 5th Int. Conf. Eur. Assoc. Fish Pathol. Budapest (Hungary).

Baxter, E.J., G. Albinyana, A. Girons, M.M. Isern, A.B. García, M. Lopez, A. Canepa, A. Olariaga, J.M. Gili and V. Fuentes. 2011. Jellyfish-inflicted gill damage in marine-farmed fish: an emerging problem for the Mediterranean. *In*: Proceedings of the 13th Congreso Nacional Acuicultura, Castelldefells (Spain).

Beck, M., R. Evans, S.W. Feist, P. Stebbing, M. Longshaw and E. Harris. 2008. *Anisakis simplex* sensu lato associated with red vent syndrome in wild adult Atlantic salmon *Salmo salar* in England and Wales. Dis. Aquat. Org. 82: 61–65.

Ben Abdallah, L.G. and F. Maamouri. 2005. The life cycle of *Bucephalus labracis* Paggi and Orecchia, 1965 (Digenea, Bucephalidae), a parasite of *Dicentrarchus labrax* in Tunisia. Bull. Eur. Assoc. Fish Pathol. 25: 297–302.

Bernardet, J.F. 1998. *Cytophaga, Flavobacterium, Flexibacter* and *Chryseobacterium* infections in cultured marine fish. Fish Pathol. 33: 229–238.

Bernardet, J.F., B. Kerouault and C. Michel. 1994. Comparative study on *Flexibacter maritimus* strains isolated from farmed sea bass (*Dicentrarchus labrax*) in France. Fish Pathol. 29: 105–111.

Bernardi, C. 2009. Preliminary study on prevalence of larvae of Anisakidae family in European sea bass (*Dicentrarchus labrax*). Food Control 20: 433–434.

Bernardi, C., A. Gustinelli, M.L. Fioravanti, M. Caffara, S. Mattiucci and P. Cattaneo. 2011. Prevalence and mean intensity of *Anisakis simplex* (sensu stricto) in European sea bass (*Dicentrarchus labrax*) from Northeast Atlantic Ocean. Int. J. Food Microbiol. 148: 55–59.

Berthe, F.C.J., C. Michel and J.F. Bernardet. 1995. Identification of *Pseudomonas anguilliseptica* isolated from several fish species in France. Dis. Aquat. Org. 21: 151–155.

Blanch, A.R., R.M. Pintó and J.T. Jofre. 1990. Isolation and characterization of an *Edwardsiella* sp. strain, causative agent of mortalities in sea bass (*Dicentrarchus labrax*). Aquaculture 88: 213–222.

Bouboulis, D., F. Athanassopoulou and A. Tyrpenou. 2004. Experimental treatments with diflubenzuron and deltamethrin of sea bass, *Dicentrarchus labrax* L., infected with the isopod, *Ceratothoa oestroides*. J. Appl. Ichthyol. 20: 314–317.

Bovo, G., T. Nishizawa, C. Maltese, F. Borghesan, F. Mutinelli, F. Montesi and S. De Mas. 1999. Viral encephalo-retinopathy of farmed fish species in Italy. Virus Res. 63: 143–146.

Bragoni, G., B. Romestand and J.P. Trilles. 1983. Cymothoadian parasitosis of the sea bass *(Dicentrarchus labrax* Linnaeus 1758) during breeding. II. Parasitic ecophysiology in the Diana pond (upper Corsica). Ann. Parasitol. Hum. Comp. 58: 593–609.

Bragoni, G., B. Romestand and J.P. Trilles. 1984. Cymothoadian parasites in the sea bass, *Dicentrarchus labrax* (Linnaeus, 1758) during rearing. 1. Parasite ecology in the case of Diana Pond (Upper Corsica) (Isopoda, Cymothoidae). Crustaceana 47: 44–51.

Breuil, G., J.F. Pepin, J. Castric, C. Fauvel and R. Thiery. 2000. Detection of serum antibodies against nodavirus in wild and farmed adult sea bass: Application to the screening of broodstock in sea bass hatcheries. Bull. Eur. Assoc. Fish Pathol. 20: 95–100.

Breuil, G., O. Moucel, C. Fauvel and J.F. Pepin. 2001. Sea bass *Dicentrarchus labrax* nervous necrosis virus isolates with distinct pathogenicity to sea bass larvae. Dis. Aquat. Org. 45: 25–31.

Breuil, G., J.F.P. Pépin, S. Boscher and R. Thiéry. 2002. Experimental vertical transmission of nodavirus from broodfish to eggs and larvae of the sea bass, *Dicentrarchus labrax* (L.). J. Fish Dis. 25: 697–702.

Buller, N.B. 2004. Bacteria from fish and other aquatic animals: a practical identification manual. CABI Publisihing, Wallingford, UK.

Cabral, P., F. Coste and A. Raibaut. 1984. The life cycle of *Lernanthropus kroyeri* Vanbeneden, 1851, a hematophagous copepod of the gills of the sea bass in wild hosts and experimental infections. Ann. Parasitol. Hum. Comp. 59: 189–207.

Caffara, M., F. Quaglio, F. Marcer, D. Florio and M.L. Fioravanti. 2010. Intestinal microsporidiosis in European seabass (*Dicentrarchus labrax* L.) farmed in Italy. Bull. Eur. Assoc. Fish Pathol. 30: 237–240.

Çağırgan, H. 2009. The use of veterinary drugs and vaccines in Turkey. pp. 29–34. *In*: C. Rodgers and B. Basurco [eds.]. The Use of Veterinary Drugs and Vaccines in Mediterranean Aquaculture. Options Méditerranéennes, A/86, CIHEAM/ FAO Publications, Zaragoza, Spain.

Çağırgan, H. and E. Tokşen. 1996. The first *Amyloodinium ocellatum* (Dinoflagellata) infestation of cultured sea bass juvenile (*Dicentrarchus labrax*) in Turkey. Pendik Veteriner Mikrobiyoloji Dergisi 27: 197–205.

Caillot, C., S. Morand, C.M. Mullergraf, E. Faliex and B. Marchand. 1999. Parasites of *Dicentrarchus labrax, Anguilla anguilla*, and *Mugil cephalus* from a pond in Corsica, France. J. Helminthol. Soc.Wash. 66: 95–98.

Candan, A., M. Kucker and S. Karatas. 1996. Pasteurellosis in cultured sea bass (*Dicentrarchus labrax*) in Turkey. Bull. Eur. Assoc. Fish Pathol. 16: 150–153.

Chen, Q., H.Q. Yu, Z.R. Lun, X.G. Chen, H.Q. Song, R.Q. Lin and X.Q. Zhu. 2008a. Specific PCR assays for the identification of common anisakid nematodes with zoonotic potential. Parasitol. Res. 104: 79–84.

Chen, Q., H. Zhang, H.Q. Song, H.Q. Yu, R.Q. Lind and X.Q. Zhu. 2008b. Prevalence of Anisakid larvae in maricultured sea fish sold in Guangzhou, China. J. Anim. Vet. Adv. 7: 1078–1080.

Chen, M.F., D. Henry-Ford and J.M. Groff. 1995. Isolation and characterization of *Flexibacter maritimus* from marine fishes of California. J. Aquat. Anim. Health 7: 318–326.

Cognetti-Varriale, A.M., A. Castelli, S. Cecchini and M. Saroglia. 1991. Therapeutic Trails Against The *Diplectanum aequans* (monogenea), Parasite of Seabass (*Dicentrarchus labrax* L.) in Intensive Farming. Bull. Eur. Assoc. Fish Pathol. 12: 204–206.

Cognetti-Varriale, A.M., A. Castelli, S. Cecchini and M. Saroglia. 1992. Distribution of *Diplectanum aequans* (Monogenea) on the gills of intensively reared sea bass (*Dicentrarchus labrax* L.). Bull. Eur. Assoc. Fish Pathol. 13: 13–14.

Colorni, A. 1987. Biology of *Cryptocaryon irritans* and strategies for its control. Aquaculture 67: 236–237.

Colorni, A. 1992. A systemic mycobacteriosis in the European seabass *Dicentrarchus labrax* cultured in Eilat (Red Sea). Isr. J. Aquacult. Bamidgeh 44: 75–81.

Colorni, A., M. Ankaoua, A. Diamant and W. Knibb. 1993. Detection of mycobacteriosis in fish using the polymerase chain reaction technique. Bull. Eur. Assoc. Fish Pathol. 13: 195–198.

Colorni, A., M. Ucko and W. Knibb. 1996. Epizootiology of *Mycobacterium* spp. in seabass, seabream and other commercialfish. Seabass and Seabream Culture: Problems and Prospects. Eur. Aquacult. Soc. Spec. Publ., Verona, Italy. pp. 259–261.

Colorni, A., A. Diamant, A. Eldar, H. Kvitt and A. Zlotkin. 2002. *Streptococcus iniae* infections in red seacage-cultured and wild fishes. Dis. Aquat. Org. 49: 165–170.

Čož-Rakovac, R., I. Strunjak-Perovič, N. Topič Popovič, M. Hacmanjek, B. Šimpraga and E. Teskeredžič. 2002. Health status of wild and cultured sea bass in the northern Adriatic Sea. Vet. Med. Czech 47: 222–226.

Crespo, S., C. Zarza and F. Padrós. 2001. Epitheliocystis hyperinfection in sea bass, *Dicentrarchus labrax* (L.): light and electron microscope observations. J. Fish Dis. 24: 557–560.

Cruz e Silva, M.P., M.L. Orge, M.M. Afonso-Roque, M.S. Grazina-Freitas and M. Carvalho-Varela. 2000. *Diplectanum aequans* (Wagener 1857) Diesing, 1858 (Monogenea, Diplectanidae) in sea bass (*Dicentrarchus labrax* (L.) 1758) from freshwater culture. Acta Parasitol. Port. 7: 53–56.

Cuadrado, M., G. Albinyana, F. Padrós, M.J. Redondo, A. Sitjà-Bobadilla, P. Álvarez-Pellitero, O. Palenzuela, A. Diamant and S. Crespo. 2007. An unidentified epi-epithelial myxosporean in the intestine of gilthead sea bream *Sparus aurata* L. Parasitol. Res. 101: 403–411.

Culurgioni, J., S. Mattiucci, M. Paoletti and V. Figus. 2011. First report of *Anisakis pegreffii* larvae (Nematoda, Anisakidae) in wild European sea bass, *Dicentrarchus labrax* (L.) from Mediterranean waters (Southern Sardinia), 58. In: Proceedings of the XVII Congresso Nazionale SIPI, Ostuni (Italy).

Cutrín, J.M., C.P. Dopazo, R. Thiéry, P. Leao, J.G. Olveira, J.L. Barja and I. Bandín. 2007. Emergence of pathogenic betanodaviruses belonging to the SJNNV genogroup in farmed fish species from the Iberian Peninsula. J. Fish Dis. 30: 225–232.

Daly, J.G. and T. Aoki. 2011. Pasteurellosis and other bacterial diseases. pp. 632–668. In: P.T.K. Woo and D.W. [eds.]. Fish Diseases and Disorders, vol. 3. CABI Publishing, U.K.

Daoudi, F. and A. Marques. 1987. *Eimeria bouixi* n. sp. et *Eimeria dicentrarchi* n. sp. (Sporozoa-Apicomplexa). Coccidies parasites du loup *Dicentrarchus labrax* (Linne 1758), en region languedocienne. Ann. Sci. Nat. Zool. Biol. Anim. 8: 237–242.

Dezfuli, B.S., L. Giari, E. Simoni, R. Menegatti, A.P. Shinn and M. Manera. 2007. Gill histopathology of cultured European sea bass, *Dicentrarchus labrax* (L.), infected with *Diplectanum aequans* (Wagener 1857) Diesing 1958 (Diplectanidae: Monogenea). Parasitol. Res. 100: 707–713.

Diamant, A., G. Issar, A. Colorni and I. Paperna. 1991. A pathogenic *Cryptocaryon*-like ciliate from the Mediterranean Sea. Bull. Eur. Assoc. Fish Pathol. 11: 122–124.

Dickerson, H.W. 2006. *Ichthyophthirius multifiliis* and *Cryptocaryon irritans* (Phylum Ciliophora). pp. 116–153. In: P.T.K. Woo [ed.]. Fish Diseases and Disorders, vol. 1: Protozoan and Metazoan disorders. 2nd edn. CABI Publishing, Cambridge, MA.

Dragesco, A., J. Dragesco, F. Coste, C. Gasc, B. Romestand, J.C. Raymond and G. Bouix. 1995. *Philasterides dicentrarchi*, n. sp. (Ciliophora, Scuticociliatidia), a histophagous opportunistic parasite of *Dicentrarchus labrax* (Linnaeus 1758), a reared marine fish. Eur. J. Protistol. 31: 327–340.

Duarte, N., N. Rosa, M. Santos and J. Rebelo. 2000. Study of ectoparasites of the sea bass gills (*Dicentrarchus labrax* L.) of Ria de Aveiro. Rev. Biol. (Lisbon) 18: 59–68.

Dyková, I., A. Figueras and Z. Peric. 2000. *Neoparamoeba* Page, 1987: light and electron microscopic observations on six strains of different origin. Dis. Aquat. Org. 43: 217–223.

Fanouraki, E., C.C. Mylonas, N. Papandroulakis and M. Pavlidis. 2011. Species specificity in the magnitude and duration of the acute stress response in Mediterranean fish in culture. Gen. Comp. Endocrinol. 173: 313–322.

Fernandes, D., C. Porte and M.J. Bebianno. 2007. Chemical residues and biochemical responses in wild and cultured European sea bass (*Dicentrarchus labrax* L.). Environ. Res. 103: 247–256.

Fernández, J.P., M. Muñoz, V.M. Orts, J. Raga and E. Carbonell. 1989. Sobre la presencia de *Cainocreadium labracis* (Dujardin, 1845) Nicoll 1909 (Trematoda: Opecoelidae) parasito intestinal de *Dicentrarchus labrax* (L.) (Pisces: Serranidae) en las costas de la Peninsula Ibérica. Rev. Iber. Parasitol. 48: 393–394.

Ferreira, M., M. Caetano, P. Antunes, J. Costa, O. Gil, N. Bandarra, P. Pousão-Ferreira, C. Vale and M.A. Reis-Henriques. 2010. Assessment of contaminants and biomarkers of exposure in wild and farmed seabass. Ecotoxicol. Environ. Saf. 73: 579–88.

Fioravanti, M.L., M. Caffara, D. Florio, A. Gustinelli and F. Marcer. 2004. *Sphaerospora dicentrarchi* and S. *testicularis* (Myxozoa: Sphaerosporidae) in farmed European sea bass. Folia Parasitol. 51: 208–210.

Fioravanti, M.L., M. Caffara, D. Florio, A. Gustinelli and F. Marcer. 2006. A parasitological survey of European sea bass (*Dicentrarchus labrax*) and gilthead sea bream (*Sparus aurata*) cultured in Italy. Vet. Res. Commun. 30: 249–252.

Fouz, B., R.F. Conchas, B. Magariños and A.E. Toranzo. 1992. *Vibrio damselae* strain virulence for fish and mammals. FHS Newsletter 20: 3–4.

Fouz, B., M.D. Esteve-Gassent, R. Barrera, J.L. Larsen, M.E. Nielsen and C. Amaro. 2001. Field testing of a vaccine against eel diseases caused by *Vibrio vulnificus*. Dis. Aquat. Org. 45: 183–189.

Fouz, B., A. Llorens, E. Valiente and C. Amaro. 2010. A comparative epizootiologic study of the two fish-pathogenic serovars of *Vibrio vulnificus* biotype 2. J. Fish Dis. 33: 383–390.

Franco-Sierra, A., A. Sitjà-Bobadilla and P. Alvarez-Pellitero. 1997. *Ichthyophonus* infections in cultured marine fish from Spain. J. Fish Biol. 51: 830–839.

Frans, I., C.W. Michiels, P. Bossier, K.A. Willens, B. Lievens and H. Rediers. 2011. *Vibrio anguillarum* as a fish pathogen: virulence factors, diagnosis and prevention. J. Fish Dis. 34: 643–661.

Frerichs, G.N., H.D. Rodger and Z. Peric. 1996. Cell culture isolation of piscine neuropathy nodavirus from juvenile sea bass, *Dicentrarchus labrax*. J. Gen. Virol. 77: 2067–2071.

Frerichs, G.N., A. Tweedie, W.G. Starkey and R.H. Richards. 2000. Temperature, pH and electrolyte sensitivity, and heat, UV and disinfectant inactivation of sea bass (*Dicentrarchus labrax*) neuropathy nodavirus. Aquacult. 185: 13–24.

Gamperl, A.K. and A.P. Farrell. 2004. Cardiac plasticity in fishes: environmental influences and intraspecific differences. J. Exp. Biol. 207: 2539–2550.

Ghittino, C., M. Latini, F. Agnetti, C. Panzieri, L. Lauro, R. Ciappelloni And G. Petracca. 2003. Emerging pathologies in aquaculture: effects on production and food safety. Vet. Research Com. 27: 471–479.

Giavenni, R., 1988. Some parasitic and other diseases occurring in seabass (*Dicentrarchus labrax* L.) broodstocks in Italy. Bull. Eur. Assoc. Fish. Pathol. 8: 45–46.

Golomazou, E., F. Athanassopoulou, E. Karagouni, S. Vagianou, H. Tsantilas and D. Karamanis. 2006. Efficacy and toxicity of orally administered anti-coccidial drug treatment on *Enteromyxum leei* infections in sharpsnout seabream (*Diplodus puntazzo* C.). Isr. J. Aquacult. Bamidgeh 58: 157–169.

Gómez, D.K., J. Sato, K. Mushiake, T. Isshiki, Y. Okinaka and T. Nakai. 2004. PCR-based detection of betanodaviruses from cultured and wild marine fish with no clinical signs. J. Fish Dis. 27: 603–608.

Gómez, D.K., G.W. Baeck, J.H. Kim, C.H. Choresca Jr. and S.C. Park. 2008. Molecular detection of betanodaviruses from apparently healthy wild marine invertebrates. J. Inv. Pathol. 97: 197–202.

Gómez-Casado, E., A. Estepa and J.M. Coll. 2011. A comparative review on European-farmed finfish RNA viruses and their vaccines. Vaccine 29: 2657–2671.

González, S.F., C.R. Osorio and Y. Santos. 2004. Evaluation of the AQUARAPID-Va, AQUAEIA-Va and dot-blot assays for the detection of *Vibrio anguillarum* in fish tissues. J. Fish Dis. 27: 617–621.

González-Lanza, C., M.P. Alvarez-Pellitero and A. Sitjà-Bobadilla. 1991. Diplectanidae (Monogenea) infestations of sea bass, *Dicentrarchus labrax* (L.), from the Spanish Mediterranean area. Histopathology and population dynamics under culture conditions. Parasitol. Res. 77: 307–314.

Gravningen, K., R. Thorarinsson, L.H. Johansen, B. Nissen, K.S. Rikardsen, E. Greger and M. Vigneulle. 1998. Bivalent vaccines for sea bass (*Dicentrachus labrax*) against vibriosis and pasteurellosis. J. Appl. Ichthyol. 14: 159–162.

Grizzle, J.M. and A.E. Goodwin. 2010. Neoplasms and related disorders. pp. 19–84. *In*: J.F. Leatherland and P.T.K. Woo [eds.]. Fish Diseases and Disorders Vol. 2: Non-infectious Disorders, 2nd edition. CABI Publishing, Oxfordshire.

Harriff, M.J., L.E. Bermudez and M.L. Kent. 2007. Experimental exposure of zebrafish, *Danio rerio* (Hamilton) to *Mycobacterium marinum* and *Mycobacterium peregrinum* reveals the gastrointestinal tract as the primary route of infection: a potential model for environmental mycobacterial infection. J. Fish Dis. 30: 587–600.

Hauck, A. and E. May. 1977. Histopathologic alterations associated with *Anisakis* larvae in Pacific herring from Oregon. J. Wildl. Dis. 13: 290–293.

Herbst, L.H., S.F. Costa, L.M. Weiss, L.K. Johnson, J. Bartell, R. Davis, M. Walsh and M. Levi. 2001. Granulomatous skin lesions in moray eels caused by a novel *mycobacterium* species related to *Mycobacterium* triplex. Infect. Immun. 69: 4639–4646.

Hodneland, K., R. García, J.A. Balbuena, C. Zarza and B. Fouz. 2011. Real-time RT-PCR detection of betanodavirus in naturally and experimentally infected fish from Spain. J. Fish Dis. 34: 189–202.

Holmes, J.M.C. and J. O'Connor. 1999. Parasitic copepods (Crustacea) from bass *Dicentrarchus labrax* and flounder *Pleuronectes flesus*, including two species new to Ireland. Ir. Nat. J. 26: 267–269.

Horton, N.T. and B. Okamura. 2003. Post-haemorrhagic anaemia in sea bass, *Dicentrarchus labrax* (L.), caused by blood feeding of *Ceratothoa oestroides* (Isopoda: Cymothoidae). J. Fish Dis. 26: 401–406.

Horton, T. and B. Okamura. 2001. Cymothoid isopod parasites in aquaculture: a review and case study of a Turkish sea bass (*Dicentrarchus labrax*) and sea bream (*Sparus auratus*) farm. Dis. Aquat. Org. 47: 181–188.

ICTV. International Committee on Taxonomy of Viruses (ICTV index of viruses). 2009. http://www.ncbi.nlm.nih.govIICTVdblIclvlindex.html.

Jacobs, J.M., C.S. Stine, A.M. Baya and M.L. Kent. 2009. A review of mycobacteriosis in marine fish. J. Fish Dis. 32: 119–130.

Kai, Y.H. and S.C. Chi. 2008. Efficacies of inactivated vaccines against betanodavirus in grouper larvae (*Epinephelus coioides*) by bath immunization. Vaccine 26: 1450–1457.

Katavic, I., J. Jug-Dujakovic and B. Glamuzina. 1989. Cannibalism as a factor affecting the survival of intensively cultured sea bass (*Dicentrarchus labrax*) fingerlings. Aquaculture 77: 135–143.

Knibb, W., A. Colorni, M. Ankaoua, D. Lindell, A. Diamant and H. Gordon. 1993. Detection and identification of a pathogenic *Mycobacterium* from European sea bass *Dicentrarchus labrax* using polymerase chain reaction and direct sequencing of 16S rRNA sequences. Mol. Mar. Biol. Biotechnol. 2: 225–232.

Kolygas, M.N., E. Gourzioti, I.N. Vatsos and F. Athanassopoulou. 2012. Identification of *Tenacibaculum maritimum* strains from marine farmed fish in Greece. Vet. Rec. 170: 623.

Korsnes, K., M. Devold, A. Nerland and A. Nylund. 2005. Viral encephalopathy and retinopathy (VER) in Atlantic salmon *Salmo salar* after intraperitoneal challenge with a nodavirus from Atlantic halibut *Hippoglossus hippoglossus*. Dis. Aquat. Org. 68: 7–15.

Kumar, S.R., V. Parameswaran, V.P.I. Ahmed, S.S. Musthaq and A.S.S. Hameed. 2007. Protective efficiency of DNA vaccination in Asian sea bass (*Lates calcarifer*) against *Vibrio anguillarum*. Fish Shellfish Immunol. 23: 316–326.

Laganà, P., G. Caruso, E. Minutoli, R. Zaccone and S. Delia. 2011. Susceptibility to antibiotics o *Vibrio* spp. and *Photobacterium damsela* ssp. *piscicida* strains isolated from Italian aquaculture farms. New Microbiol. 34: 53–63.

Lambert, A. and C. Maillard. 1975. Repartition branchiale de deux monogenes: *Diplectanum aequans* (Wagener 1857) Diesing 1858 et *Diplectanum laubieri* Lambert et Maillard 1974 (Monogenea, Monopisthocotylea) parasites simultanes de *Dicentrarchus labrax* (teleosteen). Ann. Parasit. Hum. Comp. 50: 691–699.

Le Breton, A.D. 1995. Field investigations on *Pasteurella piscicida* in Greece: Detection of the pathogen, evolution of the epidemiology, antibioresistances. *In*: Proceedings of 7th Int. Conf. Diseases of Fish and Shellfish, EAFP, Palma de Mallorca (Spain).

Le Breton, A.D. 1999. Mediterranean finfish pathologies: Present status and new developments in prophylactic methods. Bull. Eur. Assoc. Fish Pathol. 19: 250–253.

Le Breton, A., L. Grizez, E. Sweetman and F. Ollivier. 1997. Viral nervous necrosis (VNN) associated with mass mortalities in cage-reared seabass, *Dicentrarchus labrax*. J. Fish Dis. 20: 145–151.

Lehane, L. and G.T. Rawlin. 2000. Topically acquired bacterial zoonoses from fish: a review. Med. J. Aust. 173: 256–259.

Lemarié, G., C.D. Hosfeld, G. Breuil and S. Fivelstad. 2011. Effects of hyperoxic water conditions under different total gas pressures in European sea bass (*Dicentrarchus labrax*). Aquaculture 318: 191–198.

Levy, M.G., M.F. Poore, A. Colorni, E.J. Noga, M.W. Vandersea and R.W. Litaker. 2007. A highly specific PCR assay for detecting the fish ectoparasite *Amyloodinium ocellatum*. Dis. Aquat. Org. 73: 219–226.

Lewis, F.M., B. Marsh and C. von Reyn. 2003. Fish tank exposure and cutaneous infections due to *Mycobacterium marinum*: Tuberculin skin testing, treatment, and prevention. Clinical Infect. Dis. 37: 390–397.

Lewis, S. and S. Chinabut. 2011. Mycobacteriosis and Nocardiosis. pp. 397–423. *In*: P.T.K. Woo and D.W. Bruno [eds.]. Fish Diseases and Disorders, Vol. 3, CABI Publishing, UK.

Lima dos Santos, C.A.M. and P. Howgate. 2011. Fishborne zoonotic parasites and aquaculture: A review. Aquaculture 318: 253–261.

Lin, C.C., J.H.Y. Lin, M.S. Chen and H.L.Yang. 2007. An oral nervous necrosis virus vaccine that induces protective immunity in larvae of grouper (*Epinephelus coioides*). Aquaculture 268: 265–273.

Lokanathan, Y., A. Mohd-Adnan, K. Lian Wan and S. Nathan. 2010. Transcriptome analysis of the *Cryptocaryon irritans* tomont stage identifies potential genes for the detection and control of cryptocaryonosis. BMC Genomics 11: 1.

Lozano, C., J.M. Ubeda, M. de Rojas, C. Ariza and D.C. Guevara. 2001. Estudio de digénidos de peces marinos del sur de la Península Ibérica. Rev. Iber. Parasitol. 61: 103–116.

Macrì, F., G. Lanteri, G. Rapisarda, A. Costa and F. Marino. 2012. *Anisakis pegreffii* experimental challenge in *Dicentrarchus labrax*: An endoscopic study. Aquaculture 338: 297–299.

Magariños, B., J.L. Romalde, I. Bandín, B. Fouz and A.E. Toranzo. 1992. Phenotypic, antigenic and molecular characterization of *Pasteurella piscicida* isolated from fish. Appl. Environ. Microbiol. 58: 3316–3322.

Magariños, B., J.L. Romalde, J.L. Barja and A.E. Toranzo. 1994. Evidence of a dormant but infective state of the fish pathogen *Pasteurella piscicida* in seawater and sediment. Appl. Environ. Microbiol. 60: 180–186.

Magariños, B., N. Couso, M. Noya, P. Merino, A.E. Toranzo and J. Lamas. 2001. Effect of temperature on the development of pasteurellosis in carrier gilthead seabream (*Sparus aurata*). Aquaculture 195: 17–21.

Maillard, C. 1976a. Distomoses de poissons en milieu lagunaire. These Dc. Sci. Nat. Acad. Montpellier, USTL, France.

Maillard, C. 1976b. *Labratrema lamirandi* (Carrere, 1937) (Trematoda, Bucephalidae) parasite de *Dicentrarchus labrax* (L., 1758). Creation du genre *Labratrema*. Cycle evolutif 192. Bull. Mus. Natn. Hist. Nat. Paris (Zool.) 193: 69–79, 1975.

Maillard, C., L. Euzet and P. Silan. 1988. Creation of the Genus *Serranicotyle* (Monogenea, Microcotylidae)—*Serranicotyle labracis* (Vanbeneden et Hesse 1863) n. comb., Ectoparasite of *Dicentrarchus labrax* (Teleostei). Ann. Parasitol. Hum. Comp. 63: 33–36.

Marino, F., B. Chiofalo, G. Mazzullo and A. Panebianco. 2011. Multicentric infiltrative lipoma in a farmed Mediterranean seabass *Dicentrarchus labrax*: a pathological and biochemical case study. Dis. Aquat. Org. 96: 259–264.

Mattiucci, S. and G. Nascetti. 2008. Advances and trends in the molecular systematics of anisakid nematodes, with implications for their evolutionary ecology and host–parasite co-evolutional processes. Adv. Parasitol. 66: 47–148.

Mattiucci, S., P. Abaunza, S. Damiano, A. Garcia, M.N. Santos and G. Nascetti. 2007. Distribution of *Anisakis* larvae, identified by genetic markers, and their use for stock characterization of demersal and pelagic fish from European waters: an update. J. Helminthol. 81: 117–127.

McCarthy, U., N.A. Steiropoulos, K.D. Thompson, A. Adams, A.E. Ellis and H.W. Ferguson. 2005. Confirmation of *Piscirickettsia salmonis* as a pathogen in European sea bass *Dicentrarchus labrax* and phylogenetic comparison with salmonid strains. Dis. Aquat. Org. 64: 107–119.

Mennerat, A., F. Nilsen, D. Ebert and A. Skorping. 2010. Intensive Farming: Evolutionary Implications for Parasites and Pathogens. Evol. Biol. 37: 59–67.

Mladineo, I. 2002. Prevalence of *Ceratothoa oestroides* (Risso 1826), a cymothoid isopode parasite, in cultured sea bass *Dicentrarchus labrax* L., on two farms in middle Adriatic Sea. Acta Adriat. 43: 97–102.

Mladineo, I. 2003. Life cycle of *Ceratothoa oestroides*, a cymothoid isopod parasite from sea bass *Dicentrarchus labrax* and sea bream *Sparus aurata*. Dis. Aquat. Org. 57: 97–101.

Mladineo, I. 2004. Monogenean parasites in Adriatic cage-reared fish. Acta Adriat. 45: 65–73.

Mladineo, I. 2006. Check list of the parasitofauna in Adriatic sea cage-reared fish. Acta Vet. 56: 285–292.

Mladineo, I., T. Segvic and L. Grubisic. 2009. Molecular evidence for the lack of transmission of the monogenean *Sparicotyle chrysophrii* (Monogenea, Polyopisthocotylea) and isopod *Ceratothoa oestroides* (Crustacea, Cymothoidae) between wild bogue (Boops boops) and cage-reared sea bream (*Sparus aurata*) and sea bass (*Dicentrarchus labrax*). Aquaculture 295: 160–167.

Mladineo, I., M. Petric, T. Segvic and N. Dobricic. 2010. Scarcity of parasite assemblages in the Adriatic-reared European sea bass (*Dicentrarchus labrax*) and sea bream (*Sparus aurata*). Vet. Parasitol. 174: 131–138.

Morsy, K., A.R. Bashtar, F. Abdel-Ghaffar, H. Mehlhorn, S.A. Quraishy, M. El-Mahdi, A. Al-Ghamdi and N. Mostafa. 2012. First record of anisakid juveniles (Nematoda) in the European seabass *Dicentrarchus labrax* (family: Moronidae), and their role as bio-indicators of heavy metal pollution. Parasitol. Res. 110: 1131–1138.

Munday, B.L., J.S. Langdon, A. Kyatt and J.D. Humphrey. 1992. Mass mortality associated with a viral-induced vacuolating encephalopathy and retinopathy of larvae and juvenile barramundi, *Lates calcarifer* Bloch. Aquaculture 103: 197–211.

Munday, B.L., J. Kwang and N. Moody. 2002. Betanodavirus infections of teleost fish: a review. J. Fish Dis. 25: 127–42.

Nishizawa, T., M. Furuhashi, T. Nagai, T. Nakai and K. Muroga. 1997. Genomic classification of fish nodaviruses by molecular phylogenetic analysis of the coat protein gene. Appl. Environ. Microbiol. 63: 1633–1636.

Nitzan, S., B. Shwartsburd, R. Vaiman. and E.D. Heller. 2001. Some characteristics of *Photobacterium damselae* ssp. *piscicida* isolated in Israel during outbreaks of pasteurellosis in hybrid bass (*Morone saxatilis* x *M. chrysops*). Bull. Eur. Assoc. Fish Pathol. 21: 77–80.

Noga, E.J. 2011. Fish and Disease. Diagnosis and treatment. 2nd Ed. Wiley & Sons, Inc., Publication, USA.

Noga, E.J. 2012. *Amyloodinium ocellatum*. pp. 19–29. *In:* P.T.K. Woo and K. Buchmann [eds.]. Fish Parasites: Pathobiology and Protection. CABI Publishing, UK.

Noya, N., B. Magarinños and J. Lamas. 1995. Interactions between peritoneal exudate cells (PECs) of gilthead seabream (*Sparus aurata*) and *Pasteurella piscicida*. A morphological study. Aquaculture 131: 11–21.

Öktener, A. and J.P. Trilles. 2004. Report on cymothoids (Crustacea, Isopoda) collected from marine fishes in Turkey. Acta Adriat. 45: 145–154.

Öktener, A., A. Alas and K. Solak. 2009. Occurence of *Diplectanum aequans* (Wagener 1857) on the cultured sea bass, *Dicentrarchus labrax* (Linnaeus, 1758) from the Black Sea of Turkey. Bull. Eur. Assoc. Fish Pathol. 29: 102–103.

Olivier, G. 1977. Pathogenic effect of fixation of *Diplectanum aequans* (Wagener 1857) Diesing, 1858 (Monogenea, Monopisthocotylea, Diplectanidae) on gills of *Dicentrarchus labrax* (Linnaeus, 1758) (Pisces, Serranidae). Parasitol. Res. 53: 7–11.

Ortega, J.E., A. Valero, M. Wolff and González-Crespo. 1991. Contribución al estudio de la helmintofauna en tres poblaciones de lubina *Dicentrarchus labrax* (Linnaeus 1758), pp. 68. *In:* S. Mas-Coma, J.G. Esteban, M.D. Barques, M.A. Valero and M.T. Galán-Puchades J. [eds.]. Proceedings of the ICASEP I, Parasitología en el Sur-Oeste de Europa, Aguilar, S.L., Valencia, Spain.

Osorio, C. and A.E. Toranzo. 2002. DNA-based diagnostics in sea framing. pp. 253–310. *In:* M. Fingerman and R. Nagabhushanam [eds.]. Recent Advances in Marine Biotechnology Series, Seafood Safety and Human Health, vol. 7. Science Publishers, Inc., Plymouth, UK.

Palenzuela, O., F.Agnetti, G. Albinana, P. Alvarez-Pellitero, F. Athanassopoulou, S. Crespo, A. Diamant, C. Ghittino, E. Golomazou, A. Le Breton, A. Lipshitz, A. Marques, F. Padres, S. Ram and J. Raymond. 2004. Applicability of PCR screening for the monitoring of *Enteromyxum leei* (Myxozoa) infection in Mediterranean aquaculture: an epidemiological survey in sparids facilities, pp. 639–640. *In:* S. Adams and J.A. Olafsen [comp.]. Proceedings of Biotechnologies for Quality. European Aquaculture Society Special Publication No. 34. European Aquaculture Society, Barcelona, Spain.

Papapanagiotou, E.P., J.P. Trilles and G. Photis. 1999. First record of *Emetha audouini*, a cymothoid isopod parasite, from cultured sea bass *Dicentrarchus labrax* in Greece. Dis. Aquat. Org. 38: 235–237.

Paperna, I. 1980a. *Amyloodinium ocellatum* (Brown 1931) (dinoflagellida) infestations in cultured marine fish at Eilat, Red Sea—epizootiology and pathology. J. Fish Dis. 3: 363–372.

Paperna, I. 1980b. Study of *Caligus minimus* (Otto 1821), (Caligidae Copepoda) infections of the sea bass *Dicentrarchus labrax* (L.) in Bardawil lagoon. Ann. Parasitol. 55: 687–706.

Paperna, I. 1984. Review of diseases affecting cultured *Sparus aurata* and *Dicentrarchus labrax*. pp. 465–482. *In:* G. Barnabé and R. Billard [eds.]. L'Aquaculture du Bar et des Sparidés. INRA Publ., Paris.

Paperna, I. and F. Baudin Laurencin. 1979. Parasitic infections of sea bass, *Dicentrarchus labrax*, and gilthead sea bream, *Sparus aurata* in mariculture facilities in France, Aquaculture 16: 173–175.

Paperna, I., I. Sabnai and A. Zachary. 1981. Ultrastructural studies in piscine epitheliocystis: evidence for a pleomorphic developmental cycle. J. Fish Dis. 4: 459–472.

Papoutsoglou, S., M.J. Costello E. Stamou and G. Tziha. 1996. Environmental conditions at sea-cages, and ectoparasites on farmed European sea-bass, *Dicentrarchus labrax* (L.), and gilt-head sea-bream, *Sparus aurata* (L.), at two farms in Greece. Aquac. Res. 27: 25–34.

Pasnik, D.J. and S.A. Smith. 2005. Immunogenic and protective effects of a DNA vaccine for *Mycobacterium marinum*. Vet. Immunol. Immunopathol. 103: 195–206.

Pazos, F., Y. Santos, A.R. Macías, S. Núñez and A.E. Toranzo. 1996. Evaluation of media for the successful culture of *Flexibacter maritimus*. J. Fish Dis. 19: 193–197.

Pedersen, K., L. Grisez, R. van Houdt, T. Tiainen, F. Ollevier and J.L. Larsen. 1999. Extended serotyping scheme for *Vibrio anguillarum* with the definition of seven provisional Oserogroups. Curr. Microbiol. 38: 183–189.

Péducasse, S., J. Castric, R. Thiéry, J. Jeffroy, A. Le Ven and F. Baudin Laurencin. 1999. Comparative study of viral encephalopathy and retinopathy in juvenile sea bass *Dicentrarchus labrax* infected in different ways. Dis. Aquat. Org. 36: 11–20.

Peñalver, J., E. María Dolores and P. Muñoz. 2010. Absence of anisakid larvae in farmed European sea bass (*Dicentrarchus labrax* L.) and gilthead sea bream (*Sparus aurata* L.) in Southeast Spain. J. Food Prot. 73: 1332–1334.

Person-Le Ruyet, J. and N. Le Bayon. 2009. Effects of temperature, stocking density and farming conditions on fin damage in European sea bass (*Dicentrarchus labrax*). Aquat. Liv. Resour. 22: 349–362.

Pert, C.C., K. Urquhart and I.R. Bricknell. 2006. The sea bass (*Dicentrarchus labrax* L.): a peripatetic host of *Lepeophtheirus salmonis* (Copepoda: Caligidae)? Bull. Eur. Assoc. Fish Pathol. 26: 163–165.

Petrini, B. 2006. *Mycobacterium marinum*: ubiquitous agent of waterborne granulomatous skin infections. European J. Clin. Microbiol. Infect. Dis. 25: 609–613.

Piñeiro-Vidal, M., D. Gijón, C. Zarza and Y. Santos. 2012. *Tenacibaculum dicentrarchi* sp. nov., a marine bacterium of the family *Flavobacteriaceae* isolated from European sea bass. Internat. J Systematic Evolutionary Microbiol. 62: 425–429.

Pintó, R.M., J. Jofre and A. Bosch. 1991. Viral erythrocytic infection in sea bass: virus purification and confirmative diagnosis. Arch. Virol. 120: 83–96.

Poppe, T.T., R. Johansen, G. Gunnes and B. Torud. 2003. Heart morphology in wild and farmed Atlantic salmon *Salmo salar* and rainbow trout *Oncorhynchus mykiss*. Dis. Aquat. Org. 57: 103–108.

Priya, T., Y. Hong Lin, Y. Chi Wang, C. Shen Yang, P. Shing Chang and Y. Ling Song. 2012. Codon changed immobilization antigen (iAg), a potent DNA vaccine in fish against *Cryptocaryon irritans* infection. Vaccine 30: 893–903.

Pujalte, M.J., A. Sitjà-Bobadilla, M.C. Macián, C. Belloch, P. Alvarez-Pellitero, J. Pérez-Sánchez, F. Uruburu and E. Garay. 2003.Virulence and Molecular typing of *Vibrio harveyi* strains isolated from cultured dentex, gilthead sea bream and European sea bass. Syst. Appl. Microbiol. 26: 284–292.

Ragias, V., D. Tontis and F. Athanassopoulou. 2004. Incidence of an intense *Caligus minimus* Otto 1821, *C. pageti* Russel, 1925, *C. mugilis* Brian 1935 and *C. apodus* Brian 1924 infection in lagoon cultured sea bass (*Dicentrarchus labrax* L.) in Greece. Aquaculture 242: 727–733.

Raibaut, A., F. Coste and O.K. Benhassine. 1979. *Colobomatus labracis* Delamare Deboutteville and Nunes, 1952 (Copepoda, Philichthyidae), a parasite of the sea bass *Dicentrarchus labrax* (Linne 1758) from Western Mediterranean Sea. Parasitol. Res. 59: 79–85.

Ramos, P. and M. da Conceição Peleteiro. 2003. Three cases of spontaneous neoplasia in fish. Rev. Port. Cien. Vet. 98: 77–80.

Rigos, G., P. Christophilogiannis, M. Yiagnisi, A. Andriopoulou, M. Koutsodimou, I. Nengas and M. Alexis. 1999. Myxosporean infections in Greek mariculture. Aquacult. Int. 7: 361–364.

Rigos, G., M. Pavlidis and P. Divanach. 2001. Host susceptibility to *Cryptocaryon* sp. infection of Mediterranean marine broodfish held under intensive culture conditions: a case report. Bull. Eur. Assoc. Fish Pathol. 21: 33–36.

Romalde, J.L. and B. Magariños. 1997. Immunization with bacterial antigens: pasteurellosis. pp. 167–177. *In*: R. Gudding, A. Lillehaug, J. Midtlyng and F. Brown [eds.]. Fish Vaccinology. Karger, Basel, Switzerland.

Romalde, J.L., B. Magarinos, B. Fouz, I. Bandin, S. Núñez and A.E. Toranzo. 1995. Evaluation of bionor mono-kits for rapid detection of bacterial fish pathogens. Dis. Aquat. Org. 21: 25–34.

Rosales, M.J., C. Mascaró, C. Fernández, F. Luque, M. Sánchez Moreno, L. Parras, A. Cosano and J.R. Muñoz. 1999. Acute intestinal anisakiasis in Spain: a fourth-stage *Anisakis* simplex larva. Mem. Inst. Oswaldo Cruz 94: 823–826.

Salati, F., C. Cubadda, I. Viale and R. Kusuda. 2005. Immune response of sea bass *Dicentrarchus labrax* to *Tenacibaculum maritimum* antigens. Fish. Sci. 71: 563–567.

Sanjuán, E. and C. Amaro. 2007. Multiplex PCR assay for detection of *Vibrio vulnificus* biotype 2 and simultaneous discrimination of serovar E strains. Appl. Environ. Microbiol. 33: 2029–2032.

Santos, M.J. 1996. Observations on the parasitofauna of wild sea bass (*Dicentrarchus labrax* L.) from Portugal. Bull. Eur. Assoc. Fish Pathol. 16: 77–9.

Santos, M.J., F. Cavaleiro, P. Campos, A. Sousa, F. Teixeira and M. Martins. 2010. Impact of amoeba and scuticociliatidia infections on the aquaculture European sea bass (*Dicentrarchus labrax* L.) in Portugal. Vet. Parasitol. 171: 15–21.

Santos, Y., F. Pazos and J.L. Barja. 1999. *Flexibacter maritimus*, causal agent of flexibacteriosis in marine fish. pp. 1–6. *In*: G. Olivier [ed.]. ICES Identification Leaflets for Diseases and Parasites of Fish and Shellfish. No. 55. International Council for the Exploration of the Sea. Copenhagen, Denmark.

Sarusic, G. 1990. Bolesti lubina (*Dicentrarchus labrax* L.) u uvjetima intenzivnog uzgoja. Vet. Stanica 21: 159–64.

Sarusic, G. 1999. Preliminary report of infestation by isopod *Ceratothoa oestroides* (Risso 1826), in marine cultured fish. Bull. Eur. Assoc. Fish Pathol. 19: 110–112.

Siau, Y. and N.G. Sakiti. 1981. Absence de réaction histologique de défense à la présence de deux espèces de Myxosporidies parasites du poisson téléostéen *Dicentrarchus labrax* (Linne 1758). Bull. Off. Natn. Pêche (Tunisie) 5: 37–40.

Silan, P. and C. Maillard. 1986. Modalites de l'infestation par *Diplectanum aequans*, monogene ectoparasite de *Dicentrarchus labrax* en aquiculture, elements d'epidemiolgie et de prophylaxie, pp. 139–152. *In*: C.P. Vivarès, J.R. Bonami and E. Jaspers [eds.]. Pathology in Marine Aquaculture, European Aquaculture Society, Special Publication 9, Bredene, Belgium.

Silan, P. and C. Maillard. 1989. Biology of *Serranicotyle labracis*, ectoparasite of *Dicentrarchus labrax* (Teleostei): contribution to the study of its populations. Mar. Biol. 103: 481–487.

Sitjà-Bobadilla, A. 2009. Can Myxosporean parasites compromise fish and amphibian reproduction? Proc. R. Soc. B 276: 2861–2870.

Sitjà-Bobadilla, A. and P. Álvarez-Pellitero. 1990. *Sphaerospora testicularis* sp. nov. (Myxosporea, Sphaerosporidae) in wild and cultured sea bass, *Dicentrarchus labrax* (L.), from the Spanish Mediterranean area. J. Fish Dis. 13: 193–203.

Sitjà-Bobadilla, A. and P. Álvarez-Pellitero. 1992a. Light and electron microscopic description of *Sphaerospora dicentrarchi* n. np. (Myxosporea: Sphaerosporidae) from wild and cultured sea bass, *Dicentrarchus labrax* L. J. Protozool. 39: 273–281.

Sitjà-Bobadilla, A. and P. Álvarez-Pellitero. 1992b. Effect of fumagillin treatment on sea bass *Dicentrarchus labrax* parasitized by *Sphaerospora testicularis* (Myxosporea: Bival-vulida). Dis. Aquat. Org. 14: 171–178.

Sitjà-Bobadilla, A. and P. Alvarez-Pellitero. 1993a. Population dynamics of *Sphaerospora dicentrarchi* Sitjà-Bobadilla and Alvarez-Pellitero, 1992 and *S. testicularis* Sitjà-Bobadilla and Alvarez-Pellitero 1990 (Myxosporea: Bivalvulida) infections in wild and cultured Mediterranean sea bass (*Dicentrarchus labrax* L.). Parasitology 106: 39–45.

Sitjà-Bobadilla A. and P. Álvarez-Pellitero. 1993b. Pathologic effects of *Sphaerospora dicentrarchi* Sitjà- Bobadilla and Alvarez-Pellitero 1992 and *S. testicularis* Sitjà-Bobadilla and Alvarez-Pellitero, 1990 (Myxosporea: Bivalvulida) parasitic in the Mediterranean sea bass *Dicentrarchus labrax* L. (Teleostei: Serranidae) and the cell-mediated immune reaction: a light and electron microscopy study. Parasitol. Res. 79: 119–129.

Sitjà-Bobadilla, A. and O. Palenzuela. 2012. *Enteromyxum* species. pp. 163–176. *In*: P.T.K. Woo and K. Buchmann (eds.). Fish Parasites: Pathobiology and Protection. CABI Publishing, UK.

Sitjà-Bobadilla, A., A. Diamant, O. Palenzuela and P.Álvarez-Pellitero. 2007. Host factors and experimental conditions on the horizontal transmission of *Enteromyxym leei* (Myxozoa) to gilthead sea bream (*Sparus aurata* L.) and European sea bass (*Dicentrarchus labrax* L.). J Fish Dis. 29: 1–8.

Skliris, G., . Krondiris, D. Sideris, A. Shinn, W. Starkey and R. Richards. 2001. Phylogenetic and antigenic characterization of new fish nodavirus isolates from Europe and Asia. Virus Res. 75: 59–67.

Southgate, P.J. 2008. Welfare of Fish During Transport. pp. 185–194. *In*: E.J. Branson [ed.]. Fish Welfare. Blackwell Publishing Ltd., Oxford, UK.

Steiropoulos, N.A., S.A. Yuksel, K.D. Thompson, A. Adams and H.W. Ferguson. 2002. Detection of *Rickettsia*-like organisms (RLOs) in European sea bass (*Dicentrarchus labrax*) by immunohistochemistry. Bull. Eur. Assoc. Fish Pathol. 22: 338–343

Sterud, E. 2002. Parasites of wild sea bass *Dicentrarchus labrax* from Norway. Dis. Aquat. Org. 48: 209–212.

Talaat, A.M., R. Reimschuessel and M. Trucksis. 1997. Identification of mycobacteria infecting fish to the species level using polymerase chain reaction and restriction enzyme analysis. Vet. Microbiol. 58: 229–237.

Tokşen, E. 2007a. Treatment trials of gill parasite *Diplectanum aequans* (Monogenea: Diplectanidae) of cultured gilthead sea bass (*Dicentrarchus labrax*) in Aegean Sea. Ekoloji 16: 66–71.

Tokşen, E. 2007b. *Lernanthropus kroyeri* van Beneden 1851 (Crustacea: Copepoda) infections of cultured sea bass (*Dicentrarchus labrax* L.). Bull. Eur. Assoc. Fish Pathol. 27: 49–53.

Tokşen, E., E. Nemli, E. Koyuncu and M. Cankurt. 2012. Effect of trichlorfon on *Diplectanum* aequans (Monogenea: Diplectanidae) infestations in cultured sea bass, *Dicentrarchus labrax*. Bull. Eur. Assoc. Fish Pathol. 32: 103–108.

Toledo-Guedes, K., P. Sanchez-Jerez, J. Mora-Vidal, D. Girard and A. Brito. 2012. Escaped introduced sea bass (*Dicentrarchus labrax*) infected by *Sphaerospora testicularis* (Myxozoa) reach maturity in coastal habitats off Canary Islands. Mar. Ecol. 33: 26–31.

Toranzo A.E., B. Magariños and J.L. Romalde. 2005. A review of the main bacterial fish diseases in mariculture systems. Aquaculture 246: 37–61.

Toyama, T., K.K. Tsukamoto and H. Wakabayashi. 1996. Identification of *Flexibacter maritimus, Flavobacterium branchiophilum* and *Cytophaga columnaris* by PCR targeted 16S ribosomal DNA. Fish Pathol. 31: 25–31.

Trilles, J.P., B.M. Radujkovic and B. Romestand. 1989. Parasites des poissons marins du Monténégro: Isopodes (Fish parasites from Montenegro: Isopods). Acta Adriat. 30: 279–306.

Vagianou, S., F. Athanassopoulou, V. Ragias, D. Di Cave, L. Leontides and E. Golomazou. 2006a. Prevalence and pathology of ectoparasites of Mediterranean sea bream and sea bass reared under different environmental and aquaculture conditions. Isr. J. Aquacult. Bamidgeh 58: 78–88.

Vagianou, S., K. Bitchava and F. Athanassopoulou. 2006b. Sea lice (*Ceratothoa oestroides*), (Risso, 1826), infestation in Mediterranean aquaculture: new information. J. Hell. Vet. Med. Soc. 57: 223–229.

Vagianou, S., C. Bitchava, M. Yagnisi and F. Athanassopoulou. 2009. Estimation of seasonality and prevalence of the isopod parasite *Ceratothoa oestroides*, Risso 1836, in Greece. J. Hell. Vet. Med. Soc. 60: 14–25.

Varvarigos, P. 2003. Economically important pathologies of the marine fish cultured in Greece and the Aegean Sea., 69. *In*: Proceedings of the 10th National Conference of the Italian Society of Fish Pathologists (SIPI), Teramo, Italy.

Varvarigos, P. 2007. The parasitic copepod *Lernaeolophus sultanus (Pennellidae)* found on farmed sea bass *(Dicentrarchus labrax)* and sharp snout sea bream *(Diplodus puntazzo)* in coastal marine fish farms in Greece. *In*: Proceedings of the 13th International EAFP Conference on Fish and Shellfish Diseases, Grado, Italy.

Varvarigos, P. 2008. Mediterranean Farmed Fish Welfare, How good are we? *In*: Proceedings of the UK Fish Veterinary Society, Edinburgh.

Vigneulle, M., F. Baudin-Laurencin, A. Le Breton, A. Abiven and A. Leven. 1992. A case report of Nocardial infection in sea bass *Dicentrarchus labrax*. *In*: Proceedings of the Fifth International Colloquium on Pathology in Marine Aquaculture (PAMAQ) Montpellier.

Wakabayashi, H., H. Hikida and K. Masumura. 1986. *Flexibacter maritimus* sp. nov., a pathogen of marine fishes. Int. J. Syst. Bacteriol. 36: 396–398.

Yamashita, H., Y. Fujita, H. Kawakami and T. Nakai. 2005. The efficacy of inactivated virus vaccine against viral nervous necrosis (NNV). Fish Pathol. 40: 15–21.

Yambot, A.V., Y.L. Song and H. Sung. 2003. Characterization of *Cryptocaryon irritans*, a parasite isolated from marine fishes in Taiwan. Dis. Aquat. Org. 54: 147–156.

Yanagida, T., M.A. Freeman, Y. Nomura, I. Takami, Y. Sugihara, H. Yokoyama and K. Ogawa. 2005. Development of a PCR-based method for the detection of enteric myxozoans causing the emaciation disease of cultured tiger puffer. Fish Pathol. 40: 23–28.

Zlotkin, A., H. Hershko and A. Eldar. 1998. Possible transmission of *Streptococcus iniae* from wild fish to cultured marine fish. Appl. Environ. Microbiol. 64: 4065–4067.

Current Knowledge on the Development and Functionality of Immune Responses in the European Sea Bass (*Dicentrarchus labrax*)

Jorge Galindo-Villegas[a],* and Victoriano Mulero[b]

Introduction

A potential concern of fish culturists is to reduce any economic loss. In European sea bass (*Dicentrarchus labrax* L.) aquaculture rearing facilities, mostly at those located in southern European countries, infectious diseases are a leading cause of mortality. Bacterial and viral diseases of sea bass (the most recurrent, namely vibriosis, pasteurellosis, and viral erythrocytic infection) are now a major challenge for biological studies on this species and further intensive fish production. To fight against diseases, preventive strategies should be implemented in every single farm. Best management practices, which may include improved sanitary conditions, clean water supplies and vector control are by far the most effective measures to reduce the incidence of infectious diseases. However, the development of specific target strategies like vaccines, disease-resistant breeds and therapeutics are

Department of Cell Biology and Histology, University of Murcia, Campus Universitario de Espinardo, 30100. Murcia, Spain.
[a] Email: jorge-galindo@usa.net
[b] Email: vmulero@umu.es
* Corresponding author

much more important in achieving protection, but this requires sufficient understanding of fish immune system (Zapata et al. 2006). The sea bass belongs to the Teleostei which is one of three infraclasses in the class Actinopterygii. This diverse group, which arose later in evolution during the Triassic period, roughly makes up half of the extant vertebrate species and occupies a key evolutionary position in the development of immune system evolution. Teleostei is the earliest class of vertebrates sharing similar immune system organization with other vertebrates, and thus, possesses the elements of both innate and adaptive immunity (Vadstein et al. 2012). Therefore, the sea bass has become very attractive as a fish model in regards to developmental, anatomical and functional comparative studies of the immune system. So far, it has been recognized that fish immune system is characterized by a multi-layered organization that provides immunity to infectious organisms, and each layer can be considered to have an increasing complexity. In this way the central challenge of fish immunology, which includes the sea bass model, is now to fully characterize the occurring immune responses in as much tissues as possible, and determine the functionality of each layer. Foundations of present knowledge on innate and adaptive immunity in sea bass have been established during the two previous decades. Remarkable milestones were provided during the early 90s by several workers. Such studies included the isolation of sea bass innate immune cells and the description to the ultrastructural characterization of the main features displayed by each cell group (Meseguer et al. 1991, Esteban and Meseguer 1994, Meseguer et al. 1996). And, extensive reports on the role of macrophages, granulocytes, lymphocytes, plasmocytes and fibroblast-like cells after myxosporean parasites invasion on host testes (Sitja-Bobadilla and Alvarez-Pellitero 1993). Some years later, Mulero et al. (1994) conducted an interesting experiment in which functional properties of host leucocytes were shown. The authors of this paper characterized the changes followed by effector sea bass leucocytes incubated *in vitro* with HeLa or B16 melanoma cell targets. They observed that effector cells established spot contacts with targets, and characterized morphological cell changes like smooth surfaces, appearance of cell processes or surface blebs, suggesting that changes in targets are similar to those described to be mediated by mammalian cytotoxic cells (Mulero et al. 1994). On the adaptive capacity of sea bass immune system, two findings made great improvements in the study of fish immunology. These contributions were provided by Scapigliati et al. (Scapigliati et al. 2000, Romestand et al. 1995), who produced and characterized specific markers for T and B lymphocytes respectively—the monoclonal antibody DLT15 against thymocytes and the DLIg3 antibody directed against immunoglobulin (Ig)-bearing B cells. In this regard, Picchietti and colleagues (Picchietti et al. 1997), applying both antibodies, showed through functional analyses that gut-associated lymphoid tissue (GALT) developed earlier than other lymphoid compartments. This was a highly remarkable finding which provides evidence of the abundance and predominance of T cells in the gut. Together, these studies along with several more conducted in carp (dos Santos et al. 2000, Huttenhuis et al. 2006, Rombout et al. 1985, 1993), brought to light the importance of the gut as an immunocompetent organ in fish defense against invader bacteria. Recent studies have complemented the basic sea bass immune research with steadily increasing work conducted mainly at the cellular and molecular level (Scapigliati et al. 2002, do Vale et al. 2002, Buonocore et al. 2008, Buonocore and Scapigliati 2009, Pallavicini et al. 2010, Marozzi et al. 2012, Sarropoulou et al. 2012). However, the former global vision on the study of vertebrate immunology has just recently changed and new players from the environment, the microbes, have begun to be recognized as vital key modulators of the immune response (Yu et al. 2012, Galindo-Villegas et al. 2012, Rautava et al. 2012). The sea bass model of immunity is not the exception and new efforts have been directed towards discovering the mechanisms of microbial pathogenesis and fish-microbe symbiosis through the characterization of gut microbiota in conventional or gnotobiotically raised fish (Dierckens et al. 2009, Rekecki et al. 2012). Indeed, the diverse and unique immune tools available for European sea bass, its taxonomic position and commercial relevance, has turned this species into an excellent model to study developmental and comparative immunological processes, including basic and functional aspects. Therefore, in this chapter we briefly summarize and update the emergence, phylogenetic distribution and functionality of *D. labrax* innate and adaptive immunity, as well as the most relevant interactions between them towards host defense against pathogens.

Evolutionary Aspects

The inflammatory response as an effective mediator of immunity

Every living organism possesses an immune system consisting of a sophisticated, complex, highly redundant, and multilevel network of various defensive mechanisms (Danilova 2006). So far, most of what we know about the composition, functions and regulation of the two fundamental branches of the immune system (innate and adaptive), comes from studies on mice and humans. Nevertheless, fish immunity is fairly well characterized and so it has been observed that innate immune response plays a continuous, important role in orchestrating quick immune responses. It protects developing embryos and further larval stages against the hostile environment and has been considered an essential component in combating disease incidents. Indeed, the fish poikilothermic nature hamper their early adaptive immune response through particular constrains such as slow lymphocytes proliferation, limited antibody diversification and affinity maturation. Therefore, retarding their overall immunological memory capacity (Magnadottir 2006). In summary, the innate immune defense mechanisms of sea bass could immediately counteract infectivity through a complex network which includes, but is not limited to physical barriers, humoral factors and cellular defenses that working together as a system shall trigger inflammatory processes (Vadstein et al. 2012) against any pathogenic microbe attempting to assault the fish body (see: Scapigliati et al. 2002). As a general rule, defense mechanisms restrict invasion of the body by foreign components and inhibit their persistence within the tissue through physical/structural hindrance and clearance (epithelial linings, mucus, peristalsis); chemical factors (pH of body fluids, antimicrobial peptides and proteins), and phagocytic cells contributing to inflammation. Additionally, innate immunity has a powerful ally represented as an early fish protection achieved through a mixed passive immunity transmitted from maternal sources to offspring during vitellogenesis and oogenesis (Mulero et al. 2007, Swain and Nayak 2009). In most fish species, passive immunity is characterized by the presence of lytic enzymes and adaptive factors such as hormones and immunoglobulins. Nevertheless, most maternally transferred factors usually persist only for a very limited period and completely disappear thereafter at the late larval stage (Hanif et al. 2004), when the adaptive system can correctly operate. In this regard is important to stress that in most marine fish, a delayed maturation of lymphomyeloid organs is observed (Hanif et al. 2004, Scapigliati et al. 1995). In sea bass, a delayed production of thymocytes makes it unable to synthesize antibodies until several weeks after hatching (Swain and Nayak 2009). To overcome this immunological limitation, the evolutionary processes have provided sea bass innate immunity with wide capacities and mechanisms of protection that are activated just after egg fecundation, and become fully functional by the time of hatching. The literature contains many review articles on the functions and modulation of most immune parameters and immune defense mechanisms so far described in the European sea bass (Scapigliati et al. 2002, Buonocore and Scapigliati 2009, Chistiakov et al. 2007). Thus, here we will only describe the new findings reported so far. Mostly, we will emphasize such studies that follow the actual holistic perspective in which microbes must be addressed as fundamental key players on the host decision to mount most immune responses and then trigger proper defense mechanisms.

PAMPs, MAMPs, DAMPs and its receptors

Complex responses of innate immunity cells are driven by a diverse array of pattern recognition receptors (PRRs) that bind molecular motifs, conserved within all classes of microbes, which induce subsequent host immunity through multiple signaling pathways that contribute to the eradication of the pathogen (Hanif et al. 2004). Recognized structures are widely known as "pathogen-associated molecular patterns" (PAMPs) because traditionally they have been associated only with pathogens. However, one puzzling issue on the functional and orchestration capacities of the immune system is that supported microbial partners have the same conformational, molecular, or locomotive structures as their closely related pathogens, indicating further that microbe-associated molecular patterns (MAMPs) are not limited to pathogens (Galindo-Villegas et al. 2012). MAMPs may encode enough capacities to trigger an inflammatory response mediated through classical immune recognition receptors described in all classes of fish studied so far. Therefore, the wide diversity of bacterial molecular motifs present in the aquatic environment have pushed fish to evolve, diversify and transmit to upper vertebrate taxa several classes of PRRs, such as Toll-like receptors (TLRs), NOD-like receptors (NLRs), C-type lectin receptors (CLRs), complement and scavenger receptors, that have been characterized in all classes of fish studied so far. An interesting new approach, has demonstrated that in mammals, cellular injury can release endogenous "damage-associated molecular patterns" (DAMPs) which activate innate immunity through similar mechanisms as MAMPs (Zhang et al. 2010a). This activation is explained through the study of evolutionary processes in which mitochondria were evolved from endosymbionts, which in turn were derived from bacteria and so might bear bacterial, non-pathogenic, molecular motifs (e.g., the CpG repeats contained in mitochondrial DNA). Then, it would be expected that any cellular disruption by trauma releases mitochondrial DAMPs into the circulation with potential to achieve a high concentration and trigger an uncontrolled downstream innate immune signaling response collectively known as sepsis. Taking advantage of the multiple immunological tools available in sea bass, it might be interesting to test whether the recognition of endogenous danger signals is conserved, and if so determine how it affects its recognition mechanisms. This task is not easy because is well recognized that fish innate immunity signaling pathways show a higher level of complexity than those in mammals. For example, zebrafish genes corresponding with many mammalian pattern-recognition-receptors (PRRs) have been identified and are often present with several paralogs. Zebrafish show about 50% more diversity in TLRs than mammals, most likely a consequence of the early genome duplication of the teleost lineage. As a consequence, this may allow for the recognition of a larger repertoire of MAMPs/DAMPs, or other microbial conserved molecular motifs. Furthermore, there is evidence that not all zebrafish TLRs have the same MAMP specificity or signaling activity as their mammalian counterparts. To point out this feature, it has recently been reported, for the first time, the positive identification of TLR4 as a negative regulator of TLR signaling in fish, explaining in part the strange resistance of fish to the toxic effects of bacterial lipopolysaccharide (LPS) and the high concentrations of this substance required to activate the immune cells of these animals (Sepulcre et al. 2009). In sea bass, this specificity has never been addressed, but we should anticipate a similar response as the one described above for the closely related perciform, the Mediterranean gilthead sea bream (*Sparus aurata*).

TOLL-like receptors: an important mechanism of bacterial recognition

Toll-like receptors (TLRs) are part of the ancient innate arm of the immune system which possesses a skillful system that detects microbial invasion. TLRs are a highly conserved multigenic family which comprises several transmembrane and endosomal proteins first described in Drosophila (Anderson et al. 1985). The TLR signal transduction cascade has been extensively analyzed in mammals, thus they are the best understood of the innate immune receptors which recognize conserved microbial structures to induce immune effector molecules (O'Neill 2008a, 2008b, 2008c, 2008d, Oda and Kitano 2006). However, not all features of TLR signaling have been conserved throughout all vertebrate lineages. The human TLR family, for example, comprises the additional TLR6 and 10 factors, which to date have not been found in fish (Rebl et al. 2010). On the other hand, TLR21 and TLR22 appear to be fish-specific TLRs and may have been lost in mammals (Roach et al. 2005). Therefore, fish-specific adaptations may be anticipated in both, structure and function of the diverse factors making up that signaling cascade. In fish, as in mammals these receptors are mainly involved in host defence, and TLR family members share the same structure, defined by the presence of leucine-rich repeats in their extracellular domain and by the presence of a Toll/interleukin-1 (IL-1) receptor (TIR) domain in the C-terminal, the cytosolic part of the protein that initiates signal transduction after MAMPs recognition (Sepulcre et al. 2009, Anderson et al. 1985, O'Neill 2008a, 2008b, 2008c). The TIR domain containing adaptor myeloid differentiation marker 88 (MyD88) is essential for the downstream signaling of various TLRs, with the exception of TLR3. In mammals, TLR3 detects viral double-stranded (ds) RNA in the endolysosome which recruits the adaptor protein called TRIF that is totally independent of MyD88 (Takeuchi and Akira 2010). Interestingly, both pathways at the bottom of the signaling cascade, translocate into the nucleus through the master regulator nuclear factor kappa b (NFkB) or the interferon regulatory factor 7 (IRF7) and activate expression of pro-inflammatory cytokine or interferon genes respectively (Takeuchi and Akira 2010). Thus, dynamic changes in the transcriptional networks activated in response to inflammatory stimuli are likely to be highly complex. To date, 17 teleost TLRs (TLR1, 2, 3, 4, 5, 5S, 7, 8, 9, 13, 14, 18, 19, 20, 21, 22, 23) have been identified by genome and transcriptome analysis and remarkably distinct features of the TLR cascades among fish species have been reported. Additionally, the activation of antimicrobial host defense mechanisms such as the production of reactive nitrogen and oxygen radicals and antimicrobial peptides has been documented resulting from TLRs-ligand recognition. Most of this information is summarized in four major reviews dealing with piscine TLRs (Bricknell and Dalmo 2005, Rebl et al. 2010, Takano et al. 2011, Palti 2011). Unfortunately, to the best of our knowledge, so far no information for sea bass TLRs has been published, but information is expected to be released soon.

Innate Immune Cells

Professional phagocytes

Phagocytosis is an ancient cellular function. However, professional phagocytes have evolved only in higher organisms, where they play an important role in host defence. Mammalian professional phagocytes include neutrophils, monocytes, macrophages, dendritic cells and mast cells. But in most fish species the basic innate immune system consists of cell components mainly represented by phagocytes and non-specific cytotoxic cells (Galindo-Villegas and Hosokawa 2004). Fish professional phagocytes and mast cells, have been both reported as capable of engulfing relatively large microorganisms and killing them with a combination of various microbicidal systems (Krause 2000). In the European sea bass, only mononuclear phagocytes (circulating monocytes and tissue macrophages) and neutrophils have been described so far (Meseguer et al. 1991, do Vale et al. 2002). Sea bass macrophages, at the monocyte stage, possess an oval or kidney-shaped heterochromatinic nucleus and a few granules (Mulero et al. 1994). Upon localization at specific tissue, an important function of macrophages is the regulation of the immune response through the production of a variety of cytokines, such as interleukins, interferons, tumor necrosis factors and inflammatory prostanoids (Mulero et al. 2008a). In sea bass, using the monoclonal antibody Mcsfr, a preliminary study revealed that macrophages gather mainly in the head kidney and spleen of this species. We speculate that a major organ in which professional macrophages would be severely specialized is the gut of sea bass, but further experiments using the Mcsfr antibody in addition to other molecular tools must be conducted to verify this hypothesis. Neutrophils form the major type of leucocytes in peripheral blood, with counts ranging from 40% to 70% of total leucocytes under normal conditions. Ultra structurally, sea bass neutrophils were found to have abundant cytoplasmic granules positive for peroxidase and arylsulphatase, and were negative for alpha-naphthyl butyrate (ANB) esterase (do Vale et al. 2002). Fish neutrophils do not differ much from their mammalian counterpart. They are elegantly adapted to perform a variety of antimicrobial functions such as bactericidal, phagocytic and chemotactic activities. Therefore, is quite obvious that neutrophils are equipped with a complex molecular machinery to sense the site of an infection, to crawl toward the invading microorganisms, and to kill and ingest them. Upon neutrophil activation through the contact with invading microbes, reactive oxygen species (ROS) are massively generated in an oxidative burst by nicotinamide adenine dinucleotide phosphate (NADPH) oxidase consisting of a multi-subunit enzyme complex that is assembled on the membrane (Petry et al. 2010). To test the importance of this killing mechanism, pharmacological inhibition of NADPH oxidase through interference with the redox metabolism by diphenylene iodonium (DPI) has been demonstrated (Galindo-Villegas et al. 2012). An additional feature of fish neutrophils is degranulation of primary granules, releasing antimicrobial factors into the extracellular medium (Rodrigues et al. 2003). Just recently, a previously unknown defense mechanism of fish neutrophils has been described. Upon activation, they generate extracellular fibers composed of granular nuclear constituents that form neutrophil extracellular traps (NETs) which are complex structures composed of DNA, histones and proteins from granules for trapping and extracellular killing of microbes (Palic et al. 2007). So far, the only fish species in which NETs has been demonstrated is the fathead minnow (*Pimephales promelas*) but, it is expected that NETs should be demonstrated soon in sea bass too, due immune cells displaying NETs-like pattern have been recently observed after luminal uptake of *V. anguillarum* in larval stages of this fish species (P. Bossier 2012 per. comm.). Altogether, evidences demonstrate that neutrophils are uniquely capable of forming massive amounts of reactive oxygen species, peroxidase activity and other toxic molecules that effectively protect the fish body against bacterial, viral and fungal infections. Thus, to execute a large number of different functions, sufficient neutrophils must be generated at a high rate and released from the head kidney. However, since neutrophil products could be potentially harmful to the sea bass tissue, many immunological safeguards exist to prevent unwanted side effects. Therefore, both responses are worthy of deep study in order to fully understand the protective mechanisms mediated by this outstanding cellular group (Table 1).

Table 1. Summary of immune cells identified to date in the European sea bass (Body location, function, molecules released and recognizing antibodies are presented).

Cell	Type	Main Location	Main Function	Main Molecules	Antibody for recognition
Monocytes	Mononuclear phagocytes	Blood stream	Surveillance Phagocytosis		
	Tissue macrophages	Head kidney Spleen Gut	Regulation of immune response Antigen presentation	Cytokines Prostanoids	Mcsfr
Neutrophils		Blood stream All tissues	Phagocytosis	ROS, iNOS AMPs NETs?	
Mast Cells	Basophilic Acidophilic/ Eosinophilic	Connective tissue	Neutrophil attraction Macrophage activation	AMPs Lytic enzymes Serotonin Histamine	Anti-histamine Anti-piscidin
Thymocytes	T-lymphocytes	Blood stream Thymus Gut	Promotes activation of B-cells	TCRb CD4 CD8 RAG MHCII Cytokines	DLT15
	Intestinal T cells	GALT	Intraepithelial lymphocytes	TCR⁺ T cells without a prior local expansion selection	DLT15
B-lymphocytes	B-cells	Peripheral blood Spleen Head-Kidney Gut	Immune surveillance Antigen recognition	MHCI BcR	DLIg3 (WDI-1; -3)
	Plasmocytes	All tissues	Antibody production	Immunoglobulins BcR	DLIg3 (WDI-1; -3)

Abbreviations: Reactive oxygen species (ROS), Anti-microbial peptides (AMPs), Neutrophil extracellular traps (NETs), T-cell receptor (TCR), Cluster of differentiation (CD), B-cell receptor (BcR), Macrophage colony stimulating factor receptor (Mcsfr), Antibody recognizing antigenic determinants expressed by T-cells (DLT15), Antibody recognizing Ig-bearing cells (DLIg3), Monoclonal antibodies characterized as specific to sea bass Ig heavy (WDI-1) and light (WDI-3) chains.

Mast cells

Recent evidence from several laboratories has demonstrated that mast cells (MCs) are a key cell type of the hematopoietic lineage that has evolutionary conserved functions in pathogen surveillance (Abraham and St. John 2010). Presence of mast cells has been reported in all classes of vertebrates, including fish, amphibians, reptiles, birds, and mammals. Furthermore, current observations in several fish species point out the crucial location of MCs in connective tissue like the skin, gill filaments, brain and the intestinal submucosa layer, suggesting these cells have an active task on pathogens recognition or other signs of infection (Mulero et al. 2007, Reite and Evensen 2006, Dezfuli et al. 2010). Just recently, MCs have been noted to respond by migration and degranulation (Mulero et al. 2008a), giving support to previous suppositions that MCs were involved in host defense mechanisms in teleosts and shall be critical in fighting many infectious diseases (Reite

1997). Cytochemical characteristics and involvement in pathological conditions have convinced most investigators that fish MCs are analogous to mammalian ones (Dezfuli et al. 2010). However, a definite functional characterization of fish MCs has not yet been published. Fish MCs constitute a heterogeneous cell population exemplified by their diverse morphology, granular content, sensitivity to fixatives, and response to drugs (Crivellato and Ribatti 2010). Besides basic similarities among MCs, basophilic granular cells or acidophilic/eosinophilic granular cells, it is now generally accepted that MCs are involved in the induction of inflammatory responses by their effects on vasodilatation, neutrophil attraction and macrophage activation (Crivellato and Ribatti 2010, Vallejo and Ellis 1989). Several investigators have demonstrated how MCs degranulate in response to exposure to a variety of known degranulating agents and pathogens with release of their contents (Powell et al. 1991, Ellis 1985). Granules of fish MCs were believed until a few years ago to contain components common to their mammalian counterparts like antimicrobial peptides, lytic enzymes or serotonin, but lack the presence of histamine. Recently the contents of granules in fish MCs was identified using antibodies and immunostaining techniques (Table 1), and the study revealed that histamine is stored in fish mast cells, but surprisingly this was true only in those species belonging to the largest and most evolutionarily advanced order of teleosts, the Perciformes, group in which sea bass is included (Mulero et al. 2007). Furthermore, functional studies indicated that fish professional phagocyte function may be regulated by the release of histamine from MCs upon H1 and H2 receptor engagement. Whatever the explanation, evidence that histamine is biologically active and can regulate the inflammatory response of Perciformes, acting on professional phagocytes signaling against pathogens was demonstrated. Recently, in mammalian MCs, many signaling mediators upstream of degranulation have been identified and these converge on a common requirement for generating a Ca^{2+} flux in the responding cell (Kalesnikoff and Galli 2008). So far, in fish a similar pathway has not been reported but applying the different tools available for sea bass could give possible clues on the evolutionary relations among MCs and fish.

Major Proteins Involved on Immune Response

Cytokines: main orchestrators of immunity

Cytokines are small cell-signaling protein molecules that are secreted by numerous cells. They evolved from the earliest forms as intracellular molecules before the appearance of receptors and signaling cascades. Cytokines have been extensively described in mammals, but cytokine-like activities have been also demonstrated in invertebrates, such as star fish (Legac et al. 1996) and *shrimp* (Wang et al. 2012), where they seems to be involved on immune response to pathogenic infections and tissue repair. Therefore, evidences suggest a vertical transmission of these effector molecules among different taxa with an increasing functionality accordingly. Cytokines can be divided into functional classes; some trigger lymphocyte growth factors, and others function as pro-inflammatory or anti-inflammatory molecules, whereas one more type polarizes the immune response to antigen. Thus, is clear that cytokines have become recognized as important regulators of the immune system in most phylogenetic classes, and targeting these genes in fish has a potential application for development of therapeutic agents in aquaculture fish disease. Efforts in finding fish homologues of mammalian cytokines led to the cloning and characterization of important molecules related to several immune responses (Secombes et al. 2001, García-Castillo et al. 2002, Scapigliati et al. 2006). In sea bass, many trials have been conducted to characterize, under basal and infective conditions, cytokines such as TNFα, IL-1β, IL-8 or IL-18 (see: Buonocore and Scapigliati 2009). The best characterized cytokine in sea bass has been the pro-inflammatory IL-1β (Scapigliati et al. 2006). To characterize sea bass IL-1β biological activities by pioneer researchers, alignment tools were used and the site at the sequence of a putative mature peptide which allows the production of a recombinant IL-1β (r IL-1β) molecule was predicted (Buonocore et al. 2003, 2005). Interestingly, throughout previous years it was assumed that fish IL-1β lacks the cleavage sequence coding for interleukin-1β-converting enzyme (ICE), also known as caspase-1, which is a specific intracellular cysteine protease required for the processing of some cytokines lacking a signal peptide (Zou et al. 1999). Then, it has been a matter of speculation how the mature protein is released from the intracellular compartment. Recently, this issue was partially clarified by Reis et al. (2012). Using the sea bass model, they found that pro-caspase-1 auto-processing occurs through a series of intermediates, yielding active p24/p10 and p20/p10 heterodimers in a similar way to that described previously for human caspase-1. Moreover, they reported also the existence of alternative spliced variants of caspase-1, suggesting that caspase-1 isoforms have been evolutionarily maintained and therefore likely to play a regulatory role in the inflammatory response. Thus, if sea bass IL-1β is cleaved by caspase-1 as it happens in mammals, we further anticipate that soon several functional analyses must give solid evidence of this claim, and may help in searching for therapeutic targets.

Acute-phase proteins: effectors of innate immunity

Recent understanding reveals that the vertebrate innate immune system relies on a variety of antimicrobial and antiviral molecules that conspire to limit infection. Also, it is universally accepted that the innate immune system is a necessary antecedent for the development of a sustained adaptive immune response. However, in this section we focus on an essential component of the immune system, that is, non-antibody-mediated response against pathogens. These components of the innate repertoire are particularly important in fish immunity. The aquatic ecosystem, which for obvious reasons, encompasses fish is one of the most aggressive habitats due the impressive number and diversity of microorganisms coexisting per area. Therefore, fish surfaces which comprise a large area of delicate epithelium, starting from hatching constantly are exposed to potentially harmful pathogens. To overcome this limitation, innate cells, located strategically in subepithelial sites, are capable of immediately producing and secreting several endogenous peptides that as well as having direct effect over most invaders, could act as inflammatory mediators and have therefore been described as alarmins (Bianchi 2007).

Lectins

Animal lectins are proteins or glycoproteins that bind polysaccharides, glycoconjugates and membrane glycoproteins and some of them are involved in innate and adaptive immunity. Their ability to bind terminal sugars on glycoproteins and glycolipids expressed by pathogenic bacteria, yeasts, parasites and virus makes them play an important role in innate immunity in pathogen recognition and neutralization. These molecules can also trigger effector mechanisms such as activation of the complement system (Lu et al. 1990), increasing opsonisation (Camby et al. 2006), phagocytosis (Salerno et al. 2009), and apoptosis (Davey et al. 2011), which leads to pathogen destruction. Lectins can also function as regulators of adaptive immunity, leading to apoptosis of selected T cells subsets (Perillo et al. 1995), or enhancing the production of cytokines in non-activated and activated T cells, among others (van der Leij et al. 2004). Lectins have been classified into several families based on the presence of conserved amino acid sequences motifs in their carbohydrate recognition domain, structural fold, and calcium requirements (C-, P-, I-, and L-type lectin, galectins pentraxins, etc.) (Kilpatrick 2000). So far, a novel lectin family, the fucose-binding lectins (F-type), have been identified and characterized in the serum from fish (Odom and Vasta 2006, Cammarata et al. 2007). However, prior to the identification of the F-lectin family, a 34 kDa fucose-binding Ca^{2+}-independent serum lectin had been purified from the sea bass and named DIFBL (Cammarata et al. 2007). At that time DIFBL was not fully characterized and it is just recently that molecular properties and phylogenetic relationships of DIFBL have been further analysed in detail. Although lectins from the various families differ greatly in the domain architecture, they are all characterized by a carbohydrate-recognition domain (CRD), which confers the protein its carbohydrate-binding activity (Zelensky and Gready 2005). Ca^{2+}-independent soluble lectins, called galectins are widely distributed in vertebrates, invertebrates (sponges, worms, and insects), and protists (Barondes et al. 1994). In sea bass experimentally infected with nodavirus, up-regulation of expression of Sbgalectin-1 was observed as soon as 4h post infection, indicating the importance this molecule could have in fish defence and immune mechanisms (Poisa-Beiro et al. 2009). Functional *in vitro* assays conducted with recombinant sea bass galectin-1 provided further insights in its anti-inflammatory activity, suggesting a potential role for this protein against the infection mediated by nodavirus in sea bass. However, further detailed studies using those identified lectins in sea bass, still remain to be conducted to elucidate the detailed mechanisms in this species immunity and protection against pathogens.

Antimicrobial peptides

An important group of alarmins collectively termed antimicrobial peptides (AMPs) constitute a primitive immune defense mechanism found in a wide range of eukaryotic organisms, from humans to plants to insects (Meseguer et al. 1996). AMPs are evolutionarily ancient weapons, which along with regulatory protein families have provided complex multicellular organisms with defenses needed to effectively compete in a world dominated by microbes. Most of these gene-encoded peptides, systemic or inducible, share several common properties and are mobilized shortly after pathogenic infection to neutralize a broad range of microbes that can include bacteria, fungi, parasites and viruses. Therefore, during the last years an increasing number of AMPs have been isolated from a wide number of fish species (Noga and Silphaduang 2003, Bao et al. 2006). Most are amphipatic, linear, antibacterial peptides, and include pleurocidin from the skin mucus of winter flounder (*Pleuronectes americanus*) (Cole et al. 1997), pardaxin from the skin secretion of the Moses sole (*Pardachirus marmoratus*) (Oren and Shai 1996), chrysophsins from red sea bream (*Chrysophyrs major*) (Iijima et al. 2003), misgurin from the loach (*Misgurnus anguillicaudatus*) (Park et al. 1997) or moronecidin from hybrid striped bass (*Morone saxatilis*). Among fish, a widely distributed AMP is the piscidin (Silphaduang and Noga 2001). Piscidin is one of the most extensively studied AMPs. Piscidins are thought to permeabilize the membrane of pathogens by tiroidal pore formation in which lipids of the membrane are inserted between the α-helices (Campagna et al. 2007). So far, it has been isolated as five different isoforms from several species which are not exclusive among them (Peng et al. 2012). Its specific location among the immune cells has been found to be in the granules of professional phagocytic granulocytes of several fish species, like the Atlantic cod (*Gadus morhua* L.) (Ruangsri et al. 2012), the sea bass (*D. labrax*) (Mulero et al. 2008b), or tilapia (*Oreochromis niloticus*) (Peng et al. 2012). And, just recently the presence of a new AMP named chionodracine was identified in the Antartic teleost (*Chinodraco hamatus*). Analyses of the structural features of chionodracine lead to the inclusion of this peptide into the piscidins family (Buonocore et al. 2012). Genomic evidence for the presence of hepcidin (Bao et al. 2005), LEAP-2 (Bao et al. 2006) and NK-lysin (Wang et al. 2006), cathelicidins (Chang et al. 2005, 2006) and defensins (Zou et al. 2007) have been reported. Of particular interest in this book, a wide set of AMPs have been reported for the sea bass so far. Among them, different piscidins have been demonstrated to be present in mast cells and professional phagocytic granulocytes (Mulero et al. 2008b) and have been detected via bug blot, Western blot, ELISA and/or immunochemistry in gill extract (Corrales et al. 2010). Also, lysozyme (Cotou et al. 2012) gene transcripts have been detected at high amount in leukocytes underlying most immunocompetent tissues. And, the presence of a species specific piscidin (dicentracin) was recently supported by the cloning of a piscidin gene and confirmation of its antimicrobial activity (Salerno et al. 2007). However, no expression was observed in thrombocytes or in lymphocytes. Therefore, the gene expression of these key AMPs seems to be mostly related to granulocytes where these leukocytes store the AMPs in its granules (Lauth et al. 2002). The AMPs when released from the granules has potent antimicrobial activity but also promote the inflammatory process through neutrophil recruitment, amplifying the phagocytic activity, stimulating the cellular apoptosis, decreasing de cytokine production and neutralizing several substances coming from surrounding bacteria (Plouffe et al. 2005).

Behavioral Diversity of the Mucosal Tissues of Sea Bass

The mucosa-associated lymphoid tissue (MALT)

The first line of humoral defense, where any invading microbe can be effectively blocked or neutralized, is the mucosal surface. Teleost fish live in a microbe rich aquatic environment, so this diverse group developed mucosal surfaces as their strategy to protect themselves from the aggressions of the environment. Formerly, the mucosal structures were addressed as relatively simple physical barriers that may delay the infective capacities of microbes and give time to mount a specific response. But, in the last decades, robustness of these structures and the mechanisms attached has attracted attention of researchers in several fields, and massively emerged as a strong issue to be substantially explored. The mucosa-associated lymphoid tissue

(MALT) operates by two adaptive non-inflammatory mechanisms, contains B cells and immunoglobulins, which pay a pivotal role in the maintenance of mucosal immunity (Brandtzaeg et al. 2009). Until recently, there has been a general belief that IgM was the only functional immunoglobulin in teleosts, both in systemic and mucosal compartments. Recent breakthroughs in the field of fish immunoglobulins have added two new players to the scene, IgD (113) and IgT/IgZ (Danilova et al. 2005, Hansen et al. 2005). Significantly, IgT has been reported to be an immunoglobulin specialized in the gut mucosal immunity (Zhang et al. 2010b), a novel finding that makes the field of mucosal immunity in fish even more appealing. Recently, Galindo-Villegas and Mulero got the sequence for IgT in sea bass and demonstrated upon antigen recognition, an up regulation at the gut of this species (Galindo-Villegas et al. 2013). Further studies may characterize its behavior under different scenarios to clarify the role of this important immunoglobulin in sea bass.

Gut-associated lymphoid tissue (GALT): a resident player

The Teleost gut, like the mammalian intestine, is composed of very unique tissue along with the presence of commensal bacteria (Honda and Takeda 2009). To maintain beneficial relationships, the host immune system should remain hypo-responsive to commensal microorganisms and food antigens. At the same time, the intestinal immune system has to combat invasive pathogenic bacteria. Thus, the immune system in the intestinal mucosa has very complicated features that are distinct from that in other tissues. In mammals, the intestinal immune system consists both of organized lymphoid tissue, and of populations of T cells and innate immune cells dispersed throughout the lamina propria (Coombes and Maloy 2007). In mammals, an essential follicle associated epithelium above the induction sites encompasses the M cells, which are active transporters of exogenous antigens to the underlying lymphoid tissue. M cells do not present antigen, but are likely to pass it on to dendritic cells (DC) located near the epithelium, which subsequently interact with naive T cells found in the Peyer's patches (PP). The presence of DC in a number of locations throughout the intestine allows for the sampling of antigen by several routes. DC located in the subepithelial dome of the PP are likely to pick up antigen transported from the gut lumen by M Cells. These processes result in IgA class switching and the secretion of high amounts of dimeric IgA at the effector sites (Cerutti 2008). The secreted IgA is subsequently bound by the polymeric Ig receptor (pIgR) and transcytosed to the intestinal lumen or to the bile in the liver. The extracellular part of the receptor is then cleaved off and secreted as secretory component (SC) together with the IgA at mucus site. In the last decade it has been shown that dendritic cells (DC) can also take up antigen directly from the lumen and this new induction alternative is now well established (Brandtzaeg et al. 2008). Therefore, uptake of small numbers of commensal bacteria by DC through this route has been shown to induce production of IgA, which limits bacterial penetration and neutralizes LPS (Macpherson and Uhr 2004). In teleost fish, numerous studies have established the presence of a true GALT. The fish GALT is composed of lymphocytes, plasma cells, granulocytes and macrophages, of which some are located between the epithelial cells while others are disseminated throughout the lamina propria (Zapata 1979, Davina et al. 1980, Temkin and McMillan 1986, Hart et al. 1988). Studies on the carp, *Cyprinus carpio*, using MAbs against Igs, or leukocytes, have clearly shown that all cell types thought to be necessary for a local immune response, are present in the intestinal mucosa (Rombout et al. 1993). In this species (Rombout et al. 1985, 1986, 1989) and in the trout, *Salmo gairdneri* (Georgopoulou et al. 1986), it has been shown that antigens are transported from enterocytes to macrophages that process and present them to lymphoid cells. Thus, in addition to their absorptive role, certain segments of the intestine are specialized for an immunological function. Further studies are needed to clarify the development, comparative histology and physiology of the GALT in teleosts. In the sea bass, using MAbs that identify different subpopulations of lymphocytes, the existence of a GALT has been established (Scapigliati et al. 1996, Abelli et al. 1997). MAb DLT15 recognizes antigenic determinants expressed by thymocytes and T-lymphocytes (Scapigliati et al. 1995, Abelli et al. 1996, Romano et al. 1997) and MAb DLIg3 recognizes the light chain of serum IgM (Scapigliati et al. 1996, Abelli et al. 1996). These were used to study the development of the intestinal lymphoid cells in the sea bass, with the aim of gaining new information on the establishment of gut immunity. Unfortunately, so far, Peyer's patches, M cells, IgA and J-chain have not been reported in sea bass or any other teleost fish, and, there is hardly evidence for specific homing of mucosal lymphocytes through lymph nodes (Brandtzaeg et al. 2008, 2009). However, homologous isoforms dealing against fish pathogens would be expected to be present in the gut of several teleosts.

Appearance and Functionality of Adaptive Immunity

Maternal transfer and prospective fate of IgM at the first developmental stages

Most teleost fish embryos develop externally so they have developed a completely and fully functional immune system to cope with an impressive number and diversity of coexisting microorganisms (Galindo-Villegas et al. 2012). Furthermore, maturation of immunocompetence in fish develops quite late, although both lymphoid organs and T- and B-lymphocytes may appear early in embryogenesis (Zapata et al. 2006). Therefore, developing fish embryos and larvae have little or only limited ability to synthesize immune-related molecules endogenously (Wang et al. 2008). How developing fish embryos and larvae survive microbial attacks is one of the central problems for reproductive and developmental immunology, but as such, little information is available to date. The ability of fish to resist pathogens and to balance the interactive effects of stress and disease resistance capacity have been repeatedly reported to change seasonally and through ontogeny (Schreck et al. 1995). Among the diverse ontogeny steps of the immune system, passive immunity during early development may be initiated by vertical transfer from maternal sources into the egg (Picchietti et al. 2004). It is known that mature eggs of fish species contain IgM (Seppola et al. 2009), although the presence of antibody molecules related to a transfer of specific immunity is still a matter of speculation (Hanif et al. 2004, Haines et al. 2005). Previous studies revealed that in sea bream, fluctuations of serum IgM concentration relates to the onset of the reproductive cycle and thus to gender, and could be detected in the released eggs (Pichietti et al. 2001). In mature European sea bass eggs, Ig protein was detected corresponding to $6 \mu g\ g^{-1}$ egg weight and immunocytochemical analysis of paraffin sections from 5-day post-hatched embryos revealed an accumulation of material immunoreactive with the anti-IgM monoclonal antibody DLIg3 in the yolk sac (Scapigliati et al. 1999), as well as IgM gene transcripts in ovarian follicles throughout oogenesis (Pichietti et al. 2004). After fertilization, the mRNA coding for IgM disappears from sea bass embryos after 3–4 days, and IgM transcripts start to be detected again after 45 days. From these studies, it is evident that sea bass and sea bream transfer IgM to the egg through active processes, and that IgM undergo a rapid decay after fertilization. There is no clear understanding on the IgM presence in fish eggs, and considering their amount reported above, it is reasonable to speculate that unspecific IgM could be useful as antigen-opsonizing molecules during early stages of sea bass development. To reinforce this latter hypothesis, it should be remembered that fish eggs also contain complement system components (Hustenhuis et al. 2006, Lovoll et al. 2006).

Cell-mediated immunity: T cell-receptor beta, co-receptors CD4 and CD8, and MHC class II

In the last years, several papers (Buonocore and Scapigliati 2009) have reported in detail on structural and functional aspects of European sea bass lymphocytes (T and B cells). Therefore, here we will just describe briefly some of their more relevant features. Most knowledge on lymphocytes ontogenic development has been studied so far in carp (Huttenhuis et al. 2006, Rombout et al. 1993). However, sea bass has been the choice as representative of a marine species due to availability of monoclonal antibodies that have facilitated the study of thymocytes and peripheral T cells (Scapigliati et al. 1995), and IgM and B cells expression (Romestand and Breuil 1995). The combined use of these cellular probes permitted defining the timing of appearance of lymphocytes during sea bass ontogenic development (Table 1). These works have been extensively reviewed in a previous work (Rombout et al. 2005). Indeed, by using these antibodies in immunohistochemistry of sea bass larval sections, a difference in the timing of the appearance of lymphocytes emerged, with T cells preceding B cells by about 20 days (at 18°C). Recently, Picchietti and colleagues (Picchietti et al. 2008, 2009) applied RT-PCR, to investigate in sea bass, the developmental appearance of TCRb, CD8a and CD4, three important genes involved in T cell functions. For this purpose, total RNA was extracted from released eggs and pools of specimens sampled at different time points from 2 to 92 days post-hatch (dph) and from the thymus of one year old specimens, reared at 15°C. RT-PCR detected the first appearance of TCRb in larvae between 21 and 25 days post-hatch, which is well before the first expression of CD8a and CD4 between 40 and 50 (dph). The early TCRb expression confirmed data obtained in other Teleost species such as rainbow trout (*Oncorhynchus mykiss*) (Hansen and Zapata 1998, Fischer et al. 2005) and was in concert with the early development of the thymus, which is the first organ to become lymphoid in Teleosts (Abelli et al. 1997, Padrós and Crespo 1996). Further, the expression of CD8a and CD4 was detected in sea bass between 40 and 50 (dph) (Piecchietti et al. 2008, 2009), when the thymus is characterized by a distinctive lymphoid composition (Abelli et al. 1996) whose normal development is dependent upon cell interactions with thymicstroma and secreted molecules (Zapata et al. 1997), as previously reported in mammals (Maddox et al. 1987). In sea bass, real time PCR also demonstrated that transcripts of TCRb, CD8a and CD4 increased until 92 dph. At this stage, the TCRb and CD8a transcripts were significantly increased compared with all previous stages, while CD4 transcripts were significantly higher when compared with day 25 and day 51 post-hatch, suggesting that the critical events of differentiation and selection of T-lymphocytes are strongly correlated to the histological maturation of the sea bass thymus. In addition, in the thymus of 1-year-old sea bass, the amounts of CD4 and CD8 transcripts are not statistically different, while CD4 transcripts are significantly lower than TCRb transcripts, suggesting that a higher number of TCRb[+] thymocytes than CD4[+] subpopulations could be located in the thymus, or that a fraction of CD8[+] thymocytes could express different TCR chains. *In situ* hybridization with digoxigenin (DIG)-labeled RNA probes showed that at 51 days post-hatch the TCRb, CD8a and CD4 mRNAs were localized in thymocytes of the outer and lateral zones of the thymic paired glands that were not yet lobulated, while the parenchyma was regionalized. Already at day 51 and from day 75 post hatch on, an active lymphopoiesis was observed in the gland, resulting in a cortex/medulla demarcation in which the signal was restricted to cortical thymocytes. At 92 days post-hatch, the histology of the thymus was well established and TCRb[+], CD8a[+] and CD4[+] cells were mainly localized in the cortex at the cortico-medullary junction, providing new data on the understanding of the thymic microenvironment and evidence relating to the positive selection and differentiation of T cells. In teleosts, little is known about the origin of lymphoid progenitors that colonize the thymic rudiment. To increase this knowledge, a positive selection and differentiation of T cells was recently proposed in sea bass juveniles (Picchietti et al. 2008, 2009). At this stage, the thymus retained a superficial localization and was more extensively lobulated when compared with previous developmental stages. TCRb, CD8a and CD4 expression was detected by *in situ* hybridization in the cortex and at the cortico-medullary junction, in each lobe of juveniles, evidencing the compartmentalization of the parenchyma in which a network of stromal epithelial cells was established and in the meshes of which thymocytes are located. The regions housing CD8[+] and CD4[+] cells were largely overlapping. In fact, their density was nearly equal, suggesting that the cortex could be the site of double positive cells. CD4 and CD8 thymocytes were significantly less numerous in the medulla than in the cortex, while their relative numbers were not different in the medulla. In this district, large cords of CD8[+] cells extended in the gland, whereas CD4 transcripts were detected in isolated thymocytes, reflecting the occurrence of SP single positive thymocytes in the medulla. Furthermore, the presence of a CD8a, CD4 and TCRb[+] subcapsular lymphoid zone is worth noting, where the thymocytes do not express the two co-receptors and thus can be regarded as double negative. This pattern of localization of thymic subsets in distinct anatomical compartments is consistent with an out and in pattern of thymocyte migration. As evidenced in mammals, each of the T-cell maturation events takes place in a discrete region of the thymus and relies on interactions with specialized resident cells found in each of these anatomical regions (Ladi et al. 2006). Although thymocyte migrations are not defined in teleosts, some data suggest high similarity in the morpho-functional organization of the teleost thymus compared with mammals. An interesting finding was the lack of a CD8a lymphoid zone in the cortical thymic parenchyma contacting the pharyngeal epithelium, which is related to the absence of a connective capsule (Abelli et al. 1996) and hence to the lack of limiting epithelial cells that likely play a role in attracting double negative thymocytes. Taken together the data available in the literature, it can be evidenced that each fish species possesses its own ontogenetic developmental pattern, likely related to evolutionary-acquired habits (Zapata et al. 2006, Huttenhuis et al. 2006, Langenau and Zon 2005, Corripio-Miyar et al. 2007, Patel et al. 2009) and, importantly, it has been confirmed in fish that T cell development precedes B cell development. Based on the available data it is conceivable to assume that in sea bass the acquired immune system, in its main known components, namely lymphocytes bearing TcRs or IgM, MHC, RAGs and T cell co-receptors could be ready to perform full activities from 45 (dph) onwards. Thus, functionality of adaptive immune memory operation described above, might establish the correct time line in which disease resistance therapies based on the recognition, presentation and assimilation of antigen, namely vaccines, shall be started to assure success disease resistance in sea bass.

Role of Microbes in Induction of Innate Immune Protection

Gnotobiotic sea bass larvae as a model organism for studying host-bacterial interactions

The raising of marine fish larvae has only become feasible on a commercial scale because of major advances in the knowledge of nutritional requirements. However, the high and unpredictable mortality in the first weeks after hatching remains a challenging problem that needs to be solved. Different factors have been proposed to cause the observed symptoms, including the quality of gametes, inadequate nutrition, suboptimal physicochemical conditions and detrimental fish–microbe interactions (Vadstein et al. 2012). Traditionally, the focus has been mostly on the first three factors. Considerable progress has been achieved related to these factors during the past decades, and for some species, such as the European sea bass, this progress has made the industry viable. However, performance still varies considerably for identical treatments, even between replicates of the same sibling group; this indicates that egg quality, nutrition and physiochemical conditions which are fairly well controlled and kept identical, are adequate. As these conditions are kept constant, they cannot explain the variability between replicates and thus, one may hypothesize that detrimental fish–microbe interaction is the only proposed factor that can explain the lack of reproducibility between replicate treatments with full sibling groups. So far, in an effort to understand how microbial commensals are host-supported, powerful germ-free (GF) experimental approaches have been developed for several species of different taxonomical levels. Just recently, an innovative research of host-microbe interactions using GF conditions in fish has proven effective for studying developmental processes including immunity (Galindo-Villegas et al. 2012). Former GF studies in poikilothermic fish, namely salmonids received first considerations when Trust (Trust 1974) demonstrated the feasibility of collecting, fertilizing, incubating and hatching eggs aseptically of coho salmon (*Oncorhynchus kisutch*), chum salmon (*Oncorhynchus keta*), kakone salmon (*Oncorhynchus nerka*), golden trout (*Salmo aquabonito*), rainbow trout (*Oncorhynchus mikiss*) and brook trout (*Salvelinus fontanelis*). All these species fail to yield bacterial growth at least 40 days after fecundation, and whole testes from rainbow trout were demonstrated to be free from bacterial contamination in the same period. Interestingly, in the same study, a population of viable aerobic bacteria, estimated in 10^4 to 10^5 CFU g^{-1} (wet weight) of fish, was quantified in a control population of alevins at the yolk sac stage, maintained under normal hatchery conditions. These data suggests that fish reared following standard conditions are subjected not only to classic environmental pressures but include those coming from the presence of microorganisms in the rearing medium. However, at that time, no one figured out the particular characteristics, impact and magnitude exerted by microbes on every functional aspect of fish. Actually, almost 40 years later, the zebrafish (*Danio rerio*) is the choice as for most laboratories as it has several unique characteristics that make it an attractive model organism for the study of vertebrate diseases. Among the striking features displayed by zebrafish at the larval stage are transparency and fast growth and that host defense mechanisms could be totally focused on the role of innate, rather than acquired immune responses due to the the lack of functionality of the former until several days after fecundation. Additionally, to decipher the molecular foundations of host-bacterial relationships methods for raising GF zebrafish has been developed (Rawls et al. 2004), and reverse genetic analyses are possible using antisense morpholino oligonucleotide (Nasevicius and Ekker 2000). However, despite all the powerful characteristics displayed by this fish, it is not representative for all studies in marine aquaculture due to the existence of key differences among species. Zebrafish belongs to the *Cyprinidae*, can be fed artificial diets from start feeding (Carvalho et al. 2006) and has a short larval grow period, while the marine fish European sea bass belongs to the Percomorpha (Dahm and Geisler 2006), needs live zooplankton supply for several weeks and has a considerable longer larval period. Thus, to investigate the specific impact of host-microbial interactions on European sea bass biology under aquaculture conditions, a gnotobiotic larvae test system was developed (Dierckens et al. 2009). In order to obtain bacteria-free sea bass larvae, the eggs were aseptically raised and fed axenic Artemia (Marques et al. 2006) from 7 days after hatching onwards. Thereafter, gnotobiotically raised vs. conventional sea bass larva were challenged against one of the three opportunistic pathogens (*Aeromonas hydrophila*, *Listonella anguillarum* serovar O1 and O2a) strains to determine interactions. Results from two out of three challenge experiments are not clear at all, mostly due to technical constrains. However, main merit of this work is the standardized mortality observed by *L. anguillarum* challenge. Thus, this gnotobiotic sea bass system has been regarded as a powerful tool to study several host-microbial interactions in marine fish which may include the immunological competence of host versus specific pathogenic or commensal bacterial strains. The first functional study conducted using gnotobiotically raised sea bass established the interactions between sea bass larvae and the pathogenic *L. anguillarum* strain HI-610 which expressed green fluorescent protein (GFP) as reporter dye (Rekecki et al. 2012). Microscopic images revealed that after 48 hours post exposure (hpe), single or grouped bacteria are in close contact to the apical brush border of midgut and hindgut enterocytes. Onwards, bacteria were visible either within the swim bladder, or sporadically on the skin. At the same exposure time, ultrastructural analyses demonstrate that bacteria, before causing any damage to host, are located at the interspace (100–200 nm) between the microvilli of mid- and hind-gut enterocytes. Consistent with this finding, in mammalian samples using fluorescent *in situ* hybridization (FISH) it was also determined that bacteria were not in direct contact with the intestinal surface (Vaishnava et al. 2011). Rather, it was maintained ~50 μm from the villus tip through the segregation of the antibacterial lectin RegIIIγ by epithelial cells. And, this mechanism was mediated through an intrinsic TLR/MyD88 signaling, providing essential insights on how immunity regulates the physical location of the microbiota. Due to similarity in the process observed between mammals and fish, we speculate that the proposed bacterial mechanism described is evolutionarily conserved in all vertebrate classes, starting at least from teleostei. Recently, Rekecki and colleagues (Rekecki et al. 2013) using immunogold labelling and transmission electron microscopy techniques, evaluated the adhesion and translocation of pathogenic *V. anguillarum* in gut enterocytes of gnotobiotic sea bass. Findings revealed that intact bacteria were in close contact with the apical brush border in the gut lumen. And, those apical portions of the enterocytes contained secondary lysosomes positive for protein A gold particles, suggesting intracellular elimination of bacterial fragments. Thus, immunogold positive tread-like structures secreted by *V. anguillarum* witnessed the presence of outer membrane vesicles (MVs), suggesting that MVs are potent transporters of active virulence factors to sea bass gut cells (Fig. 1). Additionally, a more interesting result reported that exfoliated enterocytes were not terminally apoptotic or oncotic and were very active engulfing immunogold positive bacteria into the lumen without any damage to the epithelial lining. Collectively, these finding indicate that the route of entry of *L. anguillarum* in sea bass shall be gathered through mucosal tissue. However, much more studies using this powerful gnotobiotic sea bass set are urgently needed to clearly assess the impact of pathogenic bacteria over the immune system of this species. Thus, focused therapies could be developed to further increase standardized larval survival at any aquaculture facility.

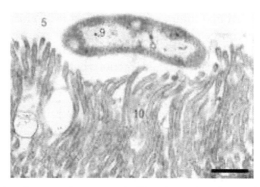

Figure 1. Transmission electron microscopy of midgut in sea bass larvae challenged with a high dose (10^8 cfu mL^{-1}) of *Vibrio anguillarum* GFP-HI-610. Putative attachment of bacteria to midgut microvilli (× 22 000). Legend: (5. lumen; 9. bacterium; 10. microvilli). Scale bar 500 nm. (Copyrighted image from (Rekecki et al. 2013) reproduced by kind permission of John Wiley and Sons).

Concluding Remarks and Future Directions

The research on sea bass immunology has risen impressively over the last years. Information on how sea bass immune system recognizes and eliminates invading pathogens is crucial to elucidate evolutionary processes and boost application-based research. Huge advances on fish immunology have been gathering until now, but a greater understanding of the sea bass immune system throughout the different developmental stages are vital to aid aquaculture production greatly through the prevention and/or treatment of diseases. Additionally, standardized experimental protocols which may allow a systematic evaluation of various immune responses shall be established as well. Thus, among urgent necessities in this research field, we propose the development of monoclonal antibodies which may recognize specific cytokines, IgT at the mucosal sites, and as much of the several and diverse lymphocyte sub-populations. Further advancements in these areas may allow the development of simple flow cytometry or ELISA based methodology. Improvements on these immunological tools might be immediately used as an effective and relatively inexpensive way to measure unequivocally Th1 or Th2 responses against specific pathogens affecting sea bass in aquaculture facilities. Also, the study in deep of sea bass mucosal immunity and several interactions with the microbiota shall reveal many clues to target specifically on effective, massive fish vaccination strategies and immune stimulation targeting oral administration or through simple immersion systems. At last, recent development of a powerful gnotobiotic sea bass system may be helpful on the study of commensal or pathogenic host-microbe interactions. Therefore, application of this system shall be essential on determining how probiotics does trigger disease resistance; however, much attention has to be paid to the immune mechanisms behind these effects.

References

Abelli, L., V.P. Gallo, A. Civinini and L. Mastrolia. 1996. Immunohistochemical and ultrastructural evidence of adrenal chromaffin cell subtypes in sea bass *Dicentrarchus labrax* (L.). Gen. Comp. Endocrinol. 102: 113–122.

Abelli, L., S. Picchietti, N. Romano, L. Mastrolia and G. Scapigliati. 1997. Immunohistochemistry of gut-associated lymphoid tissue of the sea bass *Dicentrarchus labrax* (L.). Fish Shellfish Immunol. 7: 235–245.

Abraham, S.N. and A.L. St John. 2010. Mast cell-orchestrated immunity to pathogens. Nat. Rev. Immunol. 10: 440–452.

Anderson, K.V., G. Jorgens and C. Nisslein-Volhard. 1985. Establishment of dorsal-ventral polarity in the *Drosophila* embryo: Genetic studies on the role of the Toll gene product. Cell 42: 779–789.

Bao, B., E. Peatman, P. Li, C. He and Z. Liu. 2005. Catfish hepcidin gene is expressed in a wide range of tissues and exhibits tissue-specific upregulation after bacterial infection. Dev. Comp. Immunol. 29: 939–950.

Bao, B., E. Peatman, P. Xu, P. Li, H. Zeng, C. He and Z. Liu. 2006. The catfish liver-expressed antimicrobial peptide 2 (LEAP-2) gene is expressed in a wide range of tissues and developmentally regulated. Mol. Immunol. 43: 367–377.

Barondes, S.H., D.N. Cooper, M.A. Gitt and H. Leffler. 1994. Galectins. Structure and function of a large family of animal lectins. J. Biol. Chem. 269: 20807–20810.

Bianchi, M.E. 2007. DAMPs, PAMPs and alarmins: all we need to know about danger. J. Leukoc. Biol. 81: 1–5.

Brandtzaeg, P., H. Kiyono, R. Pabst and M.W. Russell. 2008. Terminology: nomenclature of mucosa-associated lymphoid tissue. Mucosal Immunol. 1: 31–37.

Brandtzaeg, P., E. Isolauri and S.L. Prescott. 2009. 'ABC' of Mucosal Immunology. Nestlé Nutr. Inst. Workshop Ser Pediatr. Program 64: 23–43.

Bricknell, I. and R.A. Dalmo. 2005. The use of immunostimulants in fish larval aquaculture. Fish Shellfish Immunol. 19: 457–472.

Buonocore, F. and G. Scapigliati. 2009. Immune defence mechanisms in the sea bass *Dicentrarchus labrax* L. pp. 185–219. *In*: G. Zaccone, J. Meseguer, A. García-Ayala and B.G. Kapoor (eds.). Fish Defenses, Vol. 1: Immunology. Science Publishers, New Hampshire.

Buonocore, F., D. Prugnoli, C. Falasca, C.J. Secombes and G. Scapigliati. 2003. Peculiar gene organisation and incomplete splicing of sea bass (*Dicentrarchus labrax* L.) interleukin-1beta. Cytokine 21: 257–264.

Buonocore, F., M. Forlenza, E. Randelli, S. Benedetti, P. Bossu, S. Meloni, C.J. Secombes, M. Mazzini and G. Scapigliati. 2005. Biological activity of sea bass (*Dicentrarchus labrax* L.) recombinant interleukin-1beta. Mar. Biotechnol. (NY) 7: 609–17.

Buonocore, F., E. Randelli, D. Casani, L. Guerra, S. Picchietti, S. Costantini, A.M. Facchiano, J. Zou, C.J. Secombes and G. Scapigliati. 2008. A CD4 homologue in sea bass (*Dicentrarchus labrax*): molecular characterization and structural analysis. Mol. Immunol. 45: 3168–3177.

Buonocore, F., E. Randelli, D. Casani, S. Picchietti, M.C. Belardinelli, D. de Pascale, C. De Santi and G. Scapigliati. 2012. A piscidin-like antimicrobial peptide from the icefish *Chionodraco hamatus* (Perciformes: Channichthyidae): molecular characterization, localization and bactericidal activity. Fish Shellfish Immunol. 33: 1183–1191.

Camby, I., M. Le Mercier, F. Lefranc and R. Kiss. 2006. Galectin-1: a small protein with major functions. Glycobiology 16: 137R–157R.

Cammarata, M., G. Benenati, E.W. Odom, G. Salerno, A. Vizzini, G.R. Vasta and N. Parrinello. 2007. Isolation and characterization of a fish F-type lectin from gilt head bream (*Sparus aurata*) serum. Biochim. Biophys. Acta 1770: 150–155.

Campagna, S., N. Saint, G. Molle and A. Aumelas. 2007. Structure and mechanism of action of the antimicrobial peptide piscidin. Biochemistry 46: 1771–1778.

Carvalho, A.P., L. Araujo and M.M. Santos. 2006. Rearing zebrafish (*Danio rerio*) larvae without live food: evaluation of a commercial, a practical and a purified starter diet on larval performance. Aquac. Res. 37: 1107–1111.

Cerutti, A. 2008. The regulation of IgA class switching. Nat. Rev. Immunol. 8: 421–434.

Chang, C.-I., O. Pleguezuelos, Y.-A. Zhang, J. Zou and C.J. Secombes. 2005. Identification of a novel cathelicidin gene in the rainbow trout, *Oncorhynchus mykiss*. Infection and immunity 73: 5053–5064.

Chang, C.-I., Y.-A. Zhang, J. Zou, P. Nie and C.J. Secombes. 2006. Two cathelicidin genes are present in both rainbow trout (*Oncorhynchus mykiss*) and Atlantic salmon (*Salmo salar*). Antimicrob. Agents Chemother. 50: 185–195.

Chistiakov, D.A., B. Hellemans and F.A. Volckaert. 2007. Review on the immunology of European sea bass *Dicentrarchus labrax*. Vet. Immunol. Immunopathol. 117: 1–16.

Cole, A.M., P. Weis and G. Diamond. 1997. Isolation and Characterization of Pleurocidin, an Antimicrobial Peptide in the Skin Secretions of Winter Flounder. J. Biol. Chem. 272: 12008–12013.

Coombes, J.L. and K.J. Maloy. 2007. Control of intestinal homeostasis by regulatory T cells and dendritic cells. Semin. Immunol. 19: 116–126.

Corrales, J., I. Mulero, V. Mulero and E.J. Noga. 2010. Detection of antimicrobial peptides related to piscidin 4 in important aquacultured fish. Dev. Comp. Immunol. 34: 331–343.

Corripio-Miyar, Y., S. Bird, K. Tsamopoulos and C.J. Secombes. 2007. Cloning and expression analysis of two pro-inflammatory cytokines, IL-1b and IL-8, in haddock (*Melanogrammus aeglefinus*). Mol. Immunol. 44: 1361–1373.

Cotou, E., M. Henry, C. Zeri, G. Rigos, A. Torreblanca and V.-A. Catsiki. 2012. Short-term exposure of the European sea bass *Dicentrarchus labrax* to copper-based antifouling treated nets: Copper bioavailability and biomarkers responses. Chemosphere 89: 1091–1097.

Crivellato, E. and D. Ribatti. 2010. The mast cell: an evolutionary perspective. Biol. Rev. Camb. Philos. Soc. 85: 347–360.

Dahm, R. and R. Geisler. 2006. Learning from small fry: the zebrafish as a genetic model organism for aquaculture fish species. Mar. Biotechnol. (NY) 8: 329–345.

Danilova, N. 2006. The evolution of immune mechanisms. J. Exp. Zool. B Mol. Dev. Evol. 306: 496–520.

Danilova, N., H.L. Saunders, K.K. Ellestad and B.G. Magor. 2005. The zebrafish IgH locus contains multiple transcriptional regulatory regions. Dev. Comp. Immunol. 35: 352–359.

Davey, G.C., J.A. Calduch-Giner, B. Houeix, A. Talbot, A. Sitjà-Bobadilla, P. Prunet, J. Perez-Sanchez and M.T. Cairns. 2011. Molecular profiling of the gilthead sea bream (*Sparus aurata* L.) response to chronic exposure to the myxosporean parasite Enteromyxum leei. Mol. Immunol. 48: 2102–2112.

Davina, J., G. Rijkers, J. Rombout, L. Timmermans and W. Van Muiswinkel. 1980. Lymphoid and non lymphoid cells in the intestine of cyprinid fish. pp. 129–140. *In*: J.D. Horton (ed.). Development and Differentiation of the Vertebrate Lymphocytes. Elsevier/North-Holland Biomedical Press, Amsterdam.

Dezfuli, B.S., F. Pironi, M. Campisi, A.P. Shinn and L. Giari. 2010. The response of intestinal mucous cells to the presence of enteric helminths: their distribution, histochemistry and fine structure. J. Fish Dis. 33: 481–488.

Dierckens, K., A. Rekecki, S. Laureau, P. Sorgeloos, N. Boon, W. Van den Broeck and P. Bossier. 2009. Development of a bacterial challenge test for gnotobiotic sea bass (*Dicentrarchus labrax*) larvae. Environ. Microbiol. 11: 526–533.

do Vale, A., A. Afonso and M.T. Silva. 2002. The professional phagocytes of sea bass (*Dicentrarchus labrax* L.): cytochemical characterisation of neutrophils and macrophages in the normal and inflamed peritoneal cavity. Fish Shellfish Immunol. 13: 183–198.

dos Santos, N.M., N. Romano, M. de Sousa, A.E. Ellis and J.H. Rombout. 2000. Ontogeny of B and T cells in sea bass (*Dicentrarchus labrax* L.). Fish Shellfish Immunol. 10: 583–596.

Ellis, A.E. 1985. Eosinophilic granular cells (EGC) and histamine responses to Aeromonas salmonicida toxins in rainbow trout. Dev. Comp. Immunol. 9: 251–260.

Esteban, M. A. and J. Meseguer. 1994. Phagocytic defence mechanism in sea bass (*Dicentrarchus labrax* L.): an ultrastructural study. Anat. Rec. 240: 589–597.

Fischer, U., J.M. Dijkstra, B. Kollner, I. Kiryu, E.O. Koppang, I. Hordvik, Y. Sawamoto and M. Ototake. 2005. The ontogeny of MHC class I expression in rainbow trout (*Oncorhynchus mykiss*). Fish Shellfish Immunol. 18: 49–60.

Galindo-Villegas, J. and H. Hosokawa. 2004. Immunostimulants: towards temporary prevention of diseases in marine fish. Trends in Aqua. Nut. VII: 279–319.

Galindo-Villegas, J., D. Garcia-Moreno, S. de Oliveira, J. Meseguer and V. Mulero. 2012. Regulation of immunity and disease resistance by commensal microbes and chromatin modifications during zebrafish development. Proc. Natl. Acad. Sci. USA 109: E2605–2614.

Galindo-Villegas, J., I. Mulero, A. Garcia-Alcazar, I. Munoz, M. Penalver-Mellado, S. Streitenberg, G. Scapigliati and V. Mulero. 2013. Recombinant TNFa as oral vaccine adjuvant protects European sea bass against vibriosis: Insights into the role of the CCL25/CCR9 axis. Fish Shellfish Immunol. 35: 1260–1271.

Garcia-Castillo, J., P. Pelegrin, V. Mulero and J. Meseguer. 2002. Molecular cloning and expression analysis of tumor necrosis factor alpha from a marine fish reveal its constitutive expression and ubiquitous nature. Immunogenetics 54: 200–207.

Georgopoulou, U., M.F. Sire and J.M. Vernier. 1986. Immunological demonstration of intestinal absorption and digestion of protein macromolecules in the trout (*Salmo gairdneri*). Cell Tissue Res. 245: 387–395.

Haines, A.N., M.F. Flajnik, L.L. Rumfelt and J.P. Wourms. 2005. Immunoglobulins in the eggs of the nurse shark, *Ginglymostoma cirratum*. Dev. Comp. Immunol. 29: 417–430.

Hanif, A., V. Bakopoulos and G.J. Dimitriadis. 2004. Maternal transfer of humoral specific and non-specific immune parameters to sea bream (*Sparus aurata*) larvae. Fish Shellfish Immunol. 17: 411–435.

Hansen, J.D. and A.G. Zapata. 1998. Lymphocyte development in fish and amphibians. Immunol. Rev. 166: 199–220.

Hansen, J.D., E.D. Landis and R.B. Phillips. 2005. Discovery of a unique Ig heavy-chain isotype (IgT) in rainbow trout: Implications for a distinctive B cell developmental pathway in teleost fish. Proc. Natl. Acad. Sci. USA 102: 6919–6924.

Hart, S., A.B. Wrathmell, J.E. Harris and T.H. Grayson. 1988. Gut immunology in fish: A review. Dev. Comp. Immunol. 12: 453–480.

Honda, K. and K. Takeda. 2009. Regulatory mechanisms of immune responses to intestinal bacteria. Mucosal Immunol. 2: 187–196.

Huttenhuis, H., N. Romano, C. Van Oosterhoud, A. Taverne-Thiele, L. Mastrolia, W. Van Muiswinkel and J. Rombout. 2006. The ontogeny of mucosal immune cells in common carp (*Cyprinus carpio* L.). Anat. Embryol. 211: 19–29.

Iijima, N., N. Tanimoto, Y. Emoto, Y. Morita, K. Uematsu, T. Murakami and T. Nakai. 2003. Purification and characterization of three isoforms of chrysophsin, a novel antimicrobial peptide in the gills of the red sea bream, *Chrysophrys major*. Eur. J. Biochem. 270: 675–686.

Kalesnikoff, J. and S.J. Galli. 2008. New developments in mast cell biology. Nat. Immunol. 9: 1215–1223.

Kilpatrick, D.C. 2000. Mannan-binding lectin concentration during normal human pregnancy. Hum. Reprod. 15: 941–943.

Krause, K.H. 2000. Professional phagocytes: predators and prey of microorganisms. Schweiz Med. Wochenschr. 130(4): 97–100.

Ladi, E., X. Yin, T. Chtanova and E.A. Robey. 2006. Thymic microenvironments for T cell differentiation and selection. Nat. Immunol. 7: 338–343.

Langenau, D.M. and L.I. Zon. 2005. The zebrafish: a new model of T-cell and thymic development. Nat. Rev. Immunol. 5: 307–317.

Lauth, X., H. Shike, J.C. Burns, M.E. Westerman, V.E. Ostland, J.M. Carlberg, J.C. Van Olst, V. Nizet, S.W. Taylor, C. Shimizu and P. Bulet. 2002. Discovery and characterization of two isoforms of moronecidin, a novel antimicrobial peptide from hybrid striped bass. J. Biol. Chem. 277: 5030–5039.

Legac, E., G.L. Vaugier, F. Bousquet, M. Bajelan and M. Leclerc. 1996. Primitive cytokines and cytokine receptors in invertebrates: the sea star *Asterias rubens* as a model of study. Scand. J. Immunol. 44: 375–380.

Lovoll, M., T. Kilvik, H. Boshra, J. Bogwald, J.O. Sunyer and R.A. Dalmo. 2006. Maternal transfer of complement components C3-1, C3-3, C3-4, C4, C5, C7, Bf, and Df to offspring in rainbow trout (*Oncorhynchus mykiss*). Immunogenetics 58: 168–179.

Lu, J.H., S. Thiel, H. Wiedemann, R. Timpl and K.B. Reid. 1990. Binding of the pentamer/hexamer forms of mannan-binding protein to zymosan activates the proenzyme C1r2C1s2 complex, of the classical pathway of complement, without involvement of C1q. J. Immunol. 144: 2287–2294.

Macpherson, A.J. and T. Uhr. 2004. Induction of protective IgA by intestinal dendritic cells carrying commensal bacteria. Science 303: 1662–1665.

Maddox, J.F., C.R. Mackay and M.R. Brandon. 1987. Ontogeny of ovine lymphocytes. I. An immunohistological study on the development of T lymphocytes in the sheep embryo and fetal thymus. Immunology 62: 97–105.

Magnadottir, B. 2006. Innate immunity of fish (overview). Fish Shellfish Immunol. 20: 137–151.

Marozzi, C., F. Bertoni, E. Randelli, F. Buonocore, A.M. Timperio and G. Scapigliati. 2012. A monoclonal antibody for the CD45 receptor in the teleost fish *Dicentrarchus labrax*. Dev. Comp. Immunol. 37: 342–353.

Marques, A., F. Ollevier, W. Verstraete, P. Sorgeloos and P. Bossier. 2006. Gnotobiotically grown aquatic animals: opportunities to investigate host-microbe interactions. J. Appl. Microbiol. 100: 903–918.

Meseguer, J., M.A. Esteban and B. Agulleiro. 1991. Stromal cells, macrophages and lymphoid cells in the head-kidney of sea bass (*Dicentrarchus labrax* L.). An ultrastructural study. Arch. Histol. Cytol. 54: 299–309.

Meseguer, J., M.A. Esteban and V. Mulero. 1996. Nonspecific cell-mediated cytotoxicity in the seawater teleosts (*Sparus aurata* and *Dicentrarchus labrax*): ultrastructural study of target cell death mechanisms. Anat. Rec. 244: 499–505.

Mulero, I., M.P. Sepulcre, J. Meseguer, A. Garcia-Ayala and V. Mulero. 2007. Histamine is stored in mast cells of most evolutionarily advanced fish and regulates the fish inflammatory response. Proc. Natl. Acad. Sci. USA 104: 19434–19439.

Mulero, I., M.P. Sepulcre, F.J. Roca, J. Meseguer, A. Garcia-Ayala and V. Mulero. 2008a. Characterization of macrophages from the bony fish gilthead seabream using an antibody against the macrophage colony-stimulating factor receptor. Dev. Comp. Immunol. 32: 1151–1159.

Mulero, I., E.J. Noga, J. Meseguer, A. Garcia-Ayala and V. Mulero. 2008b. The antimicrobial peptides piscidins are stored in the granules of professional phagocytic granulocytes of fish and are delivered to the bacteria-containing phagosome upon phagocytosis. Dev. Comp. Immunol. 32: 1531–1538.

Mulero, V., M.A. Esteban, J. Munoz and J. Meseguer. 1994. Non-specific cytotoxic response against tumor target cells mediated by leucocytes from seawater teleosts, *Sparus aurata* and *Dicentrarchus labrax*: an ultrastructural study. Arch. Histol. Cytol. 57: 351–358.

Nasevicius, A. and S.C. Ekker. 2000. Effective targeted gene 'knockdown' in zebrafish. Nat. Genet. 26: 216–220.

Noga, E.J. and U. Silphaduang. 2003. Piscidins: a novel family of peptide antibiotics from fish. Drug News Perspect. 16: 87–92.

Oda, K. and H. Kitano. 2006. A comprehensive map of the toll-like receptor signaling network. Mol. Syst. Biol. 2: 2006 0015.

Odom, E.W. and G.R. Vasta. 2006. Characterization of a binary tandem domain F-type lectin from striped bass (Morone saxatilis). J. Biol. Chem. 281: 1698–1713.

O'Neill, L.A. 2008a. Bacteria fight back against Toll-like receptors. Nat. Med. 14: 370–372.

O'Neill, L.A. 2008b. 'Fine tuning' TLR signaling. Nat. Immunol. 9: 459–461.

O'Neill, L.A. 2008c. The interleukin-1 receptor/Toll-like receptor superfamily: 10 years of progress. Immunol. Rev. 226: 10–18.

O'Neill, L.A. 2008d. When signaling pathways collide: positive and negative regulation of toll-like receptor signal transduction. Immunity 29: 12–20.

Oren, Z. and Y. Shai. 1996. A class of highly potent antibacterial peptides derived from pardaxin, a pore-forming peptide isolated from Moses sole fish *Pardachirus marmoratus*. Eur. J. Biochem. 237: 303–310.

Padrós, F. and S. Crespo. 1996. Ontogeny of the lymphoid organs in the turbot *Scophthalmus maximus*: A light and electron microscope study. Aquaculture 144: 1–16.

Palic, D., J. Ostojic, C.B. Andreasen and J.A. Roth. 2007. Fish cast NETs: neutrophil extracellular traps are released from fish neutrophils. Dev. Comp. Immunol. 31: 805–816.

Pallavicini, A., E. Randelli, M. Modonut, D. Casani, G. Scapigliati and F. Buonocore. 2010. Searching for immunomodulatory sequences in sea bass (*Dicentrarchus labrax* L.): transcripts analysis from thymus. Fish Shellfish Immunol. 29: 571–578.

Palti, Y. 2011. Toll-like receptors in bony fish: from genomics to function. Dev. Comp. Immunol. 35: 1263–1272.

Park, C.B., J.H. Lee, I.Y. Park, M.S. Kim and S.C. Kim. 1997. A novel antimicrobial peptide from the loach, Misgurnus anguillicaudatus. FEBS Letters 411: 173–178.

Patel, S., E. Sorhus, I.U. Fiksdal, P.G. Espedal, O. Bergh, O.M. Rodseth, H.C. Morton and A.H. Nerland. 2009. Ontogeny of lymphoid organs and development of IgM-bearing cells in Atlantic halibut (*Hippoglossus hippoglossus* L.). Fish Shellfish Immunol. 26: 385–395.

Peng, K.C., S.H. Lee, A.L. Hour, C.Y. Pan, L.H. Lee and J.Y. Chen. 2012. Five different piscidins from Nile tilapia, *Oreochromis niloticus*: Analysis of their expressions and biological functions. PLoS One 7: e50263.

Perillo, N.L., K.E. Pace, J.J. Seilhamer and L.G. Baum. 1995. Apoptosis of T cells mediated by galectin-1. Nature 378: 736–739.

Petry, A., M. Weitnauer and A. Gorlach. 2010. Receptor activation of NADPH oxidases. Antioxid. Redox Signal. 13: 467–487.

Picchietti, S., F.R. Terribili, L. Mastrolia, G. Scapigliati and L. Abelli. 1997. Expression of lymphocyte antigenic determinants in developing gut-associated lymphoid tissue of the sea bass *Dicentrarchus labrax* (L.). Anat. Embryol. (Berl) 196: 457–463.

Picchietti S., S.M. Fanelli, F. Barbato, S. Canese, L. Mastrolia, M. Mazzini and L. Abelli. 2001. Sex-related variations of serum immunoglobulin during reproduction in gilthead sea bream and evidence for a transfer from the female to the eggs. J. Fish Biol. 59: 1503–1511.

Picchietti, S., A.R. Taddei, G. Scapigliati, F. Buonocore, A.M. Fausto, N. Romano, M. Mazzini, L. Mastrolia and L. Abelli. 2004. Immunoglobulin protein and gene transcripts in ovarian follicles throughout oogenesis in the teleost *Dicentrarchus labrax*. Cell Tissue Res. 315: 259–270.

Picchietti, S., L. Guerra, L. Selleri, F. Buonocore, L. Abelli, G. Scapigliati, M. Mazzini and A.M. Fausto. 2008. Compartmentalisation of T cells expressing CD8alpha and TCRbeta in developing thymus of sea bass *Dicentrarchus labrax* (L.). Dev. Comp. Immunol. 32: 92–99.

Picchietti, S., L. Guerra, F. Buonocore, E. Randelli, A.M. Fausto and L. Abelli. 2009. Lymphocyte differentiation in sea bass thymus: CD4 and CD8-alpha gene expression studies. Fish Shellfish Immunol. 27: 50–56.

Plouffe, D.A., P.C. Hanington, J.G. Walsh, E.C. Wilson and M. Belosevic. 2005. Comparison of select innate immune mechanisms of fish and mammals. Xenotransplantation 12: 266–277.

Poisa-Beiro, L., S. Dios, H. Ahmed, G.R. Vasta, A. Martinez-Lopez, A. Estepa, J. Alonso-Gutierrez, A. Figueras and B. Novoa. 2009. Nodavirus infection of sea bass (*Dicentrarchus labrax*) induces up-regulation of galectin-1 expression with potential anti-inflammatory activity. J. Immunol. 183: 6600–6611.

Powell, M.D., G.M. Wright and J.F. Burka. 1991. Degranulation of eosinophilic granule cells induced by capsaicin and substance P in the intestine of the rainbow trout (*Oncorhynchus mykiss* Walbaum). Cell Tissue Res. 266: 469–474.

Rautava, S., R. Luoto, S. Salminen and E. Isolauri. 2012. Microbial contact during pregnancy, intestinal colonization and human disease. Nat. Rev. Gastroenterol. Hepatol. 9(10): 565–576.

Rawls, J.F., B.S. Samuel and J.I. Gordon. 2004. Gnotobiotic zebrafish reveal evolutionarily conserved responses to the gut microbiota. Proc. Natl. Acad. Sci. USA 101: 4596–4601.

Rebl, A., T. Goldammer and H.M. Seyfert. 2010. Toll-like receptor signaling in bony fish. Vet. Immunol. Immunopathol. 134: 139–150.

Reis, M.I., A. do Vale, P.J. Pereira, J.E. Azevedo and N.M. Dos Santos. 2012. Caspase-1 and IL-1beta processing in a teleost fish. PLoS One 7: e50450.

Reite, O.B. 1997. Mast cells/eosinophilic granule cells of salmonids: staining properties and responses to noxious agents. Fish Shellfish Immunol. 7: 567–584.

Reite, O.B. and O. Evensen. 2006. Inflammatory cells of teleostean fish: a review focusing on mast cells/eosinophilic granule cells and rodlet cells. Fish Shellfish Immunol. 20: 192–208.

Rekecki, A., R.A. Gunasekara, K. Dierckens, S. Laureau, N. Boon, H. Favoreel, M. Cornelissen, P. Sorgeloos, R. Ducatelle, P. Bossier and W. Van den Broeck. 2012. Bacterial host interaction of GFP-labelled Vibrio anguillarum HI-610 with gnotobiotic sea bass, *Dicentrarchus labrax* (L.), larvae. J. Fish. Dis. 35(4): 265–273.

Rekecki, A., E. Ringo, R. Olsen, R. Myklebust, K. Dierckens, O. Bergh, S. Laureau, M. Cornelissen, R. Ducatelle, A. Decostere, P. Bossier and W. Van den Broeck. 2013. Luminal uptake of *Vibrio (Listonella) anguillarum* by shed enterocytes—a novel early defence strategy in larval fish. J. Fish. Dis. 36(4): 419–426.

Roach, J.C., G. Glusman, L. Rowen, A. Kaur, M. K. Purcell, K.D. Smith, L. E. Hood and A. Aderem. 2005. The evolution of vertebrate Toll-like receptors. Proc. Natl. Acad. Sci. USA 102: 9577-9582.

Rodrigues, A., M.A. Esteban and J. Meseguer. 2003. Phagocytosis and peroxide release by seabream (*Sparus aurata* L.) leucocytes in response to yeast cells. Anat. Rec. Part A 272A: 415.

Romano, N., L. Abelli, L. Mastrolia and G. Scapigliati. 1997. Immunocytochemical detection and cytomorphology of lymphocyte subpopulations in a teleost fish *Dicentrarchus labrax*. Cell Tissue Res. 289: 163–171.

Rombout, J., C.H.J. Lamers, M.H. Helfrich, A. Dekker and J.J. Taverne-Thiele. 1985. Uptake and transport of intact macromolecules in the intestinal epithelium of carp (*Cyprinus carpio* L.) and the possible immunological implication. Cell Tissue Res. 239: 519–530.

Rombout, J., A. Taverne-Thiele and M. Villena. 1993. The gut associated lymphoid tissue (GALT) of carp (*Cyprinus carpio* L.): an immunocytochemical analysis. Dev. Comp. Immunol. 17: 55–66.

Rombout, J.H., H.B. Huttenhuis, S. Picchietti and G. Scapigliati. 2005. Phylogeny and ontogeny of fish leucocytes. Fish Shellfish Immunol. 19: 441–455.

Rombout, J.H.W.M., A.A. van den Berg, C.T.G.A. van den Berg, P. Witte and E. Egberts. 1989. Immunological importance of the second gut segment of carp. III. Systemic and/or mucosal immune responses after immunization with soluble or particulate antigen. J. Fish Biol. 35: 179–186.

Rombout, J.W.H.M., L.J. Blok, C.H.J. Lamers and E. Egberts. 1986. Immunization of carp (*Cyprinus carpio*) with a *Vibrio anguillarum* bacterin: Indications for a common mucosal immune system. Dev. Comp. Immunol. 10: 341–351.

Romestand, B., G. Breuil, A.A. F. Bourmaud, J.L. Coeurdacier and G. Bouix. 1995. Development and characterisation of monoclonal antibodies against seabass immunoglobulins *Dicentrarchus labrax* Linnaeus 1758 Fish Shellfish Immunol. 5: 347–357.

Ruangsri, J., J.M. Fernandes, J.H.W. Rombout, M. Brinchmann and V. Kiron. 2012. Ubiquitous presence of piscidin-1 in Atlantic cod as evidenced by immunolocalisation. BMC Vet. Res. 8: 46.

Salerno, G., N. Parrinello, P. Roch and M. Cammarata. 2007. cDNA sequence and tissue expression of an antimicrobial peptide, dicentracin; a new component of the moronecidin family isolated from head kidney leukocytes of sea bass, *Dicentrarchus labrax*. Comp. Biochem. Physiol. Part B: Biochem. Mol. Biol. 146: 521–529.

Salerno, G., M.G. Parisi, D. Parrinello, G. Benenati, A. Vizzini, M. Vazzana, G.R. Vasta and M. Cammarata. 2009. F-type lectin from the sea bass (*Dicentrarchus labrax*): purification, cDNA cloning, tissue expression and localization, and opsonic activity. Fish Shellfish Immunol 27: 143–153.

Sarropoulou, E., J. Galindo-Villegas, A. Garcia-Alcazar, P. Kasapidis and V. Mulero. 2012. Characterization of European sea bass transcripts by RNA SEQ after oral vaccine against *V. anguillarum*. Mar. Biotechnol. (NY) 14: 634–642.

Scapigliati, G., M. Mazzini, L. Mastrolia, N. Romano and L. Abelli. 1995. Production and characterisation of a monoclonal antibody against the thymocytes of the sea bass *Dicentrarchus labrax* (L.) (Teleostea, Percicthydae). Fish Shellfish Immunol. 5: 393–405.

Scapigliati, G., N. Romano, S. Picchietti, M. Mazzini, L. Mastrolia, D. Scalia and L. Abelli. 1996. Monoclonal antibodies against sea bass *Dicentrarchus labrax* (L.) immunoglobulins: immunolocalisation of immunoglobulin-bearing cells and applicability in immunoassays. Fish Shellfish Immunol. 6: 383–401.

Scapigliati, G., S. Meloni and M. Mazzini. 1999. A monoclonal antibody against chorion proteins of the sea bass *Dicentrarchus labrax* (Linnaeus 1758): studies of chorion precursors and applicability in immunoassays. Biol. Reprod. 60: 783–789.

Scapigliati, G., N. Romano, L. Abelli, S. Meloni, A.G. Ficca, F. Buonocore, S. Bird and C.J. Secombes. 2000. Immunopurification of T-cells from sea bass *Dicentrarchus labrax* (L.). Fish Shellfish Immunol. 10: 329–341.

Scapigliati, G., N. Romano, F. Buonocore, S. Picchietti, M.R. Baldassini, D. Prugnoli, A. Galice, S. Meloni, C.J. Secombes, M. Mazzini and L. Abelli. 2002. The immune system of sea bass, *Dicentrarchus labrax*, reared in aquaculture. Dev. Comp. Immunol. 26: 151–160.

Scapigliati, G., F. Buonocore and M. Mazzini. 2006. Biological activity of cytokines: an evolutionary perspective. Curr. Pharm. Des. 12: 3071–3081.

Schreck, C.B., L. Jonsson, G. Feist and P. Reno. 1995. Conditioning improves performance of juvenile chinook salmon, *Oncorhynchus tshawytscha*, to transportation stress. Aquaculture 135: 99–110.

Secombes, C.J., T. Wang, S. Hong, S. Peddie, M. Crampe, K.J. Laing, C. Cunningham and J. Zou. 2001. Cytokines and innate immunity of fish. Dev. Comp. Immunol. 25: 713–723.

Seppola, M., H. Johnsen, S. Mennen, B. Myrnes and H. Tveiten. 2009. Maternal transfer and transcriptional onset of immune genes during ontogenesis in Atlantic cod. Dev. Comp. Immunol. 33: 1205–1211.

Sepulcre, M.P., F. Alcaraz-Perez, A. Lopez-Munoz, F.J. Roca, J. Meseguer, M.L. Cayuela and V. Mulero. 2009. Evolution of lipopolysaccharide (LPS) recognition and signaling: fish TLR4 does not recognize LPS and negatively regulates NF-kappaB activation. J. Immunol. 182: 1836–1845.

Silphaduang, U. and E.J. Noga. 2001. Peptide antibiotics in mast cells of fish. Nature 414: 268–269.

Sitjà-Bobadilla, A. and P. Alvarez-Pellitero. 1993. Pathologic effects of *Sphaerospora dicentrarchi* Sitjà-Bobadilla and Alvarez-Pellitero, 1992 and *S. testicularis* Sitjà-Bobadilla and Alvarez-Pellitero 1990 (Myxosporea: Bivalvulida) parasitic in the Mediterranean sea bass *Dicentrarchus labrax* L. (Teleostei: Serranidae) and the cell-mediated immune reaction: a light and electron microscopy study. Parasitol. Res. 79: 119–129.

Swain, P. and S.K. Nayak. 2009. Role of maternally derived immunity in fish. Fish Shellfish Immunol. 27: 89–99.

Takano, T., H. Kondo, I. Hirono, M. Endo, T. Saito-Taki and T. Aoki. 2011. Toll-like receptors in teleosts, pp. 197–208. *In*: M.G. Bondad-Reantaso, J.B. Jones, F. Corsin and T. Aoki (eds.). Diseases in Asian Aquaculture VII. Fish Health Section, Asian fisheries Society, Selangor, Malaysia. 385 pp.

Takeuchi, O. and S. Akira. 2010. Pattern recognition receptors and inflammation. Cell 140: 805–820.

Temkin, R.J. and B. McMillan. 1986. Gut-associated lymphoid tissue (GALT) of the goldfish, *Carassius auratus*. J. Morphol. 190: 9–26.

Trust, T.J. 1974. Sterility of salmonid roe and practicality of hatching gnotobiotic salmonid fish. Appl. Microbiol. 28: 340–341.

Vadstein, O., Ø. Bergh, F.-J. Gatesoupe, J. Galindo-Villegas, V. Mulero, S. Picchietti, G. Scapigliati, P. Makridis, Y. Olsen, K. Dierckens, T. Defoirdt, N. Boon, P.D. Schryver and P. Bossier. 2012. Microbiology and immunology of fish larvae. Rev. Aquaculture 4: 1–25.

Vaishnava, S., M. Yamamoto, K.M. Severson, K.A. Ruhn, X. Yu, O. Koren, R. Ley, E.K. Wakeland and L.V. Hooper. 2011. The antibacterial lectin RegIIIgamma promotes the spatial segregation of microbiota and host in the intestine. Science 334: 255–258.

Vallejo, A.N., Jr. and A.E. Ellis. 1989. Ultrastructural study of the response of eosinophil granule cells to *Aeromonas salmonicida* extracellular products and histamine liberators in rainbow trout *Salmo gairdneri* Richardson. Dev. Comp. Immunol. 13: 133–148.

van der Leij, J., A. van den Berg, T. Blokzijl, G. Harms, H. van Goor, P. Zwiers, R. van Weeghel, S. Poppema and L. Visser. 2004. Dimeric galectin-1 induces IL-10 production in T-lymphocytes: an important tool in the regulation of the immune response. J. Pathol. 204: 511–518.

Wang, P.H., D.H. Wan, L.R. Pang, Z.H. Gu, W. Qiu, S.P. Weng, X.Q. Yu and J.G. He. 2012. Molecular cloning, characterization and expression analysis of the tumor necrosis factor (TNF) superfamily gene, TNF receptor superfamily gene and lipopolysaccharide-induced TNF-alpha factor (LITAF) gene from *Litopenaeus vannamei*. Dev. Comp. Immunol. 36: 39–50.

Wang, Q., Y. Wang, P. Xu and Z. Liu. 2006. NK-lysin of channel catfish: Gene triplication, sequence variation, and expression analysis. Mol. Immunol. 43: 1676–1686.

Wang, Z., S. Zhang, G. Wang and Y. An. 2008. Complement activity in the egg cytosol of zebrafish *Danio rerio*: evidence for the defense role of maternal complement components. PLoS One 3: e1463.

Yu, L.C., J.T. Wang, S.C. Wei and Y.H. Ni. 2012. Host-microbial interactions and regulation of intestinal epithelial barrier function: From physiology to pathology. World J. Gastrointest. Pathophysiol. 3: 27–43.

Zapata, A. 1979. Ultrastructural study of the teleost fish kidney. Dev. Comp. Immunol. 3: 55–65.

Zapata, A.G., M. Torroba, A. Varas and A.V. Jimenez. 1997. Immunity in fish larvae. Dev. Biol. Stand. 90: 23–32.

Zapata, A., B. Diez, T. Cejalvo, C. Gutierrez-de Frias and A. Cortes. 2006. Ontogeny of the immune system of fish. Fish Shellfish Immunol. 20: 126–136.

Zelensky, A.N. and J.E. Gready. 2005. The C-type lectin-like domain superfamily. FEBS J. 272: 6179–6217.

Zhang, Q., M. Raoof, Y. Chen, Y. Sumi, T. Sursal, W. Junger, K. Brohi, K. Itagaki and C.J. Hauser. 2010a. Circulating mitochondrial DAMPs cause inflammatory responses to injury. Nature 464: 104–107.

Zhang, Y.A., I. Salinas, J. Li, D. Parra, S. Bjork, Z. Xu, S. E. LaPatra, J. Bartholomew and J.O. Sunyer. 2010b. IgT, a primitive immunoglobulin class specialized in mucosal immunity. Nat. Immunol. 11: 827–835.

Zou, J., P.S. Grabowski, C. Cunningham and C.J. Secombes. 1999. Molecular cloning of interleukin 1beta from rainbow trout *Oncorhynchus mykiss* reveals no evidence of an ice cut site. Cytokine 11: 552–560.

Zou, J., C. Mercier, A. Koussounadis and C. Secombes. 2007. Discovery of multiple beta-defensin like homologues in teleost fish. Mol. Immunol. 44: 638–647.

The Response to Stressors in the Sea Bass

Lluis Tort,[1,]* Josep Rotllant,[2] Michalis Pavlidis,[3] Daniel Montero[4] and Genciana Terova[5]

Stress in Fish, Stress in the Sea Bass

Stress is an important physiological response that all animals, including fish, experience when they are subjected to alarm situations.

The stress impact in fish and its consequences was early detected and assessed in relation to fish transport in aquaculture (Barton 1997). Later studies on fish stress response have involved all potential stressors, such as the natural environmental factors, but in particular those related to husbandry practices, including the social stressors. On the other side there is extensive scientific literature on the chemical stressors in water (pollutants, heavy metals, pesticides). In parallel to the study of the effects of stressors, a number of studies on stress physiology and stress responses have been developed during the last few decades.

[1]Univ Autonoma Barcelona, Dept Cell Biol Physiol & Immunol, Cerdanyola Del Valles 08193, Spain.
[2]CSIC, Inst Invest Marinas, Aquat Mol Pathobiol Lab, Eduardo Cabello 6, ES-36208 Vigo, Spain.
[3]Univ Crete, Dept Biol, POB 2208, Iraklion 71409, Crete, Greece.
[4]Univ Las Palmas Gran Canaria, Grp Invest Acuicultura, Transmontana S-N, Las Palmas Gran Canaria 35416, Canary Islands, Spain.
[5]Univ Insubria, Dept Biotechnol & Life Sci DBSV, I-21100 Varese, Italy
*Corresponding author

Stress, stressors and stress response

From the first definition of the General Adaptation System by Hans Selye (1956), i.e., "the non-specific response of the body to any demand placed upon it", many definitions about stress have been proposed by several authors. Regarding fish, authors such as Brett (1958) proposed that stress is the state which extends the adaptive responses developed of an animal beyond the normal range or the normal function; Schreck 2000 included the physiological cascade of events that occurs when the organism is attempting to resist death or re-establish homeostatic norms in the face of insult. It is not well known whether fish develop complete stress responses when the danger is not real, i.e., the perception of danger or symbolic stress, as it occurs in mammals, but some studies have shown that a characteristic physiological response would be evoked if the fish experiences fright or discomfort (Schreck 1981) and some reports have described the characteristic increase of cortisol and glucose in untouched fish from the same tank than others to which the stressor is directly applied (Molinero et al. 1997). Altogether, there is increasing evidence that fish have complex behavioural patterns and responses to non-physical stressors; this has been clearly demonstrated in episodes of dominance, aggressiveness, choice of coping strategies, response to conspecifics under different social contexts, and competition for food sources, etc. (Mackenzie et al. 2009).

From an applied point of view, the issue of stress in aquaculture is a very clear one, as a number of consequences, often undesired but always relevant for fish performance such as growth, disease resistance, behaviour and energetics are clearly observed.

Stressors may be common in nature where extreme natural conditions for fish regarding oxygen concentration, salinity, temperature or a rapid change in these variables (Altimiras et al. 1994, Tort et al. 1998) may induce an alarm response. Other factors such as the presence of predators or competitors may be included as normal stressors for fish. However, in captivity, fish are subjected to non-natural stressors, which can be of a wide variety. Thus, stressors can be related to the physical conditions of water (low oxygen, temperature change, salinity change), the chemical components of water (levels of ammonia or nitrite, presence of chemicals, metals or contaminants), or they may be related to the condition of artificial environment itself (confinement, density) and the consequences that this situation generates such as social interactions with dominance or subordinate relationship. A key feature of these imposed situations is that fish cannot escape from this environment and therefore natural avoidance mechanisms such as migration, protection in sand bottoms, and change of habitat cannot be displayed and therefore the stress reaction becomes uncontrollable. Other than severity and time, the factors of uncontrollability and unpredictability are the relevant ones regarding the induction of a consistent stress response (see Fig. 1).

There are two main types of stress, i.e., the acute and the chronic stress that are defined in terms of intensity and time. While the acute stress is high in intensity and short in time, the chronic stress is normally of lower intensity but longer time. The acute stress response is normally an adaptive response (as if the fish survives to the stressor, i.e., the intensity does not surpass the capacity of response) which can overcome the next stressor of the same nature with higher success. During this period, important changes in most physiological systems are engaged such as energy, signalling systems, sensorial awareness, behaviour, and metabolism, which are all directed to overcome the situation. This potent response after acute stress involves important quantitative changes in the energetics and metabolism as medium and long term processes will be delayed due to the diversion of resources.

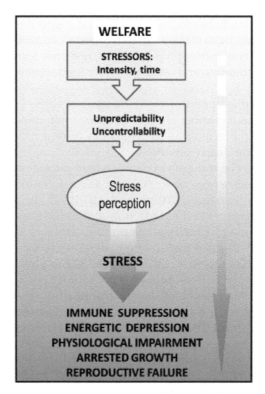

Figure 1. Stress reaction after stressor occurrence characterized by intensity and time but finally induced if uncontrollability and unpredictability are present.

Chronic stress is less common in nature and therefore it can easily become maladaptive. In aquaculture practice, fish are confined in captivity. This circumstance makes it more likely that physical or chemical characteristics of the water or the interaction between individuals become chronic stressors, as the stressor cannot be avoided. In these situations the stressors may generate a maladaptive response, with a continued decrease in energy resources, susceptibility to diseases and an overall reduced performance.

Other than the classical acute and chronic stressors, other models of stressors such as repeated acute stress or combinations like an acute stressor on chronically stressed fish, have to be taken into consideration as they involve a more complex regulatory work of neuroendocrine systems, including the effects of the combined stressors plus the feedback regulations over the endocrine, immune and neural systems.

The nature of the stress response in the sea bass

The European sea bass (*Dicentrarchus labrax*) has often been described as a stress sensitive species, as demonstrated by the fact that both basal and post-stress plasma cortisol levels are much higher in sea bass (Cerda-Reverter et al. 1998, Rotllant et al. 2003, Fanouraki et al. 2008, Pavlidis et al. 2011) compared with other phylogenetically close species, such as sea bream, *parusaurata* (Tort et al. 2001), red sea bream, *Pagrus major* (Biswas et al. 2006) or red porgy, *Pagrus pagrus* (Fanouraki et al. 2007). Thus, it has also been shown that stressors such as rearing density severely affects sea bass performance (Paspatis et al. 2003, D'Orbcastel et al. 2010) as well as the fish response to several acute stress challenges (Di Marco et al. 2008). This statement was confirmed in a study by Rotllant et al. 2006, where it was demonstrated that fish stress response to an identical stressor and analyzing samples with the same analytical method (radioimmunoassay) was clearly high. In addition to this, data obtained with superfusion techniques on isolated head kidney cells showed that the basal release of cortisol from sea bass head kidneys is much higher than that released from other fish species (Rotllant et al. 2000, 2003). Moreover, the level of stress hormones in sea bass blood is in a high range among fish. It is also known that fish farmers prefer not to touch the fish in the sea cages for grading purposes because significant mortalities are commonly recorded after this handling practice. Similarly, other authors report that the mere presence of humans can significantly reduce feeding activity (Rubio et al. 2010). These two facts (hormone levels and aquaculture stress ability) are not found in other close species cultured under very similar husbandry conditions. The basis for this differential stress response in the sea bass may have anatomical and physiological explanations that are reviewed in the next sections.

The Integrated Stress Response

The stress response is a highly conserved mechanism in all animals. It involves a number of alterations in cells, tissues, organs and systems, generating responses at gene, molecular, cellular, physiological and behavioural levels. Thus, the stress response is an integrated mechanism in which both the reaction and the consequences of this response and therefore its regulation, is exerted at all levels in an organism. In this picture, there are a number of molecules that have a principal role as the main mediators for activating both the response and its regulation (see Fig. 2).

Some hypotheses have been developed on the response to cognitive stimuli that would be related to an evolution from a more primitive and ancestral stress response. Some authors propose that this early response could be related to the development of basic functions of immune cells that later on could be capable of producing hormones and neuropeptides other than resources devoted to build up an immune reaction (Borghetti et al. 2009). This diversified role can be identified in fish in which cytokines and neuropeptides perform neuroendocrine and immune functions. Thus, these key molecules and the head kidney play a central role in organizing the stress response involving close communication between regulatory systems. This explains why stress deeply involves central and peripheral immune and neuroendocrine cells and the respective messengers.

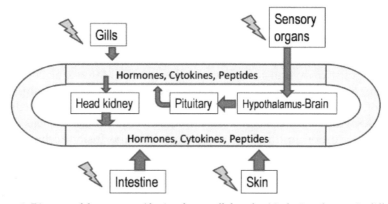

Figure 2. Diagram of the stressors (depicted as small thunders) inducing changes in different tissues and releasing chemical messengers in the blood. In addition, the brain-pituitary-head kidney axis is also included in the diagram.

Genomic response

A number of genes have been studied in sea bass by several authors with the aim of testing whether the expression of these genes would become a representative or useful indicator of the stress response in this species.

Some of these genes are related to the effects caused by pollutants and contaminants and the consequent cellular and molecular response, among them, metallothioneins, heat shock proteins and a number of other specific genes.

MT are cysteine-rich metal binding proteins that increase both in sea bass liver and brain after exposure to high rearing density, indicating that MT is not a specific biomarker, but that it can be useful to monitor fish stress conditions (Gornati et al. 2004).

Heat shock proteins (HSP) mRNA level too increases after crowding exposure. Moreover, in sea bass, there are constitutive and inducible members of HSP70 and HSP90 and it is known that the constitutive forms of HSP play an important chaperoning role in unstressed cells, whereas the inducible forms are important for protein machinery, in particular after acute stressors. HSP90 is present in brain, where it is greatly upregulated after a thermal stress, but is hardly detectable in liver of unstressed animals. Cytochrome P450 has been identified in most vertebrates, and it is widely used as a biomarker when assessing exposure to environmental contaminants. Cytochrome P450 is also overexpressed in fish reared at high density which shows that cytochrome P450 can be an unspecific biomarker to physiological stressors, too (Gornati et al. 2004).

Chemical stressors such as heavy metals may alter the structure of cell membranes by stimulating the lipid peroxidation process, a complex sequence of biochemical reactions, and oxidative deterioration of polyunsaturated fatty acids. The cellular defense system against toxicity originating from active oxygen forms includes induction of scavenging superoxide anion radicals and elimination of hydrogen peroxide. Copper and cadmium injected into *Dicentrarchus labrax* caused an enlargement of the lysosomal membrane of the kidney pronephros, Cu being more toxic than Cd. Following injection, metal uptake, measured in muscle, liver and kidney, was much higher with cadmium treatment than with copper, the kidney being the main accumulating organ of cadmium. In this organ, metal accumulation was correlated with increased zinc level, suggesting metallothionein induction. *In vivo* exposure to metal decreased the lysosomal membrane stability of pronephros. When added *in vitro*, Cu significantly raised lipid peroxidation, expressed as malondialdehyde equivalents, Cd having a lower effect. Catalase activity was significantly reduced by Cd whereas Cu showed lower effects. Hence, although both metals caused in vivo damage, Cu activated the redox process by generating oxygen radicals but it did not affect *in vitro*, the protective catalase activity, unlike Cd which appeared to weakly participate in oxygen radical generation but altered in vitro protective catalase activity (Romeo et al. 2000).

Another approach has been the management of genetic tools to assist in reducing inbreeding, accelerating growth rate, lowering losses of farmed fish from infections and improving resistance to suboptimal environmental conditions such as under stress. A work by Chistiakov et al. (2005) produced a genetic linkage map of the sea bass (*Dicentrarchus labrax*) constructed from 174 microsatellite markers including stress markers (Chistiakov et al. 2005).

Glucocorticoid receptor expression

One of the molecules with a higher number of expression studies performed in sea bass exposed to stress is the Glucocorticoid Receptor (GR). GR is a key molecule since it is the main cortisol receptor that mediates corticosteroid signaling and generating effects in many types of cells.

Although most vertebrates possess one GR, all teleosts studied to date, apart from the zebrafish (*Danio rerio*) and sea bream (*Sparus aurata*), have been shown to display two GR genes that encode different receptor protein isoforms (Alsop and Vijayan 2008, Acerete et al. 2007). In the European sea bass, *Dicentrarchus labrax*, two different GR genes have also been cloned and sequenced (Terova et al. 2005, Vizzini et al. 2007). These genes encode different GR isoforms (GR-1 and GR-2) with high amino acid sequence identity, but no significant sequence similarity at the transcriptional activation domain. Terova et al. (2005), cloned *D. labrax* GR. This isoform shows a 9 amino acid insertion between the two zinc fingers within its DNA binding domain which is different from that of mammals. This insertion was already identified in trout, tilapia and flounder. Such additional exon, in comparison with mammalian GRs, may be present in all teleosts and, as it promotes greater DNA affinity in the GR, could have been selected to serve the large spectrum of cortisol functions in fish (Lethimonier et al. 2002). Down regulation of GR-1 in accordance wth high blood cortisol levels has been reported in the liver of adult sea bass exposed to very high stocking densities (Terova et al. 2005). However, no correlation between high cortisol plasma levels and GR-1 and GR-2 liver expression levels was observed in slow release cortisol implanted adult sea bass (Di Marco et al. 2008). Work by Terova et al. (2005) found downregulation of GR by cortisol in fish, similarly with what has been found in other fish species such as *Oncorhynchus kisutch* or *Salmo salar*. In these species, GR concentration in gills decreased inversely with blood cortisol levels (Shrimpton and McCormick 1999).

Vizzini et al. (2007) cloned and sequenced a 2592 base pairs (bp) cDNA encoding a glucocorticoid receptor (DLGR1) from sea bass (*Dicentrarchus labrax*) peritoneal leukocytes. The DLGR1 functional domains presented homologies with glucocorticoid receptors of other vertebrate species. Dose-dependent cortisol inhibitory effects and significant competitive activity of a receptor blocker (a low concentration of mifepristone RU486) confirmed that cortisol–GR interaction is involved in modulating phagocytes chemiluminescent response via a genomic pathway.

Di Bella et al. 2008 showed initial specific GR2 expression only in the cells of the lateral portion of *optic tectum* in post-hatching European sea bass larvae, whereas in the subsequent stages of development, expression was found in other tissues (see Pavlidis et al. 2011).

Vazzana et al. (2008) showed glucocorticoid receptor (DGR1 or GR2) mRNA expression and tissue immunohistochemical localization in sea bass head kidney, spleen, gills, intestine, heart, and liver tissues in support of the wide physiological role of cortisol that is the main glucocorticoid hormone, responsible for fish homeostasis maintenance. GR2 mRNA is localized in the nucleus and cytoplasm of spleen and head kidney white pulp cells, in the nucleus of branchial epithelium undifferentiated cells and chondrocytes, of intestinal columnar cells, of cardiac muscular fibers, and of hepatocytes. The cytolocalization of mRNA and GR protein in the cell may suggest various phases of the cellular activity as indicated in previous histological studies on this species (Grassi-Milano et al. 1997). In the nucleus, the presence of ligand-bound GR2 is related to GR2 transcriptional regulation activity, whereas unliganded GR2 is mainly localized in the cytoplasm (Bury and Sturm 2007). In the gills, GR2 mRNA and protein are located in the nucleus of undifferentiated cells, thus supporting the potential role of cortisol in the functional differentiation of the branchial epithelium functional differentiation. In fish intestine, GR2 mRNA and protein is expressed in the nucleus and cytosol of the columnar cells, which can be related to an osmoregulatory function. The fact that GR2 mRNA was found in the nucleus of the hepatocytes, and the protein was expressed in the cytosol (Vazzana et al. 2008), supports a role of cortisol in modulating the hepatic glucose metabolism. The expression of GR2 in the head kidney and spleen parenchyma may indicate an important role of glucocorticoids in the immune response, in particular in relation to the immune-suppressive effect of cortisol in lymphocytes and macrophages of these lymphoid organs (Tort 2011).

Physiological response

The physiological stress response in European sea bass is similar to that of other marine teleosts. It starts with the activation of the Brain-Sympathetic-Chromaffin axis (BSC-axis) and the Hypothalamus-Hypophysis-Interrenal axis (HPI-axis). This primary response results in the secretion of catecholamines (BSC-axis) and cortisol (HPI-axis). Data on circulating or brain catecholamine concentrations or other important neurotransmitters are lacking for European sea bass. On the contrary, data on cortisol are available in terms of ontogeny, response to acute or chronic stressors, and in relation to changes in abiotic (light and photoperiod; 4.3 temperature) and biotic (feeding and dietary influences) conditions (see later).

European sea bass following exposure to acute stressors show a quick increase in plasma cortisol concentrations to reach a maximum around one hour post-stress (Fanouraki et al. 2011). However, the magnitude of the cortisol response is among the highest reported for fish, and in particular around two- to four-fold higher than in sharpsnout sea bream (*Diplodus puntazzo*), gilthead sea bream (*Sparus aurata*) and common Pandora (*Pagellus erythrinus*) and 20-fold higher than of dusky grouper (*Epinephelus marginatus*), meagre (*Argyrosomus regius*) and common dentex (*Dentex dentex*) (Fanouraki et al. 2011). Size seems to affect the timing, at 0.5 hours post-stress for a mean (± SD) body weight) of 30 (± 7.1) g and at one hour for 125 (±13.9) g or larger fish, but not the magnitude or the duration of the peak cortisol concentrations. Sea bass also show a typical secondary stress response following exposure to acute stressors with elevated levels of plasma glucose and lactate. However, elevated glucose levels show a longer duration than in other species from 0.5 hours to eight hours post-stress, with a peak at two hours (Fanouraki et al. 2008). Hyperlactaemia is also observed at 0.5 hours and remains until two to four hours post-stress (Fanouraki et al. 2008).

Most, if not all studies concerning the chronic stress response in sea bass are related to husbandry procedures in aquaculture. Sea bass are farmed at high densities to reduce the relative production costs. Common on-growing stocking density for sea bass on commercial farms may reach around 15 kg–30 kg/m₃ (Paspatis et al. 2003), but very high densities (40 kg–150 kg/m₃) can be reached in high technology land-based farms, where water recirculation systems, oxygenation with pure oxygen and continuous monitoring ensure high water quality standards (Terova et al. 2005). The effect of stocking density on the secondary and tertiary stress response has been studied extensively in sea bass. Stocking density, at a range of 50 to 200 individuals l^{-1} does not have any significant effect on survival, growth rate and feed intake of larvae reared under controlled conditions in 50^{-1} cylindroconical tanks (Hatziathanasiou et al. 2002). However, stocking density at the post-larvae ontogenetic phase (5 to 20 individuals l^{-1}) seems to affect significantly both survival (higher in the lower densities of 5 and 10 individuals l^{-1}), due to cannibalism, and growth (highest at 5 individuals l^{-1}) (Hatziathanasiou et al. 2002). On the contrary, Papoutsoglou et al. (2006) reported that juvenile fish (mean body weight of 6.6 g) reared at high stocking densities (325 and 650 fish m^{-3}) displayed the highest specific growth rate and the lowest food conversion ratio than that reared at lower densities (80 and 165 fish m^{-3}). No differences were observed in growth, daily feed intake and survival of adult sea bass (mean BW of approximately 140 g) reared in recirculating tanks under different stocking densities (10 to 70 kg m^{-3}) for a period of six to 10 weeks (Di Marco et al. 2008, Sammouth et al. 2009). In another study, sea bass (mean initial weight of 82.1 g) reared at 50 or 75 kg m^{-3} showed reduced feed intake and growth and alterations in swimming behavior than fish reared at lower stocking densities (8.1 and 25.2 kg m^{-3}) (Santos et al. 2010). This feed intake reduction compensated, although partially, a decrease in the maintenance requirements involving differences in the nitrogen and energy partitioning. Thus, sea bass seems to tolerate relatively high stocking densities with important differences, however, related to the ontogenetic phase and the life-history of the fish.

Varsamos et al. (2006) showed that juvenile sea bass with different early life husbandry-related stress experience displayed differences in both stress and health status. Sea bass juveniles reared during early life in high and constant temperature perform better in terms of stress-related indicators. Variable water temperature triggered dramatic changes in terms of stress parameters and susceptibility to nodavirus. This suggests a strong and prolonged activation of the HPI axis, which corresponds to an allostatic overload. Scrubbing induced some disturbances typical for mild short-term acute stress, but did not affect plasma cortisol, growth or susceptibility to nodavirus of sea bass. Plasma cortisol concentration and osmolality trends were opposite to those displayed by serum IgM content and weight. The sampling rank had no effect on the experiments performed under constant temperature regimes, but a clear effect for body weight and rank has been found, which is explained by the fact that larger fish are caught more easily than smaller ones (Varsamos et al. 2006).

Differences in performance have been shown by several studies working at high densities. Thus, no significant differences have been found regarding final body weight, specific growth rate, survival, blood metabolites and chemical composition of sea bass reared at 20 or 40 kg/m3 for 18 months (Roncarati et al. 2006), but other works on high densities as a chronic stressor show reduced performance due to an increase in energy requirements for maintenance and altered energy partitioning, even if the density itself does not induce changes in cortisol. Thus bass reared at 50 or 75 kg/m^3 reduced feed intake and growth and showed alterations of behavioral traits such as swimming speed. This feed intake reduction, compensated, although partially, a decrease in the maintenance requirements, involving differences in the nitrogen and energy partitioning (Santos et al. 2010).

Stocking density has an effect on the primary and secondary stress indicators. After the stress challenge, a significant increase in glucose levels was observed in sea bass held at stocking densities of 15 kg/m^3 and 30 kg/m^3. The increase in glucose levels was less pronounced in sea bass kept at the higher density of 45 kg/m^3. Similar findings were already reported in sea bass confined for 3 hours at 10 kg/m^3, which showed higher plasma glucose levels than sea bass confined at 60 kg/m^3 for the same duration of time (Vazzana et al. 2002). These differences have been related with the higher available metabolic reserves for mobilization in un-stressed fish than in stressed fish. The non-essential fatty acids concentration increased 1 hour after crowding and these changes correlated significantly with cortisol levels. Crowding induced also a significant decrease in total proteins concentration in sea bass. These changes may occur as a consequence of peripheral proteolysis in response to hypercortisolemia. Minor changes are also detected in cholesterol levels after overcrowding, but ten days after the stress challenge, the levels of metabolites returned to pre-stress values, with no differences between experimental groups (Di Marco et al. 2008). If the crowding conditions of 45 kg/m^3 of sea bass are maintained for longer periods, they did not greatly affect fish ability to cope with a subsequent crowding stressor. The magnitude of responses to a crowding stressor was significantly larger in sea bass maintained at the density of 45 kg/m^3 even though the pre-stress levels were recovered after 24 hours–48 hours in all groups, regardless of the stocking density (Di Marco et al. 2008).

Other stress indicators were used to evaluate the effects of high stocking densities on fin erosion, appetite, growth, feed conversion efficiency, condition index, and hepatosomatic index. However, these indicators did not show a complete correlation under stress conditions, and it should be taken into account that often the results obtained with any of these indicators show variations that may be related to water quality or social interactions which may be different from different experiments, even being the same stressor and same species. In the work by Varsamos et al. (2006) changes in four stress predictors, i.e., plasma cortisol levels, plasma osmolality levels, serum IgM levels, and growth rate display different immunoendocrine signatures depending on the husbandry stressors. Scrubbing elicited a cortisol response, which resulted in an adaptation response based on previous learning. Others have reported that following five minutes of scrubbing the tank walls, sea bass juveniles displayed a secondary stress response comparable to the consequences of a severe non-hypoxic exercise, characterized by an increase in several hematological parameters including glycemia, proteinemia, hematocrit and hemoglobinemia (Hadj-Kacem et al. 1986). These data suggest a potent adrenergic function, which in teleosts is very responsive to stress. The adrenergic function in fish results in an increase of plasma catecholamines within a few minutes. *Dicentrarchus labrax* juveniles can adapt to this kind of routine husbandry stress (learning response) showing an increase in plasma osmolality and decrease in serum IgM content (Varsamos et al. 2006).

A combination of stressors may also induce significant changes in the fish response. Sea bass, previously subjected to high density stress and then infected by injection of bacteria or virus clearly showed an immunosuppression-induced effect. High density reared fish showed a lower ability to respond to the pathogen challenge, as both lysozyme and complement levels were significantly attenuated (Mauri et al. 2011a). In addition, the expression of the C3 protein, a key molecule which influences the complement response was also significantly attenuated in previously stressed fish (Mauri et al. 2011b).

Juvenile sea bass with different early life husbandry-related stress experience displayed differences in both stress and health status. Sea bass juveniles reared during early life in high and constant temperature perform better in terms of stress-related indicators. Variable water temperature triggered dramatic changes in terms of stress parameters and susceptibility to nodavirus. This suggests a strong and prolonged activation of the HPI axis, which corresponds to an allostatic overload. Scrubbing induced some disturbances typical for mild short-term acute stress, but did not affect plasma cortisol, growth or susceptibility to nodavirus in sea bass. Plasma cortisol concentration and osmolality trends were opposite to those displayed by serum IgM content and weight. The sampling rank had no effect on the experiments performed under constant temperature regimes, but a clear effect for body weight and rank has been found, which is explained by the fact that larger fish are caught more easily than smaller ones (Varsamos et al. 2006).

Rearing stress of short duration (3 hours) but high severity (10 to 60 kg/m^3) affect both plasmatic parameters and cellular innate immune responses, which do not recover to normal immune competence for the next two days. Variations in plasma glucose and cortisol levels of sea bass showed that both confinement and high population density can induce stress responses. Increased levels of cortisol, glucose and osmolarity were found in fish confined at a low density and these levels further rose after maintenance at higher density. It appears to be an immediate and transitory response to stress, since it approached basal levels during the two days post stress. An inverse correlation of physiological indicators with cytotoxicity in peritoneal cells was shown. This response appeared to be more sensitive to stress than the respiratory burst response of pronephros leukocytes. On the other hand, some immunological effects, such as depression of peritoneal leukocyte cytotoxicity, have been described in adult sea bass crowded at 60 kg/m^3 for 48 h (Vazzana et al. 2002). In addition, an increase of both plasma cortisol levels and stress genes expression levels was observed when juveniles were held at 80 and 100 kg/m^3 for three months (Gornati et al. 2004, Terova et al. 2005).

The extent and manner of cortisol involvement in the mediation of rearing density induced effects on sea bass growth remains poorly explored. Acute stress challenges can modify short-term food intake (Rubio et al. 2010). Chronic stress induced by repetition of acute stress protocols and dietary cortisol administration induced a significant attenuation of food intake and food conversion efficiency, severely impairing sea bass performance (Leal et al. 2011). In addition, both repeated stress and cortisol treatments induced modification of the feeding activity rhythms, suggesting involvement of the stress response in the temporal organization of feeding behavior. In this case, the influence of aquaculture-related stressors on the sea bass performance was very high: even a single tank cleaning process in the week was sufficient to severely impair food intake, conversion and fish growth (Cerda-Reverter et al. 2011). A more prolonged acute stress, such as 24-hour confinement in sea bass (70 kg/m^3) resulted in sustained high plasma cortisol levels associated with an increase in the interrenal cells activity (enlarged nuclear diameters measured after 1 hour and 4 hours of confinement (Rotllant et al. 2003).

The rearing density and size grading had effects on sex ratios in 30 families of sea bass reared in the same tank from the fertilization stage onwards. In this case, the differences shown by sea bass were dependent of the development stage. A slower growth at low density has already been reported in sea bass by Papoutsoglou et al. (1998). Saillant et al. (2003) reported faster growth under low density from 49 to 191 days post fertilization (dpf), but equivalent until the end of the treatments 414 dpf. Density had no effect on sex ratio, suggesting that the high densities usually applied in aquaculture are not involved in the systematic excess of males reported in farmed sea bass populations. Repeated size grading performed from 84 dpf to 199 dpf had no effect on the sex ratio of the overall studied population. This result shows that sex determination was not affected by the density treatment. Thus, it could be concluded that growth is depressed at extreme high and low densities whereas it is maximal at intermediate densities (Saillant et al. 2003).

The Anatomical Elements of the Physiological Stress Response

In fish, the integrated stress response involves the activation of two main pathways, the hypothalamus-pituitary-interrenal (HPI) and brain-sympathetic nerves-chromaffin cell axes, resulting in elevated circulating levels of glucocorticoids (cortisol) and cathecholamines in the blood. Thus, fish present the majority of the characteristics of the main functional features of the brain, pituitary and adrenal structures of the rest of vertebrates (Tort and Teles 2011).

Brain and pituitary

The anatomical and histological characteristics of the brain and hypothalamus of sea bass have been described by several authors (Cerda-Reverter et al. 2001, Cerda-Reverter and Canosa 2009). Regarding the cyto-architecture of the preoptic area, ventral and dorsal thalamus, epithalamus, and synencephalon resemble the histological pattern of other teleosts. The hypothalamus, the main area responsible for the neuroendocrine control of the stress response, seems to differ slightly from that of other teleosts. Five subdivisions of the lateral tuberal nucleus and three different nuclei around the lateral recesses have been shown in sea bass (Cerda-Reverter et al. 2001). A medial nucleus of the inferior lobe is also present in the hypothalamus of sea bass whereas it is not common in other advanced teleosts. In addition, three cell masses which have been assigned to the migrated area of the posterior tuberculum of the diencephalon, have only been identified in sea bass and in another teleost, *Sparus aurata* (Cerda-Reverter et al. 2001).

Light microscopic staining on hypothalamo-hypophysial sections, revealed the projection of different neuropeptide-immunoreactive neurons innervating the hormone-producing cell populations in the pituitary gland. In the rostral pars distalis (PD), the ACTH cells are found in close proximity to fibers immunoreactive for somatostatin (SRIF), growth hormone-releasing hormone (GRF), corticotropin-releasing hormone (CRH), vasotocin (VT), isotocin (IT), substance P (SP), neurotensin, and galanin (GAL), while the PRL cell zone seemed only innervated by nerve fibers immunopositive for GAL. In the proximal PD, fibers immunoreactive for SRIF, GRF, VT, IT, cholecystokinin, SP, neuropeptide Y, and GAL formed a close relationship with the growth hormone producing cells. The gonadotrophs were observed near nerve fibers immunostained for gonadotropin-releasing hormone, IT, and less obviously GRF and VT, while fibers positive for GRF, CRF, VT, IT, SP, and GAL penetrated between and formed a close association with the thyrotrophs. In the pars intermedia the MSH cells and the PAS-positive cells are both innervated by separate nerve fibers immunoreactive for GRF, CRH, melanin concentrating hormone (MCH), VT, IT, and SP (Moons et al. 1989).

The head kidney. The interrenal and chromaffin tissues

The head kidney (HK, or pronephros) in fish is a basic organ forming the blood cells but also contains the interrenal tissue (cortisol producing) and chromaffin (catecholamine producing) cells of the mammals adrenal gland. Therefore fish lack a distinct adrenal gland. The head kidney in sea bass is a non-compact Y-shaped organ in which the anterior branches and the anterior region of the single posterior branch consist of hematopoietic tissue devoid of renal tubules and glomeruli (Grassi-Milano et al. 1997, Gallo and Civinini 2003). The ducts of Cuvier and the anterior and posterior cardinal veins run through the anterior kidney and the more cephalic region of the posterior kidney is drained by the post cardinal veins. Chromaffin cells are clustered in large groups in the walls of the ducts of Cuvier, in the dorsal walls of the anterior cardinal veins and in the walls of the post cardinal veins. The steroidogenic cells, generally in contact with the chromaffin cells, are grouped in cords arranged exclusively in the cephalic kidney inside the walls of the posterior cardinal veins and around the confluence between the ducts of Cuvier and the cardinal veins (Grassi-Milano et al. 1997). The interrenal cells can show some variations depending on sex, season and occurrence of stressors and it has been suggested that a periodic or cyclic renewal of these cells takes place together with other cell types in the interrenal tissues (Gallo and Civinini 2003). In the sea bass, two main subtypes of interrenal cells have been described—a first elongated type and a second, irregular type filled with numerous vesicles which can match with such a cell cycle in which the first one would correspond to an intense period of synthesis and the second with storage and release (Gallo and Civinini 2003).

Production of cortisol, the major corticosteroid in fish, occurs in the interrenal tissue, in cells which are grouped in chords arranged exclusively in the cephalic kidney or head kidney (Grassi Milano et al. 1997). In sea bass, chromaffin cells line the walls of the PCV together with the cortisol producing interrenal cells, but are also present in the opisthonephros where interrenal cells are absent (Abelli et al. 1996). The location of the chromaffin cells, as well as the possibility of distinct adrenaline and noradrenaline producing cells, provides the basis for potential local paracrine interactions within the head kidney of the sea bass (Rotllant et al. 2006).

The interrenal and chromaffin function in sea bass

In sea bass catecholamines, adrenaline and noradrenaline exert a significant *in vitro* glucocorticoid secretagogue action on head kidney cells. Moreover, using selective inhibitors it has been demonstrated that the secretagogue action of catecholamines is mediated by a β-adrenoceptor stimulating the release of intracellular cAMP. The results of this study indicated that the action of catecholamines on cortisol release in sea bass was different from sea bream, as it stimulated release in the former species but had no effect in the latter species at the concentrations tested. Therefore, a paracrine control of catecholamines on cortisol secretion in sea bass has been proposed and this may provide an explanation for the higher corticotropic activity of the head kidney in this species (Rotllant et al. 2006).

There are a number of indications that suggest that sea bass may have a higher interrenal activity, when compared to other fish. It has been shown that cortisol content of head-kidney homogenates in unstressed sea bass were approximately 2.5-fold higher than values previously reported for gilthead sea bream (Rotllant et al. 2000a, 2001). These higher levels may indicate a high degree of activity of the interrenal cells or accumulation of cortisol within the head kidney. These findings were supported by the *in vitro* superfusion studies. Therefore, the basal unstimulated release of cortisol from the head kidney of sea bass was much greater than that reported for gilthead sea bream (40-fold higher; Rotllant et al. 2000a,b) and rainbow trout *Oncorhynchus mykiss* (30-fold higher; Balm and Pottinger 1995), although the stimulatory capacity of ACTH found for sea bass was similar to other teleosts. All these data suggest that the interrenal tissue of sea bass has an intrinsic high degree of activity under control conditions involving a high turnover in corticosteroid production and release. Sea bass which was subjected to confinement, showed a rapid increase of plasma cortisol levels already after six minutes, reaching peak levels of 100–120 ng/ml after one hour (Rotllant et al. 2003). This rapid increase in plasma cortisol levels may be facilitated by the high cortisol content of the head kidney tissues and high basal cortisol release. During confinement, plasma cortisol levels remain high even after 24 hours of confinement. Vazzana et al. (2002) also showed increased serum cortisol levels in sea bass confined for up to 48 hours. The prolonged increase of plasma cortisol levels during confinement may be due to an increased activity of the interrenal cells. Indeed, larger interrenal cell nuclear diameters, a reduction in the cortisol content of the head kidney homogenates, as well as both hyperplasia and hypertrophy of head kidney, have been associated with an increase of interrenal activity (Rotllant et al. 2003).

Previous studies have shown that other fish have a reduced sensitivity to ACTH stimulation after crowding or temperature reduction (Rotllant et al. 2000, 2001). However, no difference in the maximum release rate of cortisol after ACTH stimulation between control and confined sea bass has been observed. This indicates that sea bass interrenal cells do not lose their sensitivity to ACTH and, therefore, retain the capacity to respond to additional stressors. This may be another difference regarding the hormonal stress response in sea bass.

Variability in Response

Basal vs. *acute levels of corticosteroids*

The sea bass response to stressors has been shown to be very variable (see Fig. 3 on the cortisol levels of different works on stressed sea bass, modified

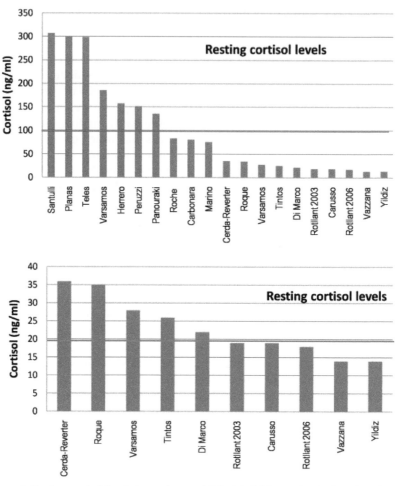

Figure 3. Resting cortisol levels in sea bass published in bibliography (right column) and most reliable resting values from selected references (see Cerda-Reverter, DiMarco, Rotllant, Varsamos, Vazzana).

from Ellis et al. 2012). Such variability may be related to different factors. Firstly, it has already been shown, both in laboratory and in aquaculture facilities, that sea bass is a species with a high sensitivity to stressors. Secondly, sea bass shows a higher basal rate of cortisol release than other fish species, as aforementioned. Thirdly, it is relevant to be accurate with fish handling before blood sampling for cortisol level measurements. Often, the variability in the cortisol response may not be a characteristic feature of the sea bass but rather, could be related to wrong blood sampling procedures.

Light and photoperiod influence

The water column acts as a chromatic filter, and the aquatic environment dramatically changes the spectral composition of incident light. There is a rapid attenuation of spectral composition with depth: blue wavelengths become predominant in all but the most shallow or turbid waters. In contrast, standard lighting systems commonly used in hatcheries create bright point light sources that are neither environment-specific nor species-specific and could potentially affect fish performance. Moreover, much of the light energy is wasted in the form of unsuitable longer yellow-red light wavelengths which are rapidly absorbed by water molecules and therefore cannot be detected by fish (Migaud et al. 2006).

European sea bass larvae are clearly affected by the different rearing conditions of photoperiod and light spectrum, displaying better growth, development, and fewer malformations under a 12h:12h cycle of blue light, which is close to the natural environmental conditions that fish larvae encounter in the wild. The onset of exogenous feeding is associated to a stress episode involving massive mortality, both in nature and in laboratory populations. Poor feeding performance under dark and red light treatments was probably due, to some extent, to a lack of appropriate photons that stimulate the larval visual system (Villamizar et al. 2009).

Temperature influence

Sea bass and seabream are classified as eurythermal fish, able to tolerate temperature ranging from 2°C to 32°C (Barnabe and de Solbe 1990). Rearing in flow-through tank systems under ambient temperatures within the range of 11°C to 25°C, thus within the normal range for the species. Therefore, in such systems as well as in RAS, water temperature is not normally a welfare issue. However, temperature has profound effects on fish metabolism and also affects the stress response of a given individual fish.

Rearing *Dicentrarchus labrax* juveniles during their early life at high and stable temperature is a good procedure as indicated by low plasma cortisol levels, normal plasma osmolality and better growth. Differences recorded between *D. labrax* juveniles subjected to constant high and low temperatures are diminished when the temperature conditions become variable. Variable temperature conditions trigger a drastic increase in plasma cortisol levels, concomitant with hemoconcentration, i.e., increased osmolality, delayed growth, and this suggests a chronic activation of the HPI axis which may represent an allostatic overload. However, within the normal temperature range for a poikilothermic species, sudden shifts in temperature can be stressful (Varsamos et al. 2006).

Temperature changes can have profound effects on the immune system. It should also be pointed out that as water temperature rises, the O_2 carrying capacity of the water decreases and therefore, the stress from temperature elevation is probably linked with hypoxic stress. This is partially confirmed by other authors (Hadj-Kacem et al. 1986) who reported that a quick temperature increase from 17°C to 26°C modified *D. labrax* haematology, reminiscent of a hypoxic situation. Temperature elicited high cortisol levels and a decrease in serum IgM content, which were related to higher susceptibility to viral challenges.

Stress is known to increase the susceptibility of fish to infectious diseases and challenges with various (bacterial and protozoan) pathogens. Stress during early life stages of sea bass has profound effects on stress response and health status of juveniles. Fish nodaviruses cause viral encephalopathy and retinopathy which are responsible for mortalities in larvae and juveniles. After applying nodavirus challenge to discriminate effects of different early life stress experiences on disease susceptibility, sea bass reared at low temperatures were more affected by virus than those reared in high temperature. Thus, fish reared at 23°C displayed the best profile of blood parameters and were less sensitive to nodavirus challenge compared to fish reared at 17°C. This result indicates that sea bass may develop a better immune response at higher temperatures, but in any case, the temperature change should not be quick, since acute shifts of temperature are known to induce a hypoxic-like stress response. Moreover, fish reared under constant but different temperature conditions were less affected by nodavirus challenge compared to those reared under variable temperature conditions (Varsamos et al. 2006).

Feeding and dietary influences

The sensitivity of sea bass to stress can also be identified in its feeding behavior. Although fish may be visually isolated to reduce the influence of human interference, feeding activity is reduced when operators are working in the culture facilities. For example, the cleaning protocol involving partial emptying and scrubbing of the tanks results in a reduction of self-feeding for two days (Rubio et al. 2010). The feeding response to acute stressors, such as sampling, showed seasonal differences. The compensatory intake observed in fish sampled during the late spring was more pronounced than in fish sampled during the winter season. This difference may not be related to the stress response *per se* but could reflect seasonal differences in energy expenditure (Rubio et al. 2010).

Variations in the diet or dietary regimes have been demonstrated to be modulators of the stress response in fish (Montero et al. 2003, Montero and Izquierdo 2010, Ganga et al. 2011a,b). Fasting and different dietary levels have been proved to be acting as chronic stressor. In a recent study on the evolution of plasma cortisol on sea bass held at different types of stressful situation, fasting within a week increased basal plasma cortisol values four-fold higher than unstressed fish (65.21 and 14.12 ng/ml, respectively) but lower that those values from sea bass confined for a week (123.92 ng/ml) (Montero et al. unpublished results). High stocking density increased the response of sea bass to an acute stressor, this effect being dependent on the feeding level and significantly reduced in well fed fish (Lupatsch et al. 2010).

Not only are variations in the feeding regime altering the physiological response to stress but also variations in the different nutrients within the diet, in terms of deficiency of a certain nutrient or imbalances in the ratios among different nutrients. At larval stages there are some studies concerning the effect of unbalanced diets in larval resistance to stressful conditions. Certain essential fatty acids such as arachidonic acid (20:4n-6, ARA) must be supplied in the diet of this species, since, as a marine fish, European sea bass is not able to elongate and desaturate 18:C fatty acids. ARA has been demonstrated to be directly related with the stress response in European sea bass (Atalah et al. 2011), as described for other species, such as the gilthead sea bream (Koven et al. 2001). Increased dietary ARA (from 0.3 to 1.2 percent of the diet) improved the stress resistance in terms of larval survival after an activity test consisting of handling the larvae out of the water in a scoop net for one minute and monitoring survival after 24 hours (Atalah et al. 2011).

There are few studies on the modulation of stress response by the dietary lipids in sea bass, but the inclusion of vegetable oils as alternative to fish oil in the diet during the on-growing period, and specifically linseed oil, has been proved to increase the plasma cortisol after one week of confinement stress (Montero et al. unpublished results), as described for other marine species such as the gilthead sea bream (Montero et al. 2003). The inclusion of vegetable oils has been demonstrated to be affecting the release of cortisol from the interrenal cells (Ganga et al. 2011a,b). The mechanism by which fatty acids are modulating this mechanism is not well known and of a complex nature, and some hypotheses have been suggested to be mediated by protein kinases (PKs) C and A, where PKA induces cortisol synthesis and PKC inhibits steroidogenesis (Lacroix and Hontela 2001) through cyclic AMP activation, triggered by the binding of ACTH to the membrane receptor. PKA phosphorilates steroidogenic acute regulatory protein (StaR), which in turn mediates the delivery of cholesterol for cortisol synthesis. PKC is activated by increased levels of $Ca2+$, which are mediated by the diacylglycerol (DGA) and inositol triphosphate (IP3) (Matthys and Widmaier 1998). These messengers are released by the activation of PKC induced by ACTH. Fatty acids have been proved to modulate this enzyme activation, free ARA being able to activate PLC, down-regulating the synthesis of cortisol. Indeed, an increase of dietary ARA levels from 0.3% to 1.2% induced a decrease in the expression of 11B hydrolase, a cytochrome involved in the last step of cortisol synthesis, and an increase in the expression of the glucocorticoid receptor gene, both measured in whole larvae (Montero et al. unpublished results).

Besides, the release of cortisol from interrenal cells has been proved to be related, at least in part, with the action of cyclooxygenase and lipoxygenase metabolites (Ganga et al. 2011a), the eicosanoids, a well-known series of hormones by the action of those enzymes on ARA and Eicosapentaenoic acid (20:5n-3: EPA). In mammals, there are some evidences that eicosanoids and specifically prostaglandins (PG) modulate the release of hypothalamic corticotrophin releasing hormone and pituitary ACTH (Nasushita et al. 1997). Although there are no evidences on how dietary type of oil and imbalances in fatty acids induce changes in the plasma cortisol levels, the use of vegetable oils has been demonstrated to alter the prostaglandins synthesis in marine fish, and specially the production of PG3 (Ganga et al. 2005). A reduction of PGE2 has been reported for sea bass fed on diets based on different blends of vegetable oils (Mourente et al. 2007).

Not only are imbalances and/or deficiencies of fatty acids modulating the stress response but also other nutrients such as certain vitamins are involved in stress resistance. Sea bass fed a diet with 300 mg of vitamin E per kg of diet had a higher survival rate after activity test when compared with larvae fed a 150 mg vit. E/kg diet (Betancor et al. 2011). In gilthead sea bream, increases of dietary vitamin E enhanced resistance to both acute and chronic stress in terms of reduced plasma cortisol after stress (Montero et al. 2001).

The use of additives in the diets, such as immunostimulants and probiotics has been demonstrated to alter some indicators of stress. Bagni and co-authors (2005) found a significant increase of HSP concentration in liver and gills of sea bass 30 days after a 15-day feeding period with alginins and glucans. However, Tovar-Ramirez and co-authors (2010) found no significant effect on dietary live yeast (*Debaryomyces hansenii*) on the expression of HSP70 in sea bass larvae. In juvenile sea bass, the inclusion of 4 mg of mannanoligosaccharides (MOS) per kg of diet significantly reduced the levels of plasma cortisol in sea bass after four hours of confinement stress and infection with *Vibrio anguillarum* (Torrecillas et al. 2012). These authors suggested that the effect of MOS on plasma cortisol could be related with the induced-increase of mucosa production, and the MOS effect on decreasing the infection through the intestine by both decreasing the translocation of pathogens through the intestine and increasing the potential of the immune barrier, consequently ameliorating the cortisol increase induced by both the stressor and the infection.

Ontogenic changes

Sea bass shows a typical ontogenetic pattern of cortisol changes, the magnitude being different to that reported in other teleosts. Cortisol content in embryos (1.71 ± 0.67 ng g^{-1}) was within the range reported for other marine fish species with short embryonic periods such as the gilthead sea bream barramundi, Japanese flounder, and red drum. Sea bass basal cortisol content at flexion and at the more advanced developmental stages is one of the highest reported for fish, showing an average of 32–48 ng/g. Peak cortisol levels in European sea bass were similar only to those reported for the chum salmon, *Oncorhynchus keta* which is a species with a different ontogenetic pattern. They were also near to the peak cortisol values reported during metamorphic climax (5–11 ng/g) in the Japanese flounder, barramundi, yellow perch, and tilapia, (see Pavlidis et al. 2011).

The exposure to a physical stressor increased the whole body cortisol above control levels, at all the developmental stages examined (Pavlidis et al. 2011). However, the magnitude of the cortisol response was low at first feeding and high in juvenile fish. These results indicate, in relation to the expression of glucocorticoid receptor genes, that European sea bass is capable of a stress-induced stimulation of cortisol production even at first feeding, as it has been shown in a number of other fish species (Pavlidis et al. 2011).

It has also been shown that both GR1 and GR2 receptors are present in sea bass embryos and hatched eggs, and in all stages of larvae examined, GR-1 being more expressed than GR-2 throughout early development. This is in accordance with results obtained in cortisol implanted adult fish, where liver GR1 mRNA was more than 100 times over expressed than GR2 (Di Marco et al. 2008). Similar results are reported for the zebrafish (*D. rerio*) in which GR transcripts were present from 49 hours until 146 hours post fertilization (Alsop and Vijayan 2008). The presence of GR2 receptor during early development has been confirmed in brain, gills, liver, head kidney and the anterior intestine of pre-larvae, larvae and post larvae of European sea bass (Di Bella et al. 2008). The highest GR1 expression was found at the end of transformation and in juveniles in accordance with high basal cortisol levels. However, down-regulation of GR1 transcripts has been found in post flexion larvae, where also high basal cortisol levels were reported.

References

Abelli, L., S. Picchietti, N. Romano, L. Mastrolia and G. Scapigliati. 1996. Immunocytochemical detection of thymocyte antigenic determinants in developing lymphoid organs of sea bass *Dicentrarchus labrax* (L.), Fish Shellfish Immunol. 6: 493–505.
Acerete, L., J.C. Balasch, B. Castellana, B. Redruello, N. Roher, A.V. Canario and L. Tort. 2007. Cloning of the glucocorticoid receptor (GR) in gilthead seabream (*Sparus aurata*).

Differential expression of GR and immune genes in gilthead seabream after an immune challenge. Comp. Biochem. Physiol. B 148: 32–43.

Alsop, D. and M. Vijayan. 2008. Development of the corticosteroid stress axis and receptor expression in zebrafish. Am. J. Physiol. Integr. Comp. Physiol. 294: R711–R719.

Altimiras, J., S.R. Champion, M. Puigcerver and L. Tort. 1994. Physiological responses of the gilthead sea bream *Sparus aurata* to hypoosmotic shock. Comp. Biochem. Physiol. 108(A)1: 75–80.

Atalah, E., C.M. Hernández-Cruz, E. Ganuza, T. Benitez, R. Ganga, J. Roo, D. Montero and M.S. Izquierdo. 2011. Importance of dietary arachidonic acid for survival, growth and stress resistance of larval European sea bass (*Dicentrarchus labrax*) fed high dietary docosahexaenoic and eicosapentaenoic acids larvae. Aquaculture Research 42: 126–128.

Bagni, M., N. Romano, M.G. Finoia, L. Abelli, G. Scapigliati, P.G. Tiscar, M. Sarti and G. Marino. 2005. Short- and long-term effects of a dietary yeast β-glucan (Macrogard) and alginic acid (Ergosan) preparation on immune response in sea bass (*Dicentrarchus labrax*). Fish shellfish Immunol. 18: 311–325.

Balm, P.H.M. and T.G. Pottinger. 1995. Corticotrope and melanotrope POMC-derived peptides in relation to interrenal function during stress in rainbow trout (*Oncorhynchus mykiss*). Gen. Comp. Endocrinol. 98: 279–288.

Barnabe, G. and J.F. de Solbe. 1990. Aquaculture. Vol. 1. Taylor and Francis, Hertfordshire, UK 528 pp.

Barton, B.A. 1997. Stress in finfish: past, present and future—a historical perspective. pp 1–33. *In*: G.K. Iwama, A.D. Pickering, J.P. Sumpter and C.B. Schreck (eds.). Fish Stress and Health in Aquaculture. Cambridge University Press, Cambridge, UK.

Betancor, M.B., E. Atalah, M.J. Caballero, T. Benítez-Santana, J. Roo, D. Montero and M.S. Izquierdo. 2011. α-Tocopherol in weaning diets for European sea bass (*Dicentrarchus labrax*) improves survival and reduces tissue damage caused by excess dietary DHA contents. Aquaculture Nutrition 17: e112–e122.

Biswas, A.K., M. Seoka, K. Takii, M. Maita and H. Kumai. 2006. Stress response of red sea bream Pagrus major to acute handling and chronic photoperiod manipulation.Aquaculture 252: 566–572.

Borghetti, P., R. Saleri, E. Mocchegiani, A. Corradi and P. Martelli. 2009. Infection, immunity and the neuroendocrine response. Vet Immunol Immunopathol. 130(3–4): 141–62.

Brett, J.R. 1958. Implications and assessment of environmental stress. pp. 69–83. *In*: P.A. Larkin (ed.). The Investigation of Fish-power Problems. H.R.MacMillan Lectures in Fisheries. Univ. British Columbia, Vancouver, Canada.

Bury, N.R. and A. Sturm. 2003. Evidence for two distinct functional glucocorticoid receptors in teleost fish. J. Mol. Endocrinol. 31(1): 141–56.

Bury, N.R. and A. Sturm. 2007. Evolution of the corticosteroid receptor signalling pathway in fish, Gen. Comp. Endocrinol. 153: 47–56.

Cerdá-Reverter, J.M. and L.F. Canosa. 2009. Neuroendocrine Systems of the Fish Brain. *In*: N.J. Bernier, G. Van Der Kraak, A.P. Farrell and C.J. Brauner (eds.). Fish Physiology, Academic Press 28: 3–74.

Cerda-Reverter, J.M., S. Zanuy, M. Carrillo and J.A. Madrid. 1998. Time-course studies on plasma glucose, insulin, and cortisol in sea bass (*Dicentrarchus labrax*) held under different photoperiodic regimes. Physiol. Behav. 64: 245–250.

Cerda-Reverter, J.M., S. Zanuy and J.A. Munoz-Cueto. 2001. Cytoarchitectonic study of the brain of a perciform species, the sea bass (Dicentrarchus labrax). II. The diencephalon. J. Morphol. 247: 229–251.

Chistiakov, D.A., B. Hellemans, C.S. Haley, A.S. Law, C.S. Tsigenopoulos, G. Kotoulas, D. Bertotto, A. Libertini,and F.A.M. Volckaert. 2005. A Microsatellite Linkage Map of the European Sea Bass *Dicentrarchus labrax* L. Genetics 170: 1821–1826.

Di Bella, M.L., M. Vazzana, A. Vizzini and N. Parrinello. 2008. Glucocorticoid receptor (DIGR1) is expressed in pre-larval and larval stages of the teleost fish Dicentrarhus labrax. Cell Tissue Res. 333: 39–47.

Di Marco, P., A. Priori, M.G. Finoia, A. Massari, A. Mandich and G. Marino. 2008. Physiological responses of European sea bass *Dicentrarchus labrax* to different stocking densities and acute stress challenge. Aquaculture 275: 319–328. doi:10.1016/j.aquaculture. 2007.12.012

D'Orbcastel, E.R., G. Lemarie, G. Breuil, T. Petochi, G. Marino, S. Triplet, G. Dutto, S. Fivelstad and J.L. Coeurdacier. 2010. Effects of rearing density on sea bass (Dicentrarchus labrax) biological performance blood parameters and disease resistance in a flow through system. Aquat Liv Res 23: 109–117.

Ellis, T.H. Yavuzcan Yildiz, J. López-Olmeda, M.T. Spedicato, L. Tort, O. Øverli and Martins, Catarina. 2012. Cortisol and finfish welfare. Fish Physiol. Biochem. 38: 163–188.

Fanouraki, E., P. Divanach and M. Pavlidis. 2007. Baseline values for acute and chronic stress indicators in sexually immature red porgy (*Pagrus pagrus*). Aquaculture 256: 294–304.

Fanouraki, E., N. Papandroulakis, T. Ellis, C.C. Mylonas, A.P. Scott and M. Pavlidis. 2008. Water cortisol is a reliable indicator of stress in European sea bass, *Dicentrarchus labrax*. Behaviour 145 Suppl. SI: 1267–1281.

Fanouraki, E., C.C. Mylonas, N. Papandroulakis and M. Pavlidis. 2011. Species specificity in the magnitude and duration of the acute stress response in Mediterranean marine fish in culture. Gen. Comp. Endocrinol. 173: 313–322.

Gallo, P.V. and A. Civinini. 2003. Survey of the Adrenal Homolog in Teleosts. *International Review of Cytology*, Academic Press. 230: 89–187.

Ganga, R., J.G. Bell, D. Montero, L. Robaina, M.J. Caballero and M.S. Izquierdo. 2005. Effect of feeding gilthead seabream (*Sparus aurata*) with vegetable lipid sources on two potential immunomodulator products: prostanoids and leptons. Comp. Biochem. Physiol. 142: 410–418.

Ganga, R., G.J. Bell, D. Montero, E. Atalah, Y. Vraskou, L. Tort, A. Fernandez, and M.S. Izquierdo. 2011a. Adrenocorticotrophic hormone-stimulated cortisol release by the head kidney interrenal tissue from sea bream (*Sparus aurata*) fed with linseed oil and soyabean oil. British Journal of Nutrition 105: 238–247.

Ganga, R., D. Montero, J.G. Bell, E. Atalah, E. Ganuza, L. Vega Orellana, L. Tort, J.M. Acerete, J.M. Afonso, T. Benitez-Santana, H. Fernández-Palacios and M.S. Izquierdo. 2011b. Stress response in sea bream (*Sparus aurata*) held under crowded conditions and fed diets containing linseed and/or soybean oil. Aquaculture 311: 215–223.

Gornati, G., E. Papis, S. Rimoldi, G. Terova, M. Saroglia and G. Bernardini. 2004. Rearing density influences the expression of stress-related genes in sea bass (*Dicentrarchus labrax* L.). *GENE*. 341: 111–118.

Grassi Milano, E., F. Basari and C. Chimenti. 1997. Adrenocortical and adrenomedullary homologs in eight species of adult and developing teleosts: morphology, histology and immunohistochemistry. Gen. Comp. Endocrinol. 108: 483–496.

Hadj-Kacem, N., J.F. Aldrin and B. Romestand. 1986. Influence immediate du brossage des bacs sur certains parametres sanguins du loup d'elevage *Dicentrarchus labrax* L.: effet du stress. Aquaculture 59: 53–59.

Hatziathanasiou, A., M. Paspatis, M. Houbart, P. Kestemont, S. Stefanakis and M. Kentouri. 2002. Survival, growth and feeding in early life stages of European sea bass (*Dicentrarchus labrax*) intensively cultured under different stocking densities. Aquaculture 205: 89–102

Koven, W.M., Y. Barr, S. Lutzky, I. Ben Atia, R. Weiss, M. Harel, P. Behrens and A. Tandler. 2001. The eVect of dietary arachidonic acid (20:4n ¡6) on growth, survival and resistance to handling stress in gilthead seabream (*Sparus aurata*) larvae. Aquaculture 193: 107–122.

Lacroix M. and A. Hontela. 2001. Regulation of acute cortisol synthesis by cAMP-dependant protein kinase and protein kinase C in a telost species, the rainbow trout (*O. mykiss*). J. Endocrinol. 169: 71–78.

Leal, E., B. Fernandez-Duran, R. Guillot, D. Ríos and J.M. Cerda-Reverter. 2011. Stress-induced effects on feeding behavior and growth performance of the sea bass (Dicentrarchus labrax): a self-feeding approach. J. Comp. Physiol. B.

Lupatsch, I., G.A. Santos, J.W. Schrama and J.A.J. Verreth. 2010. Effect of stocking density and feeding level on energy expenditure and stress responsiveness in European sea bass Dicentrarchus labrax. Aquaculture 298: 245–250.

MacKenzie, S., L. Ribas, M. Pilarczyk, D.M. Capdevila, S. Kadri and F.A. Huntingford. 2009 Screening for Coping Style Increases the Power of Gene Expression Studies. PLoS ONE 4(4): e5314.

Matthys, L.A. and E.P. Widmaier. 1998. Fatty acids inhibit adrenocorticotropin induced adrenal steroidogenesis. Horm. Metab. Res. 30: 80–83.

Mauri, I., A. Romero, L. Acerete, S. MacKenzie, N. Roher, A. Callol, I. Cano M.C. Alvarez and L. Tort. 2011a. Changes in Complement responses in Gilthead seabream (Sparus auratus) and European seabass (Dicentrarchus labrax) under crowding stress, plus viral and bacterial challenges Fish and Shellfish Immunology 30: 182–188.

Mauri, I., N. Roher, S. MacKenzie, A. Romero, M. Manchado, J.C. Balasch, J. Béjar, M.C. Álvarez and L. Tort. 2011b. Molecular cloning and characterization of European seabass (*Dicentrarchus labrax*) and Gilthead seabream (*Sparus aurata*) complement component C3. Fish and Shellfish Immunology 30: 1310–1322.

Migaud, H., J.F. Taylor, G.L. Taranger, A. Davie, J.M. Cerdá-Reverter, M. Carrillo, T. Hansen and N.R. Bromage. 2006. Pineal gland sensitivity to light intensity in salmon (Salmo salar) and sea bass (Dicentrarchus labrax): an *in vivo* and *ex vivo* study. J. Pineal Res. 41: 42–52.

Molinero, A., E. Gómez, J. Balasch and L. Tort. 1997. Stress by fish removal in the sea bream *Sparus aurata*: a time course study on the remaining fish in the same tank. J. Appl. Aquaculture 7: 1–12.

Montero, D. and M.S. Izquierdo. 2010. Welfare and health of fish fed vegetable oils as alternative lipid sources to fish oil. pp. 439–485. *In*: G. Turchini, W. Ng and D. Tocher (eds.). Fish Oil Replacement and Alternative Lipid Sources in Aquaculture Feeds. CRC Press, Cambridge, UK. ISBN: 978-1-4398-0862-7.

Montero, D., L. Tort, L. Robaina, J.M. Vergara y and M.S. Izquierdo. 2001. Low vitamin E in diet reduces stress resistance of gilthead seabream (*Sparus aurata*) juveniles. Fish & ShellFish Immunol. 11: 473–499.

Montero, D., T. Kalinowski, A. Obach, L. Robaina, L. Tort, M.J. Caballero, y and M.S. Izquierdo. 2003. Vegetable lipid sources for gilthead seabream (*Sparus aurata*): Effects on fish health. Aquaculture 225: 353–370.

Moons, L., M. Cambre, F. Ollevier and F. Vandesande. 1989. Immunocytochemical demonstration of close relationships between neuropeptidergic nerve fibers and hormoneproducing cell types in the adenohypophysis of the sea bass (Dicentrarchus labrax). Gen. Comp. Endocrinol. 73: 270–283.

Mourente, G., J.E. Good, K.D. Thompson and J.G. Bell. 2007. Effects of partial substitution of dietary fish oil with blends of vegetable oils, on blood leucocyte fatty acid compositions, immune function and histology in European sea bass (*Dicentrarchus labrax* L.). British. J. Nutr. 98: 770–779.

Nasushita, R., H. Watanobe and K. Takebe. 1997. A comparative study of adrenocorticoropin-releasing activity of prostaglandins E1, E2, F2a and D2 in the rat. Prostaglandins Leukot. Essent. Fatty Acids 56: 165–168.

Papoutsoglou, S.E., N. Karakatsouli, G. Pizzonia, C. Dalla, A. Polissidis and Z. Papadopoulou-Daifoti. 2006. Effects of rearing density on growth, brain neurotransmitters and liver fatty acid composition of juvenile white sea bream Diplodus sargus L. Aquac. Res. 37: 87–95.

Paspatis, M., T. Boujard, D. Maragoudaki, G. Blanchard and M. Kentouri. 2003. Do stocking density and feed reward level affect growth and feeding of self-fed juvenile European sea bass? Aquaculture 216: 103–113.

Pavlidis, M., E. Karantzali, E. Fanouraki, C. Barsakis, S. Kollias and N. Papandroulakis. 2011. Onset of the primary stress in European sea bass Dicentrarhus labrax, as indicated by whole body cortisol in relation to glucocorticoid receptor during early development. Aquaculture 315: 125–130.

Romeo, M., N. Bennani, M. Gnassia-Barelli, M. Lafaurie and J.P. Girard. 2000. Cadmium and copper display different responses towards oxidative stress in the kidney of the sea bass *Dicentrarchus labrax*. Aquatic Toxicology 48: 185–194.

Roncarati, A., P. Melotti, A. Dees, O. Mordenti and L. Angellotti. 2006. Welfare status of cultured sea bass (*Dicentrarchus labrax* L.) and sea bream (*Sparus aurata* L.) assessed by blood parameters and tissue characteristics. J. Appl. Ichthyol. 22: 225–234.

Rotllant, J., R.J. Arends, J.M. Mancera, G. Flik, S.E. Wendelaar Bonga and L. Tort. 2000. Inhibition of HPI axis response to stress in gilthead sea bream (*Sparus aurata*) with physiological plasma levels of cortisol. Fish Physiol. Biochem. 23: 13–22.

Rotllant, J., P.H. Balm, J. Perez-Sanchez, S.E. Wendelaar-Bonga and L. Tort. 2001. Pituitary and interrenal function in gilthead sea bream (*Sparus aurata* L., Teleostei) after handling and confinement stress. Gen. Comp. Endocrinol. 121(3): 333–42.

Rotllant, J., N.M. Ruane, M.J. Caballero, D. Montero and L. Tort. 2003. Confinement stress response in sea bass (*Dicentrarchus labrax*) is characterised by an increased biosynthetic capacity of interrenal tissue, with no effect on ACTH sensitivity. Comp. Biochem. Physiol. 136A: 613–620.

Rotllant, J., N.M. Ruane, M.T. Dinis, A.V.M. Canario and D.M. Power. 2006. Intra-adrenal interactions in fish: catecholamine stimulated cortisol release in sea bass (*Dicentrarhus labrax* L.). Comp. Biochem. Physiol. 143: 375–381.

Rubio, V.C., E. Sanchez and J.M. Cerda-Reverter. 2010. Compensatory feeding in the sea bass after fasting and physical stress. Aquaculture 298: 332–337.

Saillant, E., A. Fostier, P. Haffray, B. Menu, S. Laureau, J. Thimonier and B. Chatain. 2003. Effects of rearing density, size grading and parental factors on sex ratios of the sea bass (*Dicentrarchus labrax* L.) in intensive aquaculture. Aquaculture 221: 183–206.

Sammouth, S., E. Roque d'Orbcastel, E. Gasset, G. Lemarié, Gilles Breuil, G. Marino, J.L. Coeurdacier, S. Fivelstad and J.P. Blancheton. 2009. The effect of density on sea bass (*Dicentrarchus labrax*) performance in a tank-based recirculating system. Aquac. Engin. 40: 72–78.

Santos, G.A., J.W. Schrama, R.E.P. Mamauag, J.H.W.M. Rombout and J.A.J. Verreth. 2010. Chronic stress impairs performance, energy metabolism and welfare indicators in European seabass (*Dicentrarchus labrax*): the combined effects of fish crowding and water quality deterioration. Aquaculture 299: 73–80.

Schreck, C.B. 1981. Stress and compensation in teleostean fishes: response to social and physical factors. pp. 295–321. *In*: A.D. Pickering (ed.). Stress and Fish. Academic Press, London.

Schreck, C.B. 2000. Accumulation and long-term effects of stress in fish. *In*: Moberg, G.P., Mench (eds.). The biology of animal stress. CAB International, Wallingford pp. 147–158.

Selye, H. 1956. The Stress of Life. McGraw Hill, New York, USA.

Shrimpton, J.M. and S.D. McCormick. 1999. Responsiveness of gill Na$^+$/K$^+$-ATPase to cortisol is related to gill corticosteroid receptor concentration in juvenile rainbow trout. J. Exp. Biol. 202: 987–995.

Terova, G., R. Gornati, S. Rimoldi, G. Bernardini and M. Saroglia. 2005. Quantification of a glucocorticoid receptor in sea bass (*Dicentrarchus labrax* L.) reared at high stocking densities. *GENE.* 357(2): 144–151.

Torrecillas, S., A. Makol, M.J. Caballero, D. Montero, A.K.S. Dhanasiri, J. Sweetman and M.S. Izquierdo. Effects on mortality and stress response in European sea bass, *Dicentrarchus labrax* (L.), fed mannan oligosaccharides (MOS) after *Vibrio anguillarum* exposure. J. Fish Dis. In press. doi:10.1111/j.1365-2761.2012.01384.x.

Tort, L. 2011. Stress and immune modulation in fish. Develop. Comp. Immunol. 35: 1366–1375.

Tort, L. and M. Teles. 2011. Endocrinology of stress. A comparative view. *In*: Endocrinology. Book 3. Ed by F. Akin. Intech publisher, Rijeka (Croatia), pp:- ISBN 978-953-307-554-9.

Tort, L., F. Padros, J. Rotllant and S. Crespo. 1998. Winter syndrome in the gilthead sea bream Sparus aurata. Immunological and histopathological features. Fish and Shellfish Immunology. pp. 837–47.

Tort, L., D. Montero, L. Robaina, H. Fernandez-Palacios and M.S. Izquierdo. 2001. Consistency of stress response to repeated handling in the gilthead sea bream, *Sparus aurata.* Aquaculture Research 32: 593–598.

Tovar-Ramirez, D., D. Mazuaris, P. Gatesoupe, P. Quazuguel, C.L. Cahu and J.L. Zambonino-Infante. 2010. Dietary probiotic live yeast modulates antioxidant enzyme activities and gene expression of seabass (*Dicentrarchus labrax*) larvae. Aquaculture 300: 142–147.

Varsamos, S., G. Flik, J.F. Pepin, S.E. Bonga and G. Breuil. 2006. Husbandry stress during early life stages affects the stress response and health status of juvenile sea bass, Dicentrarchus labrax. Fish Shellfish Immunol. 20(1): 83–96.

Vazzana, M., M. Cammarata, E.L. Cooper and N. Parrinello. 2002. Confinement stress in sea bass (Dicentrarchus labrax) depresses peritoneal leukocyte cytotoxicity. Aquaculture 210: 231–43.

Vazzana, M., A. Vizzini, G. Salerno, M.L. Di Bella, M. Celi and N. Parrinello. 2008. Expression of a glucocorticoid receptor (DlGR1) in several tissues of the teleost fish *Dicentrarchus labrax*. Tissue and Cell 40: 89–94.

Villamizar, N., A. García-Alcazar and F.J. Sánchez-Vázquez. 2009. Effect of light spectrum and photoperiod on the growth, development and survival of European sea bass (Dicentrarchus labrax) larvae. Aquaculture 292: 80–86.

Vizzini, A., M. Vazzana, M. Cammarata and N. Parrinello. 2007. Peritoneal cavity phagocytes from the teleost sea bass express a glucocorticoid receptor (cloned and sequenced) involved in genomic modulation of the *in vitro* chemiluminescence response to zymosan. Gen. Comp. Endocrinol. 150: 114–123.

Index

17b-estradiol 242

About the Editors

Dr. F. Javier Sánchez Vázquez graduated in Biology in 1992 at the University of Murcia and got his PhD. degree in 1995, pioneering the investigation on sea bass feeding rhythms and their diurnal/nocturnal dualism. With the aid of an EU postdoc grant, he spent two years in Japan at Teikyo University of Science & Technology, where he further specialised in fish chronobiology. In 1997 he got a lectureship position at the University of Murcia, and finally became full Professor of Physiology in 2011. Dr. Sánchez-Vázquez has participated in many National and European research projects devoted to the circadian regulation of biological rhythms in fish, as well as collaborated in the translation of basic research into direct applications for the aquaculture industry. To date, he has published over 139 SCI scientific papers, reviews and book chapters. In addition, he has been actively involved in teaching activities (Physiology, Chronobiology and Aquaculture as major subjects) and supervised 13 PhD. Thesis on feeding, reproduction and development rhythms in fish (10 in sea bass).

Dr. José A. Muñoz-Cueto graduated in Biology in 1987 at the University of Sevilla and got his PhD. degree in 1992 in the Complutense University of Madrid. His PhD. work was done at the Cajal Institute (CSIC), and was focussed on investigating sex differences in developing mammalian central nervous system. In 1992, he joined the University of Cádiz as Assistant Professor, later as Associate Professor and finally became Full Professor of Zoology in 2010. He was postdoc in the University of Bordeaux I and the University of Rennes I (France), and Invited Professor in the University of Rennes I and the University Pierre and Marie Curie Paris VI, in Banyuls sur Mer (France). In the University of Cadiz, his research has been directed towards understanding how hormones and environmental factors control the development and reproduction in three marine species which are of special interest for aquaculture such as sea bass, sole and seabream. Dr. Muñoz-Cueto has led many European, national and regional research projects, published almost 70 SCI scientific papers and supervised 7 PhD. thesis on functional neuroanatomy, neuroendocrinology, development and chronobiology of fish. His teaching activity has been largely on Animal and Environmental Physiology, Comparative Endocrinology and Basic Biology, as well as in Postgraduate and Doctorate programs in the field of Aquaculture.

Color Plate Section

Chapter 1.1

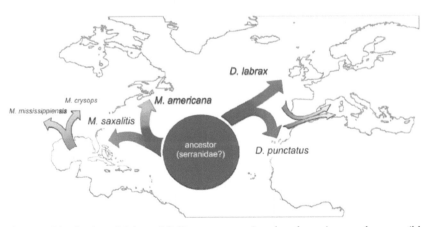

Figure 1. Distribution of Atlanto-Mediterranean species of sea bass. Arrows show possible ways of speciation and colonization or recolonization of the Mediterranean from an ancient species probably originated in the Gulf of Mexico and distributed through the North Atlantic and in the Mediterranean.

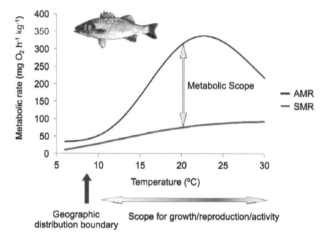

Figure 2. Metabolic scope for sea bass depending on environmental temperature according to the model from Claireaux and Lagardère 1999. AMR is the active metabolic rate and SMR the standard metabolic rate. The difference between them determines the scope for growth and reproduction effort than sea bass individuals have available. Dark arrow shows the down limit of temperature niche for the species determining the distribution range and the migratory spatial and temporal pattern.

Chapter 1.2

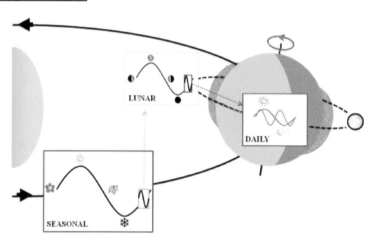

Figure 1. Seasonal, lunar and daily geophysical cycles. Geophysical phenomena, such as the Earth translation around the sun, lunar translation around the Earth and the Earth rotation, lead different periodicities cycles in the terrestrial environment: seasonal ($\tau \approx 365$ days), mensual-lunar ($\tau \approx 28$ days), daily ($\tau \approx 24$ h) and tidal ($\tau \approx 12$ h), being contained the shortest within the longest one.

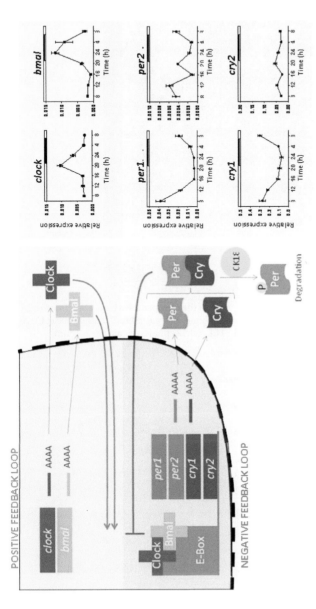

Figure 2. Sea bass molecular clock diagram based on the zebrafish one. The left panel shows the diverse sea bass molecular clock elements that are identified up to now are represented in according to the well-known zebrafish clock molecular interactions. The clock genes are organized in two feedback loop: (i) the positive one that is composed by *clock* and *bmal* genes, whose proteins form the Clock:Bmal complex to active (blue narrows) the negative loop clock genes; (ii) the negative loop with *per* and *cry* (*1* and *2*) genes, whose proteins inhibit their own transcription by means of blocking Clock:Bmal complex. The right panel show: the rhythmic daily expression of every clock gene (except for Cry2) in sea bass pituitary during summer under 14L/10D (7–21 h) at 20°C (MEAN±SEM, n=5; Cosinor p < 0.05). The relative expression is represented in the vertical axis and the time (in hours) in horizontal one. Black and white bars in the top mark the dark and photophase, respectively. Modified from Sánchez et al. 2010; Del Pozo et al. 2012a; Herrero et al. 2012.

Chapter 2.1

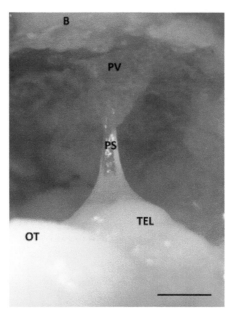

Figure 1. View of the dorsal surface of the brain of the European sea bass, *Dicentrarchus labrax*, showing the pineal gland attached to the skull bone. *Abbreviations*: B, cranial bone; OT, optic tectum; PV, pineal vesicle; PS, pineal stalk; TEL, telencephalon. Scale bar = 1 mm.

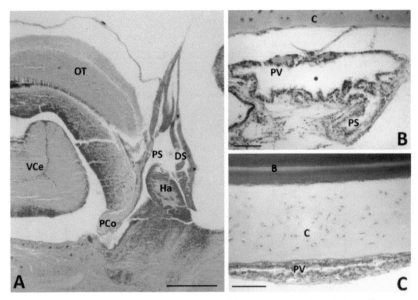

Figure 2. Histological sections of the pineal organ of the European sea bass. (**A**), Midline sagittal sections of sea bass pineal stalk emerging from the posterior commissure between the habenula and the optic tectum. (**B, C**), Transverse sections through the pineal end-vesicle of the sea bass. The pineal vesicle, whichexhibits a convoluted epithelium around the pineal lumen, appears apposed to the cranium and attached to the distal pineal stalk (**B**). At the caudal end, the pineal vesicle appears firmly attached to the skull bone and exhibits a tiny lumen (**C**). *Abbreviations:* B, bone; C, cartilage; DS, dorsal sac; Ha, habenula; OT, optic tectum; PCo, posterior commissure; PS, pineal stalk; PV, pineal vesicle; VCe, valvula of the cerebellum. The asterisk in **B** indicates the pineal lumen. Scale bar = 500 μm in (A), and100 μm in (B) and (C).

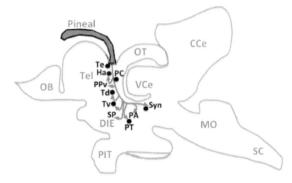

Figure 4. Schematic sagittal drawing of the brain of the European sea bass showing the efferent (red arrows) and afferent (black dots) connections of the pineal gland.*Abbreviations:* CCe, corpus of the cerebellum; DIE, diencephalon; Ha, habenula; MO, medulla oblongata; OB, olfactory bulb; OT, optic tectum; PIT, pituitary gland; PA, pretectal area;PC, posterior commissure; PPv, ventral periventricular pretectum; PT, posterior tubercle; SC, spinal cord; SP, superficial pretectum; Syn, dorsal synencephalon; Td, dorsal thalamus; Te, thalamic eminence; Tel, telencephalon; Tv, ventral thalamus; VCe, valvula of the cerebellum.

Chapter 2.2

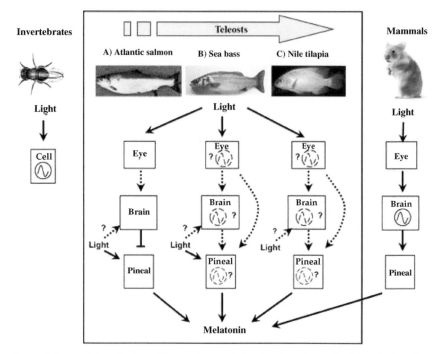

Figure 4. Suggested evolution of the regulation of pineal melatonin synthesis by the circadian axis in teleosts. In addition to the two types of circadian organisation already proposed in fish (A and B), a third type could exist where pineal light sensitivity would be dramatically reduced (C). The regulation of pineal activity would have thus evolved from an independent light sensitive pineal gland, without pacemaker activity, as seen in salmonids (A); to an intermediary state where the pineal gland remains light sensitive and could possess a circadian pacemaker, but is also regulated by photic information perceived by the retina as seen in seabass and cod (B); to reach a more advanced system closer to higher vertebrates where light sensitivity of the pineal gland would be significantly reduced and its melatonin synthesis activity primarily regulated by a circadian pacemaker (unknown location) entrained by photic information perceived by the retina (C) (adapted from Migaud et al. 2010).

Chapter 2.3

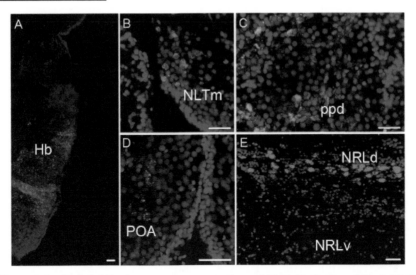

Figure 4. *In situ* hybridization on transverse sections of kiss1 (A, B, C) and kiss2 (D, E) expressing cells in the brain and pituitary gland of the European sea bass. (A) Kiss1 mRNAs were detected into the habenular nucleus (Hb) in both sexes. (B) In sea bass sacrificed during the breeding period an addition population of kiss1-expressing cells appear in the hypothalamus, at the level of the medial part of the lateral tuberal nucleus (NLTm), and (C) in the proximal pars distalis of the pituitary gland (ppd). (D) Few kiss2 expressing cells were located in the preotic area (POA), whereas the most abundant kiss2 cell population was observed in the hypothalamus, at the level of the dorsal and ventral parts of the nucleus of the lateral recess (NRLd, NRLv). Scale bar = 25 μm.

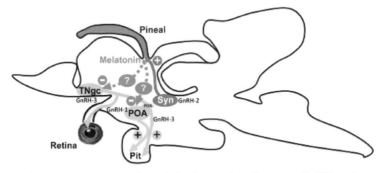

Figure 5. Schematic representation of the interactions between GnRH systems and photoreceptor organs in the European sea bass. Preoptic (POA) GnRH1 cells (light gray circle) massively project to the pituitary gland. GnRH2 cells (black oval) located in the synencephalon (Syn) and GnRH3 neurons (dark gray diamond) of the terminal nerve (TN) are connected to the pineal gland and retina, respectively. Moreover, GnRH2 stimulates (+) the pineal melatonin secretion, which in turns, inhibits (–) brain expression levels of GnRH1 and GnRH3 forms (gray dotted arrows). Cc, corpus of the cerebellum; MO, medulla oblongata; OB, olfactory bulb; Pit, pituitary gland; SC, spinal cord; Tel, telencephalon.

Chapter 2.7

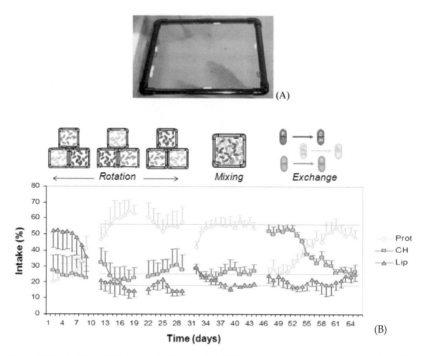

(A)

(B)

Figure 4. Illustration of dietary selection in fish. Different macronutrients encapsulated into three different colour capsules (A). Fish are able to discriminate and evaluate the quality of encapsulated diets through postingestive processes and establish a percentage of selection (B). This information can be used by the fish to develop an associative learning between the colour and nutrient content of gelatine capsules.

Chapter 2.8

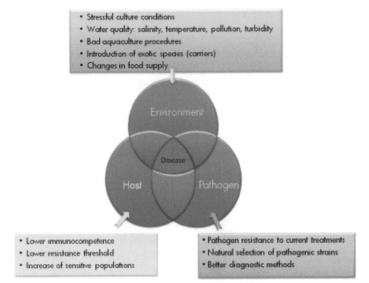

Figure 1. Venn diagram showing the three types of factors involved in a disease. For each of them, a set of possible modulators are in action.

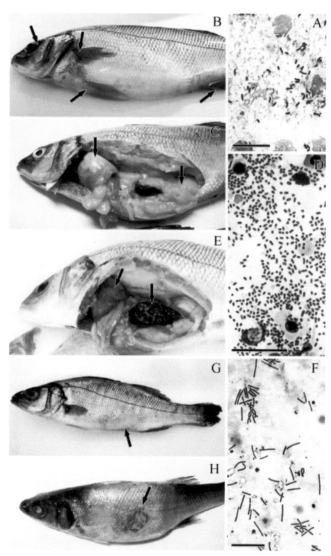

Figure 2. Pathologies due to bacteria in European sea bass. **A.** Rod shaped bacillus of *Vibrio anguillarum* in a fuchsine-stained smear. (B) External and (C) internal typical signs of vibriosis. Note the haemorrhagic fins, head (in B), liver and digestive tract (in C) (arrows). (D) Bacillus of *Photobacterium damselae* subsp. *piscicida* in a methylene blue- stained smear. (E) Internal clinical signs of photobacteriosis, with the typical yellowish nodules in the spleen and haemorrhages in the liver (arrows). (F) *Tenacibaculum maritimum* filamentous bacteria in a methylene blue-stained smear. (G-H) External lesions and ulcers typical of tenacibaculosis (arrows). Bar scales = 10 µm. All illustrations are original from the authors (CZ).

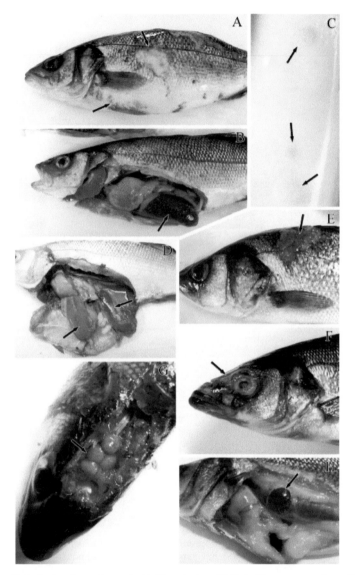

Figure 3. Pathologies due to bacteria (A-E) and virus (F-H) in European sea bass. External (**A**) and internal (B) clinical signs of mycobacteriosis, Note the external lesions and ulcers (arrows in A), and the granulomata in the spleen (arrow in B). **C.** Colonies of *Mycobacterium marinum* grown in Lowenstein-Jensen agar. (D) Clinical internal signs of streptococosis, note the haemorrhages in the digestive tract (arrows). (E) Nocardiosis external lesions, note the large ulcer (arrow). (F-H) Viral Nervous Necrosis (VNN) external (F) and internal (G-H) clinical signs, note the head erosions (F, arrow), the haemorrhages in the encephalus (G, arrow) and the inflamed swim bladder (H, arrow). All illustrations are original from author CZ.

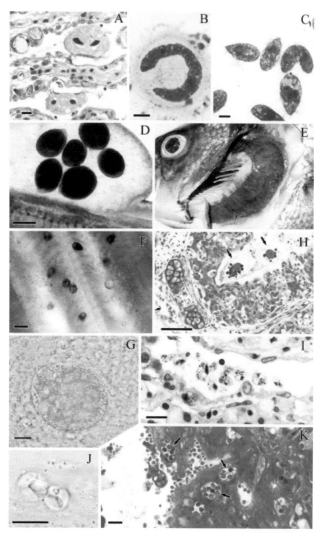

Figure 4. Protozoan parasites of European sea bass. (A-C) *Trichodina* sp. in toluidine blue stained section of gills (A, arrows), in Giemsa-stained smear (B) note the horseshoe-shaped macronucleus. (C) *Philasterides dicentrarchi* Giemsa-stained smear. (D-E) *Cryptocaryon irritans* in the gills. (F-H) *Amyloodinium ocellatum* in fresh smears of gills (F-G), and in a toluidine blue-stained section of gills (H), note the dark-stained amyloopectin granules, (arrows). (I) *Cryptobia branchialis* in a toluidine-blue stained section of gills. (J) Mature oocysts of *Eimeria dicentrarchi* in a fresh smear of intestine. (K) *Cryptosporidium molnari* in a toluidine-blue stained section of stomach invading all the epithelium. Note the groups of oocysts, arrows. Scale bars: 200 μm (D); 50 μm (F, H); 10 μm (A-C, G-K). Illustrations courtesy of: Dr. Ariel Diamant (National Center for Mariculture, Israel) (D), Dr. Panos Varvarigos (VetCare, Greece) (F), Dr. Pilar Alvarez-Pellitero (IATS-CSIC) (G, J). Illustrations original from the authors: ASB (A-C, H, I, K), CZ (E).

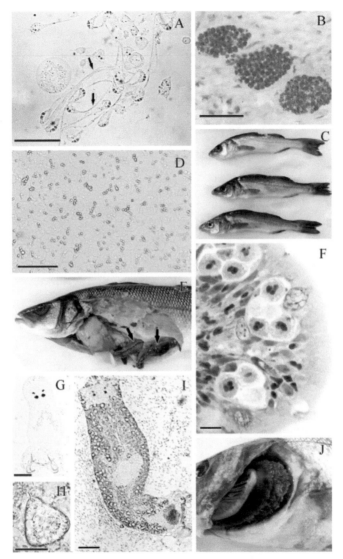

Figure 5. Metazoan parasites of European sea bass. (A) Fresh smear of bile with trophozoites and spores of *Ceratomyxa labracis*. Note the long appendages of the spores (arrows). (B) *Sphaerospora dicentrarchi* in a PAS-stained section of intestine. (C) Juveniles infected by *S. dicentrarchi*. Note the depressed abdomen and the poor dorsal musculature. (D) *Sphaerospora testicularis* in a fresh smear of seminal fluid. (E) Necrotic testes (arrows) infected by *S. testicularis* at the end of the spawning season. (F) *Enteromyxum leei* in a toluidine blue-stained section of intestine. (G-J) *Diplectanum aequans* juvenile (G), egg (H), adult (I), affected gills (J). Scale bars: 100 µm (I); 50 µm (A, D, H); 20 µm (B); 10 µm (F, G). Illustrations original from the authors: ASB (A, B, D, F-I), CZ (C, F, J).

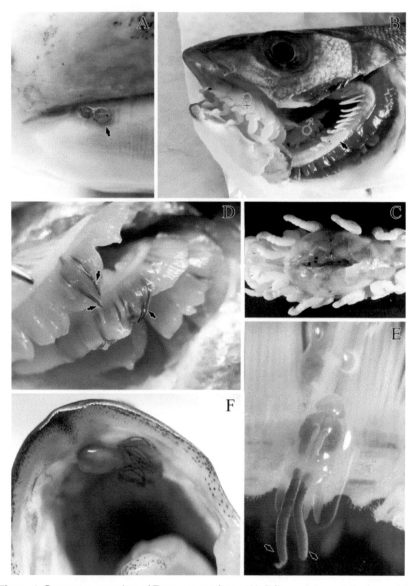

Figure 6. Crustacean parasites of European sea bass. (A) *Caligus minimus* at the base of the mouth. (B) *Cerathotoa oestroides*. Note the female (♀) and the male (♂) in the oral cavity and the larvae (arrow) in the gill arch. (C) Gravid female of *C. oestroides* full of larvae. (D-E) *Lernanthropus kroyeri*. The ovigerous sacs of females are visible out of the gill filaments (arrows). Illustrations courtesy of: Dr. Ivona Mladineo (Institute of Oceanography & Fisheries, Croatia) (C) and Dr. Panos Varvarigos (VetCare, Greece) (D-F). Illustrations original from the author CZ (A, B).

Figure 7. Non-infectious pathologies of European sea bass. (A) Skin lesions produced by bad handling during routine aquaculture procedures, like vaccination (B) or sedation, treatments and samplings in tarpaulins (C). (D) Cannibalism. (E) Hearts of fish affected by SMS (left, centre) compared with a normal heart from a wild fish (right). (F-G) Gas bubble disease, air bubbles are visible in the gill capillaries (F, arrows) and the fins (G, arrows). (H) Typical skin lesions due to nefrocalcinosis. (I) Gill affected by jellyfish. (J) Fibropapilloma. Illustrations courtesy of Dr. Panos Varvarigos (VetCare, Greece) (B, C, F-H). Illustrations original from the author CZ (A, D, E, I, J).

T - #0339 - 071024 - C15 - 234/156/19 - PB - 9780367378585 - Gloss Lamination